T0224561

Layered Structures, Epitaxy, and Interfaces

MATERIALS RESEARCH SOCIETY SYMPOSIA PROCEEDINGS

ISSN 0272 - 9172

MATERIALS RESEARCH SOCIETY SYMPOSIA PROCEEDINGS

MATERIALS RESEARCH SOCIETY SYMPOSIA PROCEEDINGS   VOLUME 37

Layered Structures, Epitaxy, and Interfaces

Symposium held November 26-30, 1984, Boston, Massachusetts, U. S. A.

EDITORS:

J. M. Gibson
AT&T Bell Laboratories, Murray Hill, New Jersey, U. S. A.

L. R. Dawson
Sandia National Laboratories, Albuquerque, New Mexico, U. S. A.

MRS MATERIALS RESEARCH SOCIETY
Pittsburgh, Pennsylvania

CAMBRIDGE UNIVERSITY PRESS
Cambridge, New York, Melbourne, Madrid, Cape Town,
Singapore, São Paulo, Delhi, Mexico City

Cambridge University Press
32 Avenue of the Americas, New York NY 10013-2473, USA

Published in the United States of America by Cambridge University Press, New York

www.cambridge.org
Information on this title: www.cambridge.org/9781107405752

Materials Research Society
506 Keystone Drive, Warrendale, PA 15086
http://www.mrs.org

© Materials Research Society 1985

First published 1985
First paperback edition 2012

Single article reprints from this publication are available through
University Microfilms Inc., 300 North Zeeb Road, Ann Arbor, MI 48106

CODEN: MRSPDH

ISBN 978-0-931-83702-9 Hardback
ISBN 978-1-107-40575-2 Paperback

Contents

SECTION B: EPITAXY OF THIN FILMS

SECTION C: INSULATORS ON SILICON

SECTION D: STRAINED-LAYER SUPERLATTICES

SECTION G: INTERFACES

SECTION I: SCHOTTKY BARRIERS

Preface

This volume contains written records of almost all of the presentations
included in the symposium: "Layered Structures, Epitaxy and Interfaces"
held at the 1984 Fall Meeting of the Materials Research Society at the
Marriot Copley Place Hotel, Boston, Massachussetts, November 26th-30th.
This symposium represented the first attempt to bring together materials
scientists involved in the growth, characterization and physical
properties of the unique thin films and layered structures which can be
fabricated by epitaxial growth. The success of this venture, which is
documented in this volume, can be attributed to the broad interdisciplinary
appeal of the Materials Research Society. The editors also extend their
gratitude to the U.S. Army Research Office and the U.S. Office of Naval
Research for financial support, to the invited speakers who set the
tone of the symposium and to the other authors whose contributed papers
were of a high standard. All those who helped in the logistics of the
symposium deserve special credit, particularly the session chairpersons
and our administrative assistant Diana Barzana.

J.M. Gibson

L.R. Dawson

January, 1985

Quantum Well Structures

PHYSICS IN QUANTUM WELLS

H. L. STORMER
AT&T Bell Laboratories, 600 Mountain Avenue, Murray Hill, N.J.   07974

EXTENDED ABSTRACT

Over the past decade Molecular Beam Epitaxy has developed into a
versatile and reliable crystal growth technique.[1] This growth process,
being inherently two-dimensional, is ideally suited to generate two-
dimensional (2D) carrier systems.[2] Employing a modulation-doping technique,[3] the mobilities in the 2D plane can well exceed 10^6 cm^2/Vsec for
electrons in GaAs-(AlGa)As systems at low temperatures. 2D hole systems
have also been prepared and show comparably high performance.[4,5] These
materials are ideally suited to study the physical properties of two-
dimensional systems.[6]

One of the most remarkable discoveries in these 2D structures is the
integral quantum Hall effect.[7,8] At low temperatures and in high magnetic
fields the Hall resistance is found to be quantized to ρ_{xy} = h/ie^2, whereby
i is an integral number within several parts in 10^7.[9] Concomitantly with
this quantization of ρ_{xy} the resistivity ρ_{xx} seem to vanish as the temperature approaches T $\to$ 0. Resistivities as low as $\rho_{xx} \leq 10^{-10}$ Ω/[] equivalent
to $\rho \leq 10^{-16}$ Ω cm have been reported.[10]

More recently, quantization of the Hall resistance and simultaneous
loss of resistivity has also been found with quantum numbers i which are
rational fractions rather than integers.[11,12]

While the integral quantum Hall effect can be understood on the basis
of the single particle density of states of a 2D system in a high magnetic
field[13,14] no such interpretation is possible for the fractional quantum
Hall effect. Various theories and model calculations indicate that the
electronic state underlying this fractional quantum effect is a novel
quantum liquid formally related to superfluid He.[14-18]

The ability to fabricate multiple-layered 2D systems via MBE and
thereby to increase the total number of 2D carriers per volume has led
to new developments.[6] The magnetic field dependence of the specific heat
of a 2D electron system has just been reported.[19] These experiments are
performed on a total of only ~10^{12} electrons. They reveal astonishingly
clear oscillation in the specific heat which allow to determine the total
density of states of the sample in a magnetic field.

Measurements of the oscillatory magnetic moment (deHaas-van Alphen
effect) are now similarly successful.[10,20] The magnetic moment of a

layered 2D electron system in a magnetic field has been determined via a very sensitive DC method. A DC method proved essential in order to avoid eddy currents which can have lifetimes of up to several hundred seconds.[10] The sensitivity of this technique approaches the limit where one can investigate a single 2D system with only a few 10^{11} electrons.

2D hole systems become of increasing importance not only technologically where the first steps towards complementary logic on GaAs are being taken,[21,22,23] but also from the point of view of physics.

The degeneracy at the top of the valence band lead to unusual 2D subband structures. Due to a relatively large g-factor of the holes one can clearly observe a surprising phenomenon. It turns out that the spin degeneracy of the valence band can be lifted away from k = 0 in the absence of a magnetic field.[24] This lifting of the degeneracy is sensitive to the symmetry of the potential well in which the carrier are trapped.[25] Potential wells having inversion symmetry lack this effect while those lacking inversion symmetry exhibit the splitting. This result demonstrates the overruling influence which the well symmetry has on the dispersion relation of the carrier in these new materials where the degree of "symmetry breaking" can be fine tuned during fabrication.

ACKNOWLEDGEMENTS

This review is only made possible due to collaborations with a large group of individuals at AT&T Bell Laboratories as well as other institutions. I would like to thank all of these colleagues, listed in the references, for their invaluable contributions.

[1] A. Y. Cho in Technology and Physics of Molecular Beam Epitaxy, edts. E.H.C. Parker and M.B. Dowsett, Plenum, N.Y. (in print)

[2] review: A. C. Gossard and A. Pinczuk in Synthetic Modulated Structures, edts. L. L. Chang and B. C. Giessen, Academic Press, N.Y. (1984) (in print)

[3] review: H. L. Stormer, Surf. Science, 132, 519 (1983).

[4] H. L. Stormer, A. C. Gossard, W. Wiegmann, R. Blondel, and K. Baldwin, Appl. Phys. Lett. 44, 139 (1984).

[5] W. I. Wang, E. E. Mendez, and F. Stern, Appl. Phys. Lab., 45, 639 (1974).

[6] review: H. L. Stormer, Surface Science 142, 130 (1984).

[7] K. von Klitzing, Festkorperprobleme, Advances in Solid State Phys. 21, 1 (1981) edt. J. Treusch, Vieweg, Braunschweig.

[8]   D. C. Tsui and A. C. Gossard, Appl. Phys. Lett. 37, 550 (1981).

[9]   D. C. Tsui, A. C. Gossard, B. F. Field, M. E. Cage, and R. F. Dziuba,
      Phys. Rev. Lett. 48, 3 (1982).

[10] T. Haavasoja, H. L. Stormer, D. J. Bishop, V. Narayanamurti, A. C.
      Gossard, and W. Wiegmann, Surface Science 142, 294 (1984).

[11] D. C. Tsui, H. L. Stormer, and A. C. Gossard, Phys. Rev. Lett. 48,1559
      (1982).

[12] review: H. L. Stormer, Festkörperprobleme. Advances in Solid State
      Phys. 24, 25 (1984) edt. P. Grosse, Vieweg, Braunschweig.

[13] R. B. Laughlin, Phys. Rev. B. 23, 5632 (1981).

[14] B. I. Halperin, Helv. Phys. Acta. 56, 75 (1983).

[15] R. B. Laughlin, Phys. Rev. Lett. 50, 1395 (1983).

[16] D. Yoshioka, B. I. Halperin, and P. A. Lee, Phys. Rev. Lett. 50,
      1219 (1983).

[17] F. D. M. Haldane, Phys. Rev. Lett. 51, 605 (1983).

[18] R. B. Laughlin, Surface Science 142, 163 (1984).

[19] E. Gornik, R. Lassning, H. L. Stormer, W. Seidenbusch, A. C. Gossard,
      W. Wiegmann, and M. V. Ortenberg, 17th Intl. Conf. Phys. Semicond.
      San Francisco 1984 (in print).

[20] J. P. Eisenstein, H. L. Stormer, V. Narayanamurti, A. C. Gossard,
      Intl. Conf. on Superlattices and Microdevices, Urbana-Champaign,
      Ill., 1984 (in print).

[21] H. L. Stormer, K. Baldwin, A. C. Gossard, and W. Wiegmann, Appl. Phys.
      Lett. 44, 1062 (1984).

[22] R. Kiehl and A. C. Gossard, IEEE Electron Device Lett. Dec. 1984
      (in print).

[23] S. Tiwari and W. I. Wang, IEEE Electron Device Lett. EDL-5, 333
      (1984)

[24] H. L. Stormer, Z. Schlesinger, A. Chang, D. C. Tsui, A. C. Gossard,
      and W. Wiegmann, Phys. Rev. Lett. 51, 126 (1983). See also for
      electrons: D. Stein, K. von Klitzing, and G. Weimann, Phys. Rev.
      Lett. 51, 130 (1983).

[25] J. P. Eisenstein, H. L. Stormer, V. Narayanamurti, A. C. Gossard,
      and W. Wiegmann (submitted)

Cyclotron Resonance Study of Modulation Doped GaAs/Al$_x$Ga$_{1-x}$As Heterojunction Superlattices

M. J. Chou and D. C. Tsui

Department of Electrical Engineering and Computer Science
Princeton University
Princeton, N.J. 08544.

ABSTRACT

Cyclotron resonance of free carriers in modulation doped GaAs/ Al$_x$Ga$_{1-x}$As heterojunction superlattices is studied, using an optically pumped far infrared laser and an 8T superconducting magnet. In the limit the free carriers in each GaAs layer form an independent two-dimensional electron gas (2DEG), the cyclotron mass is heavier than in bulk GaAs and varies from m = (0.068±0.001)m$_0$ to (0.070± 0.002)m$_0$ for 2D densities n=5.3×10^{11}cm^{-2} to 9.2×10^{11}cm^{-2}. The scattering time, obtained from fitting the resonance lineshape, differs from that obtained from low-field Hall measurements and is dependent on the laser frequency.

1. Introduction

Cyclotron resonance (CR) probes directly the free carriers in a semiconductor. It does not require ohmic contacts to the semiconductor, and it can measure simultaneously the effect mass (m*) of the carriers and their scattering time (τ) at finite frequencies [1]. Recently, there is great interest in the electronic properties of GaAs/Al$_x$Ga$_{1-x}$As superlattice structures. Experimental investigations to date are focused on DC transport [2] and optical measurements [3][4], and no CR experiment has been reported in the literature. The purpose of this paper is to report our cyclotron resonance investigations made on several samples of GaAs/Al$_x$Ga$_{1-x}$As heterojunction superlattice, in the limit that each GaAs layer form an independent two-dimensional electron gas (2DEG). Special attention was paid to the effect of free carriers in the (AlGa)As layers.

We observed an effective mass varying from (0.068±0.001)m$_0$ to (0.070±0.002)m$_0$ with 2D electron densities n=5.3×10^{11} cm^{-2} to 9.2 ×10^{11} cm^{-2}. This variation in m* is attributed to the nonparabolicity of the conduction band in GaAs. Owing to the fact that absorption in these multiple layer structures can be as high as 45%, the scattering time must be extracted from a computer fitting of the CR lineshape. In the low mobility samples, the extracted τ is in reasonable agreement with that from low-field Hall measurements. However, large discrepancies are seen in the high mobility samples. In particular, τ shows strong dependences on the resonance magnetic field (B$_r$), suggestive of the oscillatory behavior seen recently in single interface samples.[5]

2. Experimental results

The superlattice samples used in this work were similar to those in Ref. 2 and Ref. 4. They were grown by molecular beam epitaxy and the high mobilities were achieved by introducing an undoped Al$_x$Ga$_{1-x}$As spacer, of thickness d$_i$ in the structure illustrated in Fig. 1. Table 1 gives the sample parameters, where n$_0$ is the density in the ground state subband, obtained from the Shubnikov-deHaas (SdH) oscillations, and n is the total density in each well, obtained from the low-field Hall measurements.

Fig. 2 shows a schematic diagram of the experimental arrangement used in our far infrared (FIR) CR measurements. The FIR laser is pumped by a pulsed CO$_2$ laser. The laser radiation of the desired wavelength is divided by using a Si beam

SAMPLE	x	d'_1 (Å)	d''_1 (Å)	d_2	P	$n_0(10^{11}$cm$^{-2})$	$n(10^{11}$cm$^{-2})$	μ(cm^2/v-sec)
					PERIODS			MOBILITY
0	0.12	302	0	255	23	6.3	9.2	11,000
1	0.12	290	51	245	19	5.8	9.5	15,700
2	0.12	287	99	244	17	5.4	5.7	58,000
3	0.12	293	151	250	15	4.0	5.3	54,000

Table 1. Sample parameters

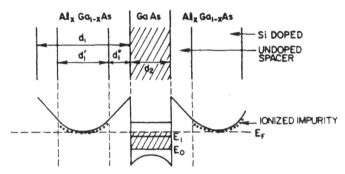

Fig. 1. A GaAs/Al$_x$Ga$_{1-x}$As heterojunction superlattice structure and the energy band diagram.

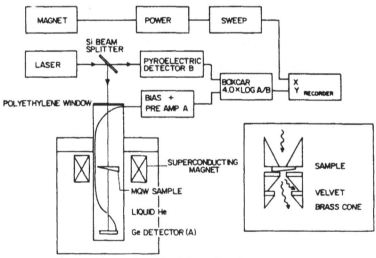

Fig. 2. A schematic diagram of the far infrared cyclotron resonance experimental set up.

splitter into two beams. One, for the reference, is directed to a pyroelectric detector at room temperature; the other, guided by a 1/2- inch light pipe through a polyethylene window, is made incident upon the sample. In order to prevent interference, the sample is made into a 4° wedge to separate the first-order beam, which passes through the sample once, from the higher-order beams. The high-order beams are trapped between the two cones, placed below the sample, and are absorbed by the black velvet (inset of Fig. 2). Only the first-order beam is transmitted to the detector, which is Ga-doped Ge, placed at 8 inches below the center of the magnet. The magnetic field (B) up to 8 tesla is provided by a superconducting solenoid and is parallel to the direction of the incident beam.

Both the reference signal and the transmitted signal are fed into a boxcar, which normalizes the signal by the reference to eliminate the influence of the fluctuations in the laser output. The output of the boxcar, which is the logarithm of the normalized signal, is plotted on an X-Y recorder. It is clear from the recorder trace, shown as the solid curve in Fig. 3, that the data allow unambiguous identification of the resonance position (B_r), which yields directly m^* through $m^* = eB_r/\omega$ (ω is the laser frequency). In Fig. 4, we plot the laser energy against B_r for all four samples. The data follow straight lines whose slopes yield m^* varying from (0.068 ± 0.001) m_o to (0.070 ± 0.002) m_o for 2D densities from $5.3\times10^{11}\text{cm}^{-2}$ to $9.2\times10^{11}\text{cm}^{-2}$. The mass is heavier than 0.0665 m_o [6] in bulk GaAs and the enhancement can be accounted for by the nonparabolicity of the GaAs conduction band.

The 2DEG density in our samples varies from $n= 5.3\times10^{11}\text{cm}^{-2}$ to $9.2\times10^{11}\text{cm}^{-2}$ in a GaAs layer thickness $d=250\text{Å}$, equivalent to a three dimensional density $N=n/d$ from $2.1\times10^{17}\text{cm}^{-3}$ to $3.7\times10^{17}\text{cm}^{-3}$. Owing to this high density of the electrons and their high mobility in these heterojunction structures, the absorption can be as high as 45%. This strong absorption makes the extraction of τ from the full width at half maximum of the resonance [7] unreliable. We follow a classical description of the 2DEG and calculate the CR absorption. Both τ and m^* are extracted from best fits to the observed CR data.

3. Calculations

We consider a linearly polarized wave, shown in Fig. 5(a), normally incident on a thin film. The electric (E) and magnetic (H) fields at boundaries 1 and 2 must satisfy the following relationship [7]:

$$\begin{bmatrix} E_1 \\ H_1 \end{bmatrix} = \begin{bmatrix} \cos(\vartheta_\pm) & i\sin(\vartheta_\pm)/n_\pm \\ in_\pm\sin(\vartheta_\pm) & \cos(\vartheta_\pm) \end{bmatrix}\begin{bmatrix} E_2 \\ H_2 \end{bmatrix} = M_1 \begin{bmatrix} E_2 \\ H_2 \end{bmatrix}$$

Here, M_1 is the characteristic matrix, $n_\pm$ is the complex index of refraction, and $\vartheta_\pm = k_\pm \times d$. The film thickness is d and the complex wave number is $k_\pm$. The $\pm$ sign refers to left and right circularly polarized waves, respectively. In addition,

$$k_\pm=(\omega/c)n_\pm \quad , \quad n_\pm=(\varepsilon-i\sigma_\pm(B)/\omega\varepsilon_0)^{1/2} \quad ,$$

$$\sigma_\pm(B)=\sigma_0(1+i\tau(\omega\pm eB/m^*))^{-1} \quad , \text{ and } \quad \sigma_0=Ne^2\tau/m^*$$

Here, ω is the FIR frequency, ε is the d.c. dielectric constant, $\sigma(B)$ is the a.c. conductivity in the presence of B, and N is the carrier density per unit volume $(N=n/d)$.

Since M relates the fields at two adjacent boundaries, it follows, for the case shown in Fig. 5(b),

(1) $\qquad \begin{bmatrix} E_1 \\ H_1 \end{bmatrix} = M_1 M_2 M_3 \dots\dots M_{2p+1}\begin{bmatrix} E_{2p+2} \\ H_{2p+2} \end{bmatrix}$

M_{2p+1} with p = 0,1,2,...., which relates the fields at the boundaries (2p+1) and

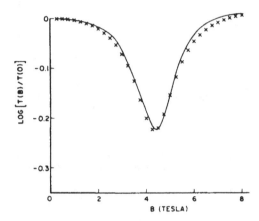

Fig. 3. Cyclotron resonance transmittance as a function of B. Solid curve : experimental; crosses : calculated values.

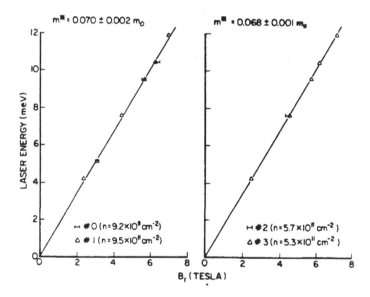

Fig. 4. Laser energy vs. resonance magnetic field

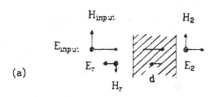

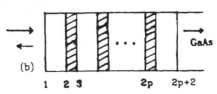

(a)

Fig. 5 An illustration for normally incident light on (a) a thin film and (b) a GaAs/Al_xGa_{1-x}As heterojunction superlattice, dashed area is GaAs, blank area is Al_xGa_{1-x}As.

(b)

(2p+2) in the Al_xGa_{1-x}As, is given by

$$(2) \qquad M_{2p+1} = \begin{bmatrix} \cos(k_{1,\pm}d_1) & i\sin(k_{1,\pm}d_1)/n_{1,\pm} \\ i\,n_{1,\pm}\sin(k_{1,\pm}d_1) & \cos(k_{1,\pm}d_1) \end{bmatrix}$$

M_{2p} with p = 1,2,...., which relates the fields at the boundaries (2p) and (2p+1) in GaAs, is given by

$$(3) \qquad M_{2p} = \begin{bmatrix} \cos(k_{2,\pm}d_2) & i\sin(k_{2,\pm}d_2)/n_{2,\pm} \\ i\,n_{2,\pm}\sin(k_{2,\pm}d_2) & \cos(k_{2,\pm}d_2) \end{bmatrix}$$

If there are no carriers in the (AlGa)As layers, the characteristic matrix for (AlGa)As can be simplified to :

$$(4) \qquad M_{2p+1} = \begin{bmatrix} 1 & 0 \\ 0 & 1 \end{bmatrix} \qquad \text{when } k_1d_1 \ll 1$$

In our experiment, the FIR wavelengths are from 100 μm to 300 μm, that is, k_1 is in the order of 10^5 m^{-1}, and d_1 is from 300 Å to 600 Å. Thus $k_1d_1 \ll 1$. The transmission coefficient ($t_\pm$) and the transmittance (T) can be expressed as follows:

$$(5) \qquad M = M_2M_4 \ldots\ldots M_{2p} = M_2^p = \begin{bmatrix} m_{1,\pm} & m_{2,\pm} \\ m_{3,\pm} & m_{4,\pm} \end{bmatrix}$$

$$(6) \qquad t_\pm = \frac{E_{2p+2,\pm}}{E_{input,\pm}} = \frac{2}{m_{1,\pm} + m_{2,\pm}n_{GaAs} + m_{3,\pm} + m_{4,\pm}n_{GaAs}}$$

$$(7) \qquad T = (\,|\,t_+\,|^2 + |\,t_-\,|^2\,)/2$$

Here $n_{GaAs} = \sqrt{\varepsilon} = 3.6$, and $m_{1,\pm}$, $m_{2,\pm}$, $m_{3,\pm}$, and $m_{4,\pm}$ are the elements of M. In case free carriers are present in the (AlGa)As layers, constituting an additional conducting channel parallel to the GaAs channel,

$$(8) \qquad M = M_1^p M_2^{p+1} = \begin{bmatrix} m_{5,\pm} & m_{6,\pm} \\ m_{7,\pm} & m_{8,\pm} \end{bmatrix}$$

And the matrix elements $m_{5,\pm}$, $m_{6,\pm}$, $m_{7,\pm}$, and $m_{8,\pm}$ substitute $m_{1,\pm}$, $m_{2,\pm}$, $m_{3,\pm}$, and $m_{4,\pm}$, respectively, in Eq. (6) and Eq. (7) for $t_\pm$ and T.

4. Discussions

We analyzed our data following the model with no carriers in the (AlGa)As layers and extracted τ and m^* by fitting the data to Eqs. (5), (6), and (7), using p and $N=n/d_2$ from the sample parameters given in Table 1. Fig. 3 shows an example of the fit and Fig. 6 shows the extracted τ and m^* plotted against the laser energy. Two points worth emphasizing. First, in all cases, our calculated transmittance in the high B tail of the resonance lies below the experimental trace. This difference, which can be accounted for by introducing an additional conducting channel in the model, introduces an uncertainty in the values of τ and m^* of several percent (not shown in Fig. 6). Second, the quantum oscillations observed in the SdH measurements, made on our CR samples, show a large positive background magnetoresistance, which increase with increasing B. The quantized Hall plateaus expected of the 2DEG were not observed. These observations also suggest the presence of a conducting channel other than the 2DEG in the GaAs layers. The average m^* extracted for samples 0, 1, 2 and 3 are $0.071m_0$, $0.069m_0$, $0.068m_0$ and $0.0685m_0$, respectively, in good agreement with those obtained from the slopes of ω vs. B_r (Fig. 4). However, because of the large absorption in these structures, τ estimated from the full width at half maximum of the resonance can differ from the extracted value by more than a factor of 2. For the two low mobilities samples (0 and 1), the average mobilities from the extracted m^* and τ, are 12,200 and $14,600 cm^2/V$-sec in agreement with the low-field Hall mobilities, 11,000 and $15,700 cm^2/V$-sec, respectively. For the high-mobility samples (2 and 3), τ is smaller than that obtained from the low-field Hall measurement, and it shows a strong variation with the laser energy. A smaller τ observed in the CR may be indicative of inhomogeneities in our samples. The variations with respect to laser energy may reflect the oscillatory effect from impurity screening on the occupation of Landau levels [8], which was observed in $GaAs/Al_xGa_{1-x}As$ single interface heterostructures [5].

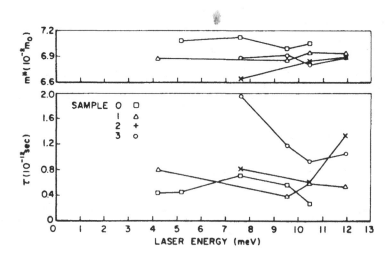

Fig. 6. τ and m^* vs. laser energy extracted from best fits to the experimental data using Eqs. (5),(6) and (7).

Acknowledgements

We thank A.C. Gossard and W. Wiegmann of AT&T Bell Labs for supplying the MBE grown heterojunction superlattices and H.L. Stormer for his helpful suggestions. This work is supported by the Office of Naval Research through contract No. N00014-82-K0450 and a grant from the National Bureau of Standards.

Reference

(1) R.J. Wagner, T.A. Kennedy, B.D. McCombe and D.C. Tsui, Phy. Rev. B22,945 (1980)

(2) H.L. Stormer, A. Pinczuk, A.C. Gossard and W. Wiegmann, Appl. Phys. Lett. 38, 691 (1981)

(3) See, for example, R.C. Miller and A.C. Gossard, Phys. Rev. B28, 3645(1983)

(4) A. Pinczuk and J.M. Worlock, Surf. Sci. 113, 69(1982)

(5) Th. Englert, J.C. Maan, Ch. Uihlein, D.C. Tsui, and A.C. Gossard, Solid State Commun. 46, 545 (1983)

(6) G.E. Stillman, C.M. Wolfe and J.O. Dimmock, Solid State Commun. 7,921 (1969)

(7) E. Hecht and A. Zajac, Optics (Addison-Wesley, MA) p.312

(8) S. Das Sarma, Solid State Commun. 36, 357 (1980)

THE EFFECT OF DOPING ON THE INTERFACE BETWEEN GaAs AND AlGaAs

W.J. Schaff, P.A. Maki and L.F. Eastman
School of Electrical Engineering
L. Rathbun
The National Research and Resource Facility for Submicron Structures
B.C. De Cooman and C.B. Carter
Material Science Department
Cornell University, Ithaca, NY 14853

ABSTRACT

The quality of the interface between (AlGa)As and GaAs is of interest for application to the inverted modulation doped structure. Results of analysis to assess the quality of interface of the inverted structure as compared to the conventional non-inverted structure are discussed. Si doping is placed on one side of a GaAs quantum well cladded by $Al_{0.3}Ga_{0.7}As$. The concentrations and thicknesses of the Si doped regions are chosen such that the structures are of the modulation doped variety for application as FET devices. Hall mobilities obtained when the inverted modulation doped structure is attempted are inferior to the non-inverted structure. Cross-section TEM measurements indicate uniform quantum well thicknesses and no extended crystal defects at either interface. Photoluminescence from the structures is discussed.

INTRODUCTION

Modulation doped heterostructures have attracted considerable interest recently as their high conductances promise substantial improvements in FET performance. While the conventional (AlGa)As grown on GaAs structure exhibits high mobility, the GaAs grown on (AlGa)As structures typically exhibit lower mobility. This has been attributed to interface roughness, impurities, or Si diffusion [1]. Recently, improvements have been made through growth at low substrate temperatures [2] or lower doping concentrations [3].

Previous work [4] where quantum wells have doping applied to the last grown confinement layer have been fabricated and exhibit good performance as modulation doped structures. Studies of undoped single quantum well (SQW) structures have shown that by careful control of growth conditions very narrow well luminescence linewidths can be obtained [5]. The photoluminescence of quantum well structures provides qualitative evidence of interface quality [6,7]. This suggests that there are no intrinsic problems with GaAs grown on undoped (AlGa)As. In this work, we show that the metallurgical interfaces are smooth and free from defects and dislocations and provide evidence that Si redistribution is in fact the predominate mechanism which influences the heterojunction quality. A quantum well structure is used to examine the influence of Si doping on the two dimensional electron gas mobility and photoluminescence behavior. Comparison is made between photoluminescence and the quantum well structure as determined by cross-sectional TEM measurements. Attempts to measure redistribution of Si impurities by SIMS are also discussed.

MATERIALS GROWTH AND HALL EFFECT MEASUREMENTS

Typical structures investigated are shown in Fig.1. They have been grown by Molecular Beam Epitaxy (MBE) in a Varian GEN II machine.

16

The Al$_{0.3}$Ga$_{0.7}$As layers and GaAs well are grown at a substrate temperature of 680°C with GaAs growth rates of 1μm/hr and a metal stable (3x1) surface reconstruction. The measured Hall mobilities and sheet concentrations for these structures is shown in Fig. 1. The high mobility modulation doped quantum well where doping is introduced after the last grown interface is typical of what is obtained for the conventional modulation doped structures. The mobility for the inverted configuration is poor by comparison.

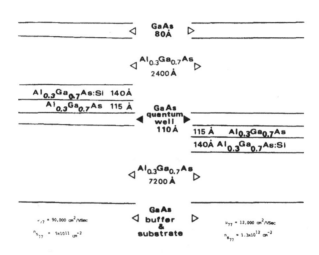

Fig.1.Structures grown
by MBE.

SIMS

Secondary Ion Mass Spectroscopy (SIMS) measurements were performed to measure the Si profiles in the two structures. The measurement of Si profiles in AlGaAs is complicated by the presence of AlH$^-$ which has AMU 28 and limits the sensitivity of Si detection. To avoid this problem, the signal for ^{30}Si$^+$ was observed for comparison to ^{28}Si. Where agreement between the two profiles is good, it can be assumed that the ^{28}Si profile is representative of the actual Si profile. The results of the measurement are seen in Figs. 2 and 3. Si appears to exist predominantly in the locations where it has been intentionally introduced. Some accumulation of Si at the surface of each structure is also observed. In the structure with Si added after the GaAs well a Si accumulation is observed in the near surface region, 2000 Å beyond the region where it was intentionally introduced, and is evidence of surface segregation of the dopant.

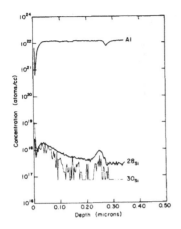

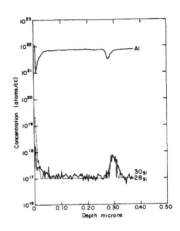

Fig.2. SIMS profile for structures with Si-doping above the SQW.

Fig.3. SIM: profile for structures with Si-doping beneath the SQW.

PHOTOLUMINESCENCE

While photoluminescence (PL) from excitons in quantum wells is a useful technique for evaluating interface quality and the presence of impurities in undoped quantum well structures, the luminescence from modulation doped wells is more difficult to correlate with structural properties. A net carrier density and band bending in the quantum well region has been shown to cause broadened luminescence linewidth while the structure still exhibits excellent electrical properties [8]. The luminescence is likely due to free carrier recombination rather than the excitonic mechanism of undoped wells. The usefulness of PL here is in a comparison between the samples rather than the evaluation of interface quality.

In Fig. 4, the luminescence of a SQW (Fig.1a) with Si doping after the well and SQW (Fig. 1b) with Si doping before the well are shown. The peaks in the region of 1.9 eV in Figs. 4a,b are from the (AlGa)As and indicate the similar high quality of the (AlGa)As cladding layers in both samples. A distinct difference between the 2 samples is observed in the main luminescence peak at lower energies (1.5-1.6eV). These peaks are shown with an expanded scale in Figs. 5a and 6a. Fig. 5a shows an asymmetric emission which is characteristic of a modulation doped SQW while Fig. 6a is broader and nearly symmetric. To identify the source of this emission, the surfaces of the samples were etched to remove approximately 1000 Å. The resulting PL spectra are shown in Figs. 5b and 6b. The spectra of Figs. 5b, 6b and 5a show very similar structure, and the emission energy suggests that they are from wells of similar metallurgical structure. The broad emission at 1.53 eV in Fig. 6a is likely due to the near surface region rather than the quantum well.

18

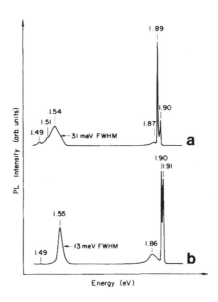

Fig.4. PL spectra for:
a) Si-doping above the SQW.
b) Si-doping beneath the SQW.

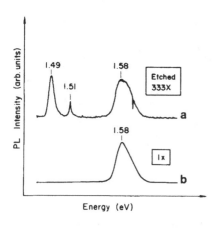

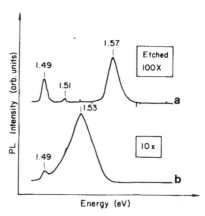

Fig.5. PL spectra of sample with Si
above the SQW.
a)as grown sample.
b)sample with etched surface

Fig.6. PL spectra of sample with
doping beneath the SQW.
a)as grown sample.
b)sample with etched surface.

TEM MEASUREMENTS

The single quantum well structures were studied by means of (002) dark-field microscopy and lattice imaging. The cross-sectional TEM results give information regarding:

a) the exact quantum well thickness and thickness homogeneity
b) the chemical abruptness of the interface at the GaAs quantum well
c) the lattice structure of the quantum well and interfaces
d) the effect of possible smearing of the Si doping profile as a result of Si segregation to the surface during MBE growth.

In Fig. 7, a (002) dark-field image of normal GaAs/n:AlGaAs hetero-junction (Fig. 1) is shown. The quantum well thickness is 93 Å. The top 7.3 nm GaAs layer is of homogeneous thickness and composition and does not show any features. For samples which contain both normal and inverted GaAs/n:AlGaAs heterojunction interfaces the situation is different.

In Fig. 8, (002) dark-field images of an inverted heterojunction is shown. The 3 nm GaAs quantum well was found to be homogeneous in thickness and composition. The top 7 nm GaAs layer however shows clear black triangular features bound by the (001) surface and the two {1$\bar{1}$1} type planes. These lattice features will be referred to as "pyramids." The density of pyramids varies (compare Figs. 8a and 8b) and is highest at the edge of the wafer. These pyramidal defects were studied by lattice imaging in the [110] cross-sectional direction. The results are summarized in Fig. 9.

Most of the surface is rough with a roughness of ~ 2-3 nm (Fig. 9a). Pyramids with cut-off tops lying entirely within the top GaAs layer (Fig. 9b) and fully developed pyramids (Fig. 6c) which extend down into the (AlGa)As are always observed in inverted heterojunctions structures. From the TEM contrast, the pyramids appear to consist of amorphous material of atomic number lower than GaAs. The conclusions of the TEM work are the following:

a. The Si-doping does not seem to influence the quality of GaAs quantum wells. There is no evidence for the presence of Al in the GaAs wells.
b. The TEM observations reveal that the surface segregation of Si gives rise to the formation of surface roughness and pyramids of amorphous material. The same results were also observed for inverted heterojunction structures grown on a superlattice buffer or with a superlattice substituted for the undoped spacer layer [9].

The observation of Si pyramids is reminiscent of the work by Dutt et al [10] on graded bandgap (AlGa)As:Si LED's grown by LPE. Although the chemistry of the pyramids could not be determined by means of SIMS, AES or AEM, previous work [10,2] and the present TEM results suggest that the observed surface features are caused by the surface segregation of the Si dopant.

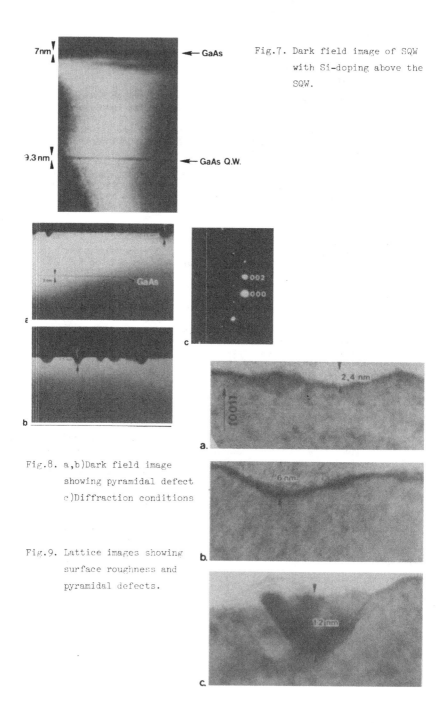

7nm ← GaAs

9.3 nm ← GaAs Q.W.

Fig.7. Dark field image of SQW with Si-doping above the SQW.

GaAs

● 002
● 000

a.

b.

c

Fig.8. a,b)Dark field image showing pyramidal defect c)Diffraction conditions

Fig.9. Lattice images showing surface roughness and pyramidal defects.

2.4 nm

[0011

a.

6 nm

b.

12 nm

c.

CONCLUSIONS

The addition of Si doping beneath a GaAs quantum well does not result in measurable reduction in AlGaAs/GaAs interface quality for the structures studied in this work. Photoluminescence measurements reveal emission from the surface of structures where doping is attempted beneath the first grown interface. Etching of the surface material removes the source of the emission and expected quantum well emission is then observed. TEM measurements of this surface reveal the presence of thin amorphous structures presumed to be Si.

The surface segregation of Si observed by SIMS (Fig. 2) in the sample with Si introduced after the SQW (Fig. 1a) is likely to occur in the sample with Si before the SQW (Fig. 1b). This suggests that the reduced mobility results from increased impurity scattering due to surface segregated Si atoms incorporated as the GaAs well growth begins [11].

ACKNOWLEDGEMENTS

The authors would like to acknowledge the support of IBM and the Office of Naval Research for this work. Discussions and technical assistance by A.R. Calawa, M.A. Hollis, G.W. Wicks, and J.D. Berry are also acknowledged.

REFERENCES

[1] H. Morkoc, T.J. Drummond, and R. Fischer, J. Appl. Phys. 53, 1030 (1982).

[2] K. Inoue and H. Sakaki, Jpn. Jour. Appl. Phys. 23, L61 (1984).

[3] S. Sasa, J. Saito, K. Nanbu, T. Ishikawa and S. Hiyamizu, Jpn. Jour. Appl. Phys. 23, L573, (1984).

[4] L.H. Camnitz, P.A. Maki, P.J. Tasker and L.F. Eastman, 11th Int. Symp. on GaAs and Related Compounds, Biarittz, France, Sept. 26-28, 1984.

[5] P.A. Maki, S.C. Palmateer, G.W. Wicks, L.F. Eastman and A.R. Calawa, Jour. Electr. Mtls. 12, 1051, (1983).

[6] C. Weisbuch, R. Dingle, A.C. Gossard, and W. Wiegmann, Inst. of Phys. Conf. Ser. 56, (IOP, London, 1981) Ch. 9., p. 711.

[7] J. Singh, K.K. Bajaj and S. Chaudhuri, Appl. Phys. Lett. 44, 805, (1984).

[8] P.A. Maki, G.W. Wicks, and L.F. Eastman, Proc. IEEE/Cornell Conf. on High-Speed Semiconductor Devices & Circuits, 209 (1983).

[9] W.J. Schaff, Ph.D. Thesis, Cornell University, Itnaca, NY (1984).

[10] B.V. Dutt, R.A. Ludwig, F. Ermans, J. Cryst. Growth 62, 21 (1983).

[11] M. Heiblum, 3rd Int. Conf. on Molecular Beam Epitaxy, San Francisco, CA, (Aug. 1-3, 1984).

SILICON REDISTRIBUTION IN MBE-GROWN GaAs AND AlGaAs DURING RAPID THERMAL PROCESSING

S. TATSUTA, T. INATA, S. OKAMURA, S. MUTO, S. HIYAMIZU, AND I. UMEBU
FUJITSU LABORATORIES LTD., 1677 Ono, Atsugi, 243-01, Japan

ABSTRACT

The diffusion coefficients of Si in MBE-grown GaAs and $Al_{0.3}Ga_{0.7}As$ have been evaluated using C-V and SIMS measurements. They have been represented empirically by the Arrhenius expression of $D_0 exp(-E_a/kT)$ over the temperature range 750-1050°C. The D_0 and E_a are 3.5×10^{-7} cm^2/s and 1.8 eV for GaAs, and 1.0×10^{-8} cm^2/s and 1.4 eV for $Al_{0.3}Ga_{0.7}As$, respectively. The diffusion coefficients of Si in GaAs are less than half of those in $Al_{0.3}Ga_{0.7}As$. Based on these results, an undoped GaAs/AlGaAs heterostructure spacer layer instead of an undoped AlGaAs spacer layer for the selectively doped GaAs/N-AlGaAs heterostructure has been developed to reduce Si redistribution, and hence, markedly reduced the degradation of mobilities of the two-dimensional electron gas in selectively doped GaAs/N-AlGaAs heterostructure caused by rapid thermal processing.

INTRODUCTION

In order to realize highly-integrated high-speed heterostructure devices, such as the High Electron Mobility Transistor (HEMT) and Heterojunction Bipolar Transistor (HBT), it is indispensable to minimize the redistribution of grown-in impurities during thermal processing following ion implantation. Though the rapid thermal processing using tungsten-halogen lamps has greatly improved it compared with furnace, there is still plenty of room for the improvement especially for the degradation of mobilities of two-dimensional electron gas (2DEG) in selectively doped GaAs/N-AlGaAs heterostructures.[1] However, there are no reports on Si redistribution in MBE-grown GaAs or AlGaAs. In this work, we have studied the diffusion of Si that is the most commonly used impurity in MBE-grown GaAs and AlGaAs, and the spacer layer structures that are effective in stopping Si redistribution during rapid thermal processing for selectively doped GaAs/N-AlGaAs heterostructures.

EXPERIMENTAL

The sample used in this work is schematically shown in Fig. 1. An undoped GaAs buffer layer and a Si-pulse-doped layer, which it consists of GaAs or $Al_{0.3}Ga_{0.7}As$, were grown on a Si-doped (100)-oriented GaAs substrate by molecular beam epitaxy (MBE). The carrier concentrations of Si doped regions are 1×10^{18}, 1×10^{18} and 5×10^{17} cm^{-3}, and their thicknesses are 50, 30 and 50 nm, respectively, for both the GaAs and the $Al_{0.3}Ga_{0.7}As$ one. Prior to thermal processing, the samples were encapsulated with a 0.1 μm-thick AlN film by the reactive sputtering method,[2] and cut into 10x10 mm

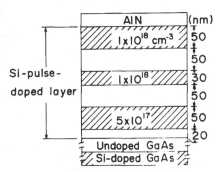

Fig. 1. Schematic diagram of the sample.

squares. The samples were sandwiched by two Si wafers polished on bothsides, and then recieved thermal processing at 900 - 1050°C for 5 - 60 seconds by tungsten-halogen lamps, as described in detail in an earlier paper.[1] Some other samples recieved thermal processing at 750 - 850 °C for 16 - 80 minutes by furnace. The redistribution of Si was investigated by the C-V method along with the step etching technique and by secondary ion mass spectroscopy (SIMS) with high mass resolution. Moreover, it was investigated through observing the degradation of 2DEG mobilities of selectively doped GaAs/N-AlGaAs heterostructures by means of Hall effect measurements at 77K.

RESULTS AND DISCUSSION

Diffusion coefficients

The Si redistribution in $Al_{0.3}$ $Ga_{0.7}As$ was found very large compared with GaAs. The carrier profiles in the as-grown and thermally processed Si-pulse-doped samples measured by the C-V method are shown in Fig. 2 for GaAs and in fig. 3 for $Al_{0.3}Ga_{0.7}As$. The main reason why the carrier profile is not rectangular even for the as-grown sample is that the majority-carrier diffuses from the doped to the undoped layers up to several times the Debye length (15 nm for a carrier concentration of 1×10^{17} cm^{-3}).[3 - 4] The SIMS profiles had the same features as the carrier profiles. The instrument resolution caused mainly by the knock-on effect during SIMS measurements is the reason for non-rectangular profiles. We have fitted gaussian curves for each peak of Si distribution profiles. The diffusion co-efficients, D's, were evaluated from the comparison of the gaussian-approximated Si-distribution profiles for the as-grown and thermally processed samples. They are caluculated from the equation:

$$\delta^2 = \delta_0{}^2 + 2Dt \qquad (1)$$

where δ_0 and δ are the standard deviations of the gaussian curves for as-grown and thermally processed samples, and t is a period of thermal processing. In Fig. 4 the evaluated diffusion coefficients of Si in GaAs and $Al_{0.3} Ga_{0.7} As$ are shown versus the reciprocal temperature of thermal processing.

The diffusion coefficients obtained from the C-V method and SIMS analysis agree very well. The

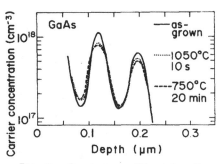

Fig. 2. Carrier profiles in the Si-pulse-doped GaAs layers for as-grown and thermally prosecced samples by tungsten-halogen lamps (1050°C, 10 s) and by furnace (750°C, 20 min).

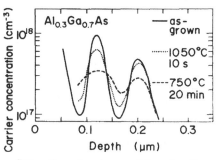

Fig. 3. Carrier profiles for the Si-pulse-doped $Al_{0.3}Ga_{0.7}As$ layers. The same thermal processing conditions in Fig. 2.

diffusion coefficients in GaAs are about one order smaller than those of Si in GaAs doped by the ion implantation technique.[5] From these results, no extrinsic defects enhancing the Si redistribution are induced during thermal processing in our experiment.

The diffusion coefficients of Si are represented by the Arrhenius expression as:

$$D_{Si} = D_0 exp(-E_a/kT) \qquad (2)$$

where the prefactor D_0 and the activation energy E_a are summarized in Table I.

The diffusion coefficients of Si in GaAs are less than half of those in $Al_{0.3}Ga_{0.7}As$. Judging from the values of the diffusion coefficients and the activation energies, the diffusion process of Si is considered to be interstitial-substitutional.

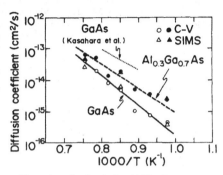

Fig. 4. Evaluated diffusion coefficients of Si in GaAs (open circles and triangles) and $Al_{0.3}Ga_{0.7}As$ (closed circles and triangles) versus the reciprocal temperature of the thermal processing. The dotted line shows those of Si in GaAs doped by ion implantation technique.[Ref. 5]

Table I. Summary of prefactors and the activation energies of the Arrhenius expression for diffusion coefficients.

	D_0 (cm^2/s)	E_a (eV)
GaAs	3.5×10^{-7}	1.8
$Al_{0.3}Ga_{0.7}As$	1.0×10^{-8}	1.4

Spacer layer structure for GaAs/N-AlGaAs

The 2DEG characteristics in a selectively doped GaAs/N-AlGaAs heterostructure degrade during thermal processing at higher than 800°C for 10 seconds.[1] We should minimize this effect to get high device performance. On the other hand, to get the higher electrical activation and the more heavily doped regions in the ion implantation process, the higher temperature thermal processing is desired. We propose a new spacer layer structure to accomodate the two factors in the IC process.

Figure 5 shows the schematic diagram of selectively doped GaAs/N-AlGaAs heterostructure and the spacer layer structures, a conventional single layer and a proposed heterostructure layer. The

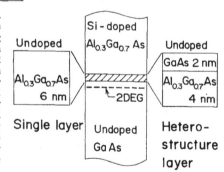

Fig. 5. Schematic diagram of the selectively doped GaAs/N-AlGaAs heterostructure (center) and its conventional spacer (left-hand) and the proposed heterostructure spacer (right-hand).

proposed spacer consists of a 2 nm-thick undoped GaAs and a 4 nm-thick undoped $Al_{0.3}Ga_{0.7}As$ layer where the GaAs layer is added in order to suppress the Si redistribution from AlGaAs to GaAs. If the Si redistribution takes place, the 2DEG mobility will decrease due to impurity scattering.

Figure 6 shows how the Hall mobility μ of 2DEG at 77K normalized by as-grown mobility μ_0 depends on the temperature of 10-second thermal processing by tungsten-halogen lamps. Here, μ_0 is 100,000 cm^2/Vs at 77K. Thus, this means the degradation of 2DEG characteristics due to thermal processing. There are two noteworthy factors. The starting temperature of degradation for the heterostructure spacer layer is about 50°C higher than that for the single spacer layer. In addition, the heterostructure spacer layer maintains more than 10% higher mobilities than the single spacer layer. These results indicate the Si redistribution in AlGaAs is suppressed in the heterostructure spacer layer. Although a small amount of electron wave function, 0.4% of 2DEG, penetrates into the GaAs quantum well in the proposed heterostructure spacer, no electrons accumulate there because of the high energy level in the well.[6] Therefore, it is concluded that the heterostructure spacer is promising to the IC process using the selectively doped heterostructures.

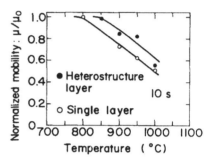

Fig. 6. Normalized 2DEG mobility, μ/μ_0, at 77K versus the temperature of 10-second thermal processing by tungsten-halogen lamps.

ACKNOWLEDGEMENTS

The authors would like to express thanks to S. Sasa for his valuable comments on the 2DEG wave function, to Y. Yoshioka of Matsushita Technoresearch Inc. for his valuable discussion on SIMS analysis, as well as to T. Misugi, O. Ryuzan, T. Kotani, O. Otsuki, and H. Nishi for their encouragement throughout the course of this work.

REFERENCES

1. S. Tatsuta, T. Inata, S. Okamura, and S. Hiyamizu, Jpn. J. Appl. Phys. 23 (1984) L147.
2. S. Okamura, H. Nishi, T. Inata and H. Hashimoto, Appl. Phys. Lett. 40 (1982) 689.
3. D. P. Kennedy, P. C. Murley, and W. Kleinfelder, IBM J. Res. Dev. 12 (1968) 399.
4. S. M. Sze, Physics of Semiconductor Devices, 2nd ed. (John Wiley & Sons Inc., 1981) p.78
5. J. Kasahara and N. Watanabe, Semi-insulating III-V Materials, Evian, 1982, edited by S. Mkram-Ebeid and B. Tuck (Shiva, Nantwich, Cheshire, 1982) 238.
6. S. Sasa, Private communication

AMORPHOUS SEMICONDUCTOR COMPOSITIONALLY MODULATED SUPERLATTICES

BENJAMIN ABELES
Exxon Research and Engineering Co., Annandale, N.J. 08801

ABSTRACT

Superlattices consisting of alternating layers of hydrogenated amorphous silicon (a-Si:H) with other hydrogenated amorphous semiconductors and insulators such as a-Ge:H, a-Si$_{1-x}$C$_x$:H, a-SiN$_x$:H, and a-SiO$_x$:H provide a powerful new probe for studying the structural and electronic properties of amorphous semiconductors and their interfaces.[1-6] An essential property that makes the new superlattice materials attractive for basic studies as well as for potential technological applications is the ability to make extremely smooth and uniform layers with atomically abrupt interfaces, as demonstrated by TEM[7] and X-ray[1,2] diffraction measurements.

An important fundamental question regarding the electronic properties of the new superlattice materials is whether they exhibit quantum size effects similar to those in the crystalline superlattices. Evidence for quantum confinement effects in a-Si:H layers less than 40Å thick comes from optical and electrical measurements of superlattices[1,3,4,6] and ultra thin films.[8] The large blue shift in the optical absorption edge and increase in the electrical resistivity in the plane of the layers with decreasing a-Si:H layer thickness observed in a-Si:H/a-SiN$_x$ superlattices has been attributed to an increase in the optical band gap in the a-Si:H layers due to confinement of the charge carriers in quantum wells between the larger band gap a-SiN$_x$:H layers. To derive the shift in the optical band gap Tiedje et al,[3] used a conduction and valence band model of free electrons and holes in a one-dimensional periodic potential. The model was also applied successfully to the a-Si:H/a-Si$_{1-x}$C$_x$:H[3,9] and a-Si:H/a-Ge:H[10] systems, where the difference in the band gaps is smaller than in a-Si:H/a-SiN$_x$:H and tunneling of carriers between wells is important.

When the a-Si:H layer thickness in the a-Si:H/a-SiN$_x$:H superlattice is larger than 40Å then the in-plane resistivity of the layered material is much lower than that of bulk a-Si:H. The low resistivity (10^3 Ω cm) persists even for relatively thick layers ($L_S = 1200$Å). The decrease in resistivity has been ascribed to transfer doping.[2,11] The underlying mechanism is pinning of the Fermi level E_F in the a-SiN$_x$:H layers.

Because of the large number of essentially identical interfaces the superlattice structure makes it possible to increase by orders of magnitude the sensitivity of measurements of interface properties. Interface defects have been studied by photoluminescence,[1,12,13] photo and dark conductivity[11] and electroabsorption.[14] Photoconductivity measurements in a-Si:H/a-SiN$_x$:H superlattices indicate that the interface state density per layer due to Si dangling bonds is about 2×10^{11} cm^{-2}. On the other hand Roxlo et al[14] deduced an interface state density of $\sim 4 \times 10^{12}$ cm^{-2} from the built in potentials in the superlattice, determined by electroabsorption measurements. Several possible explanations have been proposed for the difference in the defect densities obtained by the different techniques.[12,14] Another tool for probing the structures of the interfaces is photoemission spectroscopy. Measurements of a-Si:H/a-SiN$_x$:H and a-Si:H/a-Si$_{1-x}$C$_x$:H heterojunctions show that the interfaces are close to atomically abrupt and that the sequence in which the two layers are deposited is important.[15]

The ability to synthesize layered amorphous semiconductors with interesting and novel electronic properties is expected to have a major impact on the science and technology of amorphous semiconductors.

1. B. Abeles and T. Tiedje, Phys. Rev. Lett. 51, 2, (1983).
2. B. Abeles, T. Tiedje, K. S. Liang, H. W. Deckman, H. E. Stasiewski, J. C. Scanlon and P. M. Eisenberger, J. Non-Cryst. Solids 66, 351 (1984).
3. T. Tiedje, B. Abeles, P. D. Persans, B. G. Brooks, and G. D. Cody J. Non-Cryst. Solids, 66, 345 (1984).
4. J. Kakalios, H. Fritzsche, N. Ibaraki, and S. R. Ovshinsky, J. Non-Cryst. Solids, 66, 339 (1984).
5. M. Hirose and S. Mazaki, J. Non-Cryst. Solids, 66, 339 (1984).
6. B. Abeles and T. Tiedje in Semiconductors and Semimetals: "Hydrogenated Silicon (J. Pankove, ed) Part C", p. 407, Academic Press, NY 1984.
7. H. W. Deckman, J. H. Dunsmuir and B. Abeles, Appl. Phys. Lett. (in print).
8. H. Munekata, and H. Kukimoto, Jap. J. of Appl. Phys. 22, L542 (1983).
9. C. R. Roxlo, B. Abeles and P. D. Persans, Appl. Phys. Lett. (in print).
10. P. D. Persans, B. Abeles, J. Scanlon and H. Stasiewski, Proc. Int. Conf. on the Physics of Semiconductors, San Francisco 1984.
11. T. Tiedje and B. Abeles, Appl. Phys. Lett. 45, 179 (1984).
12. T. Tiedje, C. B. Roxlo, B. Abeles and C. R. Wronski, Proc. of 1984 Int. Conf. on Solid State Devices and Materials, Kobe, Japan (to be published).
13. T. Tiedje, B. Abeles and B. G. Brooks AIP Conference Proceedings No. 120, Optical Effects in Amorphous Semiconductors (ed. P.C. Taylor and S. G. Bishop) p. 417, 1984.
14. C. R. Roxlo, B. Abeles and T. Tiedje, Phys. Rev. Lett. 52, 1994 (1984). C. B. Roxlo, T. Tiedje and B. Abeles, ibid 13, p. 433.
15. B. Abeles, I. Wagner, W. Eberhardt, J. Stohr, H. Stasiewski and F. Sette, ibid 13, p. 394.

DOPING MODULATED AMORPHOUS SEMICONDUCTOR MULTILAYERS

J. KAKALIOS and H. FRITZSCHE
The Department of Physics and the James Franck Institute
The University of Chicago, Chicago, IL 60637

ABSTRACT

We have synthesized doping modulated amorphous silicon using a two-chamber plasma reactor. These multilayer semiconductors exhibit a photo-induced excess conductivity of one to two orders of magnitude after a few seconds illumination with white light. The steady state photoconductivity and excess conductivity exhibits a different dependence on light intensity for increasing and decreasing illumination intensity. The excess conductivity effect decreases for very small and very large layer thicknesses.

INTRODUCTION

Recently there has been great interest in the electrical and optical properties of doping modulated amorphous semiconductors.[1-5] These materials consist of alternating layers of n-type and p-type doped amorphous semiconductors, typically hydrogenated amorphous silicon (a-Si:H). These multilayer structures exhibit a large photo-induced excess conductivity at room temperature, that is, after a brief illumination with white light the dark conductivity is increased by one to two orders of magnitude over its value prior to light exposure.[1-3] This excess conductivity state is long lived; taking several days to decay at room temperature, but is removed upon annealing at 450°K. This paper describes new experimental results on the dependence of the excess conductivity on light intensity and on the thickness of the n- and p- type layers.

EXPERIMENTAL DETAILS

As described earlier,[1] our doping modulated amorphous semiconductors were prepared by rf glow discharge plasma deposition in a two chamber plasma reactor. The n-type a-Si:H was grown in one chamber by flowing silane pre-mixed with 100 ppm of phosphine while the same volume fraction of diborane was used in the second chamber to produce p-type a-Si:H. The two chambers are separated by a sealed, differential chamber, in which sits a stainless steel ball which can be rotated from the outside. A glass substrate clamped onto a flat section of this steel ball can be exposed sequentially to either chamber. By choosing the length of time the substrate faces either chamber, the individual layer thicknesses can be varied. For the samples reported here the n-doped and p-doped layer thicknesses d_n and d_p are equal. Individual layer thicknesses range from 145 Å up to 1000 Å. The total number of layers were adjusted so that all films had nearly the same total thickness of 1.5μm. In addition to the substrate on the steel ball, there are stationary substrates in both the n-type and p-type plasma chambers onto which homogeneously doped films are deposited as reference samples. All multilayer samples are grown at a substrate temperature of 490°K at a rate of ~ 1.3 Å/sec. The plasma in one chamber was extinguished before rotating the ball to the second chamber.

For conductance measurements along the film plane the multilayer film was scratched with a diamond scribe before deposition of carbon electrodes in order to make contact with all layers. This is necessary because the conductance perpendicular to the layers is very low due to the built-in potential barriers. The contacts were found to be ohmic. The samples were first annealed in darkness under vacuum at 185°C for 35 min. to remove the effects of surface adsorbates and of prior light exposure. This annealed condition is designated state A. The samples were then exposed at room temperature to heat-filtered white light ($h\nu > 1.5$ eV) from a tungsten-halogen lamp. The full thickness of the films was used for calculating the conductivity from the conductance even though most of the current is probably flowing in the n-doped layers.

EXPERIMENTAL RESULTS

Figure 1 shows the room temperature conductivity of a doping multilayer sample having $d_n = d_p = 500$ Å. After reaching state A at room temperature the sample was exposed for periods of 30 sec. to an increasing light intensity up to $F_0 \sim 50$ mW/cm^2. After each light exposure one finds first a fast decay and then a remnant excess conductivity that decreases very slowly and essentially persists for days. With increasing light intensity F both the steady state photoconductivity σ_p and the excess conductivity increase, roughly following a power law relation $\sigma_p \propto F^\gamma$ where the exponent γ lies between 0.3 and 0.5. These values are considerably smaller than $0.6 \leqslant \gamma \leqslant 0.9$ usually observed for σ_p in unlayered a-Si:H.

The saturation value of σ_E that is reached after a few minutes of light exposure is not much larger than the value obtained after the 30 sec. exposure shown in Fig. 1. The ratio σ_p/σ_E is nearly independent of F as the light intensity is increased. Figure 2 shows the equivalent results obtained with the same sample for a sequence of exposures having decreasing intensity. In this case σ_E continues to increase slightly with subsequent exposures even though the saturation values of σ_E that correspond to the smaller F (from Fig. 1) lie lower than σ_E reached by the first high intensity exposure in Fig. 2. The steady state photoconductivity σ_p decreases with F but always remains larger than σ_E. As a consequence σ_p for low F is markedly larger when the sample is in the excess conductivity state E rather than the annealed state A.

Figure 3 shows that the ratio of the conductivities in state E and A is largest near a layer thickness of $d_n = d_p = 300$ Å. We expect that $\sigma_E/\sigma_A = 1$ for zero and very large layer thicknesses because these extremes correspond to an unlayered material which does not show photo-induced excess conductivities.

DISCUSSION

Photo-induced excess conductivities, also called persistent photoconductivities because of their very long lifetimes, have been observed in numerous crystalline semiconductors.[6] They are usually attributed to internal fields which spatially separate the photo-excited electron-hole pairs and inhibit their recombination. In doping modulated semiconductors these fields can be controlled by the doping concentrations and the layer thickness d. It is easy to see that this effect disappears as d approaches zero and infinity, similar to the behavior shown in Fig. 3, because no separation is affected in the first case and the sample becomes field free in the second. Persistent photoconductivities had been predicted for doping modulated crystalline semiconductors.[7] The crystalline superlattices indeed show some of these effects but only of a small magnitude and at low temperatures. The larger band gap and higher density of deep trapping centers in a-Si:H compared to GaAs (used in crystalline doping superlattices) may explain why these effects in doping modulated a-Si:H are much more pronounced and are observed at temperatures as high as 400°K.

However, we note several observations that are difficult to explain with

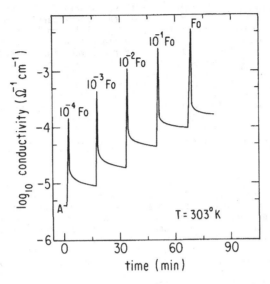

Fig. 1. Change in conductivity of doping modulated a-Si:H ($d_n = d_p = 500$ Å) as a result of 30 sec. light exposures of increasing intensity.

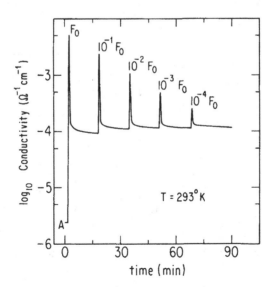

Fig. 2. Same as Fig. 1 but for light exposures having successively decreasing intensity.

the internal field model of Döhler[7] for these materials. One would expect that σ_E will saturate when the field separated charge carriers produce a counterfield that nearly compensates the original internal fields. The same σ_E should be approached for small or large light intensities contrary to our observations shown in Fig. 1. Moreover, for a given light intensity one can estimate the time needed to reach the saturation value of σ_E. For a photon flux of 10^{16} cm^{-2} s^{-1} and an absorption coefficient $\alpha = 10^4$ cm^{-1} there are about 5×10^{19} electron-hole pairs created per cm^3 and sec. An internal field E can separate photo-excited carriers over a distance $\mu\tau E$ where μ is the mobility and τ the lifetime of the carriers. Assuming the smallest value $\mu\tau = 10^{-10}$ cm^2/V measured in a-Si:H[8] one finds an average drift range of 1000 Å in a field of $E = 10^5$ V/cm. This is large compared to d_n and d_p of our samples. Therefore most photo-excited electron-hole pairs are separated; within a few milliseconds more than 5×10^{16} cm^{-3} excess electrons and holes get trapped in the n-type and p-type layers, respectively. This is more than would be needed to compensate a junction field of order 10^5 V/cm. However, instead of milliseconds we require several minutes to achieve saturation of σ_E at this light intensity. At present we cannot accomodate these discrepancies with the predictions of the present field separation model.

Alternative explanations for persistent photoconductivities in crystalline semiconductors have been proposed because of their occurance in supposedly homogeneous materials.[6] One such model involves impurity defect complexes which have the special property that lattice relaxations effectively prevent recombination of carriers released from them by optical excitation.[9] Such special centers are absent in the n-type as well as in the p-type layers of our films because there is no σ_E effect in our n-type and p-type reference samples that were prepared simultaneously with the doping modulated samples. This does not exclude the possibility that special centers having appropriate properties for creating our observed excess conductivities exist at the interfaces between the n-type and p-type layers. If this were so, the ratio σ_E/σ_A should rise with the number of interfaces. Our multilayer films are

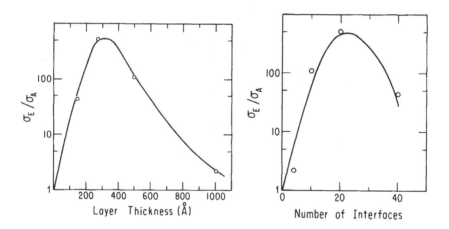

Fig. 3. Ratio of excess conductivity to annealed state conductivity as a function of doping layer thickness $d_n = d_p$.

Fig. 4. Ratio of excess conductivity as a function of number of interfaces of samples of Fig. 3.

grown to maintain a nearly constant total film thickness. Thus the sample with d_n = 275 Å has twice as many interfaces as when d_n = 500 Å. Figure 4 shows that the ratio σ_E/σ_A drops sharply between 20 and 40 interfaces contrary to the above prediction. The initial rise of the curve in Fig. 4 is an obvious consequence of any model (not only that involving interface defects) because without interfaces a-Si:H is homogeneous and σ_E/σ_A = 1.

SUMMARY

We have synthesized doping modulated a-Si:H using a two-chamber plasma reactor. These multilayer semiconducting films exhibit long-lived excess conductivities after a brief exposure to light. This effect disappears at very small and at very large layer thicknesses. The excess conductivity approaches a saturation value after a few minutes of light exposure. This illumination time is considerably longer than expected from a model that attributes σ_E to the spatial separation of electron-hole pairs by the internal junction fields. This model is also at variance with the observation that σ_E increases with light intensity.

This work was supported by Energy Conversion Devices and NSF DMR8009225. We profited very much from the materials preparation facilities of the Material Research Laboratory supported at the University of Chicago by NSF. One of us (J.K.) is an A.T.&T. Bell Laboratories Ph.D. Fellow.

REFERENCES

1. J. Kakalios and H. Fritzsche, Phys. Rev. Lett. 53, 1602 (1984).
2. J. Kakalios, H. Fritzsche and K. L. Narasimhan, in "Optical Effects in Amorphous Semiconductors" ed. by P.C.Taylor and S.G.Bishop (American Inst. Phys., New York 1984) p.425.
3. J. Kakalios and H. Fritzsche, Proc. of the 17th Intl. Conf. on the Physics of Semiconductors, San Francisco, CA., (in press).
4. M. Hundhausen, L. Ley and R. Carius, Phys. Rev. Lett. 53, 1598 (1984).
5. M. Hundhausen, J. Wagner and L. Ley, Proc. of the 17th Intl. Conf. on the Physics of Semiconductors, San Francisco, CA., (in press).
6. M. K. Sheinkman and A. Ya.Shik, Sov. Phys. Semiconductor, 10, 128 (1976).
7. G. H. Döhler, in "Advances in Solid State Physics: Festkoerprobleme", vol. 23, ed. by P. Grosse (Heyden, Philadelphia, 1983) p.207.
8. R. A. Street and J. Zesch, J. Non-Cryst. Solids, 59&60, 405 (1983).
9. D. V. Lang and R. A. Logan, Phys. Rev. Lett. 39, 635 (1977).

PROPERTIES OF $Cd_{1-x}Mn_xTe$-CdTe QUANTUM WELL STRUCTURES AND SUPERLATTICES GROWN BY MBE

R.N. Bicknell, N.C. Giles-Taylor, D.K. Blanks, R.W. Yanka,
E.L. Buckland, and J.F. Schetzina
North Carolina State University, Raleigh, NC 27695-8202

ABSTRACT

Single crystal multilayer structures of the dilute magnetic semiconductor $Cd_{1-x}Mn_xTe$ alternating with CdTe have been grown for the first time by molecular beam epitaxy. Growth techniques used to prepare these structures are described. Results of x-ray diffraction and low temperature photoluminescence film characterization experiments are also discussed.

INTRODUCTION

Dilute magnetic semiconductors (DMS) [1-9] are an interesting new class of materials in which a fraction x of the group II sites of a II-VI semiconductor are occupied by a transition metal ion, usually Mn. The presence of Mn in the DMS lattice leads to a spin-spin exchange interaction, new to the field of semiconductor physics, between localized Mn ions and conduction band electrons. This interaction gives rise to an extremely large Faraday rotation, giant negative magnetoresistance, and large positive electronic g-factors in bulk DMS materials [1-3]. Until recently these materials have only been studied in bulk form. However, the recent work of Kolodziejski et al. [8,9] and Bicknell et al. [5,6] as well as the earlier theoretical work of von Ortenberg [4] has stimulated interest in the growth of thin films and multilayers of these materials.

EXPERIMENTAL DETAILS

In the present work, the $Cd_{1-x}Mn_xTe$ layers (E_g = 1.85-2.1 eV, corresponding to x = 0.19-0.28 at 77 K) serve as barrier layers of thickness L_B between CdTe quantum wells (E_g = 1.59 eV) of thickness L_z. A number of SL's have been prepared consisting of from 14 to 240 CdMnTe-CdTe double layers ranging in thickness from 460 A to 37 A, respectively. The SL samples were grown in an ion-pumped MBE system that has been described in detail in previous publications [5-6]. Briefly, four effusion sources were employed as shown in Fig. 1. Two of these contained CdTe and served as sources of Cd and Te during the growth of the CdTe layers. These two sources also provided a stable source of Cd during the deposition of $Cd_{1-x}Mn_xTe$ layers. The other two cells contained Mn and Te, respectively, and were only employed during $Cd_{1-x}Mn_xTe$ layer depositions. For growth of $Cd_{1-x}Mn_xTe$-CdTe multilayers, the Mn and Te sources were simultaneously opened and closed while maintaining a constant Cd flux from the CdTe sources. The x-value of the $Cd_{1-x}Mn_xTe$ layers was controlled by the choice of operating temperatures of the Te and Mn sources, which were selected such as to produce the same atomic flux densities of Te and Mn, respectively, in the vicinity of the substrate. The SLs were grown on 5 um thick CdTe buffer layers of high structural perfection which were deposited onto 7 x 15 mm (0001) basal plane sapphire substrates.

X-ray diffraction techniques were employed to assess epitaxial film quality, determine the lattice constant of the SLs, and observe the positions of SL x-ray satellite peaks. An estimate of the SL period $L_z + L_B$ was obtained from the diffraction satellite spacings. This result, coupled with a knowledge of the superlattice lattice constant, allowed an estimate of individual layer thicknesses L_z and L_B to be obtained, along with the x-value for the $Cd_{1-x}Mn_xTe$ layers using easily derivable expressions [6].

Low temperature photoluminescence (PL) measurements were carried out at 20 K using 18 mW of chopped radiation (514.5 nm) from an argon ion laser. The laser beam was focused to a spot size of approximately 200 um at the SL sample surface. Further details of the photoluminescence apparatus employed in these experiments are given in an earlier publication [7].

RESULTS AND DISCUSSION

The PL spectra of all of the SLs are dominated by a narrow near-edge peak. This near-edge luminescence is extremely intense--nearly two orders of magnitude brighter than that typically observed from bulk CdTe. The position of the PL edge peak was found to shift to higher energies as the quantum well layer thickness L_z of the SL decreased, as is expected to occur due to carrier quantum confinement effects in the CdTe conduction band wells. Fig. 2 shows a PL spectrum for a sample with a double-layer period of 37 A and a CdTe well width of $L_z = 16$ A. As shown in the figure, the PL edge peak at 20.3 K occurs in the visible at 1.881 eV, corresponding to a +0.3 eV shift in energy from that of bulk CdTe, due to the quantum size effect.

Also visible in the spectrum shown in Fig 2 is a small, broad defect band centered at about 1.5 eV. This is in contrast to earlier PL experiments completed at 82 K in which the defect luminescence was found to be entirely absent from the SL spectra [5]. In the broad defect band shown there is at least one sharp feature which is seen to occur at 1.554 eV in Fig. 2. A similar feature has been seen in PL spectra from single layers of CdTe on sapphire at 1.546 eV [7]. It has been speculated that this feature may be due to donor-to-acceptor pair recombination associated with a Li impurity. The other spectral oscillations shown in this region are due to thin film interference effects associated with optical reflections which occur at the CdTe/sapphire interface when the CdTe buffer layer becomes transparent. Their occurrence is a direct indication of the highly specular nature of the surface morphology of the SL samples. There is also a second narrow and less intense peak that appears at the lower temperatures. This peak occurs at 1.765 eV for the SL shown in Fig. 2 and thus appears to be associated with the quantum well structure. However, the point defect structure associated with these new strained-layer SLs is not understood at the present time so that additional investigations are needed to clarify the origin of the 1.766 eV peak shown.

Another very interesting feature of the low temperature PL data is the dramatic intensity increase compared to PL data taken at 82 K and reported previously [6]. For several SL samples, the intensity increased by more than two orders of magnitude as the temperature was lowered from 82 K to 17 K. For comparision, the PL edge luminescence intensity from a single 5um thick CdTe layer on basal plane sapphire only increases by a factor of ten over the same temperature range. It is also interesting to note that the greatest increase in PL intensity was observed for the SLs having smallest quantum well layer thicknesses ($L_z = 16$ A). The enhanced luminescence efficiency of the SLs compared to bulk CdTe, as well as the dramatic increase in intensity which they exhibit at low temperatures may be due to modified excitonic recombination rates associated with two-dimensional carrier confinement, as has been reported for $Al_xGa_{1-x}As$-GaAs quantum well heterostructures [10]. Various physical properties of the five SLs studied at low temperatures, including the double-layer period, the well thickness L_z, the x-value of the $Cd_{1-x}Mn_xTe$ barrier layer material, the relative intensity of the luminescence, the factor by which the PL intensity changed between 82 K and 20K, and the FWHM of the PL edge peak are listed in Table 1.

In summary we have demonstrated that molecular beam epitaxy can be employed to produce high-quality multilayer structures in which the dilute magnetic semiconductor $Cd_{1-x}Mn_xTe$ serves as the barrier layer material between CdTe quantum wells. This work opens the door to a whole new area of research

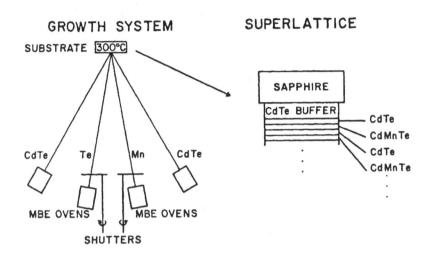

Fig. 1. Schematic of MBE growth system showing arrangement of source ovens.
Superlattice layer growth sequence is shown at the right.

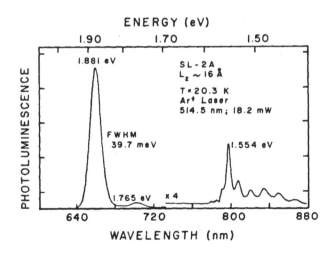

Fig. 2. Photoluminescence from a $Cd_{1-x}Mn_xTe$-CdTe superlattice at 20.3 K, with
L_z = 16 A and x = 0.23.

Table 1. Structural and Optical Properties of Cd$_{1-x}$Mn$_x$Te-CdTe Multilayers

SL Sample	Double Layers	SL Period (A)	CdTe layers L_z (A)	Cd$_{1-x}$Mn$_x$Te layers x-value	E_{1PL} (eV)	Relative Intensity	Intensity Change	FWHM (meV)	Temperature (K)
SL-1A	90	84	37	0.19	1.733	113.8	40	37.5	14.5
SL-1B	90	68	30	0.20	1.793	318.5	18	52.5	20.4
SL-2A	240	37	16	0.23	1.881	164.5	350	39.7	20.3
SL-2B	240	37	16	0.25	1.901	36.7	250	37.7	16.5
SL-3A	60	143	60	0.28	1.704	122.0	2	59.3	21.0
CdTe/sapphire epilayer					1.590	14.5	10	6.3	21.5

in which fundamental aspects of magnetism in two-dimensional structures may be studied. Such fundamental studies may in turn lead to the discovery of novel device structures which exploi⁺ the unique magnetic properties that these materials possess.

The authors wish to thank Professor H.H. Stadelmaier for providing the use of x-ray diffraction equipment. This work was supported by NSF grant DMR83-13036 through the Solid State Physics and Ceramics-Electronic Materials programs and by DARPA/ARO contract DAA29-83-0102.

REFERENCES

1. R.R. Galzka and J. Kossut, in Narrow Gap Semiconductors: Physics and Applications, Lecture Notes in Physics Series, No. 133 (Springer, Berlin, 1980), p. 245.
2. J.K. Furdyna, J. Vac. Sci. Technol. 21, 220 (1982).
3. J.K. Furdyna, J. Appl. Phys. 53, 7637 (1982).
4. M. von Ortenberg, Phys. Rev. Lett., 49, 1042 (1982).
5. R.N. Bicknell, N.C. Giles-Taylor, D.K. Blanks, E.L. Buckland, and J.F. Schetzina, Appl. Phys. Lett., 45, 92 (1984).
6. R.N. Bicknell, N.C. Giles-Taylor, D.K. Blanks, R.W. Yanka, E.L.Buckland, and J.F. Schetzina, paper presented at the 3rd International MBE Conference, San Francisco, CA (1984)-- to be published in J. Vac. Sci. Technol. B3, March-April (1985).
7. N.C. Giles-Taylor, R.N. Bicknell, D.K. Blanks, T.H. Myers, and J.F. Schetzina, (to be published in J. Vac. Sci. Technol.).
8. L.A. Kolodziejski, T.C. Bonsett, R.L. Gunshor, S. Datta, R.B. Bylsma, W.M. Becker, and N. Otsuka, Appl. Phys. Lett. 45, 440 (1984).
9. L.A. Kolodziejski, R.L. Gunshor, S. Datta, T.C. Bonsett, M. Yamanishi, R. Frohne, T. Sakamoto, R.B. Bylsma, W.M. Becker, and N. Otsuka, paper presented at the 3rd International MBE Conference, San Francisco, CA (1984) -- to be published in J. Vac. Sci. Technol. B3, March-April (1985).
10. H. Jung, A. Fischer, and K. Ploog, Appl. Phys. A 33, 97 (1984).

HIGH RESOLUTION ELECTRON MICROSCOPY OF
SEMICONDUCTOR QUANTUM WELL STRUCTURES

C.J.D.HETHERINGTON*, J.C.BARRY**, J.M.BI***, C.J.HUMPHREYS*, J.GRANGE****,
AND C.WOOD****
*Department of Metallurgy and Science of Materials, Oxford University,
Parks Road, Oxford, OX1 3PH, UK.
** Now at Department of Physics, Arizona State University, Tempe, Arizona,
85281, USA.
*** On leave from Vacuum Electron Devices Research Institute, Beijing, China.
**** GEC Hirst Research Centre, East Lane, Wembley, Middlesex, HA9 7PP, UK.

ABSTRACT

As is well known, High Resolution Electron Microscopy (HREM) can image
interfaces at the near-atomic level. However, a particular problem has
arisen in the HREM of expitaxial semiconductor multilayers composed of
similar materials (e.g. GaAs and GaAlAs), since axial HREM images of these
structures reveal little or no contrast between the layers. Hence it is
extremely difficult to see, for example, whether atomic height steps are
present at the interfaces. In this paper, we present new HREM techniques
which provide good contrast between layers of similar materials. These
techniques are applied to GaAs/GaAlAs multilayers and it is shown that
precise measurements, to Å scale accuracy, are possible of the widths of
each individual layer. In addition, interfacial steps are revealed and the
roughness of each interface may be determined.

INTRODUCTION

Very thin epitaxial layers of semiconductor materials, e.g. GaAs and
GaAlAs, of accurately controlled thickness and composition can now be
prepared by molecular beam epitaxy and other methods. The interfaces
between these layers significantly influence their electronic and optical
properties and, ultimately, device performance. Hence there is a need to
determine the interfacial structure and a wide variety of techniques are
used [1].
In this paper, we discuss some uses of conventional and high resolution
transmission electron microscopy (HREM). In particular, we propose two new
HREM techniques which reveal the layers of a GaAs/GaAlAs quantum well
structure with good contrast while lattice imaging the atomic structures
thereby enabling the interfaces to be examined.
This paper refers extensively to the orientation of the microscope
specimen with respect to the incident electron beam direction. At this
point therefore, we define the orientations in the stereogram of fig.1. The
normal to the substrate surface and layers is taken to be parallel (or
nearly parallel) to the [001] axis of the crystal. The normals to our
cross sections are either [1̄10], tilted off [1̄10] (point A on the stereogram)
or [1̄00]. The reflections in the diffraction patterns obtained with these
cross sections are given, taking forbidden reflections into account, by the
poles lying at 90° to the cross section normal. For example, the [110]
diffraction pattern includes the (1̄1̄1), (002), (1̄11), (22̄0), (1̄11̄), (002̄),
(1̄1̄1̄) and (22̄0) reflections.

CONVENTIONAL TRANSMISSION ELECTRON MICROSCOPY

The ability of dark field images to reveal III-V compound semiconductor
heterostructures by diffraction contrast is well established [2]. Fig.2
shows a dark field image of a 50 layer multiple quantum well (MQW) of

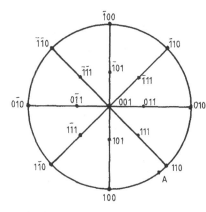

Fig.1
Stereogram showing:
[001] - substrate surface normal
[110], off-axis [110] (position A)
and [100] - directions used for
lattice images.

GaAs/Ga$_{0.7}$Al$_{0.3}$As in (110) cross section using a (002) reflection. The contrast derives from the structure factor F$_{002}$ which is the difference between the scattering factors of the group III atoms and the group V atoms. For example, for Ga$_{1-x}$Al$_x$As.

$$F_{002} = 4((1-x) f_{Ga} + x f_{Al} - f_{As}) \qquad (1)$$

The scattering factors f$_{Ga}$, f$_{Al}$ and f$_{As}$ at the scattering angle of F$_{002}$ are found in [3]. It can be seen that the values of f$_{Ga}$ and f$_{As}$ are very close so that the kinematic intensity of the (002) reflection can then be written

$$I \propto F_{002}^2 \propto 16x^2 (f_{Al} - f_{Ga})^2 \qquad (2)$$

Thus the intensity of the (002) reflection is proportional to the square of x and layers with 5% variations in Al composition from one layer of GaAlAs, to the next have been imaged [4].

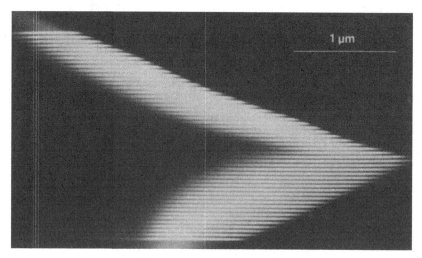

Fig.2 (002) Dark field image of 50 layer MQW

Another technique in CTEM is to image the layers of a MQW by using the superlattice reflections [2] or to study the sharpness of the interfaces by measurement of the superlattice reflection intensities [5]. These reflections have very small scattering angles at which the electron scattering factors for atoms and ions differ greatly [3]. If the bonding of the III-V compound (or even more likely the II-VI compound) is considered to be partly ionic in character, this may need to be taken into account when computing the superlattice reflection intensities and images [6].

HIGH RESOLUTION ELECTRON MICROSCOPY

CTEM dark field images of MQWs, while their interlayer contrast is high, are restricted in spatial resolution to about 5Å by the inevitable use of a small objective aperture. HREM and lattice imaging on the other hand offer near-atomic resolution and should be a more powerful technique for determining interfacial structure on an atomic scale. However, previous workers have stated that the interface between GaAs and $Ga_{1-x}Al_xAs$ is not visible in lattice images when x is small (i.e. x ~ 0.3) owing to insufficient interlayer contrast [7]. (HREM is, after all, usually called upon to provide structural rather than chemical information). We too have found this but propose here new imaging conditions which enhance the interlayer contrast to allow the interfaces to be seen in the lattice image.

Until now, HREM work on GaAs/GaAlAs layers grown on (001) substrates has only been reported when carried out on (110) cross sections with the incident electron beam parallel to the [110] axis (fig.1). We have taken two new types of lattice image; 1) looking at a (110) cross section specimen tilted a few degrees off the [110] axis and 2) looking at a (100) cross section with the incident electron beam parallel to [100]. Both of these images show clear interlayer contrast.

Experimental

The GaAs/$Ga_{0.7}Al_{0.3}As$ layers that are examined were grown on (001) GaAs substrates by MBE at GEC Hirst Research Centre, London. Cross section TEM specimens were prepared by a standard method [8] with the wafer either cleaved along the (110) plane or wire sawn along the (100) plane. After mechanically thinning to 40 μm, the specimens were ion beam thinned with Ar^i ions finishing with 3kV ions, 10 μA specimen current and 12° angle of incidence. We found that thinning specimens at liquid nitrogen (rather than room) temperature substantially reduced ion beam damage and surface roughness - which is a critical parameter particularly for the (100) cross section as discussed later. Microscopy was performed on a JEOL 200CX with a point resolution of 2.4Å, an information resolution limit of 1.8Å and 10° of specimen tilt. (45° of tilt are required to put the [100] or [010] axes of a (110) cross section parallel to the incident electron beam to allow [100], [110] and [010] lattice images of the same specimen to be taken and this has been done on a Philips 430 TEM equipped with a 60° tilting specimen holder).

Results

Figs. 3, 4 and 5 show lattice images of the same GaAs/GaAlAs multilayer. In fig.3, the incident electron beam is along [110] and 7 beams ((000), four {1Ī1} and ±(002)) are included in the objective aperture. The lattice image shows no interlayer contrast as noted above and the approximate position of the interface (arrowed) has been found by referring to a (200) dark field image. The reason fig.3 shows no detectable

Fig.3 [110] lattice image of GaAs/GaAlAs

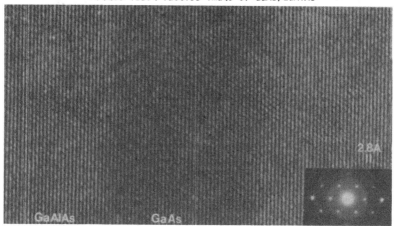

Fig.4 off-axis [110] lattice image of GaAs/GaAlAs

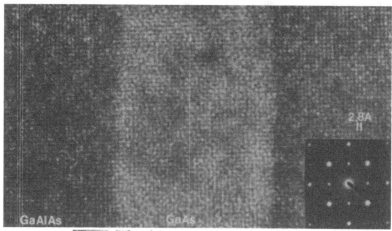

Fig.5 [100] lattice image of GaAs/GaAlAs

interlayer contrast is that the {1Ī1} reflections dominate the image and they are relatively insensitive to composition.

The ±(002) (composition sensitive) reflections in most cases make a negligible contribution to the image except for certain specific specimen thicknesses (when the {1Ī1} reflections have minimum intensity) or specific defocuses (when the contrast transfer function is zero for {1Ī1} spatial frequencies) [9]. In these particular cases the relative contributions of the {1Ī1} and ±(002) beams may be sufficiently favourable for interlayer contrast to occur (see [10] for a possible example of this). Also for multilayers where Al content in the $Ga_{1-x}Al_xAs$ layer is high, e.g. x = 1, then the intensities of the ±(002) beams, which are roughly proportional to x^2 (see equation (2)), may be sufficiently great for interlayer contrast to occur for on axis [110] lattice images [3,11].

The relative contributions of the ±(002) beams can also be increased by tilting the specimen about the [001] axis to position A in fig.1 since this results in a reduction in intensity of the four {1Ī1} beams while the ±(002) beams remain bright. Since the interface plane is parallel to (001), the incident electron beam remains parallel to the interface. As the specimen is tilted from the 110 pole, the interface becomes clearly apparent in the image (fig.4). The (002) fringes now dominate and there is good contrast between the GaAs/GaAlAs layers. The enhanced contrast results not only directly from the reduced intensity of the four {1Ī1} beams described above, but also because there is less double diffraction via {1Ī1} beams into ±(002). The (002) fringes have a spacing of 2.8 Å.

Fig.5 shows a lattice image of a (100) cross section taken with the incident electron beam exactly parallel to [100] with 5 beams in the objective aperture, (000), ±(020) and ±(002). Crossed (020) and (002) lattice fringes result and the contrast between the layers is high. This interlayer contrast can be defined as the ratio of the mean intensities of the image of GaAs and the image of GaAlAs. We have calculated these mean intensities for GaAs and $Ga_{0.7}Al_{0.3}As$ for the [110] 7 beam and [100] 5 beam lattice images taken on our microscope at Scherzer defocus and they are shown as a function of thickness of the cross section in fig.6. The [110] image intensities of GaAs and GaAlAs are similar over this range of thicknesses and the maximum "contrast" or ratio of intensities is 1:1.3 at a thickness of around 200 Å. On the other hand, the [100] image intensities are very different for GaAs and GaAlAs at certain thicknesses; at 125 Å, the "contrast" is 1:3 and at 350 Å, 1:4. The image intensities are seen to be very sensitive to thickness and this explains the need for good specimen preparation, e.g. ion beam thinning at liquid nitrogen temperatures, to ensure good smooth surfaces on the cross section. Simulated images of an interface between GaAs and $Ga_{0.7}Al_{0.3}As$ (fig.7) show the difference in contrast between the [100] and [110] lattice images for a thickness of 200 Å.

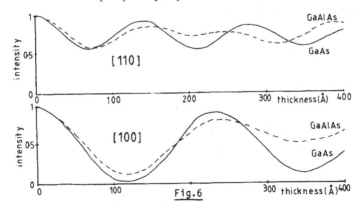

Fig.6

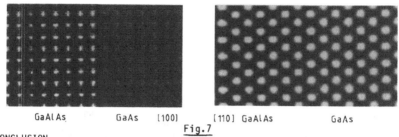

GaAlAs   GaAs   [100]   [110] GaAlAs   GaAs

Fig.7

CONCLUSION

     Using the techniques presented here, interfaces in heterostructures (not just GaAs/GaAlAs of course) can be imaged at the near-atomic level. It is then a simple matter to measure directly the width of each layer by counting the lattice fringes. The orientation of the plane of each interface may be measured (the layers are clearly not exactly parallel to (001) in figs. 4 and 5). It should be noted, though, that any one cross section, say (100), shows only one component of any misorientation of the layers from (001) and a cross section at 90°, say (010), will show a different component. We may also detect and measure interface roughness and interface steps: in fig.4, the interface is rough to about three (002) planes, i.e. ±4 Å. But it must be remembered that the finite thickness of the cross section leads to difficulties of extracting three dimensional information from a two dimensional image - for example, a step in the interface along the middle of the cross section sample would show up in the image as a diffuse interface.
     We conclude that HREM of off-axis (110) and on-axis (100) cross sections offer exciting possibilities in the study of semiconductor quantum well structures since lattice imaging is combined with the interlayer contrast that is available in the (200) dark field image.

References

1.   A.Y.Cho and J.R.Arthur, Prog.Solid State Chem. 10, 157 (1975).

2.   P.M.Petroff, J.Vacuum Sci. Technol. 14, 973 (1977).

3.   International Tables for Xray Crystallography, Vol.4, (1974).

4.   M.R.Leys, C.Van Opdorp, M.P.A.Viegers and H.J.Tulen-Van Der Mheen, J.Cryst.Growth 68, 431 (1984).

5.   T.S.Kuan, Mat.Res.Soc. Symp.Proc. 31, 143 (1984).

6.   C.J.D.Hetherington, C.J.Humphreys and J.C.H.Spence, Proc. 8th European Congress on Electron Microscopy, Budapest, Vol.1, 229 (1984).

7.   H.Ohamoto, M.Seki and Y.Horikoshi, Jpn.J.Appl.Phys. 22, L367 (1983).

8.   J.C.Bravman and R.Sinclair, J.Electron Microscopy Tech.1, 53 (1984).

9.   A.Olsen, J.C.H.Spence and P.Petroff, 38th EMSA 318 (1980).

10. T.Furuta, H.Sakaki, H.Ichinose, Y.Ishida, M.Sone and M.Onoe, Jpn.J.Appl.Phys. 23 L265 (1984).

11. S.Jeng, C.Wayman, G.Costrini and J.Coleman, Mat.Letts.2, 359 (1984).

TERRACING IN STRAINED-LAYER SUPERLATTICES

D. A. NEUMANN,* H. ZABEL* AND H. MORKOC**
* Department of Physics and Materials Research Labaoratory, University of
  Illinois at Urbana-Champaign, Urbana, Illinois 61801
** Department of Electrical Engineering and Coordinated Science Laboratory,
   University of Illinois at Urbana-Champaign, Urbana, Illinois 61801

ABSTRACT

We report an x-ray diffraction study of GaAs/GaAs$_{1-x}$Sb$_x$ superlattices
which shows superlattice reflections that do not lie on a reciprocal lattice
rod of the substrate. This result is attributed to a terraced structure
which imposes a tilt on the entire film with respect to the substrate. This
effect and a transverse broadening of the superlattice peaks are described by
a model which approximates the strain at the interfacial steps.

INTRODUCTION

Strained layer superlattices have recently become the object of much
scientific and technological interest [1-3]. These artificially grown
crystals are composed of alternating layers of two materials whose lattice
parameters may differ by several percent. If the individual layers are made
thin enough, this mismatch can be accommodated by straining the individual
layers until the lattice parameters are equal in the plane perpendicular to
the growth direction. Up to this time however, the only case considered has
been when the growth direction is parallel to a reciprocal lattice vector of
some average lattice. For strain-free superlattices, earlier x-ray
scattering experiments have shown that this is not necessarily the case [4].
A small, but non-zero, angle can exist between the chemical modulation
direction and the reciprocal lattice vector of the average lattice. This
angle has been called the superlattice terrace angle since it arises from
steps, or terraces, at the interfaces. There have also been x-ray
scattering studies of lattice mismatched, single-layer epitaxial films grown
on stepped substrates which have shown that a tilt angle develops between
the lattice planes of the film and those of the substrate due to strains at
the interfacial steps [5]. Here we report the first x-ray scattering study
of terracing in strained multilayers.

EXPERIMENTAL DETAILS

The two GaAs/GaAs$_{1-x}$Sb$_x$ superlattices used in this study were grown by
molecular beam epitaxy on GaAs substrates which were prepared as described in
reference 6. After the native oxide was desorbed, the substrate temperature
was lowered to 540°C and a 0.5 μm GaAs buffer layer was added. Twenty-five
periods of the superlattice were then grown using electronically timed
shutters to control the Sb concentration.

The x-ray scattering results were obtained with a triple-axis
spectrometer using MoK$_\alpha$ radiation from a 12 kW rotating anode x-ray source.
Flat Ge single crystals ((111) reflection) were used for the monochromator
and the analyzer. The instrumental resolution was 0.008° in both the

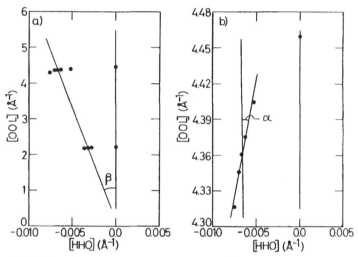

Fig. 1. a) {HHL} reciprocal lattice plane showing the (002) and (004) substrate reflections and the observed superlattice reflections for the GaAs/GaAs.7Sb.3 sample. Note that the scales are quite different resulting in the apparent distortion of the locations of the superlattice peaks. From this figure the tilt angle between the substrate and superlattice is found to be β = 0.08°. b) Expanded view of the (004) substrate reflection and the nearby superlattice reflections. The vertical and the nearly vertical lines correspond to the two lines in Figure 1a. Here one is able to determine the terrace angle α = 1.5° and the modulation length Λ = 428 Å. Note that the x and y scales are still not equal.

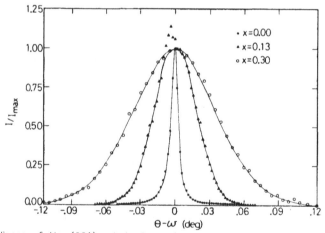

Fig. 2. Scans of the (004) substrate peak and the principal superlattice reflections showing the transverse broadening due to the strain. The x = 0.00 peak exhibits the instrumental resolution. The additional bump on the x = 0.13 reflection is due to the tail of the (004) substrate peak. All peaks have been translated to the same location for the sake of comparison.

angular and radial directions. All x-ray measurements were performed at room temperature.

RESULTS

Figure 1 shows the measured positions of both the superlattice and the substrate reflections in the reciprocal lattice plane defined by the [00L] and [HHO] rods for sample #2 (x = 0.3). Two important angles are seen. First is the angle between the chemical modulation direction and the reciprocal lattice rod of the average lattice of the film. This is the superlattice terrace angle $\alpha = 1.5°$. The other important angle is the superlattice tilt angle $\beta = 0.08°$. It describes the tilt of the average lattice with respect to the reciprocal lattice rod of the substrate. While the modulation length and the terrace angle were approximately the same for both samples, the tilt angle was only 0.02° in sample #1 (x = 0.13) due to the reduced strain. This point will be discussed in more detail later.

In addition to the shift of the superlattice reflections off the [00L] rod of the substrate, a tranverse broadening of the superlattice peaks was observed in both samples. Figure 2 shows that this broadening is smaller for the sample with the smaller strain. It also needs to be noted that there is little change of this width as a function of the order of the superlattice reflection in either sample.

DISCUSSION

The origin of these effects can be understood by considering Figure 3. Here one sees a schematic representation of a superlattice in which the lattice parameter of the white material ($GaAs_{1-x}Sb_x$) is strained to match the lattice parameter of GaAs both along the terrace and in the corner of the step, resulting in a tilt of the $GaAs_{1-x}Sb_x$ layers with respect to the GaAs layers. With x-rays one measures the average tilt of the enitre film with respect to the substrate. This average tilt β can be estimated by assuming that directly above a step, the c-axis lattice parameter in the $GaAs_{1-x}Sb_x$ layer is given by the usual Poisson expansion for a cubic unit cell compressed in two dimensions but free to expand in the third, i.e., $c' = a_x + (a_x - a_0)(1 + 2C_{12}/C_{11})$ where a_x and a_0 are the lattice parameters of $GaAs_{1-x}Sb_x$ and GaAs respectively (7). Also it will be assumed that the lattice parameters match in the corner of the step and that the c-axis lattice parameter of the $GaAs_{1-x}Sb_x$ increases linearly until it reaches the value c' which occurs at the edge of the terrace. This is clearly a simplification of the actual case, however it will give results in essential agreement with observations. With these assumptions and making use of the facts that $C_{11} \approx 2C_{12}$ for most zinc-blende semiconductors and that the thicknesses of the GaAs layers were roughly the same as those of the $GaAs_{1-x}Sb_x$ layers, one obtains

$$\beta \approx 2\alpha\left(\frac{\Delta a}{a}\right) \tag{1}$$

where $\Delta a = x \left(a_{GaSb} - a_{GaAs}\right)$.

Using this model one is also able to find that the transverse peak broadening $\gamma_{\parallel}$ due to the non-uniform strain parallel to the layers is proportional to β [8]. Several different assumptions of the actual form of

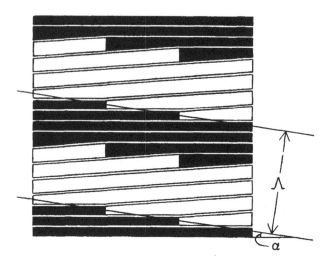

Fig. 3. Schematic diagram of terracing in a strained superlattice. Note the tilt of the white material with respect to the black material due to the "clamping" of the c-axis lattice parameter at the step.

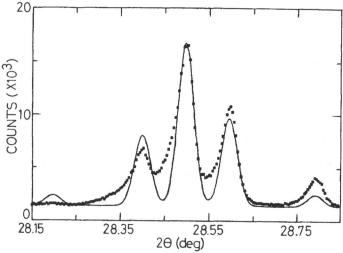

Fig. 4. Comparison of a scan of the superlattice reflections in the modulation direction and the calculated peak intensities. The calculated peak intensities have been normalized to the experimental intensity of the main peak. Also the calculated peaks have been taken to be Lorentzian in shape with peak widths matching the experimental widths. Some translation has been required to adjust for the slight discrepancy between the measured and the calculated tilt angles. Note the asymmetry of the measured intensities is larger than that predicted by the model.

the in-plane strain all give the result that the constant of proportionality is about 1. Thus one has

$$\gamma_\parallel \approx \beta. \qquad (2)$$

It's worth noting that both β and $\gamma_\parallel$ are proportional to the Sb concentration in the $GaAs_{1-x}Sb_x$ layers. It should also be pointed out that if the observed broadening were due to domain size effects related to the interfacial steps, $\gamma_\parallel$ would be the same in both samples since both displayed the same values of α and Λ. A comparison between the measured values of β and $\gamma_\parallel$ and those found from Eqs. 1 and 2 can be found in Table I. While the agreement is not complete, it is seen that the above model reproduces the observed values reasonably well despite the many approximations.

With the above assumptions one is also able to calculate the structure factor

$$F(\vec{Q}) = \sum_j f_j(Q)e^{i\vec{Q}\cdot\vec{r}_j} \qquad (3)$$

and therefore the intensities of the superlattice reflections [8]. Here $\vec{Q}$ is the scattering vector, r_j is the position of the jth atom, $f_j(Q)$ is the atomic form factor of the jth atom, and the sum is taken over the monoclinic unit cell of the superlattice. Figure 4 shows the results of such a calculation compared to an experimental scan obtained by scanning along the chemical modulation direction in the vicinity of the (004) substrate reflection. The calculated peaks have been broadened so that the widths match the experimentally determined peak widths. Also the calculated tilt angle is only 0.07° not 0.08° so the calculated peak positions have been slightly shifted for the sake of comparison. Note that the asymmetry in the experimental peak intensities is somewhat larger than the calculated asymmetry. This discrepancy is probably due to the simplifying assumptions contained in the model since the intensities of the peaks are more sensitive to the details of the strain than are the peak positions.

CONCLUSION

We have for the first time demonstrated that terracing may exist in strained-layer superlattices and that this terracing is responsible for a tilt of the film with respect to the substrate. Also a transverse broadening of the superlattice peaks due to the strain at the terraces has been observed. A simple strain model has been used to satisfactorily describe these effects and also to calculate the intensities of the super-lattice reflections.

TABLE I

	x	Λ	α	β_{meas}	β_{calc}	$\gamma_{\parallel meas}$	$\gamma_{\parallel calc}$
SAMPLE #1	0.13	430Å	1.5.°	0.08°	0.07°	0.09°	0.07°
SAMPLE #2	0.30	428Å	1.5	0.02°	0.03°	0.04°	0.04°

52

ACKNOWLEDGEMENTS

We are grateful to P. F. Miceli for valuable discussions and to J. McMillan for experimental assistance. One of us (D.A.N.) wishes to thank AT&T Bell Laboratories for fellowship support. The x-ray measurements were supported by the Department of Energy, Division of Materials Sciences, under Contract DE-AC02-76ER01198 and were carried out in the Center for Micro-analysis of Materials at the University of Illinois at Urbana-Champaign. The molecular beam epitaxy program was funded by the Joint Sciences Electronics Program.

REFERENCES

1. W. K. Chu, J. A. Ellison, S. T. Picraux, R. M. Biefeld, and G. C. Osbourn, Phys. Rev. Lett. 52, 125 (1984).
2. M. D. Camras, J. M. Brown, N. Holonyak, Jr., M. A. Nixon, R. W. Kaliski, M. J. Ludowsiez, W. T. Dietze, and C. R. Lewis, J. Appl. Phys. 54, 6183 (1983).
3. J. C. Bean, L. C. Feldman, A. T. Fiory, S. Nakahara, and I. K. Robinson, J. Vac. Sci. Technol. A2, 436 (1984).
4. D. A. Neumann, H. Zabel, and H. Morkoc, Appl. Phys. Lett. 43, 59 (1983).
5. Haruo Nagai, J. Appl. Phys. 45, 3789 (1974).
6. T. J. Drummond, H. Morkoc, and A. Y. Cho, J. Cryst. Growth 56, 449 (1982).
7. Armin Segmuller, P. Krishna, and L. Esaki, J. Appl. Cryst. 10, 1 (1977).
8. D. A. Neumann, H. Zabel, and H. Morkoc, to be published.

DOUBLE CRYSTAL X-RAY DIFFRACTOMETRY AND TOPOGRAPHY OF GaAs/GaAlAs AND GaInAs/InP MULTI-LAYERS

M.J. HILL*, B.K. TANNER* AND M.A.G.HALLIWELL**
* Durham University, Dept. of Physics, South Road, Durham DH1 3LE, England
** British Telecom Research Laboratories, Martlesham Heath, Ipswich IP5 7RE,England

ABSTRACT

GaAs/GaAlAs multi-layers grown by molecular beam epitaxy and a GaInAs/InP multi-layer grown by metallo organic vapour phase epitaxy have been examined by double crystal X-ray diffractometry. Experiments have been performed using a conventional laboratory X-ray generator and with synchrotron radiation from the Synchrotron Radiation Source at Daresbury Laboratory. Rocking curves have been simulated using a theoretical model based on the Takagi-Taupin equations for dynamical diffraction in distorted crystals. Agreement between simulated and experimental rocking curves is very good. In addition to the substrate peak a second large diffraction peak is observed which corresponds to the average lattice parameter of the epitaxial multi-layers. First order satellite reflections whose position is determined by the periodicity of the multi-layer structure are observed with approximately one hundreth and one thousandth of the intensity of the strong peaks. The observed satellites correspond well in position and intensity with those predicted theoretically. Double crystal X-ray topographs have been taken using the substrate, layer and, with synchrotron radiation, from the satellite reflections. The lattice defects observed in these topographs are described.

INTRODUCTION

Semiconductor based superlattices have been extensively studied since their growth was first reported some 14 years ago [1], when layers of GaAs and $GaAs_xP_{1-x}$ were alternately deposited on a GaAs substrate by chemical vapour deposition. Since then a number of other superlattices have been developed based on AlAs/GaAs, $Ga_{1-x}Al_xAs$/GaAs and Ge/GaAs systems using molecular beam epitaxy as the method of growth [2-5]. A variety of techniques have been utilised to study their novel properties including X-ray diffraction [2-7], electron microscopy [5], small angle X-ray scattering [8] and photoluminescence [9]. Double crystal X-ray diffraction is a particularly useful method since it is non-destructive and requires no specimen preparation.

We have used double crystal X-ray diffraction to study a number of samples of $n_1Ga_{1-x}Al_xAs/n_2GaAs$ (where n_1 and n_2 are the number of monolayers of each material in the the superlattice unit) grown by MBE on semi-insulating (001) GaAs substates and one sample of $n_1Ga_{1-x}In_xAs/n_2InP$ grown by MOCVD on a (001) InP substrate [13]. When the rocking curve is recorded in the vicinity of a substrate reflection, in addition to the substrate peak a second large diffraction peak is observed, corresponding to a lattice with the average lattice parameter of the multi-layers, along with weak 'satellite' peaks corresponding to high order reflections from the superlattice. By measuring the angular position of such satellite peaks the superlattice period, $C=2n_1d_1+2n_2d_2$, and the total number of monolayers in each unit, $L=n_1+n_2$, can be determined. It is impossible, however, to determine $n_1:n_2$ and x, the concentration of aluminium, directly from the diffraction data. Previously the kinematical theory of X-ray diffraction has been used to calculate the intensity of the satellite peaks as a function of $n_1:n_2$ and x which were then compared to the experimental values [4]. We have,[2] instead, used the dynamical theory of X-ray diffraction, based on the Takagi-Taupin equations for diffraction from non-uniform crystals, to calculate the complete rocking curve from such structures [10,11]. The dynamical theory is better suited to calculate rocking curves particularly where thick encapsulating layers are present [12]

Double crystal topographs have also been recorded using the substrate, average lattice and the first order satellite peaks utilising synchrotron radiation. Such topographs are extremely useful to check the defect concentration in the area being studied, and to check for the presence of mismatch dislocations which would indicate non-coherent layer/substrate and layer/layer interfaces.

EXPERIMENTAL TECHNIQUE

All the data was recorded using a Bragg/Bragg (+ -) double crystal arrangement utilising computer controlled double axis diffractometers in conjunction with both a conventional X-ray generator and synchrotron radiation, from the SRS at Daresbury Laboratory. For both the 002 and 004 GaAs reflections a silicon crystal was used for the first reflection. The high flux available from the synchrotron enables 100 micron beam defining slits to be used, in order to limit the area of the sample illuminated (to reduce the effect of sample curvature and area variation in layer thicknesses and compositions), while still providing enough intensity to record precise data in shorter times than with conventional sources. Double crystal topographs from the substrate and average lattice peaks are obtained in a few minutes while the topograph from the first order satellite peak required four hours (the equivalent time using a conventional generator would be about 15-20 days).

RESULTS

Sample 1

The first sample studied consisted of 100 layers of alternating GaAlAs and GaAs on a semi-insulating GaAs substrate with a 0.3 micron GaAs encapsulating layer. The X-ray rocking curves in the region of the 004 and 002 substrate reflections are shown in figs. 1 and 2 respectively. The +1 and -1 satellite peaks are visible near the 002 reflection while only the +1 is visible near the 004. The separation of the +1 and -1 satellites gives $L=n_1+n_2=71$ which corresponds to C=200.8 A. From the separation of the 004 average lattice peak and the 004 substrate peak we obtain $\bar{x}=0.185$. where $\bar{x}$ is the average concentration of aluminium. ($\bar{x} = n_1 \times /(n_1+n_2)$.) Calculating rocking curves for varying n_1 and n_2 while keeping L and $\bar{x}$ constant enables the best fit curve to be found for $n_1=43$, $n_2=28$, which is also shown in fig. 1. This corresponds to an aluminium concentration of x=0.35 which agrees well with the expected composition from the growth conditions.

004 double crystal topographs are shown in fig. 3, recorded using synchrotron radiation from the SRS operating at 2.0 GeV and 250 mA. The topograph recorded from the substrate peak shows a large concentration of defects, typical of semi-insulating GaAs. In the topographs recorded using the layer and satellite peaks the defect density does not show an increase over the substrate defect density. This implies crystal quality has been maintained across the heterointerfaces. In all three topographs only a restricted area has been imaged, due to residual distortion in the

substrate plus an overall curvature due to presence of epitaxial layers with a lattice parameter about 400ppm larger than the GaAs substrate.

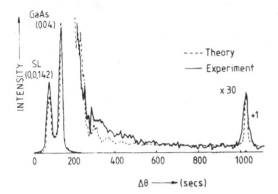

Fig.1. 004 double crystal rocking curve showing average lattice peak and first order satellite peak. Wavelength 1.54 Å.

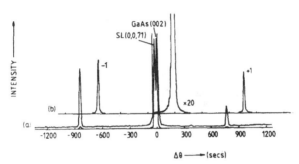

Fig. 2. (a) Experimental 002 double crystal rocking curve from sample 1 showing substrate peak, average lattice peak and the +1 and -1 satellite peaks. Wavelength 1.5 Å. (b) Theoretical 002 rocking curve as a best fit to the experimental curve (a).

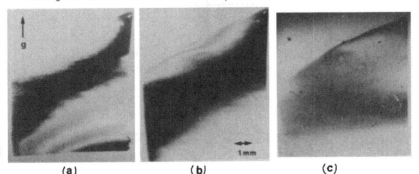

(a)   **(b)**   **(c)**

Fig. 3. 004 double crystal topographs of sample 1, wavelength 1.5 Å. Recorded using (a) substrate (b) layer and (c) first order satellite reflections. g is the direction of the incident beam with respect to the surface of the sample.

Sample 2

The second sample consisted of a multi-layer region of 100 layers of GaAs and 99 layers of GaAlAs with an encapsulating layer of 1 micron of GaAlAs and 200 A of GaAs. A 1 micron layer of GaAlAs separated the first multi-layer and the substrate. The X-ray rocking curve in the region of the 002 substrate reflection is shown in fig. 4. Only the +1 and −1 satellite peaks are observed, with intensities approximately one thousandth of the average lattice peak. The position of the satellites gives L=69 and C=195 Å. The best fit calculated curve is also shown in fig. 4 for $n_1=35$, $n_2=34$ and x=0.32. The observed satellite peaks are broader than the theoretical ones which suggests that the thicknesses of the layers forming the superlattice have a large variation.

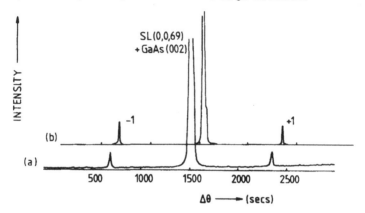

Fig. 4. (a) Experimental 002 double crystal rocking curve from sample 2, showing both first order satellites. Wavelength 1.5 Å. (b) Theoretical 002 rocking curve as a best fit to curve (a).

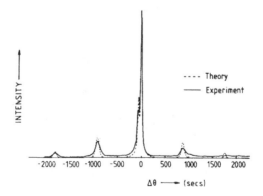

Fig. 5. 002 double crystal rocking curve from sample 3. Wavelength 1.54 Å. Both the first and second order satellite peaks are observed on both sides of the substrate peak.

Sample 3

We have recently studied a superlattice consisting of alternate layers of GaInAs and InP, grown by MOCVD, on an InP substrate. The rocking curve is shown in fig. 5, which shows satellites up to the second order. Due to the large structure factor difference between InP and GaInAs compared to that between GaAs and GaAlAs the satellite peaks are much more intense. The calculated rocking curve is also shown in fig. 5, which agrees well with the experimental curve. The superlattice period is 188 Å with $L=n_1+n_2=64$, and the best fit curve gives $n_1=38$, $n_2=26$ and $x=0.537$.

DISCUSSION

The calculated curves in general agree with the experimental data very well. There are, however, discrepancies in the intensities and widths of the satellite peaks. For the GaAs based structures the agreement between theory and experiment is best for sample 1, while the satellites are less intense and broader than predicted theoretically for sample 2. This implies that the multi-layer structures are less regular for sample 2. Independent confirmation of this result has been obtained using transmission electron microscopy at BTRL. Preliminary results show that the individual layers are less regular in sample 2 than in sample 1. Sample 3, the structure on an InP substrate, also gave broadened satellites. In this case the growth of the multi-layer was manually controlled and it is known that perfect timing was not achieved. This effect can be modelled by introducing random variations in the values of n_1 and n_2 for each repeat unit in the calculations. The effect of such variations on the satellite peaks is marked and illustrated in fig. 6, which shows the effect on the rocking curve for the InP/GaInAs sample. We note that the satellite peaks are rapidly affected where such variations exceed 10%. It is also expected that the composition profile will not be a perfect step function, with some diffusion between adjacent layers, which would also broaden and decrease the intensity of the satellite peaks starting at the highest order [2,4,7]. Additionally, all of the samples studied were curved which will cause a broadening of all peaks including the satellite peaks.

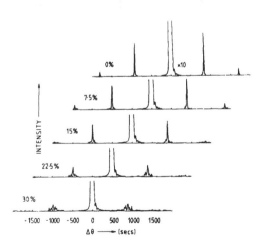

Fig. 6. A series of 002 rocking curves calculated using the dimensions of sample 4, showing the effect of an increasing random variation of the individual layer thicknesses.

ACKNOWLEDGEMENTS

Acknowledgement is made to the Director of Research of British Telecommunications plc for permission to publish this paper. One of the authors (MJH) thanks the S.E.R.C. and British Telecommunications plc for financial support through a CASE award. We thank our colleagues at British Telecom Research Laboratories, D.A.Andrews and Dr. P.C.Spurdens for the samples used in this study, and Dr. M.R.Taylor and Dr. M.Hockley for making available unpublished transmission electron microscopy results. Samples 1 and 2 were grown with financial support from the Joint Opto-electronics Research Scheme.

REFERENCES

1. L.Esaki and R.Tsu, IBM J. Res. Develop. **14**,61 (1970).

2. A.Segmuller, P.Krishna and L.Esaki, J. Appl. Cryst. **10**,1 (1977).

3. C.Chang, A.Segmuller,L.L.Chang and L.Esaki, Appl. Phys. Lett. **38**(11),912 (1981).

4. J.Kervarec, M.Baudet, J.Caulet, P.Auvray, J.Y.Emery and A.Regreney, J. Appl. Cryst. **17**, 196 (1984).

5. P.M.Petroff, A.C.Gossard, W.Weighmann and A.Savage, J. Cryst. Growth **44**, 5 (1978).

6. A.Segmuller and A.E.Blakeslee, J. Appl. Cryst. **6**, 19 (1973).

7. R.M.Fleming, D.B.McWhan, A.C.Gossard, W.Weighmann and R.A.Logan, J. Appl. Phys. **51**(1), 357 (1980).

8. L.L.Chang, A.Segmuller and L.Esaki, Appl. Phys. Lett. **28**(1), 39 (1976).

9. P.L.Gourley and R.M.Biefield, Appl. Phys. Lett. **45**(7), 749 (1984).

10. M.A.G.Halliwell, M.H.Lyons and M.J.Hill, J. Cryst. Growth **68**, 523 (1984).

11. M.A.G.Halliwell, Inst. Phys. Conf. Ser. **60**, 271 (1981).

12. M.J.Hill, B.K.Tanner, M.A.G.Halliwell and M.H.Lyons, submitted to J. Appl. Cryst.

13. R.H.Moss and P.C.Spurdens, Electronics Lett. **20**, 978 (1984).

INTERNAL STRESS AND POLARIZATION ANOMALY IN
DOUBLE HETEROSTRUCTURE SEMICONDUCTOR LASERS

J.M. LIU AND Y.C. CHEN
GTE Laboratories Incorporated, 40 Sylvan Road, Waltham, MA 02254

ABSTRACT

Temperature-dependent polarization behavior and other anomalous polarization characteristics are observed in conventional InGaAsP/InP lasers. These phenomena are attributed to internal thermal stresses in the InGaAsP active layer. These problems do not exist in conventional AlGaAs/GaAs lasers because the thermal stress in the active layer is significantly offset by the GaAs substrate. Stress analyses have been carried out for the layered structures of double-heterostructure InGaAsP/InP and AlGaAs/GaAs lasers to explain the anomalous polarization characteristics observed in InGaAsP/InP lasers and the absence of such phenomena in AlGaAs/GaAs lasers. A stress-compensated InGaAsP/InP layered structure is proposed for polarization stabilization.

INTRODUCTION

Semiconductor lasers normally operate in the TE mode (electric field parallel to the junction) because of the higher mirror reflectivity of the TE mode compared to that of the TM mode (electric field normal to the junction). It has been shown that the TM mode can be induced by applying on the active layer an uniaxial compressive stress normal to the junction [1,2] which lifts the degeneracy of the valence bands and thereby changes the relative contribution of the light and heavy hole interband transitions [2]. The same effect can be caused by an internal tensile stress along the junction plane in the active layer of a double heterostructure laser [3-6]. Several authors have recently reported observations of anomalous TM emission at room temperature from conventional double heterostructure InGaAsP/InP lasers [5,6] in which an internal tensile stress is observed and the lattice in the active layer is stretched along the junction plane. Our analysis [7] shows that a net tensile stress on the order of 10^8 dyn/cm^2 in the active layer will induce sufficient lattice deformation to promote the TM mode gain large enough to compete with the normal operating TE mode, resulting in mode transition from TE to TM or operation of the laser in a mixture of TE and TM modes. This effect has been observed in some InGaAsP/InP lasers at room temperature and in others at lower temperatures due to enhanced internal thermal stress caused by differential thermal expansion [7]. However, this effect does not exist in conventional AlGaAs/GaAs lasers with thin active layers because of their structural differences.

We have carried out the stress analyses for the layered structures of double-heterostructure InGaAsP/InP and AlGaAs/GaAs lasers to explain the polarization anomalies observed in conventional InGaAsP/InP lasers and the absence of such phenomena in conventional AlGaAs/GaAs lasers. An improved stress-compensated InGaAsP/InP laser structure is also proposed for polarization stabilization.

POLARIZATION CHARACTERISTICS

Conventional double-heterostructure lasers normally operate in the TE mode because of the higher mirror reflectivity (and thus the lower output loss) of the TE mode compared to that of the TM mode. Recently, anomalous TM emission related to internal thermal stress has been observed in conventional InGaAsP/InP lasers under various operation conditions [5-7].

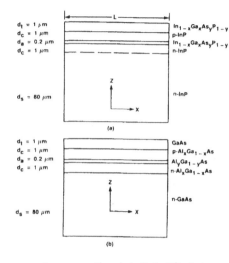

Figure 1   Schematic structure of (a) a typical conventional InGaAsP/InP laser and (b) a typical conventional AlGaAs/GaAs laser.

In conventional InGaAsP/InP laser structures, as is shown schematically in Figure 1(a), the internal stress caused by lattice mismatch and the differential thermal expansion exists mostly in the thin InGaAsP active layer. The stress is larger for lasers operating at longer wavelengths because of a larger compositional mismatch between the active layer and the surrounding InP layers which results in a larger difference in the thermal expansion coefficients. Under usual LPE growth conditions, this stress tends to be tensile along the junction plane ($\sigma_{xx} > 0$) at room temperature. For lasers operating in the wavelength range from 1.3μm to 1.55μm for optical communications, the tensile stress σ_{xx} is usually on the order of 10^8 dyn/cm^2.

A net tensile stress on the order of 10^8 dyn/cm^2 in the active layer will induce sufficient lattice deformation to promote the TM mode gain large enough to compete with the normal operating TE mode [2,4,7]. This results in some undesirable polarization characteristics of InGaAsP/InP lasers. We have observed a temperature-dependent polarization behavior in all the InGaAsP/InP lasers we have tested, which include buried heterostructure lasers and stripe-geometry lasers operating at 1.3μm wavelength. The well-behaved InGaAsP/InP lasers which operate in a pure TE mode at room temperature show operations in a pure TM mode or in a mixture of TE and TM modes at low temperatures. The onset of the TM mode occurs at a critical temperature ranging from room temperature to the lowest temperature (77°K) achievable in our apparatus, characteristic of each individual laser, and is accompanied by a kink in the power-current characteristics. A few InGaAsP/InP lasers are observed to operate in a mixture of TE and TM mode even at room temperature under usual operating conditions. Furthermore, at high injection currents, kinks in the power-current characteristics associated with the appearance of higher order TM modes are observed in many InGaAsP/InP lasers [6]. These kinks are caused by a combination of several effects. However, the stress in the active layer worsens the problem.

As an example, we show in Figure 2(a) the pulsed power-current (P-I) characteristics of a buried crescent InGaAsP/InP laser at various temperatures. At room temperature and above, the laser operates in a pure

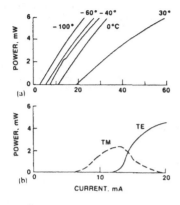

Figure 2 (a) Pulsed power vs current characteristics of an InGaAsP/InP laser at various temperatures. The pulse duration is 100 ns. (b) Polarization-resolved pulse power vs current curves measured at -40°C.

TE mode. When the temperature is lowered to a critical temperature, which is -40°C for this laser, the laser output starts to show a mixture of TE and TM modes. The TM mode gradually becomes dominant with decreasing temperature and the laser operates exclusively in the TM mode at temperatures lower than the critical temperature by 5°C or more. When the laser operates in a pure TE or TM mode, the P-I curves are kink-free. In the transition regime near the critical temperature, kinks are observed at current levels where the laser switches from the TM mode at low injection currents to the TE mode at higher currents. The detailed behavior of the mode transition process is illustrated by the polarization-resolved P-I curves in Figure 2(b). When the laser is driven by a long current pulse, the TM mode is observed in the leading edge of the laser pulse and the transition to the TE mode takes place after ≈1 µsec delay. The time delay decreases with increasing current and/or with increasing temperature. This indicates that the TM to TE switching can be explained by current-induced heating of the junction.

These problems, however, do not exist in conventional AlGaAs/GaAs lasers with thin active layers because of their structural differences, as can be seen in Figures 1(a) and (b). The thermal stress in the active layer of an AlGaAs/GaAs laser induced by the thin cladding layers is compressive in the wavelength range from 8000 Å to 8500 Å. Such compressive stress enhances the gain of the TE mode. In the longer wavelength range, the stress becomes tensile. However, the stress in the active layer of an AlGaAs/GaAs laser is always largely offset by the thick GaAs substrate whose composition is similar to that of the active layer.

STRESS ANALYSIS

It is generally true for conventional buried heterostructure lasers that the laser cavity length L is much (about 50 to 100 times) larger than the active region stripe width W. Therefore, the stress problem in this structure can be reduced to a one-dimensional one. Figure 1(a) shows the schematic multilayer structure of a typical conventional InGaAsP/InP laser. The stress in the active layer caused by thermal strain in the multilayer structure is calculated using the generalized formula in Reference 8. Throughout our analysis, we take the growth temperature to be 650°C and the difference between the growth temperature and room temperature to be -630°C. The thermal expansion coefficients of $In_{1-x}Ga_xAs_yP_{1-y}$ lattice-matched to InP have been measured [9] for y=0, 0.6, and 1 only. They are α = 4.56 x 10^{-6}/°C for InP (y=0), α = 5.42 x 10^{-6}/°C for $In_{0.74}Ga_{0.26}As_{0.60}P_{0.40}$ (y = 0.6, λ = 1.30μm), and α = 5.66 x 10^{-6}/°C for $In_{0.53}Ga_{0.47}As$ (y=1). For other compositions, we assume $\alpha(y)$ = (4.56 + 1.266y) x 10^{-6}/°C, varying linearly with y [10]. The Young's modulus, E, does not vary much with y. Throughout the calculations, we assume a (100) substrate and take a constant effective [11] $(E/1-\nu)_{(100)}$ = 9.487 x 10^{11} dyn/cm^2, where ν is the Poisson's ratio. The positive stress σ_{xx} = 10^8 dyn/cm^2 corresponds to a negative lattice mismatch $\Delta a/a \simeq -10^{-4}$ and a lattice deformation in the active layer with a lattice constant normal to the junction smaller than that along the junction.

In order to grow good expitaxial layered structures for laser devices, it is crucial to control the LPE growth conditions so that the layers are close to perfect lattice-matching at the growth temperature (typically 600°C to 650°C). However, because the $In_{1-x}Ga_xAs_yP_{1-y}$ material has a larger thermal expansion coefficient than that of InP, it is possible to have perfect lattice-match and stress-free conditions at only one temperature in the temperature range from the growth temperature to the device operation temperature. The upper part of Figure 3 shows the stress σ_{xx} at room temperature for the structure in Figure 1(a) as a function of the As content, y, in the active layer, assuming perfect lattice-matching at the growth temperature 650°C. if a lower temperature between the growth and the device operation temperatures is chosen for perfect lattice-match, as is usually done to compromise the lattice-matching requirements at the two extreme temperatures, the tensile stress in the active layer can be reduced linearly. However, under usual LPE growth conditions, the stress σ_{xx} in lasers in the

wavelength range from 1.3μm to 1.55μm for optical communications is usually tensile and on the order of 10^8 dyn/cm^2.

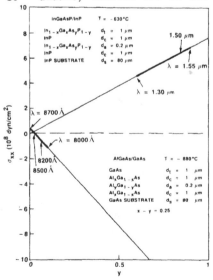

Figure 3 Active layer stress at room temperature for the structures of Figure 1, grown under perfect lattice-matching conditions at 650°C and 900°C for InGaAsP/InP and AlGaAs/GaAs, respectively. y refers to the As content in the cladding layers of the InGaAsP/InP structure and the Al content in the active layer of the AlGaAs structure, respectively.

In order to show the difference between an InGaAsP/InP laser and an AlGaAs/GaAs laser, we also perform a stress analysis for a typical conventional AlGaAs/GaAs laser structure in Figure 1(b), which has the same corresponding layer thicknesses as the InGaAsP/InP structure in Figure 1(a). We take the growth temperature to be 900°C and the difference between the growth temperature and room temperature to be -880°C. Because GaAs and AlAs have a perfect lattice-match at 900°C [12,13], the AlGaAs/GaAs structure is always grown perfectly lattice-matched at the growth temperature. The thermal expansion coefficients are $\alpha = 6.78 \times 10^{-6}/°C$ [12,13] for GaAs and $\alpha = 5.2 \times 10^{-6}/°C$ [12] for AlAs. For the ternary compositions, $Al_yGa_{1-y}As$, we assume a linear function $\alpha(y) = (6.78 - 1.58y) \times 10^{-6}/°C$. We also assume a (100) GaAs substrate and take a constant effective [11] $(E/1-\nu)_{(100)} = 12.39 \times 10^{11}$ dyn/cm^2 throughout the calculations. The composition difference between the cladding layers ($Al_xGa_{1-x}As$) and the active layer ($Al_yGa_{1-y}As$) is kept at $x-y = 0.25$. The calculated stress σ_{xx} at room temperature for the AlGaAs/GaAs structure in Figure 1(b) is shown in the lower part of Figure 3 as a function of the Al content, y, in the active layer. Substantial differences between the InGaAsP/InP and the AlGaAs/GaAs structures can be seen in the active-layer stresses in Figure 3. This is because the relationship among the layers in an AlGaAs/GaAs structure is completely different from that in an InGaAsP/InP structure. Unlike the case of the InGaAsP/InP structure, most of the stress in the AlGaAs/GaAs structure exists in the cladding layers rather than in the active layer. Therefore, a typical AlGaAs/GaAs laser is not subject to the problem of stress-induced polarization instabilities. To our knowledge, internal-stress-induced TM emission was found only in AlGaAs/GaAs lasers with very thick (>10μm) cladding layers [3].

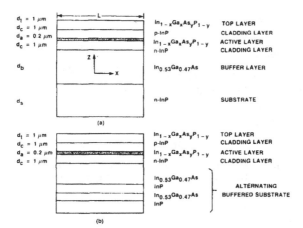

Figure 4 (a) Improved InGaAsP/InP laser structure with an InGaAs buffer layer. (b) Buffered InGaAsP/InP laser structure with alternating InGaAs and InP layers.

CONCLUSIONS

In conventional InGaAsP/InP laser structures, the internal stress caused by lattice mismatch and the differential thermal expansion is tensile along the junction plane and exists mostly in the thin InGaAsP active layer. This internal tensile stress tends to cause polarization anomalies and instabilities in the laser. This effect does not exist in the conventional AlGaAs/GaAs lasers because the internal stress in the AlGaAs/GaAs layered structure exists mostly in the AlGaAs cladding layers due to compositional similarity of the thin active layer and the thick GaAs substrate. Moreover, if a little Al content is added in the active layer, the stress can become compressive, resulting in even more stable operation of the normal operating TE mode.

In order to avoid the TM emission in the InGaAsP/InP lasers, it is very important to control the LPE growth conditions so that the epitaxial layers are perfectly lattice matched at room temperature. In case some residual stress is unavoidable, it is preferrable to have compressive stress along the junction plane rather than tensile as the tensile stress will promote the TM emission. However, this will require a large lattice mismatch at the growth temperature which is rather difficult and not favorable under usual LPE growth procedures. We therefore propose a buffered DH InGaAsP/InP structure, shown in Figure 4(a), which consists of an $In_{0.53}Ga_{0.47}As$ buffer layer for compensation of the tensile stress in the active layer. With proper choice of the layer thickness, it is possible to achieve stress-free conditions in the active layer at all temperatures. A stress-free structure at room temperature can, therefore, be grown under perfect lattice-matching conditions at the LPE growth temperature. An alternative structure with alternating buffer layers is also shown in Figure 4(b). Detailed analysis of these structures are published elsewhere [14].

REFERENCES
1. J.E. Ripper, N.B. Patel, and P. Brosson, Appl. Phys. Lett. 21, 124 (1972).
2. N.B. Patel, J.E. Ripper, and P. Brosson, IEEE J. Quantum Electron. QE-9, 338 (1973).
3. V.A. Elyukhin, V.R. Kocharyan, E.L. Portnoi, and B.S. Ryvkin, Sov. Tech. Phys. Lett. 6, 307 (1980).
4. H.D. Liu and Z.C. Feng, IEEE J. Quantum Electron. QE-19, 1016 (1983).
5. D. Akhmedov, N.P. Bezhan, N.A. Bert, S.G. Konnikov, V.I. Kuchinskii, V.A. Mishurnyi, and E.L. Portnoi, Sov. Tech. Phys. Lett. 6, 304 (1980).
6. D.C. Craft, N.K. Dutta, and W. R. Wagner, Appl. Phys. Lett. 44, 823 (1984).
7. Y.C. Chen and J.M. Liu, Appl. Phys. Lett. 45, 731 (1984).
8. Z.C. Feng and H.D. Liu, J. Appl. Phys. 54, 83 (1983).
9. R. Bisaro, P. Merenda, and T.P. Pearsall, Appl. Phys. Lett. 34, 100 (1979).
10. S. Adachi, J. Appl. Phys. 53, 8775 (1982).
11. W.A. Brantley, J. Appl. Phys. 44, 534 (1973).
12. M. Ettenberg and R.J. Paff, J. Appl. Phys. 41, 3926 (1970).
13. G.A. Rozgonyi, P.M. Petroff, and M.B. Parish, J. Crystal Growth 27, 106 (1974).
14. J.M. Liu and Y.C. Chen, IEEE J. Quantum Electron. to be published.

Epitaxy of Thin Films

THE GROWTH AND STRUCTURE OF EPITAXIAL LAYERS

D.W. PASHLEY
Imperial College, Dept. of Metallurgy & Materials Science, Prince Consort Rd.,
London, SW7 2BP, England.

ABSTRACT

A summary is given of the results of certain systematic studies of epitaxy. The influence of the mode of nucleation of an epitaxial deposit on the
subsequent growth of the layer, and on its structure at various stages of
growth, is summarised. This includes consideration of the way in which dislocations and other lattice imperfections are introduced both at the interface
between the substrate and the deposit and within the deposit. Emphasis is
given to the role of lattice misfit in the formation of these lattice defects,
as well as in the development of lattice strain in the deposits. The evidence
which is presented refers to a range of different substances, including metals,
insulators and semiconductors. An attempt is made to relate the body of this
evidence to the requirements for applying epitaxial growth techniques to the
production of electronic devices.

INTRODUCTION

The phenomenon of epitaxy was first revealed by mineralogists, who noted
a number of examples where two naturally occurring mineral species had grown
together with a unique crystal orientation relationship. The earliest known
successful laboratory experiments on epitaxy were those of Frankenheim [1] who
in 1836 grew sodium nitrate from aqueous solution on the surface of a calcite
crystal. In 1928 Royer [2] made the first systematic and detailed studies of
oriented overgrowths, and it was he who introduced the term epitaxy. In view
of the deviation in the use of the term in recent years, from that intended by
Royer, it is necessary to confirm that this review is concerned only with the
growth of highly oriented deposit layers on single crystal substrates. This
includes the growth on substrates of the same material as that of the deposit
(e.g. silicon on silicon), sometimes referred to as homoepitaxy, and growth
on substrates of a different material (e.g. gold on sodium chloride) sometimes
referred to as heteroepitaxy. Just the general term epitaxy is used in this
paper, since there is no completely sharp distinction between homoepitaxy and
heteroepitaxy in terms of the growth conditions required or the growth mechanisms operative.
    The main concept introduced by Royer is that epitaxy occurs only if the
lattice mismatch between deposit and substrate is small, say less than about
15%. This concept is still seen to have considerable significance today.
    Until about 1930, the occurrence of epitaxy had to be determined by
optical microscopy, which severely limited the range and types of deposits
which could be studied. With the discovery of electron diffraction in 1927,
the scope for surface studies was considerably widened and evidence started to
emerge that epitaxy does occur with misfits considerably in excess of 15% [3].
However, Frank and van der Merwe [4] developed a theory of epitaxy based upon
the concept of small misfit, coupled with the additional concept that the
initial deposit layers are strained so that there is a perfect lattice match
between substrate and overgrowth. This followed partly from the ideas of
Finch and Quarrell [5] who had introduced this lattice matching concept to
which they applied the term pseudomorphism, and it resulted in a theoretical
interpretation of the requirement for a small misfit. Although the Frank and
van der Merwe theory failed to produce a criterion for the occurrence of epitaxy which matched with experimental observations, it was highly successful in
introducing concepts which have considerable importance especially in the

application of epitaxy to semiconductors, as is discussed below.

One of the main reasons why the Frank and van der Merwe theory failed to explain the widespread occurrence of epitaxy is that it assumed that epitaxial growth always occurs in the form of monolayers. The developments in transmission electron microscopy (TEM) in the 1950's resulted in considerable morphological evidence being obtained about the structure of the initial deposits of epitaxial layers, and it soon became clear that, for many combinations of substrate and deposit, growth occurs by the formation of isolated three dimensional nuclei on the substrate surface. As a result, considerable efforts have been made in the last twenty years to understand the different kinds of nucleation mechanisms, and to explain how the orientations of the initial nuclei are determined. The electron microscope also allows the changes in morphological structure of deposits to be followed, as deposition proceeds, and the most powerful technique is to grow the deposits in situ in the electron microscope whilst they are being examined at high magnification. As a result, much has been learned of the detailed growth processes.

A further important advantage of TEM is its ability to provide direct images of lattice defects in the deposits, so that much has been learned of what kinds of defects occur, and how they are introduced. These observations are of real significance to the requirements for growing epitaxial layers for semiconductor devices. Of particular importance are the misfit dislocations which partially, or completely, accommodate the lattice misfit between substrate and deposit. The existence of these misfit dislocations was predicted by Frank and van der Merwe [4], and confirmed experimentally by Matthews[6][7] and others. Once such observations became possible, a sensitive technique for determining the actual lattice misfit between substrate and deposit, for cases of small misfit, became available because the spacing of an array of misfit dislocations is directly related to the lattice misfit. As a result, it was shown that pseudomorphism (i.e. the elastic straining of the deposit lattice to produce a zero misfit with the substrate) does occur under certain conditions, whereas previous attempts to obtain evidence by means of reflection electron diffraction (now called RHEED) had been largely inconclusive or negative [3]. These results have had a profound influence on the application of epitaxy to the making of semiconductor devices.

Much of the research on epitaxy, and the relevant growth processes, carried out before the 1970's involved using techniques which were liable to cause contamination of both the substrate and the deposit. Consequently there was always serious concern as to the extent to which this contamination influenced the results obtained. Some evidence was produced to show that contamination can actually improve the quality of the epitaxy [8]. This evidence was mainly confined to the growth of metal deposits on alkali halides, and since there is little evidence that similar effects occur with other systems, it demonstrates the dangers of drawing general conclusions from observations on one type of substrate deposit combination. Unfortunately the metal/alkali halide combination was by far the most extensively studied system before 1970 so that a number of effects observed with metals on alkali halides have appeared, in the past, to have more general significance than was really justified. The epitaxy of semiconductor layers must be carried out under extremely clean conditions, because of the need to control impurities at a very low level. As a result, there has been a strong stimulus for new evidence to be obtained on the characteristics of epitaxial layers grown under clean conditions, and this evidence is of general interest to the science of epitaxy, as well as being of great technological importance.

The purpose of this paper is to review the evidence on epitaxy, for a range of materials, and to consider how this evidence relates to the needs for producing epitaxial layers for semiconductor device manufacture. Some attempt is made to consider what are the current limitations on further applications of epitaxy to devices.

LATTICE MISFIT IN EPITAXY

The Royer [2] evidence was obtained for the growth of crystals from solution, for a wide range of substrates and deposits, and the most systematic evidence was obtained for alkali halides grown on cleavage surfaces of mica. This resulted in the approximately 15% rule, for the maximum allowable misfit for epitaxy. However, when Schulz [9] grew alkali halide deposits on mica by condensation from the vapour phase, he succeeded in obtaining epitaxy, as judged by RHEED, for misfits as high as 27%. Using alkali halides as substrates also, Shulz [10] obtained epitaxy with misfits ranging from -39% (LiF on KBr) to + 90% (CsI on LiF). Thus there was a complete breakdown of Royer's rule. But it is clear that lattice misfit does have an influence on epitaxy, both in terms of which orientations are preferred and in relation to the mode of growth. Schulz [9] found that for low misfits the initial deposit of an alkali halide on mica tends to be in the form of thin plate-like nuclei, or even monomolecular layers, whereas for misfits larger than about ten percent, the initial deposit consists of more equi-axed three dimensional nuclei.

Until recently, no significant distinction has been made, on either an experimental or a theoretical basis, between the treatment of positive and negative misfits. Markov and Milcher [11] have considered the consequence of there being anharmonicity in the interatomic forces of the deposit. Their theory predicts that anharmonicity should favour orientations with negative misfits (deposit periodicity less than substrate periodicity) as against orientations with a positive misfit. Further they conclude that the critical thickness for pseudomorphic growth of monolayers (see below) should be greater for negative misfits than for positive misfits. Existing experimental evidence is quoted in support of both of these conclusions.

The other important aspect of lattice matching concerns symmetry. There is often a tendency for the overgrowth and substrate crystal planes in contact to have the same symmetry, where this is possible, even when the two crystal structures involved have different symmetries. Thus for cubic materials (e.g. alkali halides or face centred cubic metals) grown on mica, which is monoclinic, it is common for (111) planes of the deposit to grown parallel to the (001) cleavage plane of the mica because both have hexagonal symmetry. In other cases, even when symmetry matching would have been possible, the orientation of the deposit is such that there is a low percentage misfit in one direction only, with the result that there is no symmetry matching. When this happens, there will sometimes be two or more deposit orientations which have equivalent matching arrangements with the substrate, and the deposits consist of a mixture of two or more orientations. This is known as multiple positioning, and means that well defined crystallographic epitaxy does not necessarily lead to a single crystal film.

Even when the lattice misfit is low, the required conditions of growth have to be determined. In particular, a minimum substrate temperature, known as the epitaxial temperature is required [3] although this temperature can also depend upon other growth parameters.

NUCLEATION OF EPITAXIAL DEPOSITS

TEM, and other techniques, have provided much information on the early stages of growth of thin surface films, and this has included much evidence on the density of nuclei, their rate of formation, their size distribution and the dependence of these on the various growth parameters. These experimental measurements, which must be carried out under suitably clean conditions to be reproducible and interpretable, have mostly been made during the last twenty years when there has been considerable theoretical work aimed at providing a quantitative theoretical framework for the understanding of epitaxy[12]. It has been found convenient to divide the nucleation process in to three classes [13], each of which has important consequences for the subsequent growth and structure of epitaxial layers. The three types of nucleation are:

(i)   nucleation in the form of monolayer islands of deposit which grow to-
        gether and form a single monolayer of deposit upon which further growth
        occurs in the same way, so that the deposit grows in a layer by layer
        manner.
(ii)   nucleation of discrete three dimensional nuclei on the substrate surface
        which increase in size and number, until a saturation density of nuclei
        is formed. These nuclei then grow in size until they intergrow with
        each other to form a continuous film. This is often known as the
        Volmer-Weber mode.
(iii)   nucleation and growth as in (i), to produce a small finite number of
        monolayers on which subsequent deposition occurs by the formation and
        growth of discrete three dimensional nuclei as in (ii). This is usually
        called the Stranski-Krastanov mode.
        The existence of these three distinct processes can be understood in
terms of the influence of surface and interface free energies, or in terms of
bond strengths. Thus when deposit atoms are more strongly bound to the substrate than they are to each other, monolayer growth occurs, at least for a
single layer. Additional deposit atoms will continue to form monolayers, provided their binding to the already covered substrate decreases monotonically
from that with the substrate to that with a bulk crystal of the deposit. If
this condition is not fulfilled, further deposit atoms can cluster together
to form three dimensional islands as in class (iii). For class (ii) nucleation
to occur, the binding of deposit atoms to each other is stronger than the binding of deposit atoms to the substrate.
        All three modes of nucleation and growth can and do result in epitaxy, and
when this does happen their main influence on the structure of films produced
after further deposition is determined by the processes and mechanisms which
take place as the nuclei grow and intergrow. Monolayer nucleation can result
in layers many atomic (or molecular) planes in thickness being formed as
successive monolayers. One of the most important features of monolayer growth
is the possibility of lattice matching between the deposit and the substrate
as postulated in the theory of Frank and van der Merwe [4]. Following the
pioneering work of Matthews and his collaborators, reliable experimental evidence for the occurrence of lattice matching has been obtained, and good comparisons between experimental observations and theoretical predictions have
been made. This is outlined below. For Volmer-Weber nucleation, lattice
matching appears to be far less common, which is partly because the natural
misfit for such three dimensional nucleation cases is generally higher than is
normal for cases of strained monolayers which are lattice matched. Thus networks of misfit dislocations occur at the interface between substrate and
deposit.

MONOLAYER GROWTH

        van der Merwe [14] has extended the original concept of lattice matched
strained monolayers to include the effect of thickening the deposit layer, by
determining the equilibrium strain as a function of thickness. As the film
thickens, it is assumed that any reduction in lattice strain is compensated by
a parallel array of edge dislocations at the interface. The spacing between
these dislocations is reduced as the deposit lattice strain decreases. For the
case of a natural misfit along one direction in the interface only, and for
various simplifying assumptions, the theory predicts [15] that for a natural
misfit of only 0.2% the perfect lattice matching will persist up to thicknesses
of 400Å. If the natural misfit is about 4%, perfect matching extends only up
to a thickness of 5Å, above which misfit dislocations are introduced. For
natural misfits of about 10% or greater, accommodation is mainly by misfit dislocations right from the smallest thickness. Experimental observations of
metals deposited on metals (Pb on Ag ; Au on Ag ; Pt on Au) are consistent with
these theoretical predictions, and support the general concepts built into the
theory.

In order for these predictions to apply, misfit dislocations must be introduced at the interface as the deposit film increases in thickness. Slip mechanisms have been proposed by Matthews [15][16], either involving the rearrangement of grown-in dislocations threading through the substrate and deposit, or involving the nucleation of dislocation loops at the deposit surface. These ideas were developed initially to explain observations on misfit dislocations present between metal deposits and substrates, for which cases the required slip systems were available. Takayanagi et al [17] studied the misfit dislocations formed between pairs of chalcogenide compounds and concluded that the observed dislocations could not be explained by a slip process, partly as a result of a detailed analysis of the Burgers vectors of the dislocations and partly because of the lack of direct evidence for slip. They concluded that the dislocations are created and moved by a vacancy climb process. Much of the detailed evidence was obtained by in situ TEM, and was more detailed and more direct than the evidence used by Matthews. Cherns and Stowell [18][19] have also used in-situ TEM to study the formation of misfit dislocations during the growth of palladium on gold in a clean vacuum system. Their detailed analysis shows that misfit dislocations nucleate at the palladium surface and that these grow into the gold palladium interface by a climb process. In a more extensive review of in-situ observations Honjo and Yagi [20] summarise results for a number of different monolayer growth systems, and indicate that slip processes cause misfit dislocations in some cases but that climb processes are more common. For both mechanisms, the driving force for bringing about the formation of misfit dislocations is the elastic stress developed in the growing film. If these are inadequate to cause one or other of the processes to operate, the elastically strained deposit layer can grow to thicknesses greater than is predicted by the van der Merwe theory, which is based upon the equilibrium configuration.

Misfit dislocations must be present whenever there is lattice mismatch not fully compensated by elastic strain. This applies also to the Volmer-Weber nucleation mode of growth, and there is some evidence that initial nuclei or islands can be at least partially elastically strained, and that misfit dislocations move into the interface, from the edge of the islands, as the islands grow [16]. Vincent [21] has observed periodic fluctuations in the measured elastic strains in tin islands grown on tin telluride, as a function of their width, and these have been associated with the introduction of additional misfit dislocations at the edge of the islands.

GROWTH VIA DISCRETE NUCLEI

When the initial growth consists of discrete three dimensional nuclei, subsequent growth occurs by the enlargement of those nuclei until they start to touch each other and merge together. After sufficient material has been deposited a continuous hole-free deposit film can be formed. The thickness at which this occurs varies from system to system, and depends upon the main parameters of growth such as substrate temperature and rate of deposition. In situ TEM is a powerful tool for studying the processes involved, and some of the early studies [22][23][24] revealed the phenomenon of liquid-like coalescence of nuclei or islands of deposit,whereby islands appear to flow together like liquid drops. Such behaviour occurs whilst the islands are solid and can be explained by surface diffusion of deposit atoms over the surfaces of the islands of deposit, the driving force for such diffusion being the associated reduction in surface area and hence total surface energy.

The early studies on in-situ TEM were carried out in microscopes with poor vacuum systems, so that contamination of the substrates and of the growing films could have had major influences on the results. However, much cleaner TEM vacuum systems have been used subsequently, as well as techniques for preparing clean substrate surfaces in situ [27], with the result that some of the basic effects have been confirmed as applying in the absence of any gross contamination. Certainly Volmer-Weber nucleation continues to apply for the same substrate/deposit combinations as before, and the liquid-like coalescence is also observed to take place in the much cleaner vacuum system. There are some effects, such as misalignment and spontaneous realignment of nuclei which do not seem to occur under the considerably cleaner conditions, although they were prominent when there was contamination present. Also, the rates of liquid-like coalescence of gold islands seem to be at least as fast in the clean conditions. Honjo and Yagi [20] deduce higher surface diffusion coefficients than do Pashley and Stowell [27], but there seems to be agreement that surface diffusion is the prominent process involved.

For some deposit materials, especially metals, the liquid-like coalescence is found to be a prominent feature of the intergrowth of nuclei and islands. For many other substances the liquid-like behaviour is either absent or much less pronounced. This applies, for example, to deposits of chalcogenide compounds and alkali halides [20]. The high surface self diffusion coefficients associated with the liquid-like behaviour have an important effect on the formation of a continuous hole-free deposit film, and so the growth sequence is different when liquid-like coalescence is absent. Thus for gold grown on molybdenite [23], the final stage of forming a continuous film is the filling-in of channels in the deposit film, with relatively little extra deposition, by rapid surface diffusion of already deposited gold to the regions of high curvature at the ends of channels. This can result in a deposit film which is fairly uniform in thickness. When the surface self diffusion coefficient is low, it means that there is little or no thickening of nuclei or islands as they grow together, so that the area of coverage of the substrate increases more rapidly in the early and middle stages of the formation of a continuous film. But the filling-in of small holes in the deposit, during the later stages, does not occur so readily because of the poor surface self diffusion of the deposit.

In addition to the liquid-like behaviour, the in-situ TEM studies [23][29] have revealed that recrystallisation of deposit islands can take place as growth proceeds. This happens by the grain boundary, formed by the coalescence of two differently oriented islands, migrating through part of the compound island. This recrystallisation process provides a mechanism for improving, or modifying, the orientation of a deposit during its growth.

STRANSKI-KRASTANOV GROWTH

Systems for which growth occurs by the Stranski-Krastanov mode have, until fairly recently, not been studied widely. Consequently there is not a large amount of experimental evidence, and not as much understanding of the processes involved as there is for the other two growth modes. A recent review [12] summarises the current position, and shows that this mode of growth is more widespread than had been supposed earlier. Wide beam Auger analysis and RHEED provide techniques for identifying the occurrence of the Stranski-Krastanov growth mode, because in both techniques a monolayer coverage of deposit considerably weakens the signal from the substrate, and further deposition in the form of nuclei on top of the monolayer causes only a small change in the strength of the substrate signal, whilst the deposit signal goes on increasing. This increase is quite significant in the case of RHEED, which also gives diffraction spots with a shape and size related to the form of the deposit nuclei.

The Stranski-Krastanov mode of growth occurs when the deposit atoms are very strongly bound to the substrate, and when one or a small number of deposited layers disturbs the smooth transition in binding from that for the first absorbed layer to that for the bulk deposit material. Stoyanov and Markov [29] have extended this approach to take into account that nucleation on top of monolayer islands can be energetically favoured, rather than the lateral growth of the monolayer islands to form a continuous monolayer. Once three dimensional nucleation occurs on top of a monolayer island, that island can re-arrange itself into one three dimensional nucleus, with the result that growth then proceeds by enlargement and coalescence of three dimensional nuclei, without the formation of any complete monolayer coverage. On this model, the Stranski-Krastanov growth is intermediate between pure monolayer growth and pure nucleation.

Many of the identified cases of the Stranski-Krastanov mode of growth result either from growth of metals on metals (especially refractory metals) or from growth of metals on semiconductors. Hence the growth mode could prove to be of some importance in semiconductor device technology.

ALLOYING AT THE SUBSTRATE DEPOSIT INTERFACE

The sharpness of the interface between an epitaxial deposit and its substrate is of both scientific and technological importance. In favourable circumstances RHEED can provide information on the diffusion of deposit atoms into the substrate during the initial stages of deposition. For metals on metals [24] it is found that condensing atoms can penetrate into the substrate surface at temperatures which would not normally result in any inter-diffusion. For copper deposited on gold at 120°C, a copper gold alloy layer is formed initially, and three dimensional copper nuclei form on top of this alloy layer. It is deduced that the kinetic energy of the arriving copper atoms is sufficient to allow some penetration into the gold surface. It is difficult to apply this idea to non-metals, in the absence of direct evidence, but it does suggest the need for similar studies with other substances.

IMPERFECTIONS IN EPITAXIAL LAYERS

Many epitaxial deposits are found to have high densities of lattice imperfections, commonly isolated dislocations, stacking faults or twins. These are introduced during the growth of the deposit, and a number of mechanisms have been identified.

With monolayer growth, the misfit dislocation network at the interface between the substrate and deposit can be linked with dislocations in the deposit film. Some of the mechanisms proposed for the formation of the misfit dislocations [15][16] involve moving or rearranging dislocations within the film, so that internal dislocation arrangements in the deposit film can be the result of the relief of epitaxial strain due to the lattice mismatch between substrate and deposit.

With Volmer-Weber growth, dislocations, and sometimes stacking faults, are introduced during the coalescence of islands. This can occur either because of some small misorientation between coalescing islands or, more generally, because of displacement mismatch [24][28]. A random displacement mismatch is inevitable if the lattice spacings in the deposit do not match exactly the lattice spacings in the substrate. Stacking faults can occur when two neighbouring islands coalesce. Dislocations can form when three or more islands coalesce to produce a hole in the film. An incipient dislocation can be trapped in the hole, and this becomes a real dislocation once the hole is filled in by further deposition. The imperfections can be eliminated, or modified, as growth proceeds but once a continuous network stage is reached dislocations become trapped and a high dislocation density can be present in a continuous epitaxial film. These processes for introducing defects do not apply to deposits formed by the strained monolayer islands if the deposit lattice is strained to match that of the substrate. Thus there is the possibility of growing deposits with very low, or even zero, dislocation densities in this case. But if monolayer growth occurs without perfect lattice matching with the substrate, the introduction of dislocations during the coalescence of monolayer islands would be expected.

Lattice defects which emerge at the surface of a substrate can be continued in an epitaxial deposit. This will almost certainly happen where the strained monolayer growth mechanism applies, and probably also where strained island growth occurs. Consequently highly perfect substrates are necessary if epitaxial layers of low defect density are to be produced by these mechanisms. For Volmer-Weber nucleation, however, it is by no means clear that either dis- locations or stacking faults are extended into the deposit.

The above mechanisms have been determined largely from studies of the epitaxy of metals. Various other mechanisms for the introduction of defects in epitaxial layers have been put forward, including stacking faults and micro- twins and especially for silicon and other semiconductor deposits. Various types of stacking fault and twin defects were observed in the earliest work on silicon epitaxy [30] and these have been associated with contamination effects [31]. Models for the formation of defects as the result of condensation of vacancies during epitaxial growth of silicon, and other substances, have been put forward [32]. More recently, techniques have been developed for preparing thin film sections perpendicular to the surface of epitaxial layers, so that TEM examination of both the deposit and the interface with the substrate can be examined. This includes the use of high resolution lattice imaging. For silicon on sapphire [33], stacking faults and twins lead away from the inter- face into the silicon layer, and these seem to provide a means of accommodating the approximately 6% lattice mismatch involved. Such silicon layers are comm- only grown to thicknesses of 0.5μm and greater, and there is inadequate evidence to determine whether the above defects are incorporated in the deposit as pure growth faults, or whether the defects result from some rearrangements caused by high stresses in the growing layers.

SEMICONDUCTOR DEVICE EPITAXY

A thorough review of the work on epitaxy of semiconductors before 1970 has been given by Holt [34]. The rapid expansion of the application of epi- taxy to the production of semiconductor devices, including opto-electronic devices, has put high demands on the requirements of epitaxy. There is now need for a wide range of semiconducting materials, especially silicon, III-V and II-VI compounds and ternary or quaternary alloys based on these compounds, to be grown as thin single crystal films. The requirements are stringent. Thickness has to be controlled, increasingly more accurately, and uniformity of thickness can be vital. Films as thin as 10Å or less are now needed. Com- position and purity must be determined within close limits, and homogeneity of these is important. For many purposes it is desirable to have no misfit dis- locations present, and to have a very sharp boundary between the substrate and deposit. For all devices the presence of lattice defects, as well as impurity defects, are harmful and for many of the potential new devices it is essential to have extremely low defect levels. Perhaps the most severe demands are being placed on the requirements for the superlattice structures, and other cases where numbers of overlapping epitaxial layers are required. It should also be noted that there are requirements for epitaxial metal layers to be grown in between layers of semiconducting compounds.

The well established application of silicon grown on sapphire (SOS) in the manufacture of devices demonstrates, as indicated above, the difficulty of producing low levels of defect densities even when there is quite a low misfit. For this reason, attempts are being made to reduce the number of defects by a post-deposition treatment of the silicon such as annealing with a laser or with an electron beam. High defect densities are likely to arise in all cases where growth occurs by the Volmer-Weber mechanism, because there is normally at least some misfit. Extremely low misfits would, in practice, be more likely to lead to monolayer growth, and this can result in defect free deposits. Thus the success of the well known lattice matching approach to the growth of heterojunctions, which is based upon arranging for a natural zero misfit so that no misfit dislocations appear at the interface, also has the important effect of inhibiting the formation of lattice defects within the epitaxial layer. Under favourable growth conditions, it should be possible to deposit defect free layers many monolayers in thickness by this means, provided there is also a fairly close matching in thermal expansion coefficients. In addition, this technique is used for the manufacture of superlattices for multi quantum well structures, which are now receiving so much attention. However, there is little direct structural evidence to confirm that the superlattices are largely free of lattice defects, although there is a rapidly growing number of examples where electronic performance is consistent with a very low defect level.

The alternative approach for producing heterojunctions with low defect densities, as applied to the growth of strained layer superlattices, is to rely on maintaining pseudomorphic layers up to the thicknesses required for the particular device. The thicknesses for which this can be achieved are predictable from the van der Merwe theory [13], and even greater thicknesses are possible if the introduction of misfit dislocations, as the layer is thickened, can be inhibited. This will tend to happen if both glide and climb are difficult, during the growth of the layer, but defects are still likely to form at the growing surface. A further factor which could be important is that it has been predicted [11] that the limiting thickness should be greater for negative misfits than for positive misfits.

In order to provide a better understanding of the mode of formation of some of the technologically important epitaxial semiconductor layers, especially with respect to the incorporation of defects in the layers, it is desirable that more attention is given to experimental studies, especially by TEM, of nucleation and growth of the layers. Without this direct evidence, reliance has to be placed on the theoretical models derived from the results of experiments carried out with epitaxial layers of other substances, particularly metals. These models have already been applied with considerable success.

There is also need to obtain more evidence on the sharpness of the boundaries at heterojunctions, especially the multiple junctions of a superlattice. The geometrical aspects of boundaries can now be examined on an atomic scale by means of high resolution TEM lattice images. The sharpness of composition changes at the boundaries is more difficult to determine. This could be influenced by: (i) the departure from perfect monolayer growth whereby the surface of one layer is not atomically smooth; (ii) diffusion of depositing species, including dopant species, into the surface during the initial stages of deposition onto the previously grown layers; (iii) inter-diffusion between adjacent layers subsequent to the initial growth. Ideally, it is desirable to apply techniques for compositional analysis to in-situ observations during growth, but it might prove more fruitful to study sections through layers after they have been grown.

The requirement for growing sandwich layers of semiconductor/metal/semiconductor presents the greatest challenge. It seems that lattice matching would be essential if high densities of lattice defects are to be avoided. In addition there are likely to be difficulties due to interdiffusion, large differences in thermal expansion coefficient, and unwanted chemical reactions, as already revealed from studies of the (Al,Ga)As/Al/(Al,Ga)As system [35]. Clearly there is a severe restriction on the systems for which success might be possible.

REFERENCES

[1] M.L.Frankenheim, Ann. Phys. 37, 516 (1836).
[2] L.Royer, Bull. Soc. franc. Min. 51, 7 (1928).
[3] D.W. Pashley, Advanc. Phys. 5, 173 (1956).
[4] F.C.Frank and J.H. van der Merwe, Proc. Roy. Soc. A 198, 205 (1949).
[5] G.I.Finch and A.G. Quarrell, Proc. Roy. Soc. A 141, 398 (1933).
[6] J.W. Matthews, Phil. Mag. 6, 1347 (1961).
[7] J.W. Matthews, Phil. Mag. 8, 711 (1963).
[8] J.W. Matthews and E. Grunbaum, Phil. Mag. 11, 1233 (1965).
[9] L.G. Schulz, Acta Cryst. 4, 483 (1951).
[10] L.G. Schulz, Acta Cryst. 4, 487 (1951).
[11] I. Markov and A. Milchev, Surface Sci. 136, 519 (1984).

[12] J.A. Venables, G.D.T. Spiller and M. Handbucken, Rep. Progr. Phys. 47, 399 (1984).

[13] E. Bauer, Z. Kristallogr. 110, 395 (1958).

[14] J.H. van der Merwe, J. Appl. Phys. 34, 117 (1963).

[15] J.W. Matthews, Physics of Thin Films 4, 137 (1967).

[16] J.W. Matthews, Phil. Mag. 13, 1207 (1966).

[17] K. Takayanagi, K. Kobayashi, K. Yagi and G. Honjo, Thin Solid Films 21, 325 (1974).

[18] D. Cherns and M.J. Stowell, Thin Solid Films 29, 127 (1975).

[19] D. Cherns and M.J. Stowell, Thin Solid Films 37, 249 (1976).

[20] G. Honjo and K. Yagi, Current Topics in Materials Science 6, 197 (1980).

[21] R. Vincent, Phil. Mag. 19, 1127 (1969).

[22] G.A. Bassett, Proc. Eur. Reg. Conf. on Electron Microscopy, Delft (De Nederlandse Vereniging Voor Electronen-microscopie : Delft),270 (1961).

[23] D.W. Pashley, M.J. Stowell, M.H. Jacobs and T.J. Law, Phil. Mag. 10, 127 (1964).

[24] D.W. Pashley, Advanc. Phys. 14, 327 (1965).

[25] K. Takayanagi, K. Yagi, K. Kobayashi and G. Honjo,J.Phys.E11, 441 (1978).

[26] R.D. Moorhead and H. Poppa, Proc. 27th A. Meeting EMSA (Baton Rouge : Claitors), 116 (1969).

[27] D.W. Pashley and M.J. Stowell, J. Vac. Sci.Techn. 3, 156 (1966).

[28] M.H. Jacobs, D.W. Pashley and M.J. Stowell, Phil. Mag. 13,129 (1966).

[29] S. Stoyanov and I. Markov, Surface Sci. 116, 313 (1982).

[30] G.R. Booker, Phil. Mag. 11, 1007 (1965).

[31] G.R. Booker and B.A. Joyce, Phil. Mag. 14, 301 (1966).

[32] M.J. Stowell, Epitaxial Growth (J.W. Matthews, Ed.) (Academic Press, New York), 437 (1975).

[33] J.L. Hutchison, G.R. Booker and M.S. Abrahams, Microscopy of Semiconducting Materials (A.Cullis,Ed.)(Institute of Physics,London),139 (1981).

[34] D.B. Holt, Thin Solid Films 24, 1 (1974).

[35] K. Okamoto, C.E.C. Wood, L. Rathbun and L.F. Eastman, J.Appl.Phys.53, 1532 (1982).

SURFACE RELIEF AND THE ORIENTATION OF VAPOR DEPOSITED FILMS

D.A. Smith, J.T. Wetzel and A.R. Taranko
IBM Thomas J. Watson Research Center, Yorktown Heights, N.Y. 10598

ABSTRACT.

It has been suggested that the topography of alkali halide substrates is a key factor in the growth of metal single crystals. The main features of the model are that small cubo-octahedral islands are mobile but preferentially adsorbed and aligned by cleavage steps where they gain a statistical advantage in the competitive growth process which occurs at coalescence. Replicas of NaCl surfaces have been made using carbon, SiO and polyimide all of which are amorphous. Gold or copper was deposited at temperatures in the range 200-350°C onto coated NaCl, and the top or bottom surfaces of the amorphous coating after removal from the substrate. The strongest preferred orientation effects were found on the insulating substrates where {100} texturing occurred. This suggests that charging of islands may be a significant factor in addition to the topography of the substrate.

INTRODUCTION

Cleaved crystals with the rocksalt structure are convenient substrates on which a variety of metals and non-metals can be grown in single crystal form by vapor deposition [1]. The cubic metals are found to grow as single crystals with {100} planes and <001> directions of deposit and substrate aligned at substrate temperatures in the range 200°C to 400°C. It is a remarkable observation that the same crystallographic features are found for fcc and bcc deposits on varying substrates irrespective of the natural misfit between substrate and deposit. It is also found that single crystal deposits are most readily formed on substrates which have had treatments which contaminate or disrupt the surface structure [2]. Single crystal gold does *not* grow on vacuum cleaved salt [3]. These observations raise two questions: (1) how does the deposit become a single crystal? and (2) how is the alignment between deposit and substrate established?

It is well known that in systems where the interaction between substrate and deposit is weak island growth occurs [4]. Diffraction patterns show that the deposit orientation evolves from a cube texture to a (twinned) {100} single crystal as the mean film thickness increases i.e. the formation of an aligned single crystal is a consequence of the island coalescence process. The grain boundary created when two misoriented islands coalesce can migrate under the capillarity driving force at the usual deposition temperatures. The resulting single crystal will have the orientation of the initially larger island [5]. This process can continue until metastable triple grain boundary junctions impede the migration process. It is this temperature dependent balance between coalescence and grain boundary movement which decides whether a single or polycrystalline deposit is produced. It is thus reasonably clear how a single orientation of deposit can be produced; the second question of the alignment mechanism of deposit and substrate is the subject of the remainder of this paper. The existing data which have been reviewed above suggest that the surface topography of rock salt structure substrates is a key factor in effecting the alignment of the deposit in contrast to the behavior of strongly interacting systems where the substrate-deposit interface is critical [4].

EXPERIMENTS

To investigate the role of substrate topography further various thicknesses in the range of 100Å-500Å of gold and copper have been deposited onto substrates which replicate the topography but not the atomic structure of rock salt type substrates. Gold was deposited onto carbon coated NaCl, SiO coated NaCl and the underside of carbon and SiO films which had been stripped off a cleaved NaCl substrate. The substrate was heated to 200°C before coating. As part of a program to investigate metal deposition on polymeric substrates copper was evaporated onto 200Å thick polyimide which had been spun onto NaCl and cured at 400°C. The trend shown by the results from these experiments was that higher temperatures during deposition and insulating substrates promoted 4-fold symmetry in the deposits. Gold on a carbon replica of a cleaved NaCl surface had a {111} fiber texture. In the case of SiO replicas, the 4-fold symmetry of the film was more marked when gold was deposited on the surface of the SiO which had been in contact with the NaCl. Similarly replicas of an NaCl surface which had been heated to above 200°C were more effective in orienting subsequently deposited films. The 4-fold symmetry evident in the diffraction patterns obtained from gold on SiO on NaCl is of course not an obvious precursor to a {100} single crystal film since intensity maxima appear in the {111} ring. However single crystal {100} oriented regions resulted from growth of gold on the underside of the SiO.

Copper deposited onto polyimide coated NaCl produced large areas of {100} single crystal. The development of the microstructure followed stages that are well known for island growth. In some cases aligned crystallites and single crystal films decorated features with the geometry characteristic of a cleaved NaCl surface. There were significant differences in the film thicknesses at which comparable microstructures resulted when copper was deposited in the same run onto bare and polyimide-coated NaCl. Island coalescence and film continuity occurred at lower thicknesses on polyimide than on NaCl. The figures show representative microstructures and diffraction patterns.

DISCUSSION

Usually deposition of an fcc metal onto an amorphous substrate results in a random distribution of orientations at low substrate temperatures and at higher substrate temperatures a {111} texture presumably to minimize surface energy [6]. Indeed if a {100} oriented gold film is annealed it recrystallizes into a {111} orientation [7]. Our observations of {100} texture provide clear evidence for an effect of surface topography on film orientation.

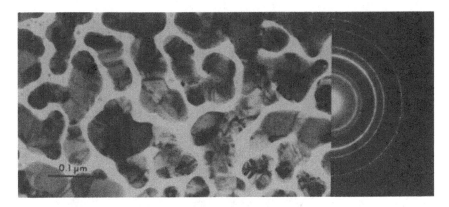

Figure 1.   The microstructure and corresponding diffraction pattern of gold deposited onto a carbon replica of an NaCl surface at 200°C.

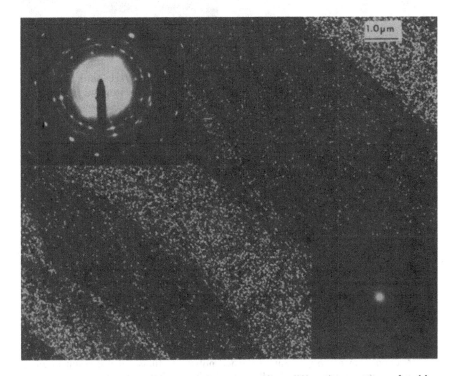

Figure 2.   The microstructure and corresponding diffraction pattern of gold deposited onto the underside of a SiO replica of an NaCl surface at 200°C. The diffraction pattern resulting from gold growth under the same conditions on the upper surface of the SiO is shown for comparison.

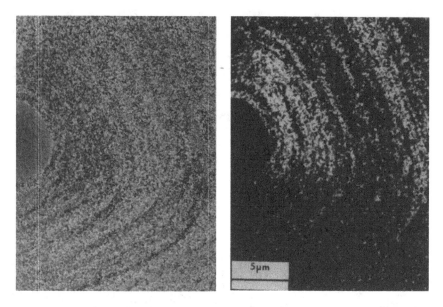

Figure 3. 150Å of copper deposited at 200°C and 20Å/sec onto polyimide coated NaCl. A large single crystal area that is only just continuous is shown in both bright field (BF) and dark field (DF) using a {200} reflection. The contrast in DF is indicative of copper following the surface relief on the polyimide which is in turn replicating the cleavage steps of NaCl.

It seems unlikely that the thermal energy $\frac{3}{2}kT$ could be sufficient to account for the mobility of islands [8] comprising more than a few atoms. However, large electrostatic forces can result from charging [9]. The enhanced {100} texture of Au on insulating SiO relative to conducting carbon may be related to this effect.

Several different island morphologies are initially produced but on an NaCl type substrate only those with at least two mutually perpendicular faces can be adsorbed on both the (100) and (010) surface and the face of a cleavage step. Such bound islands will be locked into an aligned orientation and occur in greater than random concentration. In addition, the adsorption at particular sites favors coalescence of already aligned islands which gives them an additional advantage in competitive growth.

A deposition of copper onto detached polyimide films did not result in any texture or single crystal growth. Presumably the polyimide was under tension on the TEM grids which supported them, and the sharp surface replication was lost upon heating to 300°C. We note that the glass transition for a noncrosslinking thermosetting polymer is nearly at the curing temperature (400°C in our case), so our assumption concerning relaxation of the polyimide seems reasonable.

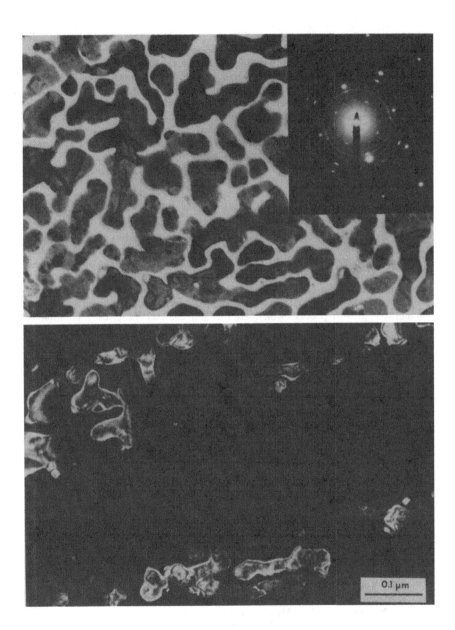

Figure 4.   150Å of copper deposited at 300°C and 20Å/sec onto polyimide coated NaCl. The DF {200} micrograph shows the sharp demarcation between between oriented single crystals, which may be adsorbed onto replicated cleavage steps, and randomly oriented polycrystalline islands.

Occasionally holes were detected in the polyimide. These could in principle provide an "epitaxial seed". However, we reject this as a possibility because of the large areas of oriented growth observed far away from holes even when the film was only just continuously connected. Large, incompletely covered cleavage steps could be a source of "seeding", but the replicating ability of polyimide was observed to completely cover steps as large as 1600Å. Holes and tears in polyimide substrates were found in regions where copper was only polycrystalline and had a random orientation. In fact it is not readily possible to distinguish between holes and tears in the polyimide film which are a result of TEM sample preparation or a result of the polyimide being initially discontinuous. Only a surface-sensitive technique such as Auger spectroscopy could give information on this point.

CONCLUSION

Cubo-octahedral islands may be preferentially adsorbed at surface steps in cube-cube alignment relative to an NaCl type substrate or an amorphous substrate with similar topography. Consequently aligned islands provide a greater than random contribution to the film texture and being localized at particular sites coalesce in a prealigned orientation again with a greater than random probability. These two effects give aligned grains a size advantage in the competitive growth process which after island coalescence gives a continuous film.

ACKNOWLEDGMENTS

We thank G. Appleby-Mougham for providing the polyimide-coated NaCl substrates, and one of us (JTW) was a Visiting Scientist supported by IBM GTD Endicott.

REFERENCES

1. Listed in E.Gruenbaum, *Epitaxial Growth Part B,* J.W. Matthews, ed., Academic Press, New York (1975)
2. F. Cosandey, Y.Komem and C.L. Bauer, *Thin Solid Films,* **59,** 165 (1979)
3. J.W. Matthews, *Phil. Mag.,* **12,** 1143 (1965)
4. Reviewed in J.A. Venables and G.R. Price, *Epitaxial Growth, Part B,* J.W. Matthews, ed., Academic Press, New York (1975)
5. D.A. Smith and M.M.J. Treacy, *Applications of Surface Science,* **11/12,** 131 (1982)
6. R.W. Vook and F. Witt, *J. Vac. Sci. Tech.,* **2,** 243 (1965)
7. S. Mader, R. Feder and P. Chaudhari, *Thin Solid Films,* **14,** 63 (1972)
8. J.J. Metois, *Surf. Sci.,* **36,** 269 (1973)
9. D.B. Dove, *J. Appl. Phys.,* **35,** 2789 (1964)

EXPERIMENTAL OBSERVATIONS ON THE GROWTH OF EPITAXIAL FILMS BY THE VOLMER-WEBER MECHANISM

K.R. MILKOVE* AND S.L.SASS**
*Cornell University, Dept. of Materials Science and Engineering,Bard Hall,
Ithaca, NY 14853 (Present Address: IBM East Fishkill, ZIP 89E, Hopewell
Junction, NY 12533)
**Cornell University, Dept. of Materials Science and Engineering, Bard Hall,
Ithaca, NY 14853

ABSTRACT

Some experimental observations are reported that are pertinent to the
nucleation and growth of epitaxial FCC metal films on alkali halide substrates (i.e., the Volmer-Weber growth mechanism). Thin epitaxial films
of Au, Ag, Pt, and Cu were deposited on the (001) surface of air cleaved
NaCl substrates under HV and UHV conditions, both with and without the use
of electron bombardment of the NaCl. Our observations contradict certain
ideas prevalent in the epitaxy literature regarding thin film growth. For
instance, the results of this study shows that vacuum pressure is <u>not</u> an
important epitaxy parameter. Instead, it is suggested that the influence
of the vacuum on thin film epitaxy was confused with various types of substrate surface modification due to the presence of water vapor and/or the
use of vacuum bakeout heat treatments. Considerable evidence was also found
which pointed to the fact that the optimum nucleus saturation density for
thin film epitaxy via the Volmer-Weber growth mechanism is 5×10^{11}
nuclei/cm^2. Several additional observations will be discussed, including
some which provide new insight regarding the use of electron bombardment.

INTRODUCTION

The growth of epitaxial FCC metals on single crystal alkali halide
substrates is governed by the Volmer-Weber growth mechanism. A nonatomistic
overview of the characteristic stages of this growth mechanism has been well
documented via in situ electron microscope studies[1]. The first observable
feature of an epitaxial thin film overgrowth layer is the appearance of a
random distribution of three-dimensional nuclei on the substrate surface.
These nuclei, which are roughly 1.5 nm across when first detected, exhibit
clearly defined geometric shapes. As the thin film deposition proceeds the
nucleation rate eventually tapers off yielding a saturation nucleus density
N_s (Figure la). Robins et al.[2-4] have shown that N_s is a function of
deposition rate R and substrate temperature T for an impurity-free substrate
surface.

Once the nucleus density saturates, the individual nuclei or islands
continue to grow (Figure 1b). Their total number remains constant until
adjacent islands start to touch one another. At the moment of contact a
rapid necking phenomenon involving considerable mass transport occurs
between the two islands, serving to reduce the radius of curvature at the
contact point. This process is referred to as liquid-like coalescence by
Pashley et al.[5] and is illustrated in Figure 1c. The island density
declines rapidly during this stage of film growth. Eventually, all islands
interconnect forming a network encompassing the entire substrate surface.
This network develops into a channeled structure which fills in as the
deposition continues (Figure 1d). As total film coalescence nears completion
the channels are reduced to small holes (Figure 1e). Ultimately, the holes
are eliminated yielding a fully coalesced foil (Figure 1f).

The key stage(s) responsible for the growth of a continuous, low defect

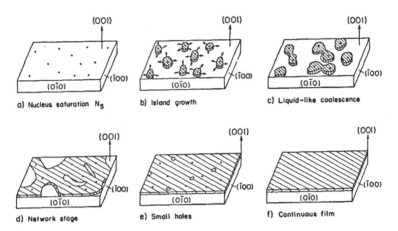

Figure 1. Thin film growth stages

density, epitaxial film is a controversial subject[6-8]. Our experiments, which involved the growth of thin epitaxial films of Au, Ag, Cu, and Pt on the (001) cleavage face of NaCl, support a hypothesis that the critical processes associated with the epitaxy of a <u>continuous</u> film occur during the nucleation stage. Specifically, it is suggested that the crystallographic orientation, size distribution, and saturation density N_s of the nucleated islands ultimately determine if an epitaxially oriented continuous foil can be grown. Evidence will be presented showing that the substrate temperature T, plus the detailed nature of the substrate surface, govern the nucleation behavior of an epitaxial overgrowth layer.

EXPERIMENTAL PROCEDURE

Barely continuous Au, Ag, and Cu films were deposited by resistive heating onto air cleaved (001) NaCl substrates. Most of the films were grown in a Varian FC-12E vacuum system which provided fast cycling ultra high vacuum capabilities in an oil-free environment. The vacuum chamber possessed two independently controlled substrate heaters, permitting an in situ comparison of film quality with the only experimental variable being the substrate temperature. The above metals were deposited over a range of pressures P extending from 1×10^{-3} to 1×10^{-7} Pa, using deposition rates R on the order of 2.4×10^{-2} to 1×10^{2} nm/sec, and substrate temperatures T between 200 to 450° C. The influence of various substrate surface treatments were also evaluated, such as electron bombardment[9-10], water vapor treatments[11-12], predeposition anneals in an air furnace[13-14], vacuum bakeouts[15], and in situ predeposition anneals[16]. The effectiveness of post deposition anneals were also evaluated as a means of reducing twin and dislocation densities [17-19], as well as surface undulations. The minimum thickness for hole-free continuous films varied from 50 to 250 nm, and exhibited a strong dependence on the choice of R,T, and predeposition substrate surface treatments.

Acceptable Pt films could not be deposited by resistive heating because the Pt alloyed with the W evaporation vessel. Deposition by electron beam, though far more suitable in the case of Pt, was adopted rather reluctantly because it required using another vacuum system. The available ranges of P and R were more limited in this evaporator, but were found to be adequate for growing epitaxial Pt films.

Figure 2. Influence of
surface treatments on twin
density. (a) Water vapor
treatment followed by
electron bombardment.
(b) Only electron bombardment.

    (a)   (b)

The epitaxial quality of all films were evaluated by TEM on a JEM 200CX
electron microscope using bright field, dark field, and selected area
diffraction techniques.

RESULTS

The optimum deposition conditions for the 70 nm thick Au films, in the
absence of a bakeout cycle, were determined to be T=282° C, R=0.2 nm/sec,
and P=6.6 X 10^{-6} Pa. It was discovered that the lowest defect density films
were obtained when freshly cleaved NaCl was briefly exposed to a high concentration of water vapor[20] followed by an in situ electron bombardment treatment using a 500-1000 volt dc accelerating potential. As shown in Figure 2,
the combination of the two surface treatments produced a lower defect density
foil than could be obtained by electron bombardment alone. This result was
unexpected, since the literature suggests that the effects of electron bombardment dominate the nucleation process[4,9-10,21], presumably overshadowing
other surface treatments that might influence nucleation. It was also found
that at constant T, a decrease in R produced a decrease in the twin density.
Conversely, with R held constant, a decrease in T produced an increase in the
twin density. Finally, the introduction of a one hour in situ post deposition
anneal at 450° C reduced the twin and dislocation densities by one to two
orders of magnitude.

The growth of Ag and Au films behaved similarly, provided the NaCl substrates were first subjected to identical surface treatments. In agreement
with the observations of Cosandey et al.[10], the defect density of an epitaxial Ag foil was about one tenth the value of a comparable Au foil. When
incorporating a vacuum bakeout as part of the experimental procedure, the
best Ag films were deposited at T=250° C, P=6.6 X 10^{-7} Pa, electron bombardment with a 1000 volt dc potential, and R=0.8 nm/sec for the first 20 nm
followed by R=4-8 nm/sec. The foils, which became hole-free at a thickness
of 85 nm, required a two step deposition procedure in order to effectively
nucleate an epitaxial film, while trying to limit the problem of agglomeration
associated with slowly deposited Ag. A 20 minute in situ post deposition
anneal at 375° C succeeded in a reduction of the twin density by a factor
of five.

The growth of a continuous epitaxial foil of Cu in a baked out vacuum
chamber was not successful until the substrate temperature was raised to
450° C. With R=1 nm/sec the films grew discontinuously due to severe
agglomeration. The problem was eventually controlled by increasing R to a
value exceeding 40 nm/sec. At these rates the Cu films became hole-free at
about 250 nm. A post deposition anneal of only 3 minutes kept the twin

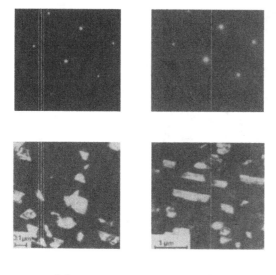

Figure 3. Comparison of selected area diffraction patterns (top) and dark field images (bottom) of epitaxial Cu films deposited using electron bombardment of the substrate. (a) HV deposit with T=280 °C, R=1.4 nm/sec, and P=1.2 X 10^{-5} Pa. (b) UHV deposit with T=450 °C, R=16.1 nm/sec, and P=6.3 X 10^{-7} Pa.

(a)   (b)

density below 1 X 10^8/cm^2. The growth behavior of the Cu was surprising because Kluge-Weiss and Bauer[22] were able to deposit epitaxial Cu foils on electron bombarded NaCl with T=280 °C, R=1 nm/sec, and P≈2.5 X 10^{-4} Pa. Using identical deposition conditions, except for a better vacuum of 6.7 X 10^{-7} Pa, our own film deposits were polycrystalline. However, as shown in Figure 3 when the bakeout was eliminated from the experiment, the Cu grew epitaxially.

Single crystal Pt foils were deposited by electron beam evaporation with T=350 °C, R=0.2 nm/sec, P=8 X 10^{-5} Pa, and a 500 volt dc electron bombardment potential. A bakeout was not employed for these experiments. The twin density of the 50 nm thick Pt foils was not reduced by a 15 minute 450 °C post deposition anneal. The inability to reduce the twin density of the Pt by an in situ anneal is consistent with the observations of Reichelt et al.[23].

ANALYSIS

Matthews et al.[15,25] have proposed that when an (001) and (111) island make contact on an (001) NaCl substrate, there is a general tendency for the smaller island to become a twin of the larger island resulting in a growth twin in the fully coalesced foil. Data reported by Cosandey et al.[10] and Matthews[15] suggests that under favorable epitaxial growth conditions the the ratio of (001) to (111) islands must exceed ~0.43 for a 2 nm mean foil thickness. Furthermore, Matthews[15] has also documented a temperature dependence for this ratio, where the (001)/(111) ratio increases with increasing temperature.

Based on the island coalescence twinning mechanism suggested by Matthews et al.[15,25], it follows that in a range of R,T values conducive to epitaxial growth, one should observe an increase in twin density with a decrease in the substrate temperature for constant R. This behavior was documented for both Au and Ag in the present study.

The difficulty of depositing Cu in ultra high vacuum (UHV) versus high vacuum (HV) may be attributable to the use of a necessary 220° C bakeout of the vacuum apparatus during all UHV experiments. Work by Lad[11] and Vermaak

and Henning[12] have shown that the nucleation density of an overgrowth layer
on air cleaved NaCl is reduced under otherwise identical deposition conditions
if the NaCl is subjected to a high temperature bakeout as a predeposition
anneal. The basis for this effect has been described by a critical accommodation model developed by Henning and Vermaak[24] which accountsfor the influence
on epitaxy of the substrate surface. Lad[11] found that under suitable conditions an air cleaved NaCl wafer acquires a NaOH surface layer via a chemical
reaction with water vapor. Calculations by Henning and Vermaak[24] show that
NaOH provides a favorable geometry for the nucleation of a stable epitaxially
oriented 4-atom overgrowth cluster for most FCC metals. On the other hand,
when an air cleaved wafer is heated up the NaOH layer is removed leaving a
bare NaCl surface. According to Henning and Vermaak's calculations a bare
NaCl surface does not provide a favorable geometry for the nucleation of stable
4-atom overgrowth clusters. We suggest that these ideas explain our inability
to grow epitaxial Cu in UHV, as opposed to HV, despite the use of electron
bombardment.

Electron bombardment has two positive effects on thin film epitaxy. It
increases both N_s and the (001)/(111) island ratio. From the observed growth
behavior of Cu, it seems that despite the use of electron bombardment, the
N_s of the Cu does not achieve what we believe is a minimum value for epitaxy
if the NaCl substrate is subjected to a bakeout. Thus, it appears that
electron bombardment alone does not totally dominate the nucleation process.
Otherwise, the use of HV versus UHV should not influence the epitaxy of the
Cu. These arguments are also consistent with the observation in Figure 2
that the combination of a water vapor treatment with electron bombardment
yields the lowest defect density Au foils.

Finally, by reviewing the available N_s data for metal/NaCl and
Metal/KCl epitaxy systems[9,12,15-16,21,26-33] the authors discovered that
there exists a range of N_s values falling between ~1 X 10^{11} to ~1 X 10^{12}
nuclei/cm^2 that appear to guarantee the growth of an epitaxial film. N_s did
not exhibit a dependence on the vacuum level, per se, but was clearly
influenced by the variety of substrate surface treatments reported in
the literature.

In summary, evidence has been presented suggesting that the ultimate
quality of an epitaxial foil is determined during the nucleation stage of
film growth. There are two key parameters:

i) The substrate temperature T, which serves a dual role of determining
   the (001)/(111) island ratio and interacting with R to establish
   the supersaturation level of the deposited metal atoms.

ii) The detailed nature of the substrate surface which in the case of
    NaCl is readily modified both morphologically[9-10,13-14] and
    chemically[11-12] in a variety of manners.

ACKNOWLEDGMENTS

The authors wish to thank James Ritchey for his able help in
depositing the Pt films used in this work.

REFERENCES

[1] D.W.Pashley, Advances in Phys. 14, 327(1965).
[2] V.N.E. Robinson and J.L.Robins, Thin Sol. Films 5, 313(1970).
[3] J.L.Robins and A.J.Donohoe, Thin Sol. Films 12, 255(1972).
[4] J.L.Robins, Surf. Sci. 86, 1(1979).
[5] D.W.Pashley, M.J.Stowell, M.H.Jacobs, T.J.Law, Phil. Mag. 10,127(1964).

[6] R.Kern, G.Lelay, and J.J.Metois, Current Topics in Materials Science 3, 133, edited by E.Kaldis, North-Holland Publishing Company, 1979.

[7] E. Bauer and H.Poppa, Thin Sol. Films 12, 167(1972).

[8] B. Lewis, Thin Sol. Films 7, 179(1971).

[9] D.J.Stirland, APL 8, 326(1966).

[10] F. Cosandey, Y. Komem, and C.L.Bauer, Thin Sol. Films 59, 165(1979).

[11] R.A.Lad, Surf. Sci. 12, 37(1968).

[12] J.S. Vermaak and C.A.O. Henning, Phil. Mag. 22, 269(1970).

[13] H. Bethge, Physica Status Solidi 2, 3(1962).

[14] H. Bethge, Surf. Sci. 3, 33(1964).

[15] J.W. Matthews, JVST 3, 133(1966).

[16] G. Gardner Sumner, Phil. Mag. 12, 767(1965).

[17] T.S. Noggle, Nuclear Instruments and Methods 102, 539(1972).

[18] L.O. Brockway and R.B. Marcus, JAP 34, 921(1963).

[19] R.W. Vook and C.T. Horng, Thin Sol. Films 18, 295(1973).

[20] T. Schober, JAP 40, 4658(1969).

[21] P.W. Palmberg, C.J. Todd, and T.N. Rhodin, JAP 39, 4650(1968).

[22] P.M. Kluge-Weiss and C.L. Bauer, Phys. Stat. Sol. A58, 333(1980).

[23] K. Reichelt, T. Schober, and J. Viehweg, J. Cryst. Growth 18, 312(1973).

[24] C.A.O. Henning and J.S. Vermaak, Phil. Mag. 22, 281(1970).

[25] J.W. Matthews and D.L. Allinson, Phil. Mag. 8, 1283(1963).

[26] C.A.O. Henning, J.C. Lombaard, and J. C. Botha, APL 14, 109(1969).

[27] B.Lewis and M.A. Jordan, Thin Sol. Films 6, 1(1970).

[28] D.J. Stirland, Thin Sol. Films 1, 447(1967/68).

[29] K. Mihama, JVST 6, 480(1969).

[30] J.W. Matthews, Phil. Mag. 12, 1143(1965).

[31] T. Inuzuka and R. Veda, APL 13, 3(1968).

[32] H. Bethge, JVST 6, 460(1969).

[33] R.F. Adamsky and R.E. LeBlanc, Trans. Symp. Am. Vac. Soc., 10th, Boston, P. 453(1963).

THEORETICAL CONSIDERATIONS OF THE VOLMER-WEBER GROWTH MECHANISM FOR THIN EPITAXIAL FILMS

K.R. MILKOVE
Cornell University, Dept. of Materials Science and Engineering, Bard Hall, Ithaca, NY 14853 (Present Address: IBM East Fishkill, ZIP 89E, Hopewell Junction, NY 12533)

ABSTRACT

An atomistic nucleation theory is proposed for the Volmer-Weber growth mechanism of thin film epitaxy. Three notable ideas have been incorporated into the theory:

    i) A general nucleation rate expression derived by Walton predicting the formation rate of small atomic clusters on a substrate.

    ii) A model, proposed by Henning and Vermaak, for the accommodation of a small cluster of overlayer atoms on a crystalline substrate.

    iii) An atomic bonding scheme for the atoms of an overlayer cluster that is consistent with experimental observations reported in the literature, as well as points i) and ii).

A cluster stability map has been derived that predicts the necessary supersaturation conditions (i.e., deposition rate per unit volume R versus substrate temperature T) for epitaxial film growth. In addition, the theory identifies three temperature transformation relations relevant to thin film epitaxy, a fundamental material constant for each metal/substrate combination, and a definite "range" of saturation nucleus densities N_S that will ultimately produce epitaxial film growth.

INTRODUCTION: THE VOLMER-WEBER GROWTH MECHANISM

The growth of epitaxial films on the single crystal surface of a dissimilar substrate has been actively investigated since the mid 1950's[1]. Considerable effort has been expended over the years trying to isolate and explain the key processes associated with epitaxy. In one of the early investigations Bauer[2,3] identified three fundamentally different growth mechanisms by which an epitaxial film can be deposited. The active mechanism associated with a specific overgrowth/substrate system is determined in part by the adhesion energy (as defined by Kern et al.[4]) between the overgrowth and its substrate[4]. When the adhesion energy of the overgrowth is less than twice the surface energy of the principal overgrowth surface layer, thin film epitaxy will proceed by the Volmer-Weber mechanism. This mode of film growth is characterized by a random distribution of overgrowth islands which tend to grow more rapidly in a direction normal to the substrate surface, rather than lateral to the surface.

One of the most intensely studied epitaxy systems that is governed by the Volmer-Weber mechanism is the growth of FCC metal films on the (001) cleavage face of alkali halide substrates (e.g. Au and Ag on NaCl or KCl). To date, a number of nucleation and growth theories have been postulated in an attempt to explain the mechanism of epitaxy in these metal/substrate systems. See, for example, reviews by Kern et al.[4], Bauer and Poppa[3], and Pashley[1]. Those efforts that have specifically concentrated on the development of a workable atomistic theory have achieved a degree of success, but typically conflict with various aspects of the experimental observations when subjected to a rigorous analysis.

The intention of this paper is to outline a new atomistic theory that incorporates a general nucleation rate expression for finite size atomic clusters derived by Walton[5,6], a semi-quantitative model proposed by

Henning and Vermaak[7] which accounts for the role of the substrate surface in epitaxy, and a novel atomic bonding formalism for critical nuclei. It will be shown that the new bonding scheme is consistent with a broad range of experimental observations reported in the literature, in addition to theoretical predictions derived from the cited works by Walton[5,6] and Henning and Vermaak[7].

WALTON'S ATOMISTIC NUCLEATION THEORY

Walton[5] recognized that two-dimensional atomic clusters as small as three and four atoms are in principle capable of exhibiting distinctive epitaxial alignment on a foreign substrate. Assuming that the epitaxial growth of a continuous film requires the initial nucleation of properly oriented particles, Walton used statistical mechanics to derive a nucleation rate expression that is applicable to metal/substrate epitaxy systems. The derived quantity $I_{n\star}$ was identified as the nucleation rate of a critical nucleus possessing $n^\star$ atoms. Walton asserted that an $n^\star$-atom critical nucleus should be a cluster of atoms which upon the receipt of one additional atom at the proper location within the cluster acquires a probability of continuing to grow that is equal to or greater than one-half. Lacking this additional atom, the probability of continued cluster growth is less than or equal to one-half.

The general nucleation rate expression derived by Walton[5] is given by

$$I_{n\star} = N_0 \ R \ \sigma_{n\star} \ d \ (\frac{R}{N_0 \nu})^{n\star} \ \exp[\frac{(n^\star+1)Q_{ad} + E_{n\star} - Q_D}{kT}] \quad , \qquad (1)$$

where $I_{n\star}$ is the nucleation rate of an "$n^\star+1$"-atom stable cluster; $n^\star$ is the number of atoms in a critical nucleus; R is the deposition rate per unit volume (i.e. the rate of incidence of metal atoms from the vapor); $\sigma_{n\star}$ is the capture width of a critical nuclues; d is the minimum diffusion jump distance on the substrate; N_0 is the density of adsorption sites; ν is the vibrational frequency of a metal atom adsorbed on the substrate; Q_{ad} is the adsorption energy of a metal atom on the substrate; $E_{n\star}$ is the dissociation energy of an $n^\star$-atom critical nucleus; Q_D is the activation energy for the diffusion of an adsorbed atom on the substrate surface; T is the substrate temperature; and k is the Boltzmann constant.

Proceeding from Walton's analysis[5,6], the successful deposition of an epitaxial overgrowth layer on a substrate exhibiting n-fold symmetry should favor the nucleation of the smallest stable atomic cluster exhibiting the same n-fold symmetry as the substrate. For instance, the (001) surface of NaCl possesses four-fold symmetry. Therefore, a likely critical nucleus for an epitaxially oriented FCC metal overgrowth film would be an "L"-shaped three atom cluster. The addition of a fourth atom at the proper location would produce a square-shaped stable cluster capable of exhibiting epitaxial alignment with the underlying substrate.

The key predictions obtained from Walton's atomistic nucleation theory[5, 6] are summarized below:

    i) For routine experimental conditions the general nucleation rate expression (equation 1) indicates that the size of a critical nucleus will be less than seven atoms.

    ii) The supersaturation level during the deposition of a thin film can be defined as the number of overgrowth atoms diffusing within the capture width of a subcritical nucleus per unit time. The supersaturation can be controlled experimentally by varying R and T. Equation 1 relates both of these parameters to the nucleation rate $I_{n\star}$, where it is noted that $I_{n\star}$ exhibits $R^{n^\star+1}$ and $\exp(1/T)$ dependences.

    iii) A relationship between R and T was derived from equation 1 predicting a transition from unoriented film growth at low levels of supersaturation to epitaxial growth as the supersaturation is raised above a critical value. At low supersaturations Walton[5] suggested that the minimum-size stable cluster consists of only two atoms, where a single metal-metal bond per atom is sufficient to maintain cluster stability. In this situation, Walton felt that the growing clusters would have the freedom to accept a new atom along their periphery without any directional constraints on the bond angle. At high supersaturation Walton proposed that the minimum-size stable cluster increases to four atoms, where each atom now possesses two metal-metal bonds in order to maintain the stability of the cluster. The two-bond per atom requirement introduces a directionality to the cluster growth that should be conducive to epitaxy.

Walton et al.[7] sought to verify the above predictions by measuring the nucleation rate of Ag deposited on vacuum cleaved NaCl as a function of T and R. A transition in the nucleation rate behavior as a function of R was clearly observed. At high supersaturation above the transition point the nucleation rate varied according to an R^4 dependence as predicted. At low supersaturation an R^2 dependence was measured. Several shortcomings with this data must be recognized, however. Milkove and Sass[8] have identified a finite range of saturation nucleus densities N_S between ~1 X 10^{11} to ~1 X 10^{12} nuclei/cm^2 that yield continuous epitaxial films for FCC metals deposited on alkali halide substrates. The nucleation rates reported by Walton et al.[7] were several orders of magnitude smaller than these values, and continuous films were apparently not deposited in conjunction with the nucleation experiments. Later work by both Matthews[9] and Ino et al.[10-11] showed that Ag deposited onto vacuum cleaved NaCl (without modifying the substrate surface with electron bombardment) initially exhibited strong (001) epitaxy at 0.1 - 1 nm in average film thickness, but as the average thickness increased the (111) orientation began to dominate the growth process. The fully continuous foils were found to be polycrystalline.

Walton et al.[7] also reported a significant inconsistency in the low supersaturation data. According to Walton's theory, when the nucleation rate is proportional to R^2, oriented islands should not be nucleated in significant numbers. Nevertheless, selected area diffraction patterns of films deposited at low supersaturations clearly exhibited a discernable (001) orientation. The authors accounted for their observations by suggesting that diffusion effects could cause some of the islands to grow epitaxially.

THE CRITICAL ACCOMMODATION MODEL OF HENNING AND VERMAAK

Henning and Vermaak[12] have proposed a semi-quantitative model, referred to herein as the critical accommodation model or CAM. This model explains the role of the substrate surface in the epitaxy of films that grow by the Volmer-Weber mechanism. Using hard-sphere concepts, Henning and Vermaak identify a set of conditions ideally suited for the nucleation of epitaxially oriented FCC metal overgrowth clusters on alkali halide substrates.

The ideal situation for (001) epitaxy is illustrated in Figure 1, where r_a and r_c are the ionic radii of the anion and cation; r_m is the atomic radius of one of the metal atoms comprising a four-atom parallel oriented stable overgrowth cluster; and x_m is the second-nearest neighbor approach distance of the metal atoms in the overgrowth cluster. In the ideal situation the atomic radius r_m of an overgrowth atom must satisfy two simultaneous conditions[12]:

    i) Each overgrowth atom in the minimum-size stable cluster lies in a potential minimum on the substrate surface.

    ii) The nearest neighbor and second nearest neighbor bulk metal-metal bond lengths are satisfied.

When these conditions are satisfied the substrate provides a critical accommodation center for the overgrowth cluster, and should represent the lowest

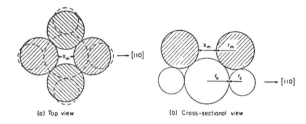

(a) Top view   (b) Cross-sectional view

Figure 1. Critical accommodation center for an oriented metal overgrowth cluster on an alkali halide substrate. After Henning and Vermaak[12].

energy orientation for an overgrowth cluster on the substrate. Henning and Vermaak[12] also note that if r_m is greater than the value depicted in Figure 1 the accommodation center becomes supercritical. The stability of the overgrowth cluster should be improved under this condition because of the likely enhancement of the metal-metal bonding. On the other hand, if r_m is less than the value in Figure 1 the substrate provides a subcritical accommodation center, which presents an unstable situation for the parallel alignment of an overgrowth cluster on the substrate.

In Walton's[5,6] atomistic nucleation theory the supersaturation level of the overgrowth atoms was considered the only experimental variable capable of influencing the size and shape of a critical nucleus. However, the CAM clearly predicts that the choice of substrate is also an important factor, particularly in the case of alkali halide substrates whose surfaces are known to be modified by exposure to water vapor[13,14], CO_2 [13,14], heat[13], and electron bombardment[15,16].

Consider for example the behavior of Au, Ag, and Cu. Both Matthews and Grünbaum[17] and Ino et al.[18] were unable to deposit Au, Ag, and Cu epitaxially on ultra high vacuum cleaved NaCl. However, if the NaCl wafers were exposed to air for an hour (and therefore water vapor), epitaxy was achieved in both high vacuum and ultra high vacuum, provided a bakeout was not employed. These results can be understood in terms of the CAM. As noted in Tables 1 and 2 of reference 12, the (001) surface of vacuum cleaved NaCl provides a subcritical accommodation center for Au, Ag, and Cu overgrowth clusters. Therefore, conditions will not be favorable for epitaxial growth. However, when the NaCl is exposed to water vapor, a chemical reaction creates an NaOH surface layer. This layer provides a supercritical accommodation center for the epitaxial alignment of the Au, Ag, and Cu overgrowth clusters, meaning that epitaxial growth should be likely. Thus, the predictions of the CAM are consistent with the noted experimental observations.

Henning and Vermaak[12] adapted their model in a semi-quantitative manner to Walton's nucleation theory. They point out that by considering the idea of critical accommodation, it can be argued that a two-atom cluster possessing a single metal-metal bond should promote epitaxial growth, contrary to Walton's original hypothesis. The reason for this conflict arises because the critical or supercritical accommodation of an overgrowth atom by the substrate will impart a directionality to a single metal-metal bond.

If the bonding scheme suggested by Henning and Vermaak is introduced into equation 1, several questionable results are predicted. For instance, although the modified nucleation rate expressions will be in general agreement with the epitaxy data available for continuous films, the new relations do not predict the change in the nucleation rate dependence from R^2 to R^4 that was documented by Walton et al.[7]. Furthermore, it is rather unsettling that in Henning and Vermaak's proposed cluster bonding formalism the choice of the proper nucleation rate expressions require an ad hoc prediction regarding whether the metal/substrate system is subcritical, critical, or supercritical. Although, this approach is not unreasonable, it lacks a certain aesthetic quality.

NEW THEORY

The cited shortcomings of Walton's[5,6] atomistic nucleation theory and Henning and Vermaak's[12] CAM are not encountered if the bonding scheme illustrated in Figure 2 is used to interpret the available experimental data. Four distinct temperature ranges are identified in the proposed bonding format. The blackened circles connected by solid lines represent the geometry of the minimum-size critical nucleus within a particular temperature range. The open circles and dashed lines illustrate how additional atoms must be added to the critical nucleus in order to stabilize it. When a sequence of growing nuclei are depicted within a temperature range, the author's intent is to demonstrate how the minimum-size stable cluster is expected to grow as it captures more atoms. Note that the number of metal-metal bonds connecting each atom in the minimum-size stable cluster changes by one at each temperature transition between adjacent temperature ranges.

In Walton's atomistic nucleation theory only one temperature transition between a two atom minimum-size stable cluster and a four atom minimum-size stable cluster was considered relevant to epitaxy. As pointed out earlier, this formulation does not account for the difference in the epitaxial behavior of Au, Ag, and Cu deposited on air cleaved versus vacuum cleaved NaCl.

Similarly, Henning and Vermaak's CAM predicts a lone temperature transition from a one atom minimum-size stable cluster to a two atom minimum-size stable cluster. In this case, the bonding scheme fails to predict the R^2 to R^4 transition in the nucleation rate that was reported by Walton et al.[7].

The bonding scheme proposed in Figure 2 avoids both of the problems encountered by the above theories without creating any new inconsistencies with the experimental data. Three transformation temperatures for critical nuclei are identified in this figure, one between each of the four bonding geometries. The transition at T_1 can be interpreted as the lower onset temperature for nucleating oriented atomic clusters on the substrate. The growth of a continuous epitaxial film is not assured for $T \gtrsim T_1$, however, because it is likely that the (001)/(111) island ratio will be too small. As T is increased from T_1 to T_2 the (001)/(111) ratio rises exponentially and eventually acquires a value that is conducive to the growth of a hole-free epitaxial film. The author suggests that the epitaxial growth temperatures reported by Sloope and Tiller[19] and Vook et al.[20-21] correspond to the temperatures which provide the minimum (001)/(111) ratios necessary for growing a continuous epitaxial film. The transition temperature at T_2 identifies the transition in the island nucleation rate from an R^2 to an R^4 dependence. Contrary to Walton et al.'s[7] interpretation, this transition should not significantly influence the growth of a continuous epitaxial film because the orientation of epitaxial islands will be the same above and below T_2. Only the number of atoms in the minimum-size critical nucleus will differ. The T_3 temperature transition identifies a high temperature limit for epitaxial growth, where the minimum-size critical nucleus changes from a single parallel oriented planar geometry with 2 metal-metal bonds per atom to a nonparallel multiple-oriented three-dimensional geometry with 3 metal-metal bonds per atom.

By equating the nucleation rate expression (equation 1) associated with

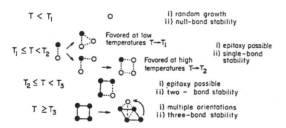

Figure 2. Minimum-size stable cluster configurations. ($T_m \equiv$ temperature above which the minimum-size stable cluster has m bonds per atom)

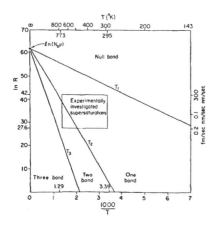

Figure 3. Cluster stability map for Ag deposited on a clean (001) NaCl Surface.

the adjacent bonding domains in Figure 2, it is possible to derive expressions for the three temperature transformations as a function of R. These equations can be applied to derive a cluster stability map which can be used to identify acceptable R,T supersaturation conditions for epitaxy. As an example, the cluster stability map for the Ag/NaCl metal/substrate system is illustrated in Figure 3. It must be recognized that a clear distinction is being made between a Ag/NaCl system and a system possessing a modified NaCl surface such as Ag/NaOH. Two particularly interesting features of the cluster stability map are:

i) The quantity $N_0 \nu$ represents a fundamental constant for cluster stability that is distinct for every metal/substrate combination. (N_0 is the density of adsorption sites on the substrate and ν is the vibrational frequency of a metal atom adsorbed on the substrate.)

ii) Along any constant temperature (constant R) tie-line there exists a limited range of acceptable R values (T values) from which an epitaxial film can be grown.

It may be implied from the tie-line behavior in the cluster stability map that equation 1 will yield a finite spread of island nucleation rates conducive to the epitaxial alignment of isolated islands as well as an even more limited range for the growth of a hole-free epitaxial film. Furthermore, Robins et al.[22-24] have observed that when a film is deposited at a constant nucleation rate, the density of nuclei ultimately reaches a saturation point N_s. It, therefore, follows that only a limited range of N_s will yield a continuous epitaxial film. As noted by Milkove and Sass[8] a careful inspection of the epitaxy literature appears to confirm this prediction.

In summary, an atomistic nucleation theory has been proposed which utilizes Walton's[5,6] general nucleation rate expression, the CAM developed by Henning and Vermaak[12], and a novel atomic bonding scheme for the nucleation of minimum-size stable atomic clusters. A cluster stability map has been derived from the theory which identifies.

i) Three temperature transformations relevant to the growth of epitaxial films by the Volmer-Weber mechanism.
ii) A fundamental cluster stability constant $N_0 \nu$.
iii) A limited range of suitable N_s values for thin film epitaxy.

In closing it is suggested that the cluster stability map in Figure 3 is the equivalent of a configurational phase diagram for the nucleation of atomic clusters possessing specific bonding geometries in a well-specified overgrowth/substrate system.

REFERENCES

[1] D.W.Pashley,"Recent Developments in the Study of Epitaxy", Recent Progress in Surface Science 3, 23(1970).
[2] E. Bauer, Z. Kristallogr. 110, 372(1958).
[3] E. Bauer and H. Poppa, Thin Sol. Films 12, 167(1972).
[4] R.Kern, G. Lelay, and J.J. Metois, Current Topics in Materials Science 3, 133, edited by E. Kaldis, North-Holland Publishing Company, 1979.
[5] D. Walton, J. Chem. Phys. 37, 2182(1962).

[6] D. Walton, Phil. Mag. $\underline{7}$, 1671(1962).
[7] D. Walton, T.N.Rhodin, and R.W. Rollins, J. Chem. Phys. $\underline{38}$, 2698(1963).
[8] K.R. Milkove and S.L.Sass, To be published in the proceedings of this conference.
[9] J.W. Matthews, JVST $\underline{3}$, 133(1966).
[10] S. Ino, D. Watanabe, and S. Ogawa, J. Phys. Soc. Jap. $\underline{17}$, 1074(1962).
[11] S. Ogawa, S.Ino, T. Kato, and H. Ota, J. Phys. Soc. Jap. $\underline{21}$, 1963(1966).
[12] C.A.O. Henning and J.S. Vermaak, Phil. Mag. $\underline{22}$, 281(1970).
[13] R.A. Lad, Surf. Sci. $\underline{12}$, 37(1968).
[14] J.S. Vermaak and C.A.O. Henning, Phil. Mag. $\underline{22}$, 269(1970).
[15] P.W. Palmberg, C.J. Todd, and T.N. Rhodin, JAP $\underline{39}$, 4650(1968).
[16] D.G. Lord and M. Prutton, Thin Sol. Films $\underline{21}$, 341(1974).
[17] J.W. Matthews and E. Grunbaum, Phil. Mag. $\underline{11}$, 1233(1965).
[18] S. Ino, D. Watanabe, and S. Ogawa, J. Phys. Soc. Jap. $\underline{19}$, 881(1964).
[19] Billy W. Sloope and Calvin O. Tiller, JAP $\underline{32}$, 1331(1961).
[20] R.W. Vook and C.T. Horng, Thin Sol. Films $\underline{18}$, 295(1973).
[21] R.W. Vook and C.H. Chung, Thin Sol. Films $\underline{37}$, 461(1976).
[22] V.N.E. Robinson and J.L.Robins, Thin Sol. Films $\underline{5}$, 313(1970).
[23] J.L. Robins and A.J. Donohoe, Thin Sol. Films $\underline{12}$, 255(1972).
[24] J.L. Robins, Surf. Sci. $\underline{86}$, 1(1979).

MISFIT DISLOCATIONS IN HgCdTe-CdTe HETEROJUNCTIONS

HIROSHI TAKIGAWA, MITSUO YOSHIKAWA, MICHIHARU ITO AND KENJI MARUYAMA
Infrared Devices Laboratory, FUJITSU LABORATORIES LTD., 1677 Ono Atsugi
243-01 Japan

ABSTRACT

Misfit dislocations in (111) HgCdTe-CdTe heterojunctions grown by
liquid phase epitaxy have been investigated. Etch pit studies showed misfit
dislocations to be distributed in the compositional transient region of the
epilayer due to interdiffusion. The density of dislocation etch pits n on
a (110) plane was found to be nearly proportional to the compositional gradient dx/dt in this area: $n = 5 \times 10^4$ dx/dt. This phenomenon can be interpreted by a model in which lattice mismatch originating from the gradual
compositional change is entirely accommodated by misfit dislocations.
These dislocations were found to move with interdiffusion during epilayer
growth. The experimental expressions which describe the compositional profile as a function of growth conditions have been presented.

INTRODUCTION

HgCdTe is a direct band-gap semiconductor having many advantages when
used in infrared detectors. Recently, a number of crystal growth techniques
have been developed [1-6]. To produce a crystal with good crystallinity,
it is generally grown by liquid phase epitaxy on a CdTe substrate [1-3].
The lattice constants of $Hg_{1-x}Cd_xTe$ and CdTe differ by 0.2% for x = 0.3.
This lattice mismatch may introduce strain-relieving misfit dislocations.
Transmission electron microscopy has shown misfit dislocations forming a
planar network in the interface between HgCdTe and CdTe [7]. However, distribution of misfit dislocations in the epilayer itself had not been investigated. It must be investigated to clarify the effects of the dislocations
on performance of infrared detectors.

This paper reports the distribution of misfit dislocations within this
heterojunction, and a model for interpreting the distribution. The conditions for epilayer growth are first briefly discussed. The observation of
dislocation etch pits and their distribution is presented. The new
relationship we found between etch pit distribution and compositional profile in the heterojunction is discussed, and a model interpreting this
relationship is proposed. Then experimental expressions describing compositional profiles of the heterojunctions are presented.

EXPERIMENTS AND RESULTS

In this study, we grew a HgCdTe epilayer having a 30% CdTe mole fraction in a Te-rich growth solution on a (111)A CdTe substrate using the
tipping liquid phase epitaxy technique. Dislocation etch pit density of a
CdTe substrate was estimated to be 10^4 - 10^5 cm^{-2} using HF:H$_2$O$_2$:H$_2$O etch
[8]. Before the epitaxial growth, the CdTe substrate was etched with a
solution of 5% by volume of bromine in methanol to remove the polishing
damage. Polishing damage was confirmed to be entirely removed. The 40-μm
epilayer was grown at growth temperature of 590 - 583°C at cooling rate of
0.02 °C/min and with growth time of 320 min.

We succeeded in finding dislocations by etching on the cleaved ($\bar{1}$10)
plane of epilayer. Figure 1 shows the cleaved, etched cross-sectional ($\bar{1}$10)
plane of the epilayer. Many dislocation etch pits appear in the epilayer,

as shown in this photomicrograph. As far as we know, this is the first time that dislocations spreading into the epilayer itself have been observed. Individual etch pits on the ($\bar{1}$10) plane were found to move only in the $\langle\bar{1}10\rangle$ direction when the surface layers of the plane were removed repeatedly by etching. This fact indicates that these etch pits originated from dislocation lines aligned in the $\langle\bar{1}10\rangle$ direction. Etch pit density varies along the growth direction of the epilayer, and is highest around the interfacial region, decreasing gradually in the direction of the growth.

The compositional profile of the heterojunction measured by X-ray microbeam analysis is shown in Figure 2. The horizontal axis of this figure shows the distance measured in the $\langle111\rangle$ direction. The horizontal axis starts at the original surface of the substrate. The gradual compositional change in the substrate and the epilayer is due to interdiffusion of Hg and Cd during epilayer growth, when Hg atoms diffused from the epilayer to the substrate and Cd atoms diffused to replace them. The dislocation etch pits were found to appear wholly in the compositional transient region on the ($\bar{1}$10) plane. The compositional gradient is highest around the interfacial region, decreasing in the direction of growth in the same way as etch pit density.

DISCUSSION

The calculations of misfit dislocation density at abrupt $Hg_{1-x(1)}Cd_{x(1)}$ $Te-Hg_{1-x(2)}Cd_{x(2)}Te$ heterojunctions are made for the (111) plane [9] where $\Delta x = \lfloor x(1) - x(2) \rfloor$ is a variable, using the theoretical treatment of Oldham and Milnes [10]. According to this study, the misfit dislocations for (111) heterojunctions may lie in the $\langle0\bar{1}1\rangle$, $\langle10\bar{1}\rangle$, and $\langle\bar{1}10\rangle$ directions with spacing h between sets as shown in Figure 3. James and Stoller observed dislocation lines aligned in each of these $\langle110\rangle$ directions and composing a planar network similar to Figure 3 in the $Hg_{1-x}Cd_xTe-CdTe$ interface, using transmission electron microscopy [7]. The $Hg_{1-x}Cd_xTe-CdTe$ interface is a special case with the variable Δx becoming $(1-x)$. However,

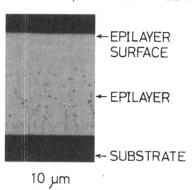

←EPILAYER SURFACE

← EPILAYER

← SUBSTRATE

10 μm

Figure 1. Photomicrograph of a cleaved, etched cross-sectional ($\bar{1}$10) plane of a HgCdTe epilayer grown on a CdTe substrate at growth temperature of 590-583°C and with growth time of 320 min. Many dislocation etch pits spreading into the epilayer itself are observed.

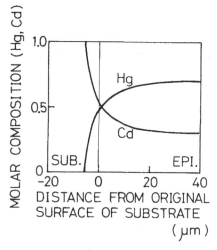

Figure 2. Compositional profile of the HgCdTe epilayer seen in Figure 1.

this method did not indicate where misfit dislocations occurred in the epilayer itself. If lattice mismatch is entirely accommodated by misfit dislocations, a mean spacing h is given theoretically by [9]

$$h = \frac{3 \ a_{x(1)}^2 \ a_{x(2)}}{\sqrt{2} \ [\ a_{x(1)}^2 - a_{x(2)}^2 \ ]} \ , \quad a_{x(1)} > a_{x(2)} \tag{1}$$

where $a_{x(1)}$ and $a_{x(2)}$ are the lattice constants for $x(1)$ and $x(2)$, respectively, on both side of the heterojunction. The lattice constants of CdTe and HgTe at room temperature are 6.4818×10^{-8} cm and 6.4620×10^{-8} cm, respectively [11]. The lattice constant of $Hg_{1-x}Cd_xTe$ varies approximately linearly with the x value across the entire composition range [11]. The linear thermal expansion coefficients for both materials is 5.0×10^{-6} /°C above room temperature [12,13]. Therefore, the mean spacing h at the growth temperature is given approximately by

$$h \ (\ cm \ ) = 2 \times 10^{-5}/\Delta x. \tag{2}$$

The distribution of dislocation etch pits on the ($\bar{1}$10) plane of a heterojunction and the compositional profile of the heterostructure are shown schematically in Figure 4. Dislocation etch pits on the cleaved ($\bar{1}$10) plane are produced by only dislocation lines aligned in the $\langle\bar{1}$10$\rangle$ direction as mentioned before. Dislocation lines aligned in the $\langle0\bar{1}$1$\rangle$ and $\langle10\bar{1}\rangle$ directions may form only invisible shallow etch pits. Therefore, if lattice mismatch originating from the compositional difference of Δx is entirely accommodated by dislocations, the number of dislocation etch pits N in the narrow area with a width of Δt and length L is given by

$$N = L/h = 5 \times 10^4 \ (\ \Delta x \ L \ ). \tag{3}$$ (a)

That is, dislocation etch pit density

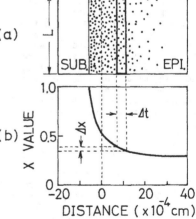

$\langle10\bar{1}\rangle \langle\bar{1}10\rangle \langle0\bar{1}1\rangle$

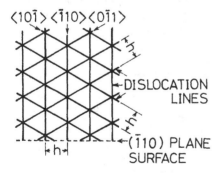

←DISLOCATION LINES

←($\bar{1}$10) PLANE SURFACE

Figure 3. Schematic diagram of misfit dislocation lines composing a planar network in the (111) plane of an abrupt $Hg_{1-x}Cd_xTe$-CdTe heterojunction.

Figure 4. Schematic diagram of distribution of dislocation etch pits on a ($\bar{1}$10) plane of HgCdTe-CdTe heterostructure: (a) distribution of dislocation etch pits; (b) compositional profile of the heterostructure.

n on the ($\bar{1}$10) plane is proportional to the compositional gradient dx/dt in the <111> direction in that area and is given by

$$n \ (\ cm^{-2}\ ) = N/(\ \Delta t\ L\ ) = 5 \times 10^4 \ dx/dt. \qquad (4)$$

Relationship between experimental data and the curve calculated from the model mentioned above is shown in Figure 5. To confirm that the previous model fits epilayers grown under different conditions and with different epilayer thicknesses, experimental data for other 40-μm and 15-μm epilayers grown under the conditions listed in Table 1 is added in Figure 5. The cleaved, etched cross-sectional ($\bar{1}$10) planes of the additional epilayers are shown in Figure 6. The experimental data for all epilayers nearly coincides with the curve calculated based on the previous model as shown in Figure 5. Therefore, it was found that the distribution of dislocation etch pits in epilayers can, on the whole, be interpreted by the previous model independent of growth conditions and the epilayer thickness. As the result, lattice mismatch originating from gradual compositional change is found to be entirely accommodated by misfit dislocations. Figure 7 shows the etch pit density as a function of the distance. The horizontal axis shows the distance from the HgCdTe-CdTe interface formed anew by interdiffusion. The solid lines were calculated from the previous model. The wider the compositional transient region becomes, the lower the etch pit density at the original surface of the substrate becomes, as shown in Figure 7. This tendency indicates that many dislocations generating in the original abrupt heterojunction at the beginning of growth move in the <111> direction with interdiffusion during epilayer growth.

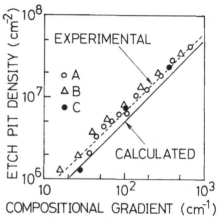

Figure 5. Experimental and calculated dislocation etch pit density on ($\bar{1}$10) planes as a function of the compositional gradient in this area for three epilayers grown under the conditions listed in Table 1.

Table 1. Growth conditions for epilayers

Symbol	Epilayer thickness (μm)	Growth temperature (°C)	Growth time (min)	Cooling rate (°C/min)
A	40	590 - 583	320	0.02
B	40	510 - 502	400	0.02
C	15	480 - 476	90	0.05

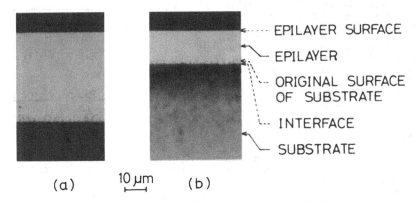

(a)   10 μm   (b)

Figure 6. Photomicrographs of cleaved, etched cross-sectional ($\bar{1}10$) planes of HgCdTe epilayers grown under the following conditions: (a) at 510 - 502°C in 400 min; (b) at 480 - 476°C in 90 min.

This high mobility of dislocations is one of the most distinctive features of this material.

The relationship between compositional profile of the heterojunction and growth conditions was investigated for epilayers grown under a variety of conditions in order to estimate etch pit distribution and compositional profile in epilayer grown under other conditions. As a result, it was found that the compositional profile of the epilayer grown at growth temperature of T K and growth time of τ s can be expressed approximately by the following experimental expressions,

$$x - X(E) = [1 - x(E)] \exp[-z/\sqrt{D(x,T)\tau}]$$

$$\cdot D(x,T) \ (\ cm^2/s \ ) = A(x) \exp[-B(x)/kT]$$

$$A(x) \ (\ cm^2/s \ ) = \exp(-8.157x + 14.51)$$

$$B(x) \ (\ eV \ ) = -0.3211x + 1.416, \tag{5}$$

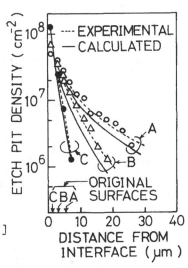

Figure 7. Experimental and calculated dislocation etch pit density on ($\bar{1}10$) planes as a function of the distance from the HgCdTe-CdTe interface for three epilayers grown under the conditions listed in Table 1.

where x(E) is the CdTe mole fraction of epitaxial surface layer, and z is the distance from the HgCdTe-CdTe interface formed anew as shown schematically in Figure 8, and k is the Boltzmann's constant. The position of the original surface of the substrate can be estimated from the compositional profile because area of the both hatched portions in Figure 8 must be same.

Relationship between the compositional profiles measured for the epilayers grown under the conditions listed in Table 1 and the curves calculated by the previous experimental expressions is shown in Figure 9. As shown in this figure, the compositional profiles for these epilayers can be approximately described by the previous experimental expressions.

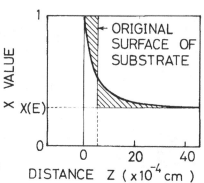

Figure 8. Schematic diagram of the compositional profile produced by interdiffusion of Hg and Cd atoms.

SUMMARY

We found by etch pit studies that misfit dislocations in a HgCdTe-CdTe heterojunction spread into the compositional transient region of epilayer due to interdiffusion of Hg and Cd atoms.

The density of dislocation etch pits n on (110) plane was found to be nearly proportional to the compositional gradient dx/dt in this area: $n = 5 \times 10^4 \, dx/dt$. This phenomenon can be fairly closely interpreted by a model in which lattice mismatch originating from gradual compositional change is entirely accommodated by misfit dislocations. These dislocations were found to move with interdiffusion during epilayer growth.

The experimental expression which can describe approximately the compositional profile as a function of growth temperature and growth time was obtained. As a result, distribution of misfit dislocations for epilayers grown under other conditions became predictable.

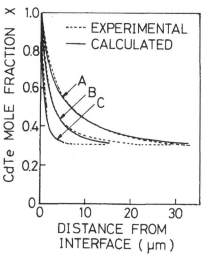

Figure 9. Experimental and calculated compositional profiles of HgCdTe-CdTe heterojunctions.

REFERENCES

[1] C.C.Wang, S.H.Shin, M.Chu, M. Lanir, and A.H.B.Vanderwyck, J.Electrochem.Soc. 127, 175 (1980).
[2] J.L.Schmit, R.J.Hager and R.A. Wood, J.Cryst.Growth 56, 485 (1982).
[3] T.C.Harman, J.Electron.Mater. 10, 1069 (1981).
[4] J.P.Faurie, A.Million and J.Piaguet, J.Cryst.Growth 59, 10 (1982).
[5] S.J.C.Irvine, J.Tunnicliffe and J.B.Mullin, J.Cryst.Growth 63, 479(1983).
[6] J.T.Cheung and D.T.Cheung, J.Vac.Sci.Technol. 21, 182 (1982).
[7] T.W.James and R.E.Stoller, Appl.Phys.Lett. 44, 56 (1984).
[8] E.P.Warekois, M.C.Lavine and A.N.Mariano, J.Appl.Phys. 33, 690 (1962).
[9] R.B.Schoolar, Proc.SPIE 409, 32 (1983).
[10] W.G.Oldham and A.G.Milnes, Solid-State Electron. 7, 153 (1964).
[11] J.C.Woolley and B.Ray, J.Phys.Chem.Solids 13, 151 (1960).
[12] Novikova, Soviet Phys.-Solid State 3, 129 (1961).
[13] Novikova and N.Kh.Abrikosov, Soviet Phys.-Solid State 5,1558 (1964).

TEM INVESTIGATION OF POLAR-ON-NONPOLAR EPITAXY:
GaAs-AlGaAs on (100) Ge

J.H. Mazur,* J. Washburn,* T. Henderson,** J. Klem,** W.T. Masselink,** R. Fischer,**
and H. Morkoç**
*Materials and Molecular Research Division, Lawrence Berkeley Laboratory, University of
California, Berkeley, CA 94720.
**University of Illinois at Urbana-Champaign, 1101 W. Springfield Ave., Urbana, IL 61801

ABSTRACT

Morphologically excellent GaAs and AlGaAs epitaxial layers have been grown on (100)
Ge substrates by molecular-beam epitaxy (MBE). Transmission electron-microscope (TEM)
studies of the thin-film cross sections revealed that defects, most probably antiphase
domains, were contained within a 20–30 nm thick initial layer near the GaAs/Ge interface,
formed with a 0.1-μm/h growth rate at 500°C. A reduction of defects occurred in the
remaining 0.1-μm thick layer, for which the growth rate and temperature had been increased
to 1 μm/h and 580°C, respectively. The interface between this GaAs layer and a subsequently grown 35-nm thick AlAs film was found to undulate with a period of 150–200 nm
and an amplitude of 5–10 nm. Such large undulations were not observed in AlGaAs-GaAs
superlattices grown after an additional deposition of 2-μm GaAs. High-resolution electron-
microscopy observations suggest that transitions between different layers are abrupt on the
atomic scale. In addition, modulation-doped field-effect transistors fabricated on these layers
had similar characteristics to those obtained with MBE layers grown on GaAs substrates.
These results demonstrate that the (100) Ge surface is suitable for MBE polar-on-nonpolar
semiconductor growth and that the integration of III-V films with silicon electronic devices
via epitaxial Ge on Si is feasible, provided that epitaxial Ge of sufficient quality grown on Si
is available.

INTRODUCTION

The growth of III-V compound semiconductor films on nonpolar substrates is of considerable interest, because of the potential application of these films for electronic and
optoelectronic devices. These applications include GaAs solar cells [1], heterojunction bipolar transistor [2], light-emitting diodes [3], and modulation-doped field-effect transistors
(MODFETs) [4,5]. The growth of GaAs and AlGaAs films on Si substrates is especially
interesting because of their good mechanical strength and low cost. In addition, such a substrate would allow monolithic integration of III-V electronic devices with well-established Si
technology. However, two materials problems, antiphase domain formation and large
(~4%) lattice mismatch [6-8], have to be overcome in order to successfully use Si substrates.
A possible solution to the problem involves epitaxial growth of Ge (which has a small,
~0.1%, lattice mismatch with GaAs) on Si followed by epitaxial growth of the III-V film, or
epitaxial growth of well-matched GaP or AlP films followed by graded transition to GaAs or
AlAs. In both cases the greatest concern would be elimination of antiphase domains. These
domains may be formed by coalescence of islands of As and Ge when epitaxial growth
begins on the nonpolar semiconductor surface, or, more probably, they may be caused by the
existence of monatomic steps on the substrate. These two possibilities are schematically
illustrated in Figs. 1a and 1b, respectively.

In this work molecular-beam epitaxy (MBE) GaAs-AlGaAs layers have been grown on
(100) Ge substrates in order to determine the viability of the first approach to solving the

*Supported by the Director, Office of Energy Research, Office of Basic Energy Sciences, Materials Sciences Division of the U.S.
Department of Energy under Contract Number DE-AC03-76SF00098.
**Supported by the Air Force Office of Scientific Research.

problem described above. The quality of the films and their interfaces have been studied by transmission electron microscopy and by the characteristics of electronic devices fabricated on these layers.

EXPERIMENTAL PROCEDURES

The III-V compound semiconductor layers were grown by MBE on (100)2° [011] oriented heavily doped p-type Ge. The substrates were first degreased in boiling tri-chloroethane, acetone, methanol, and water. They were then etched for 20 min in $H_2SO_4:H_2O_2:H_2O$ (5:1:1), rinsed in deionized H_2O, and dried with N_2 gas. The substrates were then mounted to Mo blocks using In solder, loaded into the MBE system, outgassed at 630°C (as read by a thermocouple in an auxiliary chamber), and immediately transferred into the growth chamber. Prior to the growth the substrate temperature was raised to 600°C and then lowered to 500°C. Following the procedure [7], an As prelayer was deposited, followed by 250 Å of GaAs growth with a large As overpressure and the growth rate 0.1 μm/h. The growth rate and substrate temperature were then increased to 1 μm/h and 580°C during the growth of 0.1 μm of GaAs. Different sequences of GaAs and $Al_xGa_{1-x}As$ layers were then deposited. MODFET structures were fabricated from these layers according to procedures described elsewhere [5]. Cross-sectional TEM specimens with <011> foil surface orientation were prepared using the same procedures as for cross-sectional Si specimens [9]. The obser-vations were performed at 200 kV and 100 kV in JEOL 200CX and Philips 301 electron microscopes, respectively.

EXPERIMENTAL RESULTS

TEM Results

Figures 2a and 2b show a TEM cross section of one of the MBE-grown layers and the sequence and thickness of the deposited layers, respectively. The morphology of the growth front at the top of the first layer, as revealed by the interface between GaAs and AlAs, is not planar. It undulates with a period of 0.2–0.25 μm and an amplitude of 10–15 nm. Apparently the interface consists of pyramids with facets approximately normal to <116> directions and inclined between 7–12° with respect to a planar Ge/GaAs interface, as can be seen in Fig. 3. Figure 4, which is from a different area of the same specimen, shows in greater detail the defect contrast most probably corresponding to antiphase domains at the Ge/GaAs interface. Tilting experiments did not reveal any defect contrast in other layers. It is tentatively concluded that antiphase domains were contained only within the initial region of growth and that areas further from the Ge/GaAs interface were defect-free. The transition regions between adjacent layers were atomically abrupt, as can be seen from the high-resolution electron micrographs shown in Fig. 5 and Fig. 6. The GaAs/Ge interface had good matching of the lattice planes on the two sides of the interface, as expected for materi-als with close lattice parameters. No contrast corresponding to antiphase boundaries was observed at this level of resolution, in agreement with earlier computer simulation [10]. The GaAs lattice planes continue across the AlAs layer, which is about 500 Å thick, although some AlAs, especially near the TEM specimen edge, appeared to be amorphous (probably due to rapid oxidation in the atmosphere enhanced by condensation of water vapor on the specimen after removal from the cold-stage ion mill). In the thicker regions of the specimen (Fig. 6), the lattice planes also continued across the AlAs layer, suggesting that the material was crystalline before it oxidized.

An additional interesting growth feature of these MBE layers was revealed in the bright-field image in Fig. 7, which shows a reduction of the undulation of the growth front 2 μm above the AlAs layer. This is illustrated by the flat layers of the superlattice 40× (100 Å $Al_{.25}Ga_{.75}As$/150 Å GaAs). The transition between the GaAs and $Al_{.25}Ga_{.75}As$ layers is shown in the high-resolution micrograph (Fig. 8). The contrast between layers is

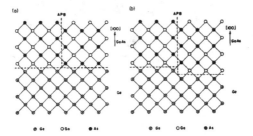

Fig. 1. Possible origin of antiphase boundaries: a) initiated on a flat surface of Ge; b) on a surface containing a monatomic step.

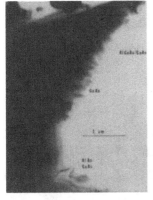

| 50 Å GaAs (Si) |
| 350 Å Al$_{.3}$Ga$_{.7}$As (Si) |
| 30 Å Al$_{.3}$Ga$_{.7}$As |
| 40 × (150Å GaAs/100Å Al$_{.25}$Ga$_{.75}$As) |
| 2 μm GaAs |
| 500 Å AlAs |
| 0.1 μ GaAs |
| 250 Å GaAs slow growth |
| (100) Ge substrate |

Fig. 2. The structure of the films grown on (100) Ge: a) TEM cross section; b) corresponding schematic of the MBE layers.

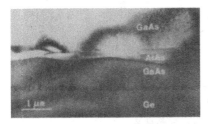

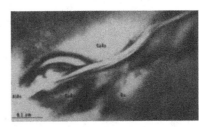

Fig. 3. TEM image of the initial layers demonstrating undulation of the MBE growth front.

Fig. 4. Defects near the Ge/GaAs interface, most probably antiphase boundaries.

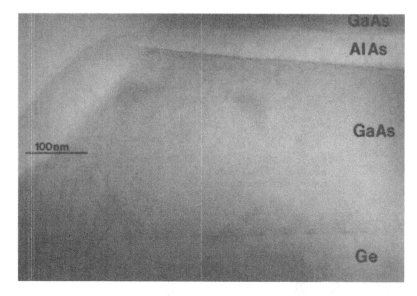

Fig. 5. High resolution image of Ge/GaAs, GaAs/AlAs interfaces.

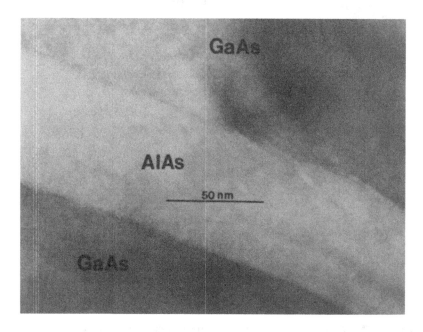

Fig. 6. High resolution image of GaAs/AlAs/GaAs showing continuity of lattice planes.

Fig. 7. Bright field image of superlattices with g = [200].

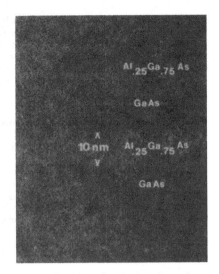

Fig. 8. High resolution image of superlattices.

Fig. 9. Surface morphology as revealed by Nomarski phase-contrast microscopy.

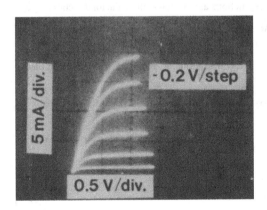

Fig. 10. MODFET I-V characteristics.

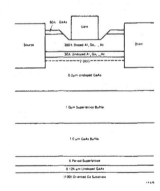

Fig. 11. MODFET structure.

due to the different atomic-scattering factor between GaAs and AlGaAs with Al randomly substituted in the Ga sublattice. This transition is abrupt on the atomic scale. Therefore, it is not surprising that the morphology of the surface as revealed by the Nomarski phase-contrast method is very good, as shown in Fig. 9. The surface flatness compares well with the surfaces of similar films grown on (100) GaAs substrates.

Characteristics of MODFETs

Modulation-doped field-effect transistors were fabricated on the layers described above using procedures described elsewhere [5]. The 300 K I-V characteristics shown in Fig. 10, were obtained from devices with 1×145 μm gates and 3μm source drain spacing fabricated from structures with the layer arrangement shown schematically in Fig. 11. The measured transconductance of the device was 160 mS/mm, and the output conductance was 3.6 mS/mm, which is comparable to results obtained from MODFETs grown on GaAs. The threshold voltage of the device shown in Fig. 11 was -0.95 V. When cooled to 77 K, the transconductance increased to 345 mS/mm, indicating existence of high-saturation-velocity two-dimensional electron gas (2-DEG). The output conductance at 77 K reduced to 2 mS/mm. The absence of looping in the I-V characteristics in Fig. 11 and the high transconductances demonstrate the high quality of the GaAs/AlGaAs grown on the Ge (100) substrates.

CONCLUSIONS

TEM studies of GaAs and GaAlAs thin films grown on (100) Ge substrates showed excellent crystalline perfection. MODFETs fabricated on these layers had characteristics similar to those obtained on MBE layers grown on GaAs substrates. These results indicate that monolithic integration of III-V and Si electronic devices on Si substrates is feasible, provided that epitaxial Ge of sufficient quality can be grown on Si.

REFERENCES

1. D.L. Miller, J.S. Harris, Appl. Phys. Lett, *37*, 1104 (1981).
2. A.G. Milnes, D.L. Feucht, "Heterojunctions and Metal-Semiconductor Junctions," New York: Academic (1972), p. 58.
3. R.M. Fletcher, D.K. Wagner, J.M. Ballantyne, Appl. Phys. Lett., *44*, 967 (1984).
4. H.K. Choi, B.-Y. Tsaur, G.M. Metze, G.W. Turner, J.C.C. Fan, IEEE Electron Device Lett., EDL-5, 207 (1984).
5. R. Fischer, J. Klem, T. Handerson, W.T. Masselink, W. Kopp, H. Morkoç, IEEE Electron Device Lett., EDL-5, 456 (1984).
6. K. Morizane, J. Cryst. Growth, *38*, 249 (1977).
7. P.M. Petroff, A.C. Gossard, A. Savage, W. Wiegmann, J. Cryst. Growth, *46*, 1972 (1979).
8. S.L. Wright, H. Kroemer, M. Inade, J. Appl. Phys., *55*, 2916 (1984).
9. J.H. Mazur, R. Gronsky, J. Washburn, in Microscopy of Semiconducting Materials, Proceedings Series, The Institute of Physics, *67*, 77 (1983).
10. T.S. Kuan, C.-A. Chang, J. Appl. Phys., *54*, 4408 (1983).

MULTILAYER GROWTH OF SILICON BY LIQUID PHASE EPITAXY

E. Bauser [+] and H.P. Strunk [++]
[+] Max-Planck-Institut für Festkörperforschung, 7000 Stuttgart 80, Germany
[++]Technische Universität Hamburg-Harburg, 2100 Hamburg 90, Germany

ABSTRACT

The growth of thin layers by liquid phase epitaxy requires rapid transport of the solutions. A growth system is briefly described, which makes use of centrifugal forces in a rotating crucible. Two microscopic growth mechanisms are discussed with respect to their applicability for growing thin, homogeneously doped layers with planar interfaces: terrace growth which proceeds via surface steps of multiple elementary height, and facet growth with growth steps of elementary height. The discussion is illustrated with results that were obtained by an optical and electron microscope study of silicon multilayers grown from various metal solutions.

INTRODUCTION

Semiconductor multilayers have become essential elements in modern device technology, and quantum well structures are currently of particular interest. Depending on the desired properties of the multilayer structures, various methods for their production may be applied. Liquid phase epitaxy (LPE), although widely used for the growth of compound semiconductors, is exceptional in silicon technology up to now /1/. Specific advantages, however, make LPE a promising candidate for Si multilayer growth. The advantages of Si-LPE result from the generally low growth temperature, and from the possibility to precisely control the microscopic kinetic growth mechanisms of the layers: Reduced interdiffusion and abrupt interfaces, in particular an excellent planarity of the interfaces, chemical homogeneity and a very low content of structural defects of the layers. The kinetic growth mechanisms will be addressed in the following, mainly in conjunction with microscopic observations of grown Si-multilayers. In addition, a growth apparatus will be outlined which utilizes centrifugal forces for a rapid transport of the solutions, thus permitting the growth of very thin layers /2-5/.

MULTILAYER GROWTH IN THE CENTRIFUGAL SYSTEM

The design of the epitaxy-crucible permits rapid transport of the solutions and avoids movable or sliding parts, because abraded particles generally disturb the defect-free growth of the layers. The crucible as a whole can be rotated, and the solutions inside the crucible are transported by utilizing centrifugal forces. The centrifugal forces are controlled by the variable rotational frequency of the crucible /2,5/. A schematic cross-section of the crucible is shown in Fig. 1. The dimensions and shapes of the flow tunnels connecting solution reservoirs and substrate chambers are designed to account for the delicate counterplay of surface tensions of the solutions at the desired growth temperature, of interfacial tensions and centrifugal forces. The construction of the crucible provides for closed cycling of each of the solutions and avoids their mixing. One charge of solutions may therefore be used in repeated epitaxial runs.

The crucible is mounted on the lower end of a rotor which is electromagnetically suspended in a thin walled vacuum tank /4/. Bearing magnets, rotor drive and position sensors are positioned outside this tank. Details of the centrifugal LPE system have been described earlier /4-6/. Single-crystalline layers are grown from solution on (111)-oriented silicon substrates. A variety of solvents are applied, Ga, In, Sn, Sb, Bi and alloys of these elements, occasionally with P or As added as dopants. The solutions are saturated with silicon and brought into contact with the substrates for epitaxial growth. Subsequent slight lowering of the temperature causes Si supersaturation and thus epitaxial growth of the layers. After predetermined time and temperature intervals the solutions are removed from the grown layers, and guided by a flow tunnel into a reservoir to get resaturated there for the growth of the next layer. Both the operation of the epitaxy centrifuge and the epitaxial growth process are computer controlled.

GROWTH MECHANISMS AND SURFACE MORPHOLOGY

A crystal generally grows by the attachment of atoms or molecules to steps that exist at its surface. The steps then move laterally along the surface /7/. This lateral microscopic growth, $L_\mu G$, adds successively new layers to the crystal and leads to the perpendicular macroscopic growth, PMG, of the crystal /8/. The morphology of the surface and the perfection of the crystal depend on the height, and hence on the origin, of these steps. Two contrasting mechanisms of growth are shown in the micrograph of a Si layer in Fig. 2a. Figure 2b shows a sketch of the surface morphology. A microscopically irregular array of steps is present in the terraced area. Step heights and distances vary in a wide range. Growth terraces develop on substrates which are oriented slightly off ($\geq$ 0.3°) a low indices plane. The adjacent and apparently flat area in Fig. 2a represents facet growth which develops in areas where the orientation of the growth interface is very close to (< 0.3°) or parallel to a low indices plane. Crystal growth then proceeds by the motion of elementary steps, which may be generated in essentially two ways. On a defect-free and atomically flat area, new monomolecular layers can arise only by two-dimensional nucleation, that is by the formation of islands of new crystal planes. This nucleation occurs at high supersaturations only and leads to an intermittent rapid lateral growth. Once an island has formed - an event which occurs statistically in time and site - it spreads with a high lateral velocity due to the high supersaturation. Steps which are the edges of two-dimensional nuclei, therefore, are likely to locally coalesce and to form steps of multiple elementary height and thus a microscopically nonplanar surface.

At the low supersaturations generally encountered in LPE, this two-dimensional nucleation is not possible at rates worth mentioning. Dislocation-controlled facet growth may occur, however, if dislocations are present in the substrates and if these dislocations act as persistent step sources.

Figure 3 shows a facet in which dislocation step sources were active. Various "growth pyramids" are visible in this facet. A direct X-ray /9/ and electron microscope analysis showed that in each apex only one dislocation emerges and acts as a persistent growth step source. An example of such a dislocation is shown in Fig. 4. The electron microscope analysis revealed /10,11/ that every dislocation may act as a step source, irrespective of the direction of its Burgers vector with respect to the growth surface /10,11/. The step generation mechanism with the Burgers vector perpendicular or oblique ('Frank step source' /12/) was comprehensively described by Frank with the spiral step model /13/, cf. Fig. 5a,b. The nucleation of layers in case of Burgers vector in the growth surface ('Bethge step source' /12/) which leads to the formation of concentric growth step patterns /14/ (Fig. 5c), is presently still under discussion.

Various explanations for the step generation at Bethge sources have been proposed. The only available microscopic experimental indications to the mechanism is the observation by TEM /11/ that such dislocations are widely split at the growth interface and thus form a stacking fault, see Fig. 4. Such a stacking fault causes a step of less than elementary height. Attach-

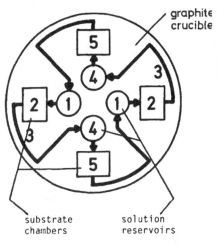

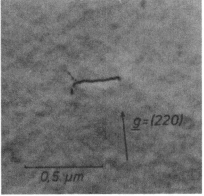

Fig.1. Schematic cross section perpendicular to the axis of rotation through the crucible used in the LPE centrifuge. Numbers give consecutive positions of the solutions. The channels are in reality inclined to the plane of the cross-section.

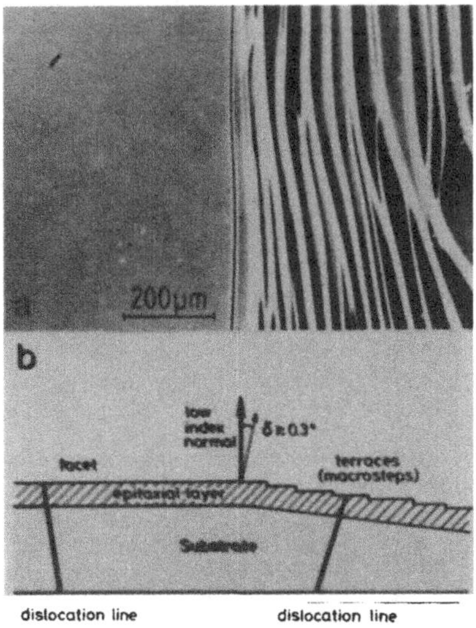

Fig.2. a) Light optical micrograph of Si layer grown from In solution. The surface morphology is seen in Nomarski differential interference contrast (NDIC).
b) Schematic cross-section through a) to depict the surface morphology.

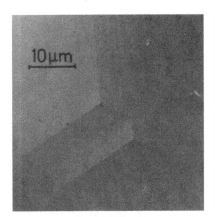

Fig.3. Growth pyramids on a (111) Si-facet grown from Ga solution. NDIC micrograph.

Fig.4. Dislocation that acted as persistent growth step source during LPE Si growth. Transmission electron micrograph. $\underline{g}$: diffraction vector. The dislocation is widely dissociated in the apex of the growth pyramid (see text).

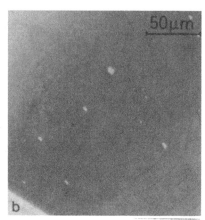

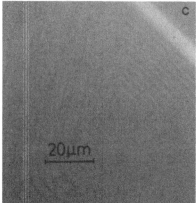

Fig.5. Growth pyramids with monomolecular steps on (111) facets.
a) Growth spiral which is poly-
gonized. Growth from In solution.
Growth temperature interval
$\Delta T = 1220 - 500$ K
b) Isotropic growth spiral, grown
from In solution. $\Delta T = 1220 - 1125$ K
c) Concentric step structure, grown
from Bi solution. $\Delta T = 1175 - 1060$ K

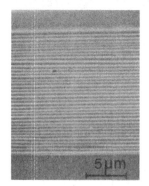
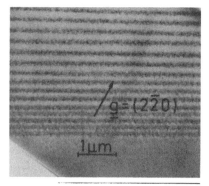

Fig.6. Multilayer grown from In and In-As solutions to produce an alternating
p-n structure.
a) Scanning electron micrograph of a photoetched cleavage face.
b) Corresponding transmission electron micrograph.

ment of elementary molecules during crystal growth cannot eliminate this step which is therefore persistent and may cause persistent, highly efficient nucleation of growth steps /6,11/. This model has recently been corroborated by the observation that in a silicon facet only widely dissociated dislocations did cause growth pyramids. The other, apparently undissociated dislocations (these are generally narrowly, say ~ 5nm /15/, dissociated, which is below the resolution of conventional TEM) did not form growth pyramids /16/. This observation is presently at variance with a model recently proposed by Giling /17/. This model bases on the existence of a surface reconstruction. This reconstruction is disturbed at the emergence point of the dislocation to form a "rough heart" /17/, where nucleation of a new crystal layer is facilitated.

GROWTH MECHANISMS AND CHARACTERISTICS OF GROWN LAYERS

The detailed discussion of the growth mechanisms is indispensable for layer growth because of two reasons: Firstly, the growth mechanisms largely determine the dopant (or impurity) distribution in the crystal and secondly, they determine the 'morphology' of the interface and thus the spatial thickness fluctuation in a grown layer.

In the terrace growth mechanism, the minimum layer thickness is limited by the height of the terrace steps, which can amount to several tens of nanometers. Growth interfaces with terraces are morphologically unstable /12/, i.e. perturbations at the growth interface tend to increase. Therefore the heights of the steps may vary during growth. Moreover, the risers of terraces are sites of enhanced or reduced dopant incorporation and thus cause distinct trails of different dopant concentration in the growing crystal, which were denoted as type II striations /8,12/. Terraces therefore cause an inhomogeneous dopant distribution.

If dislocation-controlled facet growth is applied, the thickness of epitaxial layers can in principle be reduced to one lattice distance. The growth interfaces are essentially atomically flat with a low density of equidistant atomic steps. The distance between steps may be as large as ~ 50 μm, i.e. the ratio of step distance to step height may amount to 10^5. An example is shown in Fig. 5b. Such structures represent the optimal experimental approach to an ideally atomically flat interface or surface. Further advantages of this kind of interfaces result from their morphological stability: Perturbations tend to decrease because of the steady motion of the train of equidistant monomolecular steps during growth /18/. In addition, the incorporation of dopant atoms is highly uniform at such an interface and leads to a homogeneous dopant distribution in the solid. For an optimum uniformity in the distribution of dopants in the layers, it is not yet sufficient that the growth steps are monomolecular and equidistant. The steps must be straight in addition. Hence, a train of monomolecular, equidistant and straight steps, which extends over the entire growth interface, is the condition required for the growth of thin and uniform layers and multilayers /12,18/. Strong nonuniformities may result, for example, in cases, where step trains form corners in the growth plane. Recent TEM studies of silicon LPE layers grown from Ga solution give evidence, for example, of microsegregation at the corners of growth pyramids /19/.

EXAMPLES OF MULTILAYERS

Figure 6a shows a SEM micrograph of a multilayer grown with 75 subsequent, equally thick and n-p-alternating layers. The specimen, after growth, was cleaved and photoetched to reveal the p-n-junctions. The homogeneous thickness and planarity of the layers is evident; no indication for lattice defects can be seen. Figure 6b shows a corresponding TEM cross-sectional

114

micrograph of the same multilayer. The interfaces are planar within the resolution of the micrograph. The contrast of the various layers was generated by simultaneous excitation of the two (220) diffraction vectors on both sides of the central spot. Other imaging conditions did not show any contrast at all which indicates that no lattice defects are present in the bulk and at the interfaces. The contrasts in Fig. 6b are very probably due to a different surface damage produced by the ion-milling at the n- and p-doped layers. Even thinner layers can be grown by further reduction of deposition time and selection of the facet growth mechanism.

CONCLUSIONS

The terrace and dislocation-controlled facet growth mechanisms are discussed and studied with the aid of layers grown by liquid phase epitaxy. It is shown that, by utilizing the specific features of these growth mechanisms and by using rapid solution transport in a LPE centrifuge, very thin multilayers of silicon of excellent crystalline quality can be grown. The discussion shows in particular that dislocation-controlled facet growth is capable of producing well defined thin layers.

ACKNOWLEDGMENTS
The authors are grateful to W. Appel, H. Bender, D. Käß and M. Warth for cooperation in and discussions on layer growth and helpful assistance in the preparation of the manuscript. Part of the work was performed at the Max-Planck-Institut für Metallforschung, Institut für Physik, Stuttgart. The work was supported by the Federal Ministery of Research and Technology (Contract No. NT 2597). The authors alone are responsible for the content.

REFERENCES
/ 1/ B.J. Baliga, J. Electrochem. Soc. 128, 161 (1981).
   R. Linnebach and E. Bauser, J. Crystal Growth 57, 43 (1982).
/ 2/ E. Bauser, M. Frik, K.S. Löchner and L. Schmidt, in: Gallium Arsenide and Related Compounds, Inst. Phys. Conf. Ser. No. 24, 10 (1975).
/ 3/ E. Bauser, L. Schmidt, K.S. Löchner and E. Raabe, Jap. J. Appl. Phys. 16 (1977), Supplement 16-1, pp. 457.
/ 4/ G. Schweitzer, A. Traxler, H. Bleuler, E. Bauser and P. Koroknay, Vakuum-Technik 32, 70 (1983).
/ 5/ D. Käss, M. Warth, H.P. Strunk and E. Bauser, Proc. ESSDERC 1984, to be published.
/ 6/ E. Bauser and H.P. Strunk, Report Nr. 403-7291-NT 9533, Federal Ministery of Research and Technology, Germany, June 1982.
/ 7/ W. Kossel, Naturwissenschaften 18, 131 (1930).
/ 8/ E. Bauser and G.A. Rozgonyi, Appl. Phys. Lett. 37, 1001 (1980).
/ 9/ E. Bauser and W. Hagen, J. Crystal Growth 70, 771 (1980).
/10/ E. Bauser and H. Strunk, J. Crystal Growth 51, 362 (1981).
/11/ H. Strunk and D. Käss, Thin Solid Films 81, L101 (1981).
/12/ E. Bauser and H.P. Strunk, J. Crystal Growth, in press.
/13/ F.C. Frank, Disc. Faraday Soc. 5, 49 (1949).
/14/ H. Bethge, phys. stat. sol. 2, 775 (1962) and Surface Sci. 3, 33 (1964).
/15/ J.L.F. Ray and D.J.H. Cockayne, Proc. Roy. Soc. A325, 534 (1971).
/16/ A. Kessler and H.P. Strunk, to be published.
/17/ L.J. Giling and B. Dam, J. Crystal Growth 67, 400 (1984).
/18/ E. Bauser, in: Crystal Growth of Electronic Materials, E. Kaldis Ed., North Holland, to be published.
/19/ H.P. Strunk and A. Kessler, Electron Microscopy 1984, A. Csanady, P. Röhlich and D. Szabo Eds., Programme Committee of the Eighth European Congress on Electron Microscopy, Budapest 1984, Vol. 2, p. 951.

GROWTH AND CHARACTERIZATION OF CdTe LAYERS ON InSb SUBSTRATES PREPARED BY ORGANOMETALLIC EPITAXY

I. Bhat, L. M. G. Sundaram, J. M. Borrego and S. K. Ghandhi
ECSE Department, Rensselaer Polytechnic Institute, Troy, New York 12180-3590

ABSTRACT

Organometallic vapor phase epitaxy of CdTe on InSb substrates is described, for growth under various temperatures and reactant partial pressures. Layers are characterized by photoluminescence measurements at 77°K and by electrical measurements on Au-CdTe Schottky barrier diodes. It is shown that growth of CdTe on InSb substrates results in material with better electrical and photoluminescence properties than that grown on CdTe substrates. Au-CdTe Schottky barriers on these layers resulted in diodes with an ideality factor of approximately 1.1 and a barrier height of 0.75 eV. These results are comparable to the characteristics of the devices fabricated on the best CdTe materials to date.

INTRODUCTION

Growth of high quality CdTe films is of considerable interest because of its potential application in areas of optoelectronics, solar energy conversion and X-ray detectors. In addition, its close lattice match and chemical compatibility with HgCdTe makes it an ideal substrate for the growth of this material. However, the lack of availability of high quality, large area single crystal CdTe has prompted many workers to study the growth of this material on sapphire,[1] GaAs[2], and InSb.[3] This heteroepitaxial layer can then be used as an active medium for device applications, or as a buffer layer for subsequent HgCdTe growth. InSb is of particular importance because of its close lattice match with CdTe (0.05%): moreover, it is available in the form of large diameter wafers (40 mm).

We have reported on the growth of CdTe on InSb substrates by organometallic vapor phase epitaxy.[4] This work reports, for the first time, on devices fabricated on CdTe deposited by OMVPE, either on CdTe substrates or on any other substrates.

EXPERIMENTAL

The epitaxial growth of CdTe on (100) InSb substrates was carried out at atmospheric pressure in a quartz horizontal reactor tube with a r.f. heated graphite susceptor, using palladium purified hydrogen as the carrier gas. Diethyltelluride (DETe) and Dimethylcadminum (DMCd) vapors were transported to the reaction zone by bubbling hydrogen through the respective bubblers.

Growth was carried out at temperatures ranging from 350°C to 450°C, both with excess DMCd partial pressure and with excess DETe partial pressure. The partial pressures of DETe and DMCd were in the range 1×10^{-4} atms to 8×10^{-4} atms, and the total flow was 3 ℓ/min. CdTe was also grown on (100) CdTe substrates for comparison purposes. The growth rate was typically 8 μm/hr at a temperature of 415°C.

Gold Schottky barrier diodes were formed on n-type layers by evaporating 1000°A of Au through metal masks. The area of diodes were about 1.1×10^{-3} cm^2. Since the substrate material has a low bandgap, no special ohmic contact procedure was used for the back surface.

The interface between CdTe layers and the InSb substrate was found to be ohmic.

RESULTS AND DISCUSSION

Layers of good crystalline perfection could be grown on InSb substrates at all the above reactor conditions. It was consistently found that layers grown on InSb substrates were of superior morphology compared to those grown side by side on CdTe substrates. The layers grown on CdTe substrates generally showed a series of ridged features, as opposed to the smooth surfaces obtained for layers grown in InSb substrates. This is because InSb substrates are less detected than CdTe substrates,[5] at the present state of the art.

Photoluminescence (PL) measurements were made at 77°K using a He-Ne laser as the optical excitation source. CdTe grown above 400°C and under excess DETe partial pressure exhibited one strong peak at 1.58 eV and a broad band around 1.43 eV. The 1.43 eV is generally associated with the cadmium vacancy-donor complex, and can be expected for growth under excess DETe partial pressure. The 1.58 eV peak is due to band to band transitions. The PL data for growth under excess DMCd pressure, on the other hand, shows that the broad band at 1.43 eV has been replaced by a broad band around 1.5 eV of comparable intensity. In addition, a new peak around 1.2 eV is also present in these layers. This peak has been attributed to a complex involving Te vacancies.[6]

Electrical characteristics of the layers also showed large differences depending on the growth conditions. Growth above 380°C and under excess DETe partial pressure showed n-type behavior whereas growth under excess DMCd pressure resulted in high resistivity layers. This can be explained by noting that growth under excess DETe pressure results in Cd vacancies; these should act as incorporation sites for donors such as Al, which is a common contaminant in the starting chemical sources.

Deep Level Transient Spectroscopy (DLTS) measurements were made on these devices using a capacitance meter, two box car integrators, and a pulse generator. The diode temperature was scanned from 80°K to 360°K for these experiments. For Au-n-CdTe Schottky diodes, only electron traps could be detected. One electron defect level was found at 0.6 eV below the conduction band. The capture cross section and the defect density of this level was 3.5×10^{-15} cm^{-2} and 1.5×10^{13} cm^{-3} whereas material impurity concentration was 2.3×10^{15} cm^{-3}. The above values are for a layer grown with P_{DETe}/P_{DMCd} of 6. When this ratio was increased to 8, this defect level concentration increased to 2.8×10^{13} cm^{-3}. This confirms that the basic defect in the crystal is either a tellurium interstitial or a cadmium vacancy. This level may be the same as the one commonly reported for the doubly-negative charged cadmium vacancy.[7]

The current-voltage (I-V) and 1 MHz capacitance-voltage (C-V) characteristics of the Schottky diodes were obtained to assess the quality of the layers. A photograph of the forward and reverse I-V characteristics of one of these diodes is shown in Figure 1.

The forward bias current-voltage characteristic was measured over 4-5 decades of current. More than 10 different diodes on different layers were measured; in all cases the ideality factor was found to be between 1.0 and 1.1, over 3-4 decades of current. For the diode shown in Figure 2, J_0 was found to be 3×10^{-7} A/cm^2. Since, n is close to unity, the results of thermionic emission theory were used to determine a barrier height of 0.74 eV.

Figure 3 shows the $1/C^2$ vs V characteristics of a typical diode. The curve is non-linear at voltages above 3V, indicating a slight

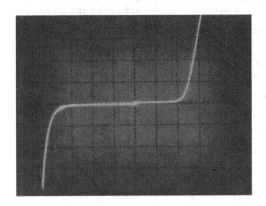

Figure 1: Current-
Voltage Characteristic
of a Au-nCdTe Schottky
Barrier Diode.

Forward: I = 0.2 mA/div
V = 0.2 V/div

Reverse: I = 20 µA/div
V = 10 V/div

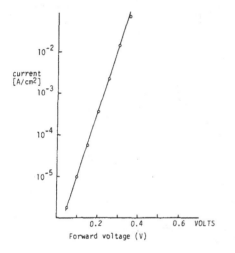

Figure 2: Forward log I
vs. V characteristic for
the Diode of Figure 1.

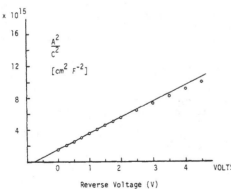

Figure 3: Reverse
$1/C^2$ vs. V characteristics for a typical Au-
Schottky barrier diode
on epitaxial CdTe.

non-uniform doping. This is true of many of the devices measured from different layers. The carrier concentration measured from the slope near the surface is about 6.7×10^{15} cm^{-3}. The value of the barrier height deduced from these curves is $\simeq 0.75$ eV which is in excellent agreement to that obtained from I-V measurement.

Small area Schottky barrier diodes with a transparent contact were fabricated to determine the diffusion length of minority carriers. Here, a 70°A Au film was first evaporated through a metal mask, followed by 1000°A Au dots. The area of thin Au was 0.06 cm^2, and the thick dots served as front contacts. The photocurrent was measured in an experimental setup consisting of a monochromator, a light chopper and a lock-in amplifier.[8] Measurements were made at 835 nm, at various reverse biases from 0 to 1.0 volt. An absorption coefficient of 1×10^3 cm^{-1} was assumed at this wavelength.

As shown in Figure 4, the best fit to the experimental photocurrent gives a hole diffusion length of 1 μm. This is measured for a layer with a carrier concentration of 8×10^{15} cm^{-3} as determined by C-V measurement. Diffusion lengths in the range 0.6 to 2.8 μm have been reported [9] in bulk CdTe grown by the Bridgman technique. Our results show that the epitaxial layer quality is comparable to that of bulk material. It should be emphasized that no attempts were made to optimize the diffusion length for this work.

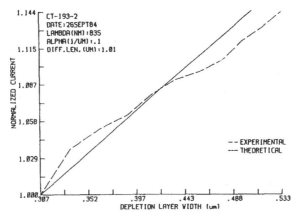

Figure 4   Normalized Photoresponse at 835 nm as a Function of Depletion Layer Width.

CONCLUSION

In conclusion, we have demonstrated that device quality CdTe can be grown on (100) InSb substrates by OMVPE. InSb is shown to be a better substrate material than CdTe for the growth of CdTe; this is because of the poor quality of presently available CdTe bulk material. Both n-type and high resistivity laers could be grown on InSb substrates by changing the reactant pressures. Detailed C-V, I-V and DLTS measurements were performed on these devices, and show that this material is equivalent in its characteristics to the best material that is presently available.

ACKNOWLEDGEMENTS

The authors wish to thank Ms. A. Hayner for manuscript preparation.

REFERENCES

1. T. H. Myers, Yawcheng Lo, R. N. Bicknell and J. F. Schetzina, Appl. Phys. Lett. $\underline{42}$, 247 (1983).

2. H. A. Mar, K. T. Chee and N. Salansky, Appl. Phys. Lett. $\underline{44}$, 237 (1984).

3. R. F. C. Farrow, G. R. Jones, G. M. Williams and I. M. Young, Appl. Phys. Lett. $\underline{39}$, 9544 (1981).

4. S. K. Ghandhi, I. Bhat, Appl. Phys. Lett. $\underline{45}$, 678 (1984).

5. R. F. C. Farrow, Extended Abstracts, 1984 U.S. Workshop on the Physics and Chemistry of Mercury Cadmium Telluride, May 15-17, San Diego, CA. pg. 5.

6. K. Zanio, "Cadmium Telluride" in Semiconductors and Semimetals, edited by R. K. Willardson and A. C. Beer (Academic, New York, 1978), Vol. 13, Chapter 4.

7. T. Takabe, T. Hirata, J. Saraie and H. Matsunami, J. Phys. Chem. Solids $\underline{43}$, pg. 5 (1982).

8. R. J. Lender, S. Tiwari, J. M. Borrego and S. K. Ghandhi, Solid State Electronics $\underline{22}$, 213 (1979).

9. P. Gaugash and A. G. Milnes, J. Electrochem. Soc., Vol. $\underline{128}$, 213 (1981).

ENHANCED GROWTH OF DOPED AMORPHOUS Si FILMS IN LATERAL SOLID PHASE EPITAXY

H. YAMAMOTO, H. ISHIWARA, AND S. FURUKAWA
Department of Applied Electronics, Tokyo Institute of Technology,
Nagatsuda, Midoriku, Yokohama 227, JAPAN

ABSTRACT

Phosphorus and boron doping effects for lateral solid phase epitaxial (L-SPE) growth in amorphous Si films, which were deposited on (100)Si substrates with SiO_2 patterns at elevated temperature and amorphized by ion implantation, were investigated. It was found from Nomarski optical microscopy that phosphorus doping is effective to enhance the L-SPE growth rate and to reduce the random nucleation rate, while boron doping is only effective to enhance the growth rate. Owing to these effects, the maximum L-SPE length of about 24μm was obtained in the film doped with 3×10^{20} P atoms/cm^3 after 8hour annealing at 600°C, which was about 6 times enhanced than that in the undoped films.

INTRODUCTION

Silicon-on-insulator (SOI) structures can be formed at low temperatures by lateral solid phase epitaxial (L-SPE) growth of amorphous Si (a-Si) films onto the SiO_2 patterns.[1-5] We have shown that a-Si films, which were deposited on (100)Si substrates with SiO_2 patterns by vacuum evaporation and then amorphized by Si^+ ion implantation, grow laterally onto the ≈40nm thick SiO_2 patterns and that the maximum L-SPE length of 5-6μm was obtained along the <010> direction after 10hour annealing at 600°C.[2] In this method, however, the L-SPE growth is essentially limited by random crystallization or polycrystallization with randomly nucleated crystal grains of the Si film and the maximum L-SPE length is determined by the competition between the L-SPE growth rate and the random crystallization time. We have shown[5] that the random crystallization time t_{poly}[sec] is expressed by the random nucleation rate n[$cm^{-2}sec^{-1}$] and the L-SPE growth rate v_L[cm/sec] as follows, under the assumptions that the nucleation occurs homogeneously[6,7] and that the poly-grain growth rate v_g[cm/sec] is proportional to v_L.

$$t_{poly} \propto (nv_L^2)^{-1/3} \tag{1}.$$

Then, the maximum L-SPE length l_L(max)[cm], which is given by the product of v_L and t_{poly}, is also expressed by n and v_L as follows.

$$l_L(max) \propto (v_L/n)^{1/3} \tag{2}.$$

We can see from eq.(2) that reduction of the random nucleation rate and enhancement of the L-SPE rate are important to enhance the L-SPE length.

The L-SPE growth rate is considered to be enhanced by incorporation of electrically active dopants such as phosphorus (P), boron (B) and arsenic (As), which are known to enhance the vertical SPE (V-SPE) growth rate.[8-10] For example, Ho et al.[10] studied about the <100>, <110> and <111> V-SPE rates at 500-550°C and found that P atoms with a concentration of about $2 \times 10^{20}cm^{-3}$ enhance the growth rate in all three orientations by a constant factor of 8.1±0.9, while a higher enhancement factor of 12.2±1.2 is obtained by B atoms with nearly the same concentration, except the case of <100>. In addition, dopant atoms may reduce the random nucleation rate, since Pai et al.[11] reported that the nucleation rate of a-Si films vacuum-deposited on sapphire substrates is reduced by P^+ ion implantation to about $1.4 \times 10^{20}cm^{-3}$. So, in this paper we investigate the L-SPE growth rate and random nucleation

Table I. Implantation conditions.

Sample	Ion	Energy / Dose [keV] [10^{15}cm^{-2}]	Concentration [10^{20}cm^{-3}]
Si-1	^{28}Si$^+$	160/2.1, 80/0.76, 40/0.38	1.0
Si-3		160/6.3, 80/2.3, 40/1.1	3.0
P-1	^{31}P$^+$	180/1.8, 90/0.79, 40/0.37	1.0
P-3		180/5.4, 90/2.4, 40/1.1	3.0
B-3	^{28}Si$^+$	150/2.0, 50/1.0	
	^{11}B$^+$	75/4.3, 45/1.4, 30/1.6	3.0

rate in P$^+$ and B$^+$ implanted a-Si films and discuss about the enhancing effect of L-SPE length by the doping.

EXPERIMENTAL PROCEDURES

Si films 230-270nm thick were deposited by e-gun evaporation on thermally-cleaned (100)Si substrates with SiO$_2$ patterns 30-40nm thick in an ultra-high vacuum chamber at the substrate temperature of 500-550°C. The pattern consists of stripe-shaped seeding regions parallel to the various directions. The deposited films were then implanted at room temperature with the conditions summarized in **Table I**, i.e., (1) the samples Si-1 and Si-3 were amorphized by Si$^+$ implantation, (2) the samples P-1 and P-3 were amor-phized and doped by P$^+$ implantation, and (3) the sample B-3 was first amorphized by Si$^+$ implantation and then doped by B$^+$ implantation. In the samples Si-1, 3 and P-1, 3, the implanted atom concentrations in the deposited films are theoretically predicted to be nearly equal to the values shown in **Table I**, except the surface regions about 40nm thick. In the sample B-3, the B concentration is predicted to be nearly equal to 3x10^{20} cm^{-3}, except the surface region about 100nm thick. These samples were finally annealed in N$_2$ atmosphere at 600°C and investigated by Nomarski optical microscopy after etching in Wright etchant.[2]

RESULTS AND DISCUSSIONS

L-SPE Characteristics in P Doped a-Si Films

In **Fig.1**, we show optical micrographs of the samples Si-1,3 and P-1,3 which were annealed at 600°C for 2hours and etched in Wright etchant. In these micrographs, the patterns are nearly parallel to the <001> axis of the (100)Si substrate. We can see from these micrographs that the L-SPE growth fronts are delineated by the difference of the etching rate between a-Si and crystalline Si. The growth lengths along the <010> direction were about 2µm in the samples Si-1 and Si-3 and they were about 6 and 8µm in the samples P-1 and P-3, respectively. These results clearly show that the L-SPE rate was enhanced by P doping. In **Fig.2**, the L-SPE lengths along the <010> direction in these samples are plotted against the annealing time at 600°C. From this figure, we can see that the all L-SPE rates in the four samples decrease during the initial growth stages and then they approach saturated values v_L(sat) of ≈1.0x10^{-8}cm/s for the samples Si-1 and Si-3, ≈3.2x10^{-8}cm/s for P-1, and $\simeq 7.1 \times 10^{-8}$cm/s for P-3, respectively. Discussions on the saturation mechanism were made elsewhere.[3] From comparison of the saturated L-SPE growth rates between the P-doped and undoped (Si-implanted) samples, we can see that the enhancement factors of the L-SPE rate by P atoms are about 3 and 7 for the concentrations of about 1 and 3×10^{20}cm^{-3}, respectively. These factors are nearly equal to those for V-SPE rate.[8]

In this figure, we also plotted experimental values of the random crystallization time, i.e., an annealing time when the volume fraction of random crystallites becomes $1-1/e$ in the unseeded region where the L-SPE growth from the pattern edge does not reach. So, we can see that the maximum L-SPE lengths, which are determined by the L-SPE length at the random crystallization time, are about 4µm in the samples Si-1 and Si-3, while they are about 8 and 24µm in the samples P-1 and P-3, respectively. That is, the maximum L-SPE lengths were enhanced by factors about 2 and 6 with the P doping to about 1 and 3×10^{20}cm^{-3}, respectively. In Fig.2, it is interesting to note that the random crystallization time of the P-doped sample P-3 is almost the same as those of the undoped samples Si-1 and Si-3, though the L-SPE rate in the sample P-3 is drastically enhanced. This result suggests that the random nucleation rate n is strongly suppressed in the sample P-3 as can be seen from eq.(1).

The random nucleation rates in the four samples were then measured at 600°C. Numbers of nucleated poly-grains were counted by optical microscopic observation[5] of the Wright-etched samples and plotted against the annealing time as shown in **Fig.3**. From this figure, we can see that the nucleation in the four samples occur with nearly constant rates. Note that, the vertical scale for the sample P-3 was expanded by a factor of 10. The nucleation rates calculated from Fig.3 are summarized in **Table II**. From this Table we can see that the random nucleation rates in the samples Si-3

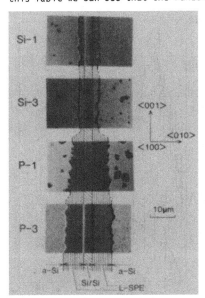

Fig.1. Nomarski optical micrographs of the samples Si-1,3 and P-1,3 annealed at 600°C for 2hours and etched by Wright etchant. The SiO$_2$ patterns are parallel to the <001> axis.

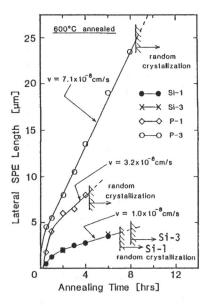

Fig.2. L-SPE growth length along the <010> direction in the samples Si-1,3 and P-1,3 as a function of annealing time at 600°C.

Table II.   Random nucleation rate and two normalized random
crystallization times (calculated and observed) at 600°C.

Sample	n [$10^2 cm^{-2} s^{-1}$]	$v_L(sat)$ [$10^{-8} cm/s$]	$\overline{(nv_L(sat)^2)^{-1/3}}$	$\overline{t_{poly}}$
Si-1	5.4	1.0	1	1
Si-3	3.5	1.0	1.2	1.1
P-1	2.8	3.2	0.6	0.6
P-3	0.15	7.1	0.9	1.2

and P-1 are not significantly reduced compared to that in the sample Si-1,
while in the sample P-3, it is reduced to about 1/40 of the rate in the
sample Si-1.   From this result we conclude that P doping to the atomic
concentration of about $3\times10^{20} cm^{-3}$ is effective to reduce the random
nucleation rate as well as to enhance the L-SPE growth rate in a-Si films
and that combination of these two effects results in enhancement of the
maximum L-SPE length.   In **Table II**, the values of $(nv_L^2)^{-1/3}$, which were
calculated with the measured values of n and $v_L(sat)$ at 600°C and normalized
with the value of the sample Si-1, are also shown, and they are compared
with the experimental values of t_{poly} at 600°C normalized with the value of
the sample Si-1.   From this Table, we can see that the above two values well
coincide each other.   This result shows that eq.(1) holds fairly well.

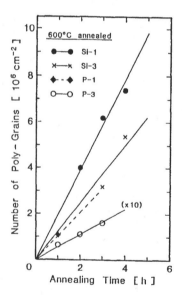

Fig.3.Numbers of nucleated poly-
grains in the samples Si-1,3 and
P-1,3 as a function of the annealing time at 600°C. The vertical scale for the sample P-3
is expanded by a factor of 10.

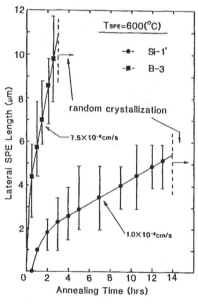

Fig.4.L-SPE growth length along
the <010> direction in the
sample B-3 as a function of
annealing time at 600°C.

L-SPE Characteristics in B Doped a-Si Films

In **Fig.4**, the L-SPE growth length along the <010> direction in the sample B is shown as a function of the annealing time at 600°C. The L-SPE length in the sample Si-1', which was formed on the same wafer using a partial mask against B+ implantation, is also plotted for comparison. From this figure, we can see that the growth rates in the both samples decrease during the initial growth stages and approach saturated values of ≈7.5 and ≈1.0x10^{-8}cm/s, respectively. That is, the enhancement factor of the saturated L-SPE growth rate is about 8 in a sample with ≈3x10^{20} B atoms/cm^3. This factor is comparable to that for the <110> and <111> V-SPE rates.[10] We can also see from Fig.4 that the random crystallization time is about 5 times reduced by the B doping and the enhancement factor of the maximum L-SPE length is only about 2. This fact indicates that the B doping is not so effective to reduce the random nucleation rate. In fact, we can calculate the reduction factor for n as ≈0.5 (i.e., about twice enhanced) by simply putting the observed enhancement and reduction factors for v_L(sat) and t_{poly} in eq.(1). One may imagine that the random crystallization in this sample is dominated by nucleation in the surface region, since it has a relatively thick surface region with lower doping concentration. However, we could observe no distinct reduction effect of the nucleation rate even in samples with nearly flat doping profiles. From these facts we conclude that B doping is only effective to enhance the L-SPE growth rate and it is not so useful as P doping to enhance the maximum L-SPE length.

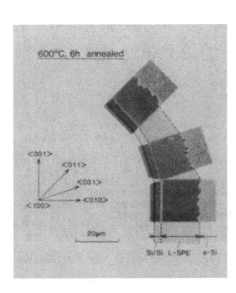

Fig.5. Nomarski optical micrographs for the sample P-3 grown along the <010>, <031> and <011> directions at 600°C for 6hours.

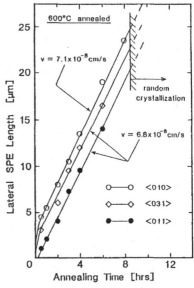

Fig.6. L-SPE growth length of the sample P-3 along the <010>, <031> and <011> directions as a function of annealing time at 600°C.

Dependence on Growth Direction

Finally, we investigate the dependence of the L-SPE characteristics on the growth direction. It was found in all samples that the L-SPE length has a maximum along the <010> direction and that it decreases monotonically to a minimum along the <011> direction, as shown in **Fig.5** for the sample P-3. However, it was also found that, the variation of the L-SPE length mainly originated from the difference of the growth rate in the initial stage and that their saturated values do not depend on the growth direction, as shown in **Fig.6** for the sample P-3.

CONCLUSION

We investigated the L-SPE growth rate and the random nucleation rate in a-Si films implanted with P^+ or B^+ ions and found that P doping to the concentration around $3 \times 10^{20} \text{cm}^{-3}$ is effective to enhance the L-SPE growth rate and to reduce the nucleation rate, while B doping to nearly the same concentration is only effective to enhance the L-SPE rate. Owing to these effects, the maximum L-SPE length of about 24µm was obtained along the <010> direction in the film doped with $\simeq 3 \times 10^{20}$ P atoms/cm^3 after 8-hour annealing at 600°C, which was about 6 times enhanced than that in the undoped film. Lightly doped SOI films necessary for LSI fabrication may be obtained by some modification of this technique, e.g., by low temperature vertical epitaxy on these doped films. So, we believe that the L-SPE technique is most promising to fabricate 3-dimensional LSIs with highly stacked structures.

ACKNOWLEDGEMENT

The authors gratefully acknowledge Dr. T.Tokuyama, Dr. M.Tamura, and Dr. M.Miyao in Central Research Lab. Hitachi Ltd. for useful discussions and preparation of patterned wafers. This work is partially supported by 1984 Grant-in-Aid for Scientific Research(B) (No.59460053) from the Ministry of Education, Science and Culture of Japan.

References

[1] Y.Kunii, M.Tabe, and K.Kajiyama, J.Appl.Phys., **54**, 2847 (1983).
[2] H.Ishiwara, H.Yamamoto, S.Furukawa, M.Tamura, and T.Tokuyama, Appl.Phys.Lett., **43**, 1028 (1983).
[3] H.Yamamoto, H.Ishiwara, S.Furukawa, M.Tamura, and T.Tokuyama, Proc. Sympo. Thin Films and Interfaces, Boston, 1983, J.E.E.Baglin, D.R.Campbell, W.K.Chu, eds. (North-Holland, New York, 1984), p.511.
[4] J.A.Roth, G.L.Olson, and L.D.Hess, Proc. Sympo. Energy Beam-Solid Interaction and Transient Thermal Processing, Boston, 1983, J.C.C.Fan, N.M.Johnson, eds. (North-Holland, New York, 1984), p.431.
[5] H.Yamamoto, H.Ishiwara, S.Furukawa, to be published in Appl.Phys.Lett.
[6] P.Germain, S.Squelard, J.Bourgoin, and A.Gheorghiu, J.Appl.Phys., **48**, 1909 (1977).
[7] K.Zellama, P.Germain, S.Squelard, J.C.Bourgoin, and P.A.Thomas, J.Appl.Phys., **50**, 6995 (1979).
[8] L.Csepregi, E.F.Kennedy, T.J.Gallagher, J.W.Mayer, and T.W.Sigmon, J.Appl.Phys., **48**, 4234 (1977).
[9] A.Lietoila, A.Wakita, T.W.Sigmon, and J.F.Gibbons, J.Appl.Phys., **53**, 4399 (1982).
[10] K.T.Ho, I.Suni, and M-A.Nicolet, J.Appl.Phys., **56**, 1207 (1984).
[11] C.S.Pai, S.S.Lau, and I.Suni, Thin Solid Films, **109**, 263 (1983).

BEAM INDUCED CRYSTALLISATION OF SILICON

J.S. WILLIAMS*, R.G. ELLIMAN+, W.L. BROWN AND T.E. SEIDEL
A.T. & T Bell Laboratories, Murray Hill, N.J. 07974
*Permanent address, Microelectronics Technology Centre, RMIT, Melbourne, 3000 Australia.
+C.S.I.R.O. Division of Chemical Physics, Clayton, Vic. 3168 Australia.

ABSTRACT

Epitaxial crystallisation of amorphous silicon layers on <100> silicon has been generated at temperatures of 200 - 450°C using MeV Ne+ irradiation. Two distinct temperature regimes have been identified : i) below 400°C growth proceeds linearly with Ne dose and is proportional to nuclear energy loss : a well defined activation energy of 0.24eV is obtained in this regime ;ii) above 400°C non-linear growth with Ne dose is observed and the apparent activation energy is higher but well below the 2.3 - 2.85 eV value normally associated with thermally-induced epitaxial growth. Dose rate and channeling studies have been employed to provide insight into the mechanism of beam annealing. Results suggest that mobile defects generated by the Ne+ beam within a few atom layers of the interface are the dominant influence on growth in the low temperature regime whereas defects migrating from greater distances play a more important role in the high temperature regime.

INTRODUCTION

Ion implanted amorphous layers on single crystal silicon are conventionally recrystallised by solid phase epitaxy in a furnace at temperatures of 500-600°C [1]. However, it has been demonstrated recently that recrystallisation can be induced by using an energetic ion beam to irradiate the silicon at temperatures in the range 200-400°C [2-8]. The first demonstrations of ion beam induced epitaxial crystallisation utilized heavy ions to produce the annealing at both low [2] and high energies [3]. More recent studies [6-8] have examined the annealing mechanism in further detail and have both eliminated local beam heating contributions and indicated that mobile point defects, generated as a result of nuclear energy loss of the impinging ion, play the dominant role in recrystallisation. In addition, it has been shown that epitaxial crystallisation can be induced by light ions (e.g. He+ [5,7]) and that doping of the silicon can enhance the growth rate for He+ beam annealing at temperatures of ∿ 440°C. Linnros et al [6] have been able to extract activation energies for ion beam induced annealing from their regrowth measurements and obtain a value of 0.36 eV compared with values of between 2.3 eV and 2.85 eV for furnace-induced epitaxial crystallisation [9]. Such a low value for beam annealing has been explained in terms of vacancy migration as the most likely mechanism for recrystallisation. In the present study, carefully controlled experimental conditions have been utilized to explore in detail the dependence of beam-induced epitaxial growth on a number of parameters including ion dose, dose rate, energy and target temperature. The observations reported in this paper provide considerable new insight into the beam annealing phenomenon.

EXPERIMENTAL

Silicon wafers of <100> orientation were first implanted with Si+ ions to generate near-surface amorphous layers. Typically, 50 keV Si+ ions were implanted to a dose of 3×10^{14} cm^{-2} with the substrates held at LN$_2$ temperatures. This produced an amorphous layer 1000Å thick, continuous to the surface. Other implant conditions and substrate orientations were employed, in particular cases, to investigate the influence of other parameters on subsequent beam-induced epitaxy.

Neon ions of energy 600 keV to 3 MeV were employed to induce epitaxial recrystallisation. The silicon samples were mounted on a goniometer providing 2 axes of rotation and 1 linear motion. Samples could be held at temperatures up to 600°C while being irradiated with Ne$^+$ ions scanned over a 3mm diameter aperture to provide dose uniformity. Up to 6 irradiated areas could be obtained on the one sample under varying dose, dose rate, beam energy or target temperature conditions.

Samples were analysed in-situ by Rutherford backscattering (RBS) and channeling using a 1.5 MeV He$^+$ beam. A glancing exit-angle detector geometry was employed to facilitate mesurements of the epitaxial growth. Conventional thermally-induced epitaxial growth was measured at 500°C and 520°C at various target positions as a check on both the linearity of growth with time for Si - generated amorphous layers and the temperature calibration of the goniometer hot stage. Thermal regrowth kinetics agreed well with the most recent data on undoped (100) Si by Olsen et al [10] and indicated a temperature accuracy and uniformity of better than $\pm$ 5°C. Prior to beam annealing the samples were preannealed at 450°C for 20 min to induce annealing of the partly amorphous (defective) transition region at the crystal-amorphous interface.

RESULTS AND DISCUSSION

Temperature, Dose and Energy Dependence

Fig. 1 illustrates the temperature dependence of 600 keV Ne+-induced recrystallisation of amorphous silicon on <100> silicon. Each of the spectra in Fig. 1 correspond to regions of the sample irradiated at selected temperatures in the range 200 to 450°C. The data clearly indicate that epitaxial crystallisation is more pronounced as the temperature increases for irradiation to a constant dose of 3×10^{16} cm^{-2}. No further thermal recovery of unirradiated parts of the sample, beyond that of preannealed samples, was observed after beam annealing. A sharp delineation in terms of extent of recrystallisation was observed between the stable amorphous layer thickness in unirradiated regions and the various Ne+-irradiated regions. The channeling spectra indicate recrystallisation to good quality single crystal silicon. For the irradiation at 450°C, which essentially resulted in complete epitaxial regrowth of the 1000Å amorphous layer, the channeling minimum yield was little different (~3%) from that obtained from virgin <100> silicon. However, the Ne$^+$ bombardment did generate damage in the silicon but this damage, most likely in the form of intermediate defect complexes and stacking faults [7], was confined to near the Ne ion range (~5000Å). This deep damage was revealed in channeling spectra taken with a 170° scattering geometry.

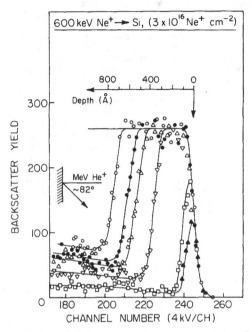

Fig. 1. Aligned RBS spectra depicting regrowth for samples irradiated at different temperatures with 600 keV Ne$^+$ to a fluence of 3 x 10^{16} cm^{-2}. The spectra represent the α-Si layer thickness following the thermal preanneal (0) and following irradiation at 200°C (●), 250°C (Δ), 318°C(▽), 425°C(□) and 450°C(▲).

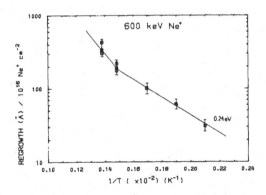

Fig. 2. Regrowth as a function of temperature for a sample irradiated with 600 keV Ne$^+$, normalised to a fluence of 1 x 10^{16}Ne$^+$ cm^{-2}.

From regrowth data similar to that shown in Fig.1, it is possible to display the temperature dependence more quantitatively on a 1/T plot as shown in Fig.2. Each of the data points in Fig. 2, corresponding to regrowth at a particular temperature, was averaged over irradiations at several doses. Only at 450°C was there any tendency for the regrowth to proceed non-linearly with Ne+ dose. At this temperature we have included individual data points which indicate more extensive growth for initial increments of Ne+ dose, where the amorphous layer is thickest. Fig. 2 high-lights two distinct temperature regimes for beam annealing (<400°C ; > 400°C). In the low temperature regime, the data are 'escribed by a well-defined activation energy of 0.24eV, whereas the non-linear regrowth with Ne+ dose at temperatures above 400°C does not allow a unique activation energy to be assigned in this temperature regime. More extensive Ne±-induced regrowth measurements at other Ne+ energies in the >400°C temperature regime confirm the trends suggested in fig.2. These data are reported elsewhere [11].

In Fig. 3 we illustrate the linearity of epitaxial growth with Ne+ dose at 318°C. The dependence of regrowth on nuclear stopping is quantified in Fig. 4. Between 0.6 and 3 MeV the regrowth increases linearly with nuclear stopping power at each discrete irradiation temperature in the range 200 - 400°C.

Regrowth Model and the Role of Mobile Defects.

The observation of an activation energy of 0.24eV is consistent with the migration energy of point defects such as vacancies. Furthermore, the observation that beam-induced regrowth is independent of the amorphous layer thickness (at 200 -400°C) suggests that the operative defects are not migrating to the interface from large distances in amorphous silicon. Our data indicate a maximum diffusion length of about 20Å. However, in the high temperature regime (>400°C) the regrowth does not proceed linearly with dose: regrowth is more extensive for the initial dose increments (Fig.2). This behaviour suggests that defects migrating from greater distances play a role in epitaxial growth in this temperature regime, an observation which is consistent with previous data [7] for He+-induced epitaxy.

Two further experiments give insight into the role of mobile defects. Firstly, the dose rate of the annealing Ne beam was varied by a factor of 4. No measurable dose rate effect was observed at low temperatures; however, a significant effect was observed in the high temperature regime. In particular, regrowth was ~ 20% less for higher dose rate irradiations. This result is again consistent with our previous data for He-induced epitaxy [7]. Secondly, a series of Ne+ bombardments were undertaken in a channeling direction to investigate the influence of reduced nuclear energy loss in the crystalline region. Although these experiments are reported in detail elsewhere [11], we show the basic effects in Fig.5. In this case, a buried amorphous layer (generated by multiple-energy Si+ implants) was first preannealed (at 450°C) and then irradiated with 1.5MeV Ne+ at 318°C with incremental doses of 5.5 x 10^16 cm^-2. Irradiations were performed sequentially in random (solid symbols) and aligned (open symbols) directions. The 1.5MeV He+ channeling spectra in Fig.6 clearly show sequential regrowth of both the front and back amorphous-crystal interfaces. Significantly, no difference in regrowth was obtained for random and aligned irradiations, suggesting that the "defects" from the crystalline side which contribute to epitaxy are generated at the amorphous-crystalline interface.

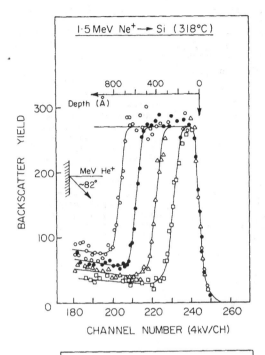

Fig. 3. Aligned RBS spectra from a sample held at 318°C and irradiated sequentially with 1.5 MeV Ne⁺ ions. Symbols depict the preannealed α-Si layer (o), and the thickness following equal fluence irradiation of $\therefore \times 10^{16}$Ne⁺ cm⁻².

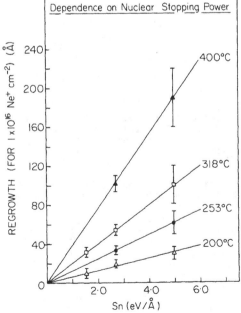

Fig. 4. Regrowth, normalised to a Ne⁺ fluence of 1×10^{16}cm⁻², as a function of nuclear stopping power for different substrate temperatures.

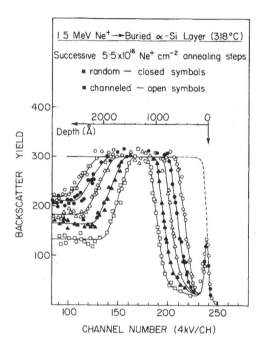

Fig. 5. Aligned RBS spectra showing the regrowth induced at the two crystal-amorphous interfaces of a buried α-Si layer following irradiation with a 1.5 MeV Ne+ ion beam. Symbols depict the α-Si layer after preannealing (o) and following sequential 5.5 x 10^{16}Ne+ cm^{-2} irradiation in random and aligned directions.

CONCLUSIONS

i) We have identified two distinct temperature regimes for ion-beam-induced epitaxy : a low temperature regime (200 - 400°C) which is characterised by a well defined activation energy of 0.24eV and a high temperature regime (>400°C) in which growth is non-linear with ion dose and displays a higher apparent activation energy.

ii) Growth is shown to scale linearly with nuclear stopping power of the annealing ions, suggesting beam-induced defects are responsible for eptiaxial growth.

iii) Channeling and dose rate studies in the low temperature regime indicate that the growth-activating defects are generated within a few atom layers of the interface. In the high temperature regime, more mobile (bulk) defects appear to contribute to epitaxial growth.

Acknowlegements

The Special Research Centres scheme is acknowledged for financial support of part of this study and one of us (R.G.E) acknowledges the CSIRO Post-doctoral fellowship award scheme.

References

[1]   L. Csepregi, E.F. Kennedy, S.S. Lau, J.W. Mayer and T.W. Sigmon Appl. Phys. Lett. $\underline{29}$, 645 (1976).

[2]   I. Golecki, G.E. Chapman, S.S. Lau, B.Y. Tsaur and J.W. Mayer, Phys. Lett. $\underline{71A}$, 267 (1979).

[3]   J. Nakata and K Kajiyama, Appl. Phys. Lett. $\underline{40}$, 686 (1982).

[4]   B. Svensson, J. Linnros and G. Holmen, Nucl. Instr. Meth. $\underline{209/210}$, 755 (1983).

[5]   R.G. Elliman, S.T. Johnson, K.T. Short and J.S. Williams Mat. Res. Soc. Symp. Proc. $\underline{27}$, 229 (1984).

[6]   J. Linnros, B. Svensson and G. Holmen, Phys. Rev. (in press).

[7]   R.G. Elliman, S.T. Johnson, A.P. Pogany and J.S. Williams, Nucl. Instr. Meth. (in press).

[8]   K.T. Short, D.J. Chivers, R.G. Elliman, J. Liu, A.P. Pogany, H.Wagenfeld and J.S. Williams Mat. Res. Soc. Symp. Proc. $\underline{27}$, 247 (1984)

[9]   A. Lietoila, A. Wakita, T.W. Sigmon and J.F. Gibbons J. Appl. Phys. $\underline{53}$, 4399 (1982).

[10]   G.L. Olson, S.A. Kokorowski, J.A. Roth and L.D Hess Mat. Res. Soc. Symp. Proc. $\underline{13}$, 141 (1983).

[11]   J.S. Williams, R.G. Elliman, W.L. Brown and T.E. Seidel, to be published

HIGH QUALITY GaAs EPITAXIAL LAYERS GROWN BY CSVT

CHAVEZ F.*, VILLEGAS D, AND MIMILA-ARROYO J.**
Centro de Investigacion y de Estudios Avanzados del IPN, Ap. Post
14-740 Mexico D.F. C.P. 07000 MEXICO
* Laboratoire de Physique des Solides - CNRS - 1, Place A. Briand -
  92190 MEUDON (France)
** Groupement de Physique des Solides de l'E.N.S., 2 Place Jussieu -
   75251 PARIS (France)

ABSTRACT

GaAs epitaxial layers were grown by the CSVT technique, using high
purity GaAs as source material and Cr-doped GaAs as substrate. Source and
substrate temperatures were in the range 750-850°C and 675-830°C
respectively. Growth rate varied between 600-10 000 Å/min and perfectly
agreed with a theoretical model proposed earlier [1] .The layer thickness
homogeneity was better than 15 % for samples of one cm^2. Resistivity and
Hall measurements showed that all the layers were N-type with electron concentration varying from 3×10^{16} - 3×10^{17} cm^{-3} and mobility in the range
of 4600 - 3000 cm^2 - V^{-1} s^{-1} respectively. Both parameters, electron concentration and their mobility follow a clear and well defined dependence on
the growth conditions. DLTS measurements made on the layers showed only a
majority carrier trap located at 0.92 eV below the conduction band with a
concentration between 10^{12} - 10^{13} cm^{-3}, without a clear relationship with
the growth conditions. The results show that high quality GaAs epitaxial layers
can be grown by the CSVT technique and owing to its excellent control of
low growth rates, it might be used for making devices and even superlattices.

INTRODUCTION

The growth of semiconductor epitaxial layers by the close spaced
vapor transport technique (CSVT) presents many advantages over other
methods. It works at moderate low temperatures, atmospheric pressure and no
toxic gases are needed. The mass transport efficiency is very large and
easily controlled. This technique has been successfully used to grow II-VI,
III-V and other semiconductors. For the particular case of GaAs the mass
transport mechanism has been studied and it is now well understood [1] .
The epitaxial layers obtained with this technique have shown very good
cristallinity [2] and preliminary results on electrical properties showed
high mobilities [3] .The influence of the growth conditions on the photoluminescence properties of this epitaxial GaAs, nas been well established.First studies
showed relatively poor and non reproductible photoluminescence spectra [4] .
Improvements in the growth reactor as well as in the procedure and growth
conditions lead to photoluminescence spectra, showing peaks
associated with excitons and other recombination processes proper to high
quality materials [5] . However little is known about electrical
properties and deep level impurities of this material and their relation
with the preparation conditions. The aim of this communication is to
present the results of a detailed study that confirms the very good
control of the growth rate, describes the behaviour of the majority
carrier concentration, their room temperature mobility and the concentration of deep levels as a function of the most important growth parameters i.e. the substrate (θ) and the source (T) temperatures.

SAMPLE PREPARATION

The reactor used to grow the samples studied in the present work as well as the technical details involved in growth process have been described before (1). Layers were grown, using a high purity GaAs as source material. This GaAs was neither intentionally doped nor compensated, with a majority carrier concentration lower than 10^{15} cm^{-3}. As a substrate, <100> oriented high resistivity GaAs : Cr was used. The source and substrate temperatures were in the range of 850 - 750°C and 830 - 675°C respectively. The layer surface was always about 1 cm^2. All the grown layers had smooth mirror like surfaces, and thickness uniformity was better than 15 %. Layer thickness was between 0.2 - 3 μm depending on the growth conditions For each condition several layers were grown and reproductibility was very good.

For the electrical properties characterization, ohmic contacts were made on the layers by indium alloying in an hydrogen atmosphere. Resistivity and Hall measurements were made using the Vander Pauw technique. DLTS measurements were made on Schottky barriers obtained by gold evaporation on the layers.

EXPERIMENTAL RESULTS

Growth law.

The mass transport mechanism has been widely studied before and it was found that it follows a rather simple law, which is a function of : the source (T) and the substrate (θ) temperatures, the distance between them, the transporting gas partial presure (g), the total gas pressure (p) and the crystal orientation. Keeping constant both pressures, the crystal orientation and the distance between the samples and under proper conditions, the growth rate follows the law given by the equation :

$$V(\theta,T) = A \exp 1/k \, (K/\theta - Q/T) \exp - \frac{\Delta H}{k} (1/T - 1/\theta) \left(1 + \exp - \frac{Q}{k} (1/T - 1/\theta)\right)^{-1} \quad (1)$$

where A is a constant depending on g and p.K, Q and ΔH are the activation energies and enthalpy variation for the reactions taking place.

For the grown layers, thickness was between 0.2 - 3 μm depending on the growth conditions Fig. 1 shows the experimental growth rate obtained for the temperature source values of 850 and 800°C and substrate temperatures between 830 - 700°C. The full lines correspond to equation (1). For the activation energy and the enthalpy variation the values given in (1) were used. A few layers grown at T = 700°C and θ between 675 - 600°C seem to follow the same law. The above results clearly show that the growth rate can be precisely controlled in a wide range and they confirm the growth law proposed in equation (1).

Electrical measurements.

Resistivity and Hall measurements were realized using the Van der Pauw technique. All the layers were N-type. Fig. 2 shows the electron concentration and their room temperature mobility. From this figure it can be seen that the electron concentration and their mobility strongly depend on the substrate and source temperature.It is also found that the layer electron concentration is always higher than the one of the source material. This higher concentration is certainly caused by doping

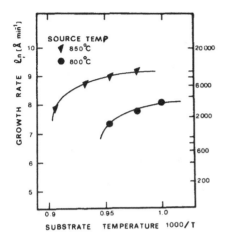

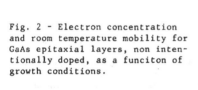

Fig. 1 - Growth rate of GaAs by CSVT as a function of experimental parameters. Substrate and source were <100> oriented.

Fig. 2 - Electron concentration and room temperature mobility for GaAs epitaxial layers, non intentionally doped, as a funciton of growth conditions.

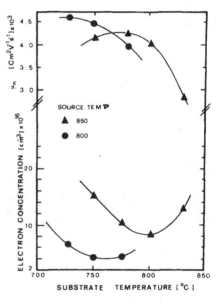

with silicon coming from the quartz reactor. Photoluminescence results seems to confirm this proposition (5). The mobility values obtained ranging from 3000 to 4600 cm^2 V^{-1} s^{-1} correspond to a reasonable low compensation level (6) and prove the high quality of the layers (studies of the mobility as a function of temperature are being realized   and results will be published).

DLTS studies.

Schottky barriers used for the DLTS studies were realized as described before. They showed good I-V characteristics. From C-V measurements made

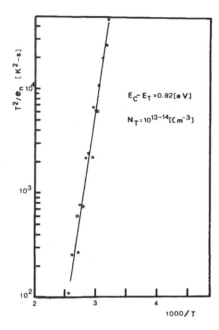

Fig. 3 -   DLTS signature
of the only deep level found in
the GaAs epitaxial layers grown by
CSVT. Its concentration was lower
than 10^{13} cm^{-3}.

on them the electron concentration was calculated through the C^{-2} - v curves.
The found value agreed well with the one obtained by Hall effect. The
DLTS measurements were made using the double lock in technique and all
the layers studied showed only one electron deep level located at 0.92 eV
below the conduction band. Fig.3 shows its signature. Its concentration was in
the range of 10^{12} - 10^{13} cm^{-3} and it does not showed a clear relation with
the epitaxial growth conditions.

CONCLUSIONS.

     The CSVT technique is able to grow GaAs epitaxial layers with excellent
control of the growth rate wich can be as low as a few hundred of angstroms
per minute. Lower growth rate has not been specifically studied but it looks
possible to be reached. Non intentionally doped layers are N type with a
carrier concentration as low as 2 x 10^{16} cm^{-3} with a room temperature
mobility of 4600 cm^2 v^{-1} s^{-1}, and only one deep level at a very low concentration, less than 10^{13} cm^{-3}, located at 0.92 eV below the conduction band.
From these results it can be concluded that high quality epitaxial GaAs can
be grown by the CSVT technique. Owing to the good control of the low growth
rate, this method can be used to make electronic devices and even superlattices.

Note added in proof : A MES-FET has been realized using GaA-CSVT as active
                      layer.

Aknowledgements : This work was supported by CONACYT of Mexico under cntract PIT/EN/OEA/82/1799 and by the Organization of American States. F. Chavez is a PhD student supported by Conacyt. Thanks to R. Legros for the carefull reading of the manuscript.

REFERENCES.

1 F. Chavez, J. Mimila-Arroyo, F. Bailly and J.C. Bourgoin, J. Appl. Phys., 54, 6646 (1983).

2 F. Chavez, M.S. Thesis, Centro de Investigacion y de Estudios Avanzados del IPN Mexico (1983).

3 J. Mimila-Arroyo, J. Reinoso, R. Legros and F. Chavez, Proc. IEEE Photovoltaic Specialists Conf. San Diego U.S.A. (1982), p. 952.

4 J. Mimila-Arroyo, L. Kratena, F. Chavez and F. de Anda, Sol. Stat. Comm. 49, 10, p. 939 (1984).

5 J. Mimila-Arroyo, R. Legros, J.C. Bourgoin and F. Chavez, to be published.

6 W. Waukiewicz, L. Lagowski, L. Jastrzebski, M. Lichtensterger and H.C. Gatos, J. Appl. Phys. 50, 899 (1979).

Insulators on Silicon

RECENT PROGRESS IN EPITAXIAL FLUORIDE GROWTH ON SEMICONDUCTORS

Julia M. Phillips

AT&T Bell Laboratories
Murray Hill, New Jersey 07974

ABSTRACT

This review considers recent progress in the epitaxial growth of alkaline earth fluorides on semiconductors. The field has grown rapidly in recent years because of the potential applications of such heteroepitaxial systems as well as the possibility which these systems present for fundamental studies on epitaxial growth. Two recent developments in the field will be treated in detail. Electrical characterization of these systems is underway. The insulating properties of fluoride films are good, and the interface transport properties are promising. The use of rapid thermal annealing to improve the epitaxial quality of as-grown fluoride films will also be discussed. This technique offers promise of relaxed growth conditions and improved film quality for epitaxial fluoride films.

INTRODUCTION

Considerable progress has been made in the research into the growth of epitaxial alkaline earth fluorides on semiconductors. Potential technological applications of such systems include the achievement of dielectric isolation of different devices on a single semiconductor substrate, the fabrication of 3-dimensional epitaxial heterostructures, and the passivation of semiconductors which lack a stable native oxide. Such possibilities, coupled with the potential that insulators grown epitaxially on semiconductors offer as models for studying the insulator-semiconductor interface have provided impetus for an increasingly wide investigation of such systems. The alkaline earth fluoride compounds, CaF_2, SrF_2, and BaF_2 and mixtures thereof have attracted attention as potential epitaxial insulators on semiconductors for a variety of reasons. They have the cubic fluorite crystal structure which is closely related to the diamond and zincblende structures of the semiconductors. In addition, the bulk lattice parameters vary over a wide range around those of the common semiconductors [1], so that one can not only obtain a close or exact lattice match to a given semiconductor but also study the role of the lattice match in these systems.

Investigation of epitaxial fluoride growth on semiconductors has proved to be a fertile ground for materials research since the first papers on the subject appeared some three years ago [2]. Much of the early work involved the identification of epitaxial systems, and the experimental parameters which resulted in epitaxial growth [3]. This sort of research has continued, and several more epitaxial systems have been discovered in the last year. The current list of epitaxial fluoride-semiconductor pairs is given in Table I. More recently, work has emphasized the study of the detailed structure of the epitaxial films, and, equally importantly, the structure of the semiconductor-insulator interface. Some of this work has involved the use of high resolution transmission electron microscopy to study the insulator-semiconductor interface. One of the most novel results from this work has been the observation of an incommensurate interface (lacking in coherence or misfit dislocations) in a very large lattice mismatch epitaxial system ($BaF_2/Ge(111)$, having a mismatch of nearly 10%) [19]. Work in this area is continuing, especially in small misfit systems, where one expects a more typical interfacial structure with either widely spaced misfit dislocations or none at all. In the last year, for the first time, several groups have begun studies of the electrical properties of the fluoride layer and the fluoride-semiconductor interface, while continuing to study the structural properties, as well. The results of these studies at present indicate that the alkaline earth fluorides have substantial promise as epitaxial insulators.

Another area of work which has recently arisen is the use of rapid thermal annealing to improve the epitaxial quality of fluoride films. This has allowed the growth conditions for the films to be relaxed substantially while still yielding the highest quality epitaxial films yet produced. The structural and insulating characteristics of films thus treated show dramatic improvement over as-grown films, indicating that this technique may offer the possibility of still better epitaxial layers.

Mat. Res. Soc. Symp. Proc. Vol. 37. ° 1985 Materials Research Society

TABLE I: Epitaxial Fluoride-Semiconductor Systems

Fluoride	Semi-Conductor	Substrate Orientations	Lattice Mismatch	References
CaF_2	Si	(100), (110), (111)	+0.61%	4,5,6
	GaAs	(100), (110)	−3.4	7,8
	Ge	(111)	−3.5	5
	InP	(100)	−6.9	9,10
SrF_2	Si	(100), (111)	+6.8	11,12
	GaAs	(100)	+2.7	13
	InP	(100)	−1.2	10
BaF_2	Si	(100), (111)	+14.2	11,12
	Ge	(111)	+9.6	14
	InP	(100), (111)	+5.5	2,5,9,11,14
	PbSe	(111)	+1.3	15
	CdTe	(100)	−4.5	2
$Ca_xSr_{1-x}F_2$	Si	(111)	+4.2	11
	GaAs	(100), (110), (111)	0	7,16,17
$Ba_xCa_{1-x}F_2$	InP	(100)	0	9
$Ba_xSr_{1-x}F_2$	Si	(111)	+10.0	12
	InP	(100)	0	10
	InAs	(100), (110), (111)	0	18

EXPERIMENTAL PROCEDURES

In the last year one of the most important changes in the experimental procedures associated with epitaxial fluoride growth has been the move towards the use of ultra high vacuum (UHV) molecular beam epitaxy (MBE) apparatus. These systems are frequently commercially available apparatus, with base pressures of $\leq 1 \times 10^{-10}$ Torr. The pressure in the growth chamber during film growth is also frequently in the 10^{-9}–10^{-10} Torr range. A schematic diagram of an MBE system used for fluoride growth is shown in Figure 1. This improvement in the growth conditions has no doubt had a great deal to do with the fact that electrical measurements on these systems are now being made. The structural quality of films grown in non-UHV conditions may have been high [3], but residual contamination prevented any sort of electrical measurements from being made.

The availability of clean growth conditions has resulted in increasing attention being paid to substrate preparation prior to film growth. In the case of fluoride growth on Si, substrates are generally cleaned using a chemical technique [20], in which a thin volatile oxide is chemically grown and stripped on a wafer several times. The wafer is introduced into the vacuum with the last grown thin oxide on it. This is stripped in situ by heating the wafer to ~850°C. Auger electron spectroscopy studies have demonstrated that the residual contamination remaining after such treatment amounts to about 0.01 monolayer of C and O [21]. The preferred cleaning technique for fluoride growth on III-V substrates has been chemical cleaning ex situ followed by final heating in situ in the presence of an overpressure of the group V element [8,10,13,16,18]. In some cases an epitaxial buffer layer of the semiconductor is grown prior to the commencement of fluoride growth [8,10,13,15,17].

A variety of analytical techniques have been applied to the study of epitaxial fluoride films. These include modulated beam mass spectrometry, reflection high energy electron diffraction (RHEED), Nomarski interference contrast optical microscopy, x-ray diffraction, Rutherford backscattering/channeling (RBS), and transmission electron microscopy (TEM). Techniques which have recently begun to be applied to these systems include secondary ion mass spectrometry (SIMS), scanning electron microscopy (SEM), and Auger electron spectroscopy (AES). Finally, a variety of electrical measurements are finding increasing use. These include measurements of breakdown voltage, capacitance-voltage, admittance-voltage, and electron mobility at the fluoride-semiconductor interface.

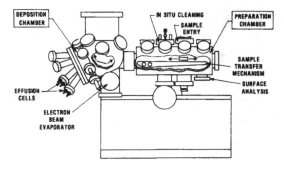

Figure 1. Schematic diagram of a molecular beam epitaxy system used for fluoride growth. The deposition and preparation chambers may be kept under vaccum while samples are introduced into the system through the sample entry. The preparation chamber is used for pre-cleaning the substrate and for surface analysis, such as Auger spectroscopy. The film growth takes place in the deposition chamber. The fluoride is evaporated from an effusion cell. Mixed fluorides can be grown using multiple cells. An electron beam evaporator allows the growth of Si or a metal on the fluoride to form a multi-layer heterostructure.

EXPERIMENTAL RESULTS

a. Electrical Characterization

Recently, several research groups have begun to report the results of investigations into the electrical characteristics of epitaxial fluoride films on semiconductors. Breakdown fields of about 5×10^5 V/cm have been reported by several groups [8,10,15,17,21,22,23,24]. The best breakdown fields reported to date are 3×10^6 V/cm at room temperature [17,21] and 4×10^6 V/cm at 77K [15]. This indicates that the films are good insulators, whose breakdown characteristics are approaching those of SiO_2 on Si, which typically has a breakdown field of $6{-}10 \times 10^6$ V/cm [28].

Metal-insulator-semiconductor (MIS) capacitors have been fabricated in a number of fluoride-semiconductor systems to study the properties of the insulator-semiconductor interface [8,10,15,17,21,22,23,24]. In the case of $CaF_2/Si(100)$ [21] the inversion capacitance is near the predicted value, and the overall behavior of the capacitance is correct. The hysteresis is generally small, especially in thin films [22]. The hysteresis tends to increase dramatically in thicker films (≥ 500 nm) or in thin films that have undergone mechanical stress. This may be a manifestation of the stress induced in the films by the larger thermal mismatch between the fluoride and semiconductor. These findings are discussed in more detail in ref. 22. The lowest interface state density reported to date is in $CaF_2/Si(100)$ where a value of 7×10^{10} eV^{-1} cm^{-2} has been observed [22]. More typical values for these systems are in the 5×10^{11} to 1×10^{12} eV^{-1} cm^{-2} range [15,21]. For comparison, a typical value for $SiO_2/Si(100)$ is $\leq 5 \times 10^{10}$ cm^{-2} eV^{-1} [28].

The fabrication of n-channel metal-epitaxial insulator-semiconductor field-effect transistors (MEISFET's) has been reported [21]. Such devices will be important for the characterization and improvement of the interface transport properties of fluoride-semiconductor systems. The MEISFET's were fabricated in the $CaF_2/Si(100)$ system. The Corbino geometry was used, as shown in Figure 2, with a 20 μm long channel and length-to-width ratio of 1:30. The drain current (I_D) as a function of drain voltage (V_D) for these devices is shown in Figure 3. The threshold voltage is 0.5V. The peak mobility at room temperature is up to 400 cm^2/Vs. Devices fabricated using a fluoride film of poorer epitaxial quality (having a ratio of backscattered 1.8 MeV ^{4}He$^+$ ions in the aligned normal $<100>$ direction to a random direction, $\chi_{min} = 0.16$ compared with $\chi_{min} = 0.09$ in the higher quality film) have somewhat worse mobility (230 cm^2/Vs), indicating that the crystalline quality of the CaF_2 film influences electron scattering at the interface. (The mobility

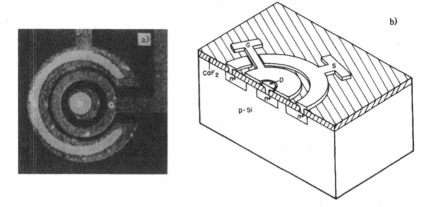

Figure 2.   a) Optical micrograph of a MEISFET fabricated in Si using CaF$_2$ as the gate insulator and Al as the metal. The source, gate, and drain regions are indicated by S, G, and D, respectivly. b) Schematic view of the MEISFET geometry. The channel is 20 μm long, and the length-to-width ratio is 1:30 (ref. 21).

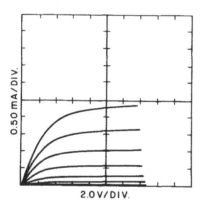

Figure 3.   Drain current as a function of drain voltage for a MEISFET fabricated on 5 Ω-cm p-Si(100) with a 640 nm CaF$_2$ film. The gate voltage is stepped between 0 and 7 V (ref. 21).

was calculated using the static dielectric constant measured from capacitance dots on the wafers. The measured value is equal to the established value of 6.8 [25].) Transmission electron microscopy indicates that the types of defects in the poorer film differ substantially from those in the better film. This no doubt accounts for the difference in interface transport in the two cases [23].

b. Rapid Thermal Annealing

While CaF₂ films of high epitaxial quality can be grown on $Si(100)$ substrates, the parameters which result in optimum film quality are quite narrowly defined. In fact, deviation of the temperature of the substrate from the optimum value in either direction by as little as 25°C results in a significant degradation of epitaxial quality, as illustrated in Figure 4. This is believed to be due to the fact that at low temperatures, the mobility of CaF₂ on $Si(100)$ is insufficient to result in good epitaxy, while at high temperatures, 3-dimensional island formation in the growing CaF₂ film takes place, in addition perhaps to reaction between the CaF₂ and Si. Therefore, there is interest in improving the epitaxial quality of poor or mediocre films after growth. Recent studies have shown that it is possible to improve dramatically the epitaxial quality of such films by rapid thermal annealing after the initial film growth. CaF₂ films 400 nm - 2 μm thick are grown under UHV conditions in an MBE system. The epitaxial quality, as measured by χ_{min} of the as-grown films is not critical. The value of χ_{min} can in fact be greater than 0.30. The samples are post-annealed in an AG 210 flash lamp annealing system in an Ar atmosphere at 1100°C for 20 sec. The time and temperature required for maximum film improvement are critical. If the temperature is too high, of if the time is too long, the CaF₂ and Si react at the interface, and the epitaxial quality of the film is degraded. If the time is too short, of if the temperature is too low, insufficient rearrangement of the molecules in the films takes place, and there is no improvement in epitaxy. These considerations are discussed in more detail in ref. 26. It is essential that all oxygen be purged from the annealing atmosphere in order to prevent reaction of oxygen with CaF₂ and Si at the fluoride-semiconductor interface. The epitaxial quality of the post-annealed films is determined by RBS/channeling. In all cases, the χ_{min} of the post-annealed films is improved over that of the as-grown films, in some cases by as much as a factor of 5 to 8. Channeling spectra for a typical film before and after annealing are shown in Figure 5. The best value of χ_{min} which we have measured in an as-grown film of CaF₂ on $Si(100)$ is 0.05. Using our post-annealing technique, films which have poor epitaxial quality after growth (typically $\chi_{min} \sim 0.10-0.20$) have, after post-annealing $\chi_{min} \sim 0.03$. This is the best published value of χ_{min} for an epitaxial fluoride grown on $Si(100)$ under any growth conditions.

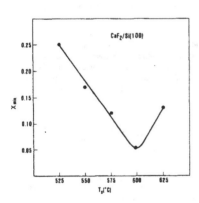

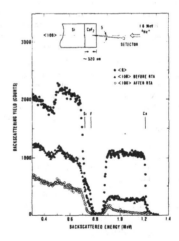

Figure 4. χ_{min} as a function of substrate temperature during film growth for 500 nm CaF₂ films on Si.

Figure 5. Channeling and random RBS spectra of a 500 nm film on $Si(100)$ before and after rapid thermal annealing (RTA). Before RTA χ_{min} = .26. After RTA χ_{min} = .03.

Nomarski interference contrast optical microscopy of the films reveals one striking difference between the as-grown and annealed films. Whereas the as-grown films are generally featureless, demonstrating smooth morphology, the annealed films have very straight lines along the {220} directions. The nature of these lines has not yet been determined. They may be either cracks in the film caused by the sizable thermal expansion mismatch between the CaF_2 and Si, or they may be slip planes in either the CaF_2 or substrate (or both).

Transmission electron microscopy reveals that an annealed film is highly single crystalline and contains no misoriented material. This contrasts with unannealed films on Si(100) which nearly always have at least a small amount of CaF_2 which is oriented with its $<111>$ direction perpendicular to the substrate surface.

Epitaxial CaF_2 films on Si(100) have generally proved to be quite resistant to chemical etching, and the annealed films are, if anything more resistant to this sort of attack than are as-grown films. Both sorts of films can withstand the chemicals associated with standard Si processing. When CaF_2 is etched in this context, HCl is the chemical of choice. It takes at least twice as long to etch contact holes in CaF_2 which has been annealed compared with CaF_2 grown under identical conditions but not annealed (> 23 minutes vs. 16 min. for $\sim$800 nm of CaF_2 in a solution of 4:1 H_2O:HCl).

Preliminary electrical measurements on post-annealed films are encouraging. In some films, the breakdown field has been improved by a factor of 5. Breakdown fields of 2×10^6 V/cm have been measured in annealed films. This is essentially equal to the breakdown field that is seen in the best unannealed films. In short, rapid thermal annealing of CaF_2 films grown on Si(100) offers the opportunity to improve predictably and reproducibly the epitaxial quality of poor to mediocre as-grown films.

DISCUSSION

The results discussed in the preceding pages are an important step in the investigation of epitaxial fluoride growth on semiconductors. Early work in the field concentrated on identifying epitaxial fluoride-semiconductor systems, on identifying epitaxial relationships, and on studying the atomic structure of the fluoride-semiconductor interface. Much of this work was done using rather simple apparatus. It is to be expected that this sort of work will continue in the future. There is still much to be learned about epitaxial growth in general, and the fluoride-semiconductor systems present an important class of systems for such study.

The increased use of UHV apparatus for fluoride film growth has enabled the advent of electrical characterization to the field. The results obtained to date are encouraging, and suggest that some of the potential applications of epitaxial insulators may soon be realized on a research scale. An important next step will be the correlation of structural and electrical properties of epitaxial fluoride films. The results on MEISFET's discussed in ref. 23, wherein a correlation is noted between the electrical properties and defect structure of the fluoride film is a first step in this direction.

Another development which has not been discussed here is the growth of epitaxial semiconductor-fluoride-semiconductor structures [8,12,16,24,27]. The semiconductor films thus grown have begun to be characterized. This is an important step in the movement toward three-dimensional integration.

Finally, the use of rapid thermal annealing to improve the epitaxial quality of fluoride films on semiconductors shows promise of allowing the initial growth conditions for the films to be relaxed while improving the epitaxial quality of the fluoride film over the best as-grown films.

ACKNOWLEDGEMENTS

The author acknowledges fruitful collaborations with J. M. Gibson, R. People, L. Pfeiffer, and T. P. Smith. W. M. Augustyniak has provided invaluable technical assistance.

REFERENCES

[1] J. M. Phillips, L. C. Feldman, J. M. Gibson, and M. L. McDonald, Thin Solid Films, *107*, 217 (1983).

[2] R. F. C. Farrow, P. W. Sullivan, G. M. Williams, G. R. Jones, and D. C. Cameron, J. Vac. Sci. Technol. *19*, 415 (1981).

[3] J. M. Phillips and J. M. Gibson, Mat. Res. Soc. Symp. Proc. *25*, 381 (1984), and references therein.

[4] H. Ishiwara and T. Asano, Appl. Phys. Lett. *40*, 66 (1982).

[5] J. M. Phillips and C. J. Yashinovitz, J. Vac. Sci. Technol. *A2*, 415 (1984).

[6] L. J. Schowalter, R. W. Fathauer, R. W. DeBlois, and L. G. Turner, J. Vac. Sci. Technol., in press.

[7] C. Fontaine, S. Siskos, and A. Munoz-Yague, Abstrasts of the 2nd European Workshop on MBE, Sussex University, 27-30 March 1983.

[8] P. W. Sullivan, G. M. Metze, and J. E. Bower, J. Vac. Sci. Technol., (in press).

[9] P. W. Sullivan, R. F. C. Farrow, and G. R. Jones, J. Crystal Growth *60*, 403 (1982).

[10] C. W. Tu, T. T. Sheng, M. H. Read, A. R. Schlier, J. G. Johnson, W. D. Johnston, and W. A. Bonnor, J. Electrochem. Soc. *130*, 2081 (1983); also in Proc. of Symp. on III-V Opto-Electronics Epitaxy and Device Related Processes, ed. V. G. Kermaides and S. Mahajan, p. 165 (Electrochem. Soc., 1983).

[11] T. Asano, and H. Ishiwara, Appl. Phys. Lett. *42*, 517 (1983).

[12] H. Ishiwara, and T. Asano, Mat. Res. Soc. Symp. Proc. *25*, 393 (1984).

[13] P. W. Sullivan, Appl. Phys. Lett. *44*, 190 (1984).

[14] J. M. Phillips, L. C. Feldman, J. M. Gibson, and M. L. McDonald, J. Vac. Sci. Technol. *B1*, 246 (1983).

[15] H. Zogg, W. Vogt, and H. Melchoir, Appl. Phys. Lett. *45*, 286 (1984).

[16] S. Siskos, C. Fontaine, and A. Munoz-Yague, Appl. Phys. Lett. *44*, 1146 (1984).

[17] C. W. Tu, S. J. Wang, R. A. Stall, J. M. Gibson, and J. M. Phillips, 1984 Fall Meeting of the Materials Research Society, Symposium on Layered Structures, Epitaxy, and Interfaces.

[18] K. Sugiyama, J. Appl. Phys. *56*, 1733 (1984).

[19] J. M. Gibson, and J. M. Phillips, Appl. Phys. Lett. *43*, 828 (1983).

[20] A. Ishizaka, K. Nakagawa, and Y. Shiraki, 2nd International Symposium on Molecular Beam Epitaxy, Tokyo (1982), p. 183.

[21] T. P. Smith, III, J. M. Phillips, W. M. Augustyniak, and P. J. Stiles, Appl. Phys. Lett. *45*, 907 (1984).

[22] R. People, T. P. Smith, III, J. M. Phillips, and W. M. Augustyniak, this volume.

[23] T. P. Smith, III, J. M. Phillips, R. People, J. M. Gibson, and P. J. Stiles, this volume.

[24] L. J. Schowalter, R. W. Fathauer, R. W. DeBlois, L. G. Turner, and J. P. Krusius, Proc. VLSI Science and Technol. Symp., Electrochem. Soc. Meeting, May, 1984 (in press).

[25] M. Wintersgill, J. Fontanella, C. Andeen, and D. Schucle, J. Appl. Phys. *50*, 8259 (1980).

[26] L. Pfeiffer, J. M. Phillips, T. P. Smith, III, W. M. Augustyniak, and K. W. West, Mat. Res. Soc. Symp. Proc. Symposium on Energy Beam-Solid Interactions and Transient Thermal Processing, November, 1984 (in press).

[27] C. W. Tu, S. R. Forrest, and W. D. Johnston, Jr., Appl. Phys. Lett. *43*, 569 (1983).

[28] E. H. Nicollian and J. R. Brews, *MOS (Metal Oxide Semiconductor) Physics and Technology* (Wiley-Interscience, New York, 1982).

HETEROEPITAXY OF CALCIUM FLUORIDE ON (100), (111),
AND (110) SILICON SURFACES

L.J. SCHOWALTER[*], R.W. FATHAUER[*+], L.G. TURNER[*], and C.D. ROBERTSON[*]
General Electric Corporate R D Center, P.O. Box 8, Schenectady, NY 12301
+School of Electrical Engineering, Cornell University, Ithaca, NY 14853

ABSTRACT

The surface morphologies of epitaxial CaF_2 films grown on (100),
(111), and (110) Si substrates by molecular beam epitaxy (MBE) have been
studied. On (100) Si surfaces, epitaxial CaF_2 can only be grown in a
fairly narrow temperature range around 550°C and is very rough with columnar structures about 20 nm across. On (110) Si substrates, (110) epitaxial
growth of CaF_2 can be obtained at substrate temperatures above 700°C. In
this case, the surface shows long troughs caused by (111) facets running
across the wafer. On (111) Si wafers, good (111) epitaxial growth of CaF_2
has been obtained at temperatures above 600°C. These films are very smooth
except for small flat-topped hillocks. These observations can be explained
in terms of the free energies of these different CaF_2 surfaces. The (100)
surface free energy of an ionic crystal with the fluorite structure is
expected to diverge with thickness, while the (110) surface free energy is
approximately 1.5 to 2 times larger. The epitaxial growth of smooth (111)
CaF_2 films on (110) Si substrates is observed below a fairly sharp transition temperature of 700°C.

INTRODUCTION

The epitaxial growth of silicon/insulator structures is of interest
for a number of applications in the semiconductor industry, including
silicon-on-insulator (SOI) and three-dimensional device structures. It is
also desirable to develop an epitaxial insulator to replace the amorphous
oxide most commonly used in FETs. We are currently studying the epitaxial
growth of CaF_2/Si heterostructures, since this material appears to be an
attractive epitaxial insulator on Si. Both CaF_2 and Si have face-centered-
cubic lattice structures and their lattice mismatch is only 0.6% at room
temperature. With a band gap of approximately 12 eV, CaF_2 is an excellent
insulator [1]. Its relatively large dielectric constant of 6.8 (compared to
3.9 for SiO_2) also makes it attractive for a gate insulator. We are using
molecular beam epitaxy (MBE) to deposit the films on Si substrates in an
ultra-high vacuum system. Calcium fluoride readily sublimes as a tightly
bound molecule so that film stoichiometry is automatically obtained [2].

In addition to the technological interest, CaF_2 presents an interesting system in which to study the heteroepitaxy of two dissimilar materials.
In particular, the question of how epitaxy is initiated across an interface
between a covalently bonded material like Si and an ionic solid like CaF_2
is fascinating. In this paper, we present the results of our study of the
epitaxial growth of CaF_2 on the (100), (110), and (111) surfaces of Si.

OBSERVATIONS OF CaF_2 HETEROEPITAXY ON (100), (111), AND (110) Si SURFACES

We are using a comercially available Si MBE system [3] to deposit CaF_2
on 3″ Si wafers. The CaF_2 is evaporated from a graphite crucible which we
heat to about 1400°C. The pressure during deposition is typically in the
mid 10^{-10} mb range.

In Fig. 1, a scanning electron microscope (SEM) micrograph and a reflection high-energy electron diffraction (RHEED) pattern of an epitaxial CaF$_2$ layer grown on a (100) Si substrate at 550°C are shown. The layer is approximately 400 nm thick. Because the CaF$_2$ is an insulator, it tends to charge up and defocus the electron beam. By evaporating a thin (10-20 nm) gold layer over the CaF$_2$ before taking the SEM micrograph, this problem can be alleviated. However, we have verified by direct viewing that the features shown here and in subsequent micrographs are not artifacts of the gold coating. The (100) CaF$_2$ surfaces always appear very rough with columnar structures typically 20-30 nm across. However, at much lower magnifications, such as those obtained with an optical microscope, the films appear mirror smooth.

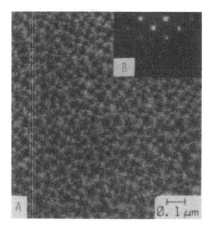

Fig. 1: Scanning electron microscope (SEM) micrograph (a) and reflection high-energy diffraction (RHEED) pattern (b) along the [011] azimuth of a CaF$_2$ epitaxial film grown on a (100) Si substrate at 550°C. The CaF$_2$ is approximately 400 nm thick.

Fig. 2: SEM micrograph of a 300-nm-thick, epitaxial CaF$_2$ film grown at 650°C on a (100) Si substrate. In this case, the (111) faceting is more obvious; however, there are many more misoriented grains which can be seen scattered through out the micrograph.

RHEED patterns, of which the pattern shown in Fig. 1b is a typical example, were used to show that the CaF$_2$ films grown on (100) Si at 550°C were single crystal films oriented in the (100) orientation. This fact was also verified with x-ray diffraction and with chemical etching techniques [4]. In Fig. 1b, the electron beam is oriented along the [011] azimuth of the wafer. The spots result because the surface is so rough that a transmission, rather than a reflection, diffraction pattern results. The streaks which can be seen connecting the dots suggest (111) and (111) facets as these would give rise to streaks perpendicular to their faces. The angle between these steaks should be 109° which is what is observed. The same diffraction pattern results when the wafer is rotated 90° so the (111) and the (111) facets are also implied. In fact, these facets can be resolved with SEM on films grown at somewhat higher substrate temperatures as shown in Fig. 2. However, in this film, which was grown at 650°C, a large number of misoriented grains are observed which can be seen scattered throughout the micrograph. This is consistent with the observations of Ishiwara and Asano [5] that the best temperature for growing CaF$_2$ on (100) Si is 550°C.

Fig. 3: SEM micrograph (a) and RHEED pattern (b) along the [112] azimuth of a 400-nm-thick, epitaxial CaF$_2$ layer on a (111) Si substrate which was grown at 700°C.

Fig. 4: SEM micrograph of (110)-oriented CaF$_2$ film on a (110) Si substrate. The film is approximately 400 nm thick and was grown at 850°C.

In stark contrast, films grown on (111) Si substrates above 600°C are quite smooth. The film shown in Fig. 3 is a typical example. It is 400 nm thick and was grown at 700°C. Small triangular hillocks can be seen which are approximately 10 nm in height. The RHEED pattern for this film shows streaks and well-defined Kikuchi lines indicating good crystal structure and a relatively smooth surface. This RHEED pattern is typical of all (111) CaF$_2$ films grown in this temperature range from a thickness of less than two monolayers up to the thickest layers we have grown (1.2 μm). It is worth noting that the quality of the RHEED pattern did not decay significantly in time although damage to the surface of the CaF$_2$ layer was clearly introduced and could be seen after the sample was removed from the chamber.

In Fig. 4, an SEM micrograph of an epitaxial CaF$_2$ film grown on a (110) Si substrate at 850°C is shown. The film is approximately 400 nm thick and is oriented in the (110) direction like the substrate. This orientation was checked by x-ray diffraction but can also be inferred from the (111) facets which can be seen forming troughs which run from one end of the micrograph to the other and are correctly aligned with the orientation of the underlying substrate. These facets result in the tilted streaks which can be seen in the RHEED pattern shown in Fig. 5. For the (110) orientation, the (111) facets should result in RHEED streaks which form an angle of 109°, which is what we observe when the electron beam is directed along the [110] azimuth (this direction is parallel to the troughs). When a RHEED pattern is taken with the beam directed along the [001] azimuth, which is perpendicular to the troughs, a spot pattern occurs since a transmission electron diffraction pattern now occurs.

If the growth temperature on a (110) substrate is lowered below 700°C, we can still obtain epitaxial films but now the orientation of the films is (111) rather than (110). The temperature at which this transition occurs appears to be sharply defined and, in fact, a film we grew at 700°C had regions of both (110) and (111)-oriented material. An SEM micrograph and RHEED pattern of a 300-nm-thick CaF$_2$ film grown on a (110) Si substrate at 600°C is shown in Fig. 6. This layer is oriented in the (111) direction as verified by x-ray diffraction. The RHEED pattern is indistinguishable from

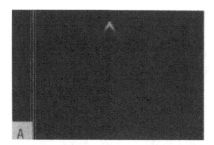 

Fig. 5: RHEED patterns from a (110)-oriented CaF_2 film on a (110) Si substrate grown under the same conditions as the film in Fig. 4. The electron
beam was directed along the [110] azimuth in (a) and along the [001] in
(b).

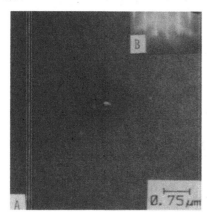

Fig. 6: SEM micrograph (a) and RHEED
micrograph (b) of a (111)-oriented
CaF_2 film grown on a (110) Si substrate. This layer was grown at a
substrate temperature of $600°C$ and
is approximately 300 nm thick.

that taken of epitaxial films grown on (111) Si substrates once the layer
is thicker than several hundred nanometers. However, for thinner films, the
RHEED patterns appear more spotty indicating that the films are somewhat
rougher.

CLASSIFICATION OF THE (100), (111), AND (110) SURFACES OF AN IONIC CRYSTAL WITH THE FLUORITE STRUCTURE

As Tasker [6] has pointed out, the surfaces of an ionic crystal can be
classified into three groups. In a type 1 surface, all the planes of atoms
which are parallel to the surface are electrically neutral. An example of
such a surface is the (110) surface of CaF_2 which is shown schematically in
Fig. 7a. In this case, the composition of all planes are identical. The
surface free energy has been calculated to be 750-1000 erg/cm^2 [7,8] for
CaF_2.

For type 2 and 3 surfaces, the individual atomic planes parallel to
the surface are charged. Of course, charge neutrality must be maintained
so the atomic planes can be grouped into identical repeat units which are
electrically neutral. For a type 2 surface, these repeat units have no net
dipole moment while for a type 3 surface, there is a net dipole moment. In

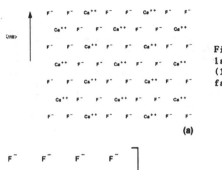

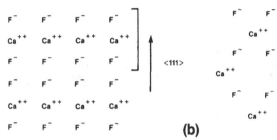

Fig. 7: Schematic diagrams of the lattice planes parallel to the (a) (110), (b) (100), and (c) (111) surfaces.

the fluorite structure, the (100) surface is a type 3 surface since the atomic planes parallel to the surface consist of alternating cation and anion planes. However, the (111) surface is a type 2 surface when the crystal surface is terminated with the anion plane. Schematics of the planes parallel to both of these surfaces are shown in Figs. 7b and 7c. The free energy of a type 3 surface must diverge with increasing thickness [6,9]. Therefore, a type 3 surface must either undergo extensive reconstruction to eliminate the dipole moment or it can facet to expose other faces which have a finite free energy. Our data suggests the latter occurs on the (100) CaF_2 surface.

The (111) surface of CaF_2, when terminated with F, has the lowest free energy for any surface of this crystal. Both experimental [10] and theoretical studies [7,8] place this energy around 450–550 erg/cm^2. Because this energy is a factor 1.5 to 2 times lower than that of the (110) surface, we can also understand why the (110) surface of the CaF_2 exhibits (111) facets.

We wish to emphasize that these arguments are completely general to any ionic crystal and we would expect the same behavior to occur in all of the ionic crystals with the fluorite structure. Other researchers [2,11] have observed spotted RHEED patterns for (100) BaF_2 and CaF_2 grown on (100) InP and no researcher has ever reported seeing streaks for a group II flouride grown on a (100) surface. However, very little high resolution microscopy has been done to determine the surface morphology of epitaxial flourides grown on semiconductors. One example where SEM microscopy of the surface morphology of $Sr_{1-x}Ba_xF_2$ on InAs has been done, has recently been published by Sugiyama [12]. In this case, the results for epitaxial (111), (100), and (110) films grown on (111), (100), and (110) substrates are very similar to our results for CaF_2 on Si. While Sugiyama does not try to explain his results, it should be expected that they would be similar to ours as $Sr_{1-x}Ba_xF_2$ also has the fluorite crystal structure.

CONCLUSIONS

We have studied the surface morphology of epitaxial CaF_2 on (100), (111), and (110) Si substrates. Very rough surfaces result on the (100) Si substrates since this is a type 3 surface for CaF_2 which has a very large free energy. At temperatures above 700°C, relatively smoother surfaces can be obtained on (110) Si but they have (111) facets since this is the lowest free-energy surface. Below 700°C on (110) Si substrates, we obtain high quality, single-crystal (111) CaF_2 growth. We do not presently understand what causes this transition temperature. Our results suggest that the (111) orientation is the best direction to grow epitaxial CaF_2 in those situations for which a very thin insulator (less than 100 nm^2 thick) is required, like gate dielectrics.

ACKNOWLEDGMENTS

We wish to thank R.W. DeBlois and R.P. Goehner for valuable technical assistance with the MBE growth and x-ray charcterization of the CaF_2 films. TEM work by N. Lewis and E.L. Hall is also gratefully acknowledged. We also thank Prof. J.P. Krusius for valuable technical discussions and L.A. Principe for her assistance with wafer cleaning. One of us (RWF) acknowledges support from a General Electric fellowship.

REFERENCES

1. G.W. Rubloff, Phys. Rev. B 5, 662 (1972); R.A Heaton and C.C. Lin, Phys. Rev. B 22, 3629 (1980).
2. P.W. Sullivan, R.F.C. Farrow, and G.R. Jones, J. Cryst. Growth 60, 403 (1982).
3. L.J. Schowalter, R.W. Fathauer, R.W. DeBlois, and L.G. Turner, General Electric CRD report No. 84CRD039 (1984).
4. L.J. Schowalter, R.W. Fathauer, R.P. Goehner, L.G. Turner, R.W. DeBlois, S. Hashimoto, J.-L. Peng, W.M. Gibson, and J.P. Krusius, to be published.
5. H. Ishiwara and T. Asano, Appl. Phys. Lett. 40, 66 (1982).
6. P.W. Tasker, Surf. Sci. 78, 315 (1979).
7. P.W. Tasker, J. Physique 41, C6-488 (1980).
8. G.C. Bensen and T.A. Claxton, Can. J. Phys. 41, 1287 (1963).
9. R.W. Fathauer and L.J. Schowalter, Appl. Phys. Lett. 45, 519 (1984).
10. J.J. Gilman, J. Appl. Phys. 31, 2208 (1960).
11. C.W. Tu, S.R. Forrest, and W.D. Johnston, Jr., Appl. Phys. Lett. 43, 569 (1983).
12. K. Sugiyama, J. Appl. Phys. 56, 1733 (1984).

THERMAL STABILITY OF HETEROEPITAXIAL CaF$_2$/Si AND Si/CaF$_2$/Si STRUCTURES

N. HIRASHITA, M. SASAKI, H. ONODA, S. HAGIWARA, AND S. USHIO
OKI Electric Industry Co., Ltd., 550-1 Higashiasakawa, Hachioji, Tokyo 193,
Japan

ABSTRACT

The stability of heteroepitaxial CaF$_2$/Si and Si/CaF$_2$/Si
structures against heat treatments has been investigated.
It is shown that an epitaxial CaF$_2$ film has two types of instability. The one is the degradation of the crystallinity
due to thermal instability. The other is the destruction of
CaF$_2$ crystal due to oxidation, which begins at the interface
between CaF$_2$ and Si and goes toward the surface. This is
related to the oxidation of Si. It is also shown that the
overgrown Si film is stable and its crystallinity is not
degraded by heat treatments.

INTRODUCTION

Heteroepitaxial growth of group II-a fluoride on semiconductor is of great
interest for formation of silicon on insulator devices and 3-dimensional
integrated circuits(ICs) [1-3]. In particular CaF$_2$ grown on Si and Si/CaF$_2$/Si
structure are promising because CaF$_2$ has a cubic structure similar to the
diamond structure and a small lattice mismatch to Si. It also has an advantage
of the lower growth temperature, which is required for fabrication of
3-dimensional ICs.
Heteroepitaxial growth of CaF$_2$/Si and Si/CaF$_2$/Si has been widely studied
[4,5], but few studies on the stability of such a structure against device
fabrication processes have been reported.
In this paper, we describe the stability of epitaxial CaF$_2$/Si and Si/CaF$_2$/Si
structures against heat treatments including oxidation.

EPITAXIAL GROWTH

The Si wafers were chemically cleaned and loaded into a UHV chamber. The
pressure was less than 10^{-7}Pa. After heatcleaning at about 800°C, CaF$_2$ and
Si films were subsequently grown by e-gun evaporation. During the deposition of
both films, the pressure was kept at less than 10^{-5}Pa. Growth rates were 10 to
20Å/sec for CaF$_2$ and less than 5Å/sec for Si. Optimum growth temperatures were
600°C for CaF$_2$ on Si(100), 800°C for CaF$_2$ on Si(111) and about 700°C for
overgrown Si. In order to improve the crystallinity of the overgrown Si film, a
thin amorphous Si layer was deposited prior to the evaporation of Si. This
technique was developed by Asano and Ishiwara [6].
For CaF$_2$ films grown on both Si(100) and Si(111), channeling minimum
yields, Xmin of Rutherford backscattering spectrometry(RBS) were less than 7%
at 200nm in thickness, which indicated its high quality single crystalline
growth. The amorphous Si predeposition technique enables us to obtain double
heteroepitaxial structures of high quality. The Xmin of overgrown Si films
were less than 8% at 500nm in thickness.

THERMAL STABILITY

ANNEALING IN OXYGEN FREE AMBIENT   Fig.1 shows RBS spectra for as-grown
and vacuum annealed CaF_2/Si(100) structure. The annealing was carried out at
800°C for 1 hour in a UHV chamber. The align yield of CaF_2 increased uniformly
throughout the film and the X_{min} of CaF_2 increased from 6% for an as-grown film
to 37% for a vacuum annealed film, which[2] indicated the crystalline degradation
of the CaF_2 film. In the spectra, neither reaction nor interdiffusion were
observed. [2] This suggests that the crystalline degradation is merely due to the
thermal instability of CaF_2 films.
    The same phenomenon was observed in Si/CaF_2/Si structure. Fig.2 shows the
X_{min} s of Si/CaF_2/Si structure as a function of post N2 annealing temperature.
The annealing was carried out in a dry N_2 ambient for 1 hour in a conventional
furnace. The X_{min} for CaF_2 films were measured after removal of overgrown Si
film by CF_4 plasma etching. It was confirmed that the CF_4 plasma etching did
not cause degradation of CaF_2 crystal. The crystallinity[4] of intermediate CaF_2
film was degraded by the annealing at the temperature higher than 800°C. The[2]
RBS spectra of CaF_2 were similar to that of vacuum annealed CaF_2 films. Neither
reaction nor interdiffusion occurred, and only the degradation of CaF_2 crystal
occurred uniformly. Epitaxial CaF_2 films have the above thermal instability,
which, we believe, can be related to stress and crystalline imperfection
characteristic of the heteroepitaxial structure.
    On the other hand, the crystallinity of overgrown Si films was not so much
degraded after the annealing up to 1000°C as shown in Fig.2. It is quite
important for applications to device fabrication that the crystallinity of
overgrown Si films is not degraded even after the crystallinity of underlying
CaF_2 films is degraded.

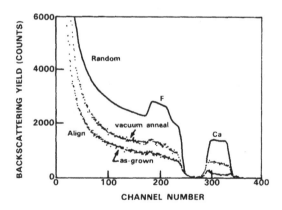

Fig. 1. RBS spectra of CaF_2/Si(100) structure for
an as-grown and a vacuum annealed films.

OXIDATION   In annealing in an oxygen ambient we observed a different
phenomenon from the former case of oxygen free annealing. Fig.3 shows the RBS
spectra for oxidized CaF_2/Si(111) structure. After the oxidation for 1 hour, the
destruction of CaF_2 crystal was observed in the vicinity of the interface
between CaF_2 and Si as shown in Fig.3(b).

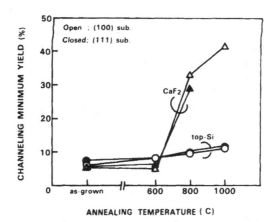

Fig. 2. The Xmin.s of Si/CaF$_2$/Si structure as a function of N$_2$ annealing temperature.

In that region, a considerable amount of silicon and oxygen were observed. This is the different result from oxygen free annealing. In the region near the surface, only a slight degradation of CaF$_2$ crystal was observed and the penetration of silicon and oxygen was not observed. This result is similar to those of oxygen free annealing.

After the oxidation for 3 hours, the crystallinity of CaF$_2$ film was entirely destroyed, and Si and oxygen were observed in the whole films as shown in Fig.3(c). These results suggest that in oxidation two types of crystalline degradation take place. The one is due to the thermal instability, observed in the sample oxidized for 1 hour. The other is the destruction of the crystal, which seems to relate to silicon and oxygen in the film. The latter occurs at the interface between CaF$_2$ and Si and goes toward the surface.

In order to examine the chemical states of the oxygen and Si in the film, X-ray photoelectron spectrometry(XPS) measurements were carried out. In the spectra from the surface of CaF$_2$ film oxidized for 1 hour, only CaF$_2$ spectra were observed but not Si and O peaks. However in the spectra for the sample oxidized for 3 hours, Ca, F, Si and O peaks were observed. Ca 2p and Si 2p spectra of them are shown in Fig.4. The Ca 2p spectrum had a satellite peak, and each was separated to two peaks, respectively. From the analysis of the chemical shift of Ca, these were assigned to CaF$_2$ and CaO. The Si 2p spectra indicate that Si in the film was oxidized. These results reveal that Si and oxygen in the film exist as SiO$_2$ and CaO.

From the results shown in Fig.3 and Fig.4, we can say that the epitaxial CaF$_2$ film on Si is oxidized from the interface, and changes into a mixture of CaF$_2$, CaO and SiO$_2$. Though CaF$_2$ is thermo-dynamically stable against oxidation [7], our experiments show that CaF$_2$ films grown on Si were oxidized and lost its crystallinity. We were not able to understand the detailed mechanism at present. We believe, however, that it is related to the oxidation of Si. The oxidation of Si destroys the CaF$_2$ crystal, which enhances diffusion of Si and oxidation of CaF$_2$. The progress of these two processes will entirely destroy the whole CaF$_2$ crystal.

It should be noted here that no cracks were observed through secondary electron microscopy even after the destruction of the whole films.

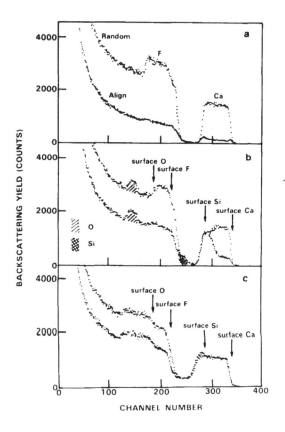

Fig. 3. RBS spectra of the oxidized CaF$_2$/Si(111)
structure for a); as-grown film and after oxidation
at 800°C for b); 1 hour and c); 3 hours.

SUMMARY

In summary, the stability of heteroepitaxial CaF$_2$/Si and Si/CaF$_2$/Si
structures against heat treatments has been investigated. It has been found
that an epitaxial CaF$_2$ film has two types of instability. The one is caused by
oxygen free annealing. The crystallinity of CaF$_2$ films is degraded, due to
thermal instability of the film. The other is caused by oxidation. By the
oxidation epitaxial CaF$_2$ films on Si lose the crystallinity and change into a
mixture of CaF$_2$, CaO and SiO$_2$. This is related to the oxidation of Si from the
interface. It has been also found that the crystallinity of overgrown Si films
is not degraded while the crystallinity of underlying CaF$_2$ films is degraded.

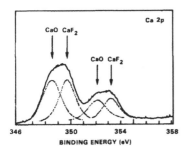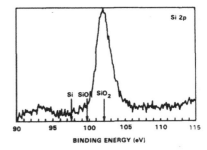

Fig. 4. XPS specra of Ca 2p and Si 2p for the oxidized $CaF_2/Si(111)$ structure at 800°C for 3 hours.

ACKNOWLEDGMENT

The authors wish to thank T. Ajioka for XPS measurements and useful discussions. The authors also wish to thank Prof. T. Inada and Dr. Y. Yamamoto of Research Center of Ion Beam Technology, Hosei University, for the use of RBS.

REFERENCES

1. H. Ishiwara and T. Asano, Appl. Phys. Lett. 40, 66 (1982).

2. T. Asano and H. Ishiwara, Thin Solid Films 93, 143 (1982).

3. R. F. C. Farrow, et al, J. Vac. Sci. Technol. 19, 415 (1981).

4. C. W. Tu, et al, J. Electrochem. Soc. 130, 2081 (1983).

5. R. W. Fathauer and L. J. Schowalter, Appl. Phys. Lett. 45, 519 (1984).

6. T. Asano and H. Ishiwara, J. Appl. Phys. 55, 3566 (1984).

7. O. Kubaschewski, E. Ll. Evans and C. B. Alcock, Metallurgical Thermochemistry (Pergamon Press., 1967).

CHARGE TRANSPORT IN CaF$_2$/Si METAL-EPITAXIAL INSULATOR-SEMICONDUCTOR FIELD-EFFECT TRANSISTORS

T. P. Smith, III

Department of Physics
Brown University
Providence, RI 02912

Julia M. Phillips, R. People, and *J. M. Gibson*

AT&T Bell Laboratories
Murray Hill, NJ 07974

P. J. Stiles

Department of Physics
Brown University
Providence, RI 02912

ABSTRACT

We have studied the material and electronic properties of metal-epitaxial insulator-semiconductor field-effect transistors fabricated by molecular beam epitaxial growth of CaF$_2$ on Si(100). We find that the mobility, interface state density, and threshold voltage can be related to the crystalline quality of the epitaxial CaF$_2$.

INTRODUCTION

Recently, the growth and material properties of epitaxial alkaline earth fluorides on semiconductors have been studied intensively.[1] However, relatively little is known about the relationship between these properties and the electronic properties of the films and transport at the semiconductor-insulator interface. Using Rutherford backscattering (RBS) and channeling, electron diffraction, and transmission electron microscopy (TEM) we have studied the material properties of metal-epitaxial insulator-semiconductor field-effect transistors (MEISFET). We have then examined our results in light of the mobility, threshold voltage, breakdown voltage, leakage current and interface state density of these devices. Our results indicate that there is a strong correlation between the material and electronic properties of these devices.

EXPERIMENTAL

n-channel MEISFETs were fabricated on two boron doped Si(100) substrates with resistivities of 5 ohm-cm (wafer A) and 10 ohm-cm (wafer B). Standard Si metal-oxide-semiconductor (MOS) cleaning and photolithography were used throughout the process. Using SiO$_2$ masks, a p$^+$ back contact was formed by thermal diffusion from a BBr$_3$ source, and n$^+$ contacts to the channel were diffused from a POCl$_3$ source. The wafers were stripped in buffered oxide etchant and a volatile oxide was grown in an HCl solution at 85°C for 10 minutes. The wafers were then loaded into the molecular beam epitaxy (MBE) system and heated to 850°C for 30 minutes to remove the volatile oxide and any impurities on the oxide.[2] In situ Auger electron spectroscopic analysis of the resultant Si surface revealed less than 0.01 monolayer of C and O contamination of the surface following this treatment.

To date most epitaxial films of alkaline earth fluorides on semiconductors have been grown under high vacuum conditions.[3] The films described here were grown under ultra-high vacuum (UHV) conditions in a VG V80H MBE system dedicated to epitaxial fluoride growth. The base pressure in the growth chamber is 5×10^{-11} Torr. Epitaxial CaF$_2$ films 640-800 nm thick were grown at rates of 4.5-6.5 nm/minute while the substrate was held at 565°C. Substrate rotation during the film growth resulted in film thickness uniformity of better than ±2% over the entire 2" wafer. Following growth, contact holes to the source and drain were etched in the CaF$_2$ using dilute HCl followed by HF. Aluminum was then evaporated over the entire wafer and patterned in a weak basic solution.

The MEISFETs were fabricated in the Corbino (circular) geometry.[4] The channel is 20 microns long and the length-to-width ratio is 1:30.

The crystalline quality of the CaF_2 was determined by Rutherford backscattering and channeling at 175 degrees using 1.8 MeV $^4He^+$ ions and by electron diffraction and transmission electron microscopy at 200 kV. Transport properties, capacitance-voltage characteristics and interface state densities were all measured at room temperature.

RESULTS

The backscattered ion yield for wafer A is shown in figure 1. The ratio of the yield in the aligned normal direction to the yield in a random direction (χ_{min}) is 0.09, indicating an epitaxial film of good, although not perfect, quality. (The χ_{min} along the <100> direction in bare bulk Si is 0.03.) The χ_{min} of wafer B is 0.16 indicating significantly poorer epitaxial quality. Electron diffraction tends to confirm these results. In figure 2 the electron diffraction pattern from wafer A indicates the presence of only a small amount of misoriented CaF_2 while the diffraction pattern from wafer B contains rings of spots indicating that there are many more polycrystallites present. However, the electron micrographs in figure 2 show that while the CaF_2 on wafer A has very few defects a significant number of (111) grains are present. The CaF_2 on wafer B has on the order of 10^{11} defects/cm^2 and a larger density of stacking faults. However, a smaller percentage of the area of this film contains misoriented grains. This may indicate that the contribution of defects and the presence of strain in the CaF_2 film on wafer B are responsible for its higher χ_{min}. We also find that the dislocation loops in film A are larger than in film B, and finally that the CaF_2 on wafer A is much more susceptible to electron beam damage.

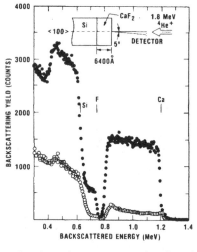

Figure 1.   Channeling and random Rutherford backscattering spectra of a CaF_2 film 640 nm thick on Si(100). The epitaxial fluoride film was used as the gate insulator in MEISFETs. $\chi_{min} = 0.09$.

Capacitance dots were evaporated on the wafers after MEISFET fabrication was complete. The capacitance-voltage characteristics for wafer B is shown in figure 3. Both wafers show some hystersis as the diodes are swept from accumulation to inversion and back to accumulation. However, the shift in the flatband voltage of wafer A is much smaller than in wafer B. The capacitance measurements yield a static dielectric constant of 6.8 for both the films. Using the AC admittance technique,[5,6] the interface state density was found to be approximately 5×10^{11}/cm^2 eV for wafer A and $5-10 \times 10^{12}$/cm^2 eV for wafer B.

Figure 4 shows the drain current (I_D) versus drain voltage (V_D) for MEISFETs from wafers A and B. The peak mobility at room temperature is about 400 cm^2/Vsec for devices fabricated on wafer A and 230 cm^2/Vsec for devices fabricated on wafer B. The mobility was calculated using the static dielectric constant measured by capacitance. There is detectable leakage current $(I_D > 0.2$ uA) when V_D is greater than 4V for devices with the poorer quality CaF_2 while there is none in the devices with the better CaF_2 for $V_D < 15V$.

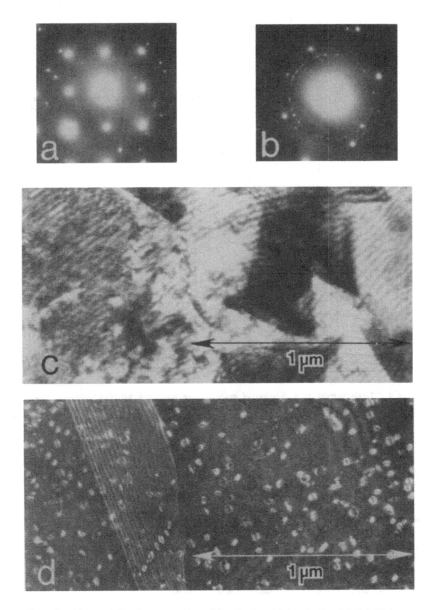

Figure 2.   Electron diffraction patterns from: (a) wafer A and (b) wafer B and dark field electron micrographs of: (c) wafer A and (d) wafer B.

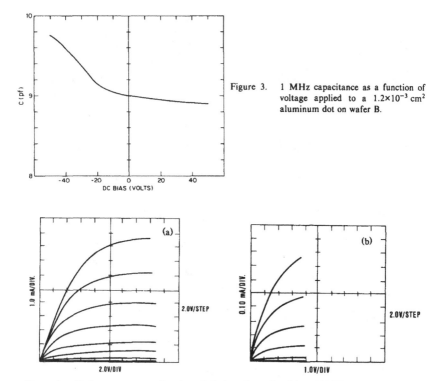

Figure 3.　1 MHz capacitance as a function of voltage applied to a 1.2×10^{-3} cm^2 aluminum dot on wafer B.

Figure 4.　Drain current as a function of drain voltage for (a) a MEISFET fabricated on 5 Ω−cm p-Si(100) with a 640 nm CaF$_2$ film and (b) a MEISFET fabricated on 10 Ω−cm p-Si(100) with an 800 nm CaF$_2$ film. The gate voltage is stepped between 0 and 14V in (a) and 0 and 10V in (b).

Consistent with the hysterisis in the capacitance voltage measurements the threshold voltages of MEISFETs fabricated on wafer A are between −0.5 and 0.5V and relatively stable. Initially, the threshold voltages of devices from wafer B are between −1.0 and 1.0V but after only a few seconds under positive bias the devices begin to turn off as the threshold voltage increases. This shift in threshold voltage occurs at temperatures as low as −100°C and the threshold voltage moves to as high as 33V before devices fail.

The catastrophic breakdown voltage of the films is about 2.5×10^5 V/cm between the gate and source on wafer A but 5.0×10^5 on wafer B. However, the breakdown voltage between Aluminum dots and the undoped substrate on wafer A is 5×10^5 but as high as 3×10^6 on wafer B. There is the possibility that the epitaxial quality of CaF$_2$ on degenerately doped silicon is different from that grown on lightly doped material. Optical microscopy reveals that the surface morphology of the CaF$_2$, above the n$^+$ regions is rougher than on the rest of the wafer.

DISCUSSION

Devices were fabricated on Si(100) substrates for several reasons. First, although RBS and channeling indicate that higher quality epitaxial films are obtained on (111) substrates, TEM reveals that in general these films have a much higher defect density than (100) films and capacitance-voltage measurements indicate that the Fermi level at the semiconductor-insulator interface on (111) wafers is often pinned.[6] Second, (111) oriented films are not chemically stable and hence not compatible with the wet chemical processes used in fabricating devices. Last, because the effective mass is smallest for electrons in inversion layers on (100) Si surfaces a higher mobility can be achieved.

As stated above there is strong correlation between the defect density of the epitaxial CaF_2 film and the interface state density. There are very few defects in wafer A and the interface state density is approximately 1/20th that of wafer B. However, the defect density is not directly related to the interface state density. Only a fraction of the defects observed in the film are located at or near the interface and so the areal defect density may be several orders of magnitude smaller than the areal interface state density. The surface state density of wafer B is close to the density of atoms at misfit dislocation sites if we assume a uniform distribution of interface states between the valance and conduction bands in the Si. This suggests that the defects in the film may allow impurities or ions to migrate to these sites and form interface traps. The fact that the room temperature mobilities of devices on these two wafers differ by less than a factor of two is not inconsistent since many of the interface traps may be neutral and not scatter electrons strongly. The large shifts in threshold voltage of devices on wafer B are probably also mediated by these defects. As a positive bias is applied to the gate mobile charges, fluorine vacancies, and impurities are able to move along these defects toward or away from the interface and trap electrons out of the inversion layer. This is in contrast to the drifting of positive mobile ions (e.g. Na^+) in SiO_2 where application of a positive gate bias increases the number of electrons in the channel but reduces the mobility and causes the threshold voltage to become more negative.

The fact that the CaF_2 on wafer A is more sensitive to electron beam irradiation may be related to its lower breakdown voltage. However, it is not clear why the defects present in the poorer quality film would not contribute more readily to electrical breakdown.

SUMMARY

In conclusion, we have studied the material and electronic properties of $CaF_2/Si(100)$ MEISFETs. We find that the mobility, threshold voltage, and interface state density are all strongly dependent on the crystalline quality of the epitaxial CaF_2. The presence of defects is reflected in the interface state density and is most likely responsible for threshold voltage shifts. The elimination of these imperfections or their effects is essential to the fabrication of high mobility, stable electronic devices.

ACKNOWLEDGEMENTS

The authors gratefully acknowledge the technical support and assistance of W. M. Augustyniak, K. Warner and E. Povilonis and helpful discussions with J. J. Rosenberg. One of us (TPS) acknowledges support from an Office of Naval Research Graduate Fellowship.

REFERENCES

1. J. M. Phillips and J. M. Gibson, Mat. Res. Soc. Symp. Proc. and references therein, *25*, 381 (1984).

2. A. Ishizaka, K. Nakagawa, and Y. Shiraki, 2nd International Symposium on Molecular Beam Epitaxy, Tokyo 1982 p. 183.

3. J. M. Phillips, L. C. Feldman, J. M. Gibson, and M. L. McDonald, Thin Solid Films *107*, 217 (1983).

4. T. P. Smith, III, J. M. Phillips, W. M. Augustyniak and P. J. Stiles, Appl. Phys. Lett. *45*, 907 (1984).

5. For a complete discussion see for example E. H. Nicollian and J. R. Brews, *MOS (Metal Oxide Semiconductor) Physics and Technology* (Wiley, New York, 1982).

6. R. People, T. P. Smith, III, J. M. Phillips, and W. M. Augustyniak, this volume.

ELECTRICAL CHARACTERIZATION OF THE CaF$_2$/Si-EPITAXIAL INSULATOR/SEMICONDUCTOR INTERFACE BY MIS ADMITTANCE

R. PEOPLE[*], T.P. SMITH, III[†], J.M. PHILLIPS[*], W.M. AUGUSTYNIAK[*], AND K.W. WECHT[*]
[*] AT&T Bell Laboratories, Murray Hill, N.J. 07974
[†] Brown University, Physics Department, Providence, R.I. 02912

ABSTRACT

Preliminary measurements of the electrical characteristics of the ultra high vacuum grown CaF$_2$/Si, epitaxial insulator on Silicon, interface are presented. MIS diode admittance has provided information on the net density of surface states and small signal time constant for growth on p-type silicon substrates. Interface state densities as low as $7 \times 10^{10} \mathrm{eV}^{-1} \mathrm{cm}^{-2}$ have been observed for some CaF$_2$/P–Si(100) diodes whereas CaF$_2$/P–Si(111) interface state densities tend to be higher by more than two orders of magnitude. In the present study, we have concentrated on characterization of CaF$_2$/P–Si(100) films having negligible hysteresis. Such films have been consistently grown, having typical thicknesses ~1,000Å and interface state densities $\leq 7 \times 10^{11} \mathrm{eV}^{-1} \mathrm{cm}^{-2}$. Measurements of the small signal time constant versus surface potential show a novel behavior in that capture and emission processes are dominated by minority carriers. These results are explained in part by the presence of interface trapping level(s) above midgap which have electron capture cross sections $\sim 3 \times 10^{-9} \mathrm{cm}^2$. The presence of interface electron traps should have a pronounced and deteriorative effect on MEISFET characteristics. Work is continuing at present in an effort to determine the origin and nature of the observed defect states.

The growth of epitaxial alkaline-earth fluorides (e.g. CaF$_2$) on silicon has potential applications ranging from high mobility surface channel (FET) devices to silicon-on-insulator (SOI) technology, with emphasis on three-dimensional integration. This system offers the prospect of achieving a perfectly ordered heteroepitaxial crystalline insulator on semiconductor interface having transport properties superior to those attainable at an amorphous-insulator on semiconductor interface. Since the first reported growth of epitaxial alkaline earth fluorides on semiconductors [1], the study of these systems has rapidly expanded [2,4]. The room temperature lattice mismatch between CaF$_2$ and Si is ~0.6%; however, due to differences in the coefficient of thermal expansion of these two materials the mismatch at the growth temperature may exceed 1-2%. The extent to which lattice mismatch affects heteroepitaxy has not been quantified; however, it is to be expected that mismatch will cause film quality to be sensitive to (i) film thickness and (ii) cool down-cycle.

In the present study, we report on the electrical characteristics of the CaF$_2$/P–Si interface as obtained by the MIS admittance technique. These measurements give the net surface state density and the small signal time constant. The variation of the small signal time constant with surface potential determines the capture cross-section for those states which dominate carrier capture and emission processes.

We begin with a brief review of the metal-insulator-semiconductor (MIS) diode admittance produced by surface states, and show the means by which surface trap density, D$_{it}$, and capture cross section are extracted. We next present data on a number of films consisting of CaF$_2$/P–Si(100). These films: (i) show low hysteresis, (ii) are $\leq 1,000$Å thickness and (iii) have typical surface state densities $\leq 7 \times 10^{11} \mathrm{eV}^{-1} \mathrm{cm}^{-2}$. It will be seen that the time constant versus surface potential data exhibit novel behavior which suggests capture and emission processes are dominated by minority carriers. Some insight into this observed behavior can be gained by analyzing the small signal trap admittance in the more general case of both minority and majority capture. Such an analysis suggests the observed trend is indicative of a dominant electron trap(s) above midgap level in which the electron capture cross section σ_n greatly exceeds the hole capture cross-section, σ_p.

Figure 1 shows the energy band diagram and charge distribution for an MIS diode structure on p-type silicon substrate. The applied gate bias V$_G$ is divided between the insulator (V$_i$) and the semiconductor surface (Ψ_S). V$_{00}$ and Ψ_{00} denote the zero bias voltage drops across the insulator and semiconductor surface and are due to work function differences between metal and silicon and the presence of charge in surface states. The following regions are of importance: (i) $\Psi_s < 0$

(accumulation), (ii) $\Psi_B > \Psi_S > 0$ (depletion) and (iii) $\Psi_S > \Psi_B$ (inversion); here $\Psi_B = \dfrac{kT}{q}$ ln $\left(\dfrac{N_A}{n_i}\right)$, and denotes the bulk surface potential. For the ideal MIS diode $Q_{it} = 0$ and the measured capacitance of the diode C_m, will be given by the series combination of the insulator capacitance ($C_{ins} = K_{ins}\cdot\epsilon_0/d_{ins}$) and the semiconductor capacitance C_S (due to depletion + inversion layer charge densities); therefore

$$\left[\frac{C_m}{C_{ins}}\right] = \left[\frac{C_s}{C_s + C_{ins}}\right] \tag{1}$$

C_s is frequency dependent, due to the finite rate of minority carrier generation and recombination [5]. In general,

$$C_s = -\left[\frac{dQ_s}{d\Psi_s}\right] \tag{2a}$$

where

$$Q_s = Q_n - q(N_A - N_D)X_d , \tag{2b}$$

Q_n denotes the inversion layer charge and X_d the effective depletion width within the semiconductor.

Fig. 1 Band diagram for MIS diode on p-type semiconductor. Lower figure shows charge density versus depth.

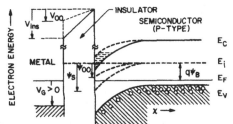

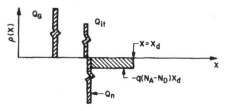

The admittance of an MIS diode gives information on interface trap densities and capture cross-sections. Admittance measures the small signal energy loss, due to interface traps, induced by application of an ac-signal. Since interface traps cannot respond instantaneously to variations in the ac-signal source, there will be a time lag between the ac source and the interface trap filling (emptying) current. This lag gives rise to ohmic losses which are detected as a conductance. The admittance of a single hole trap at the interface of an MIS diode on p-type Si has been given by Nicollian and Goetzberger [6] and will serve to illustrate how interface trap density, D_{it} and capture cross-sections are obtained. If capture and emission by majority carriers only are considered, then the real and imaginary parts of the small signal admittance are given by

$$\left[\frac{G_{it}}{\omega}\right] = D_{it}\left[\frac{q^2}{kT}\right] f_0(1-f_0)\left[\frac{\omega\tau_m}{1+\omega^2\tau_m^2}\right] , \tag{3a}$$

$$C_{it} = D_{it}\left[\frac{q^2}{kT}\right] f_0(1-f_0)\left[\frac{1}{1+\omega^2\tau_m^2}\right] , \tag{3b}$$

where

$$\tau_m = (1-f_0)/\sigma_p \bar{v} p_{so} = \frac{(1-f_0)}{\sigma_p \bar{v} N_A} \cdot \exp(q\Psi_{so}/kT) . \tag{3c}$$

The quantity f_0 is the dc-bias dependent trap occupancy fermi function, σ_p the hole capture cross-section, $\bar{v}$ denotes the thermal velocity of the carriers, Ψ_{so} the dc-bias induced semiconductor surface potential, and p_{so} denotes the surface concentration of free holes. This simple analysis ignores corrections to D_{it} due to fluctuations in surface potential. Brews [7] has shown how surface potential fluctuations affect D_{it} determination. Following his procedure we find that the stated values of D_{it} are accurate to within a factor of two.

Note that both G_{it}/ω and C_{it} contain information on D_{it}, however G_{it}/ω is more amenable in that this quantity peaks for $\omega\tau_m = 1$ giving τ_m (majority carrier time constant) directly; while the amplitude of G_{it}/ω at it's peak gives D_{it}. From (3c) it can be seen that application of a dc-bias ramp (V_G) allows variation of τ_m [through variations of $\Psi_{so}(V_G)$]. For each measurement frequency (ω_j), G_{it}/ω will peak at a unique value of surface potential $(\Psi_{so})_j$, in order that the relation $\omega\tau_m = 1$ is satisfied. Further, note that for majority carrier capture and emission, measurements at increasingly high frequencies will cause G_{it}/ω to peak at increasingly smaller values of surface potential Ψ_{so} (i.e. the G_{it}/ω peaks should move toward more negative bias values as frequency is increased).

In Figs. 2 and 3 we show measured capacitance and equivalent parallel conductance (after correcting for insulator capacitance) for a $CaF_2/P-Si(100)$ MIS diode having $d_{ins} \sim 730$Å. It will be noted that $D_{it} \sim 5 \times 10^{11} eV^{-1} cm^{-2}$ in both figures. The nonzero G_{it} at negative bias values in Fig. 2 is an artifact which simply reflects the absence of appreciable gate bias induced band-bending for $V_G \leq -1.4V$. The most striking feature of these data is the observation that the peak of the G_{it}/ω curves move toward more positive bias values as frequency is increased. This behavior is contrary to what would be expected for majority carrier dominated capture and emission, as previously described. In order to obtain further information on the interface trap(s) responsible for this behavior, we have measured the variation of the small signal time constant versus surface potential (as described by Goetzberger [8]). These results are shown in Fig. 4. It is found that the small signal time constant, τ_c, satisfies

$$\tau_c = \tau_0 \exp[\gamma(\Psi_B - \Psi_{so})q/kT] , \tag{4a}$$

where $\tau_0 = 2.3 \times 10^{-9}$s and $\gamma = 0.90$. Therefore the present capture cross-section is determined as $\sigma_n = 3 \times 10^{-9} cm^2$. The value of σ_n is not typical of donor traps in bulk silicon. The fact that $\gamma > 0$, shows that trap occupation is dominated by electron capture and emission. Further, the observation that γ is not exactly 1.0 is taken as an indication that majority carrier loss processes are contributing, but to a lesser extent.

Fig. 2 Measured capacitance and equivalent parallel conductance for CaF$_2$/P-Si(100) MIS diode. Measurement frequency f = 100 kHz, $d_{ins} \simeq 730$Å.

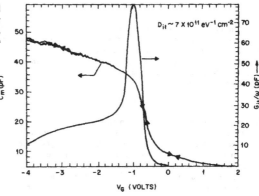

Fig. 3 Measured capacitance and equivalent parallel conductance for $CaF_2/P-Si(100)$ diode (same as Fig. 2) for $f = 1$ MHz $d_{ins} = 730\text{Å}$. Note conductance peak has shifted to a more positive voltage value at this higher frequency.

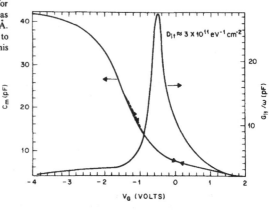

Fig. 4 Small signal time constant versus normalized difference in bulk and surface potentials for sample shown in Figs. 2 and 3. These data give capture cross section of minority carriers (electrons) $\sigma_n = 3 \times 10^{-9} cm^2$.

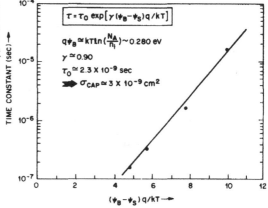

In order to determine the required properties of an interface trap, such that the present measurement trend be observed, we briefly state the results of a calculation of the small signal admittance of an interface trap considering both majority and minority carrier capture and emission. It will suffice to say that the form of the small signal admittance in this case is identical to that obtained in Eq. 3 with the exception being that the majority carrier time constant τ_m, is replaced by the small signal time constant τ_c, where

$$\tau_c = \{\overline{v}[\sigma_n \cdot n_{so} + \sigma_n \cdot n_1 + \sigma_p \cdot p_{so} + \sigma_p \cdot p_1]\}^{-1} . \qquad (4b)$$

Here $n_{so}(p_{so})$ are the surface electron (hole) concentration respectively, $\sigma_n(\sigma_p)$ denote the electron (hole) capture cross section of the interface trap, and $n_1(p_1)$ denote the free electron (hole) density that would be produced if the trap level coincided with the Fermi level (i.e. $n_1 = N_c \exp[-(E_c - E_T)/kT]$, etc.).

It can be readily shown that the present trap occupancy will be dominated by electron capture and emission if and only if: (i) $\sigma_n n_{so} \gg p_1 \sigma_p$ and (ii) $\sigma_n \cdot n_1 \gg \sigma_p \cdot p_{so}$. The first relation denotes electron dominated "filling" of the interface state ($\sigma_n n_{so}$ represents trap filling by electron capture from the conduction band whereas the term $p_1 \sigma_p$ represents trap filling via hole emission from the valence band, and similarly for $\sigma_n \cdot n_1$ and $\sigma_p \cdot p_{so}$). The second relation represents electron dominated "emptying" of the interface traps. Both of these relations are equivalent to:

$$\sigma_n \gg \left[\frac{N_A}{n_i}\right] e^{-\beta(E_T - E_i)} e^{q\Psi_m/kT} \cdot \sigma_p,$$ (5a)

which for $N_A \sim 10^{15} cm^{-3}$ can only be satisfied if $E_T \geq E_i$ (i.e. above midgap) and

$$\sigma_n \gg 10^5 \sigma_p.$$ (5b)

The relations in [5] are readily satisfied by electron traps having a Coulomb repulsive potential for holes; oxygen donors in bulk Si are one example in that the charge states of [0] in Si are "+/++".

The present results are in reasonable agreement with the requirements stated in Eq. (5b), since interfacial hole traps have $\sigma_p \sim 10^{-15}-10^{-16} cm^2$ [6]. The origin of the observed interfacial electron traps are at present unknown, however, two possibilities are immediately suggested: (i) trapping by residual charged interfacial dislocations and (ii) trapping by surface contaminants such as oxygen. The "locally" commensurate nature of the $CaF_2/P-Si$ interface has been recently shown by TEM studies [9], whereas "in-situ" Auger electron spectroscopy of the cleaned Si surfaces prior to CaF_2 film growth typically reveal less than 1/100 of a monolayer of surface oxygen ($\leq 10^{13} cm^{-2}$, more than sufficient to produce the observed interface state densities). In the absence of a measurement of the activation energy of the present interface states, reference to residual charged dislocations and/or surface oxygen as the source of the present defects remain speculative.

In conclusion, MIS admittance studies of the CaF_2/Si-epitaxial insulator on Si interface have shown that: (i) nominal interface trap densities $\sim 5 \times 10^{11} cm^{-2} eV^{-1}$ are present in low hysteresis (≤ 100 mV) structures and (ii) small signal loss is dominated by an electron trap having capture cross section $\sim 3 \times 10^{-9} cm^2$. Further, the present behavior is explained in part by the present of electron trap(s) having a Coulomb repulsive potential for hole capture; as is the case for oxygen in Si or residual dislocations having similar charge states. Until such time that an activation energy can be obtained for the present interface state(s), the supposition that either of the aforementioned mechanisms is responsible for our present observations remain speculative. The presence of an interfacial electron trap(s) is expected to have a pronounced and deteriorative effect on MEISFET operation (especially at low temperatures [10]). Work is continuing at present in an effort to determine the origin and nature of the observed defect states.

References

1. R.F.C. Farrow, P.W. Sullivan, G.M. Williams, G.R. Jones, and D.C. Cameron, J. Vac. Sci. Technol. 19, 415 (1981).

2. H. Ishiwara and T. Asano, Appl. Phys. Lett., 40, 66 (1982).

3. J.M. Phillips, L.C. Feldman, J.M. Gibson, and M.L. McDonald, Thin Solid Films, 107, 217 (1983).

4. J.M. Phillips and J.M. Gibson, Proceedings of the Materials Research Society (North-Holland, Amsterdam, 1984), Vol. 25, p. 381.

5. A.S. Grove, B.E. Deal, E.H. Snow, and C.T. Sah, Sol. St. Electron., 8, 145 (1965).

6. E.H. Nicollian and A. Goetzberger, Bell System Tech. Jrl. Vol. XLVI, 1055 (1967).

7. J.R. Brews, Sol. St. Electron., 26, 711 (1983).

8. A. Goetzberger, CRC Critical Reviews on Sol. St. Physic., 6, 1 (1976).

9. T.P. Smith, III, R. People, J.M. Phillips, W.M. Augustyniak, and P.J. Stiles; to be published in Proceedings of the 1984 Materials Research Society Symposium; Boston, MA (Nov. 26-30, 1984).

10. T.P. Smith, III, J.M. Phillips, W. Augustyniak, and P.J. Stiles, Appl. Phys. Lett., 45, 907 (1984).

GROWTH OF ORIENTED FLUORIDE AND Si/FLUORIDE FILMS ON AMORPHOUS SUBSTRATES

T. ASANO, N. KAIFU, and H. ISHIWARA
Graduate School of Science and Engineering, Tokyo Institute of Technology,
4259 Nagatsuda, Midori-ku, Yokohama 227, Japan

ABSTRACT

The crystalline structures of CaF_2, SrF_2 and BaF_2 films grown
on amorphous SiO_2, and of Si films grown on BaF_2/SiO_2
structures have been investigated. The films were grown by
vacuum deposition onto heated(500-750°C) substrates. It has
been found that preferentially (111) oriented fluoride films
grow on SiO_2 at 700°C though the crystallite orientation was
randomly distributed in the planar direction. At this growth
temperature, the spread of the crystallite orientation in a
BaF_2 film was as small as 0.45°, while that in other
fluoride films was larger. The channeling effect of MeV ions
was observed in BaF_2 films, and the minimum yield in the
film grown at 700°C was as low as 9% near the surface. The
average size of crystallites in BaF_2 films was about 400 nm.
It has also been found that Si films grow epitaxially onto
the preferentially (111) oriented BaF_2 films on SiO_2.

INTRODUCTION

Well-oriented, large-grain polycrystalline Si films on amorphous
insulating substrates are useful not only for fabrication of Si thin film
transistors[1] but also for fabrication of high-speed, high-density three-
dimensional integrated circuits[2]. Various methods such as molecular beam
deposition, graphoepitaxy, beam recrystallization and zone-melting
recrystallization have been studied to form such Si films. From the view
point of low temperature preparation, the molecular beam deposition
technique is attractive because it provides a process below 600°C to grow
fairly large-grain Si films, while the other techniques require
temperatures close to or above the melting temperature of Si. However, the
crystalline structure of molecular beam deposited Si films is essentially
similar to that of Si films prepared by the conventional chemical vapor
deposition technique at higher temperatures[3].

We have been investigating the heteroepitaxial growth of group-IIa
fluoride films on single crystal Si substrates, and have found from x-ray
diffraction analyses that preferentially (111) oriented BaF_2 films can be
grown even on amorphous SiO_2[4]. Such well oriented fluoride films on
amorphous materials may provide a new approach to the formation of oriented
semiconductor films on amorphous insulating substrates, since typical
elemental and compound semiconductors are expected to grow epitaxially on
the fluoride films due to their similar lattice constants and structures.

In this paper, we first investigate the crystalline structure of CaF_2,
SrF_2 and BaF_2 films grown by vacuum deposition onto amorphous SiO_2 at
various substrate temperatures. It is found that preferentially (111)
oriented films of these fluorides can be grown at 700°C, and that the
spread of the crystallite orientation in BaF_2 films is so small that the
channeling effect of MeV ions can be observed. Next, the crystalline
structure of Si films grown on oriented BaF_2 films is investigated.

EXPERIMENT

The growth of fluoride films was carried out in a vacuum system having a base pressure of about 1×10^{-6} Pa. Thermally grown SiO_2 on mirror polished Si(100) wafers were used as substrates. The thickness of SiO_2 films was about 500nm. Prior to deposition of fluoride films the substrates were heated to 900°C for 10 min in order to remove contaminants from the SiO_2 surface. Fluoride grains with 99.9–99.99% purity were evaporated from a Ta boat. The substrate temperature during growth of fluoride films was kept constant in the 500–750°C range at a 50°C increment. The deposition rate of fluoride films was about 0.5 nm/s. The thickness of fluoride films was about 400 nm. For the growth of Si films on BaF_2 films, Si was evaporated from an e-gun. The deposition rate of Si films was about 0.1 nm/s. The thickness of Si film was about 340 nm. During the growth of Si, the substrate temperature was kept at 700°C.

X-ray diffraction analyses were carried out in order to examine the orientation of crystallites in fluoride films and to determine the spread of the crystallite orientation in the films. Rutherford backscattering and channeling measurements with 1.5MeV ^{4}He$^+$ or 220keV H$^+$ ions were also used to characterize the crystalline quality of films. The diameter of the probe ion beam was 1 mm. The size of crystallites and the planar alignment of the crystallite orientation were examined by transmission electron microscopy(TEM).

RESULTS

All fluoride films grown on SiO_2 were first analyzed by x-ray diffraction in order to examine the crystallite orientation. CaF_2 films grown on SiO_2 at substrate temperatures below 650°C showed no strong preferred orientation. But a CaF_2 film grown at 700°C showed signals diffracted only from (111) oriented crystallites. Increase of the growth temperature up to 750°C resulted in disappearance of the strong (111) preferred orientation. SrF_2 films showed similar dependence on the growth temperature of the crystallite orientation to CaF_2 films. On the contrary, all BaF_2 films grown in the temperature range of 500–750°C were preferentially (111) oriented films. Figure 1 shows an x-ray diffraction specrtrum taken from a BaF_2 film grown onto SiO_2 at 700°C, which shows no diffracted signal other than (111) oriented BaF_2 crystallites.

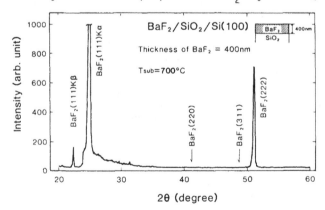

Fig. 1 An x-ray diffraction spectrum taken from a 400 nm thick BaF_2 film grown on SiO_2/Si(100) structure at 700°C.

In order to examine the crystalline quality of fluoride films more
precisely, ion channeling measurements were carried out. Figure 2(a) shows
Rutherford backscattering random and aligned spectra of a BaF$_2$ film grown
on a SiO$_2$/Si(100) structure at 700°C. The incident angle of the ^{4}He$^+$ probe
beam for taking the aligned spectrum was almost normal to the substrate
surface. The random spectrum shows that the BaF$_2$ film is stoichiometric and
uniform. The aligned spectrum shows that crystallites in the BaF$_2$ film are
oriented so well that the channeling effect of ions occurs in the film. The
channeling minimum yield x_{min} near the surface of the BaF$_2$ film is about
9%. Figure 2(b) shows dependence of the channeling minimum yield of BaF$_2$
films on the growth temperature. The channeling effect was not observed in
the films grown at 500-550°C, but it was observed in the films grown at
temperatures higher than 600°C. The channeling minimum yield of the films
decreases with increasing the growth temperature, and it shows a minimum at
700°C. Increase of the growth temperature to 750°C resulted in degradation
of the crystalline quality.

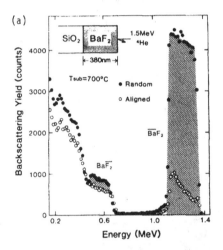

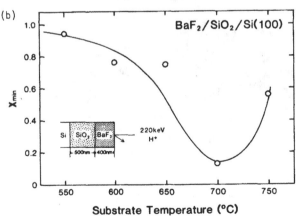

Fig. 2
(a):Rutherford backscattering
random and aligned spectra for a
BaF$_2$(380nm)/SiO$_2$(500nm)/Si(100)
structure. The BaF2 film was grown
at 700°C.
(b):Dependence on the growth temperature of the channeling minimum
yield of BaF$_2$ films grown on SiO$_2$.
x_{min}'s were measured with 220keV
protons.

Channeling measurements with 1.5MeV ^{4}He$^+$ ions were also performed on CaF$_2$ and SrF$_2$ films grown on SiO$_2$ at 700°C. However, the channeling effect was not observed in these films, although the films were preferentially (111) oriented as far as they were measured by x-ray diffraction analyses. These results imply that the <111> axis of each constituent crystallite in the CaF$_2$ and SrF$_2$ films is almost aligned to the substrate surface normal, but the spread of the crystallite orientation is wider than the channeling critical angle(about 1°). In order to confirm this, the spread of the crystallite orientation in the CaF$_2$, SrF$_2$ and BaF$_2$ films grown on SiO$_2$ at 700°C was measured by x-ray rocking curve analyses. The standard deviations of the spread of the crystallite orientation in the CaF$_2$, SrF$_2$ and BaF$_2$ films were about 4.8°, 2.2° and 0.45°, respectively. X-ray rocking curve analyses were also carried out for BaF$_2$ films grown at 550-650°C. The standard deviation of the spread of the crystallite orientation was decreased from 3.8° at 550°C to 0.50° at 650°C with increasing the growth temperature. Therefore, we consider that the observed variation of the channeling minimum yield in BaF$_2$ films with growth temperature(Fig. 2(b)) is owing mainly to the variation of the spread of the crystallite orientation in the films. From these results we conclude that the use of BaF$_2$ films grown at 700°C is suitable to grow well oriented Si films on fluoride/SiO$_2$ structures, though the lattice mismatch between Si and BaF$_2$ is relatively large(-12%).

Next, the size of crystallites and the planar alignment of the crystallite orientation in BaF$_2$ films were investigated by TEM. Figure 3 shows a TEM bright-field micrograph of the BaF$_2$ film grown at 700°C. The size of crystallites ranges from 200 to 600 nm, and it is about 400 nm on the average. The electron diffraction pattern taken from this area showed signals diffracted only from crystal planes perpendicular to the (111) plane of BaF$_2$. But, as we can see in the micrograph, the orientations of crystallites are almost randomly distributed in planar directions.

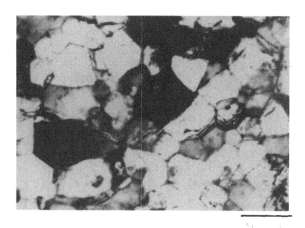

Fig. 3 TEM bright-field micrograph of a BaF$_2$ film grown on SiO$_2$ at 700°C.

Finally, the crystalline structure of Si films grown on top of BaF$_2$/SiO$_2$ structures was investigated. Figure 4 shows backscattering spectra taken from a Si/BaF$_2$/SiO$_2$/Si(100) structure. In this sample, both the BaF$_2$ and Si films were grown at 700°C. The random spectrum shows that the top Si film is uniform. The aligned spectrum shows that the channeling effect of ions occurs in both the top Si and the underlying BaF$_2$ films. The channeling minimum yield of the Si film is about 73%. X-ray diffraction analyses have shown that the top Si film is (111) oriented as well as the underlying BaF$_2$ film.

Figure 5 shows a TEM dark-field micrograph of the Si film grown on the BaF$_2$/SiO$_2$ structure at 700°C. The size of crystallites in the Si film is about 400nm on the average, which corresponds to that in the underlying BaF$_2$ film. On the other hand, a Si film grown directly onto SiO$_2$ at the same temperature(700°C) showed no strong preferred orientation of crystallites, and the size of crystallites was about 60 nm on the average. These results clearly demonstrate that Si films grow epitaxially on preferentially (111) oriented BaF$_2$ films on amorphous substrates, and that well-oriented, large-grain Si films can be grown on amorphous substrates at relatively low temperatures by the use of fluoride intermediate layers.

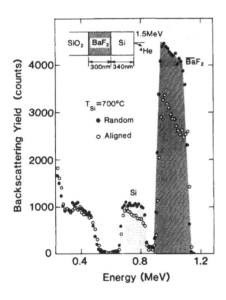

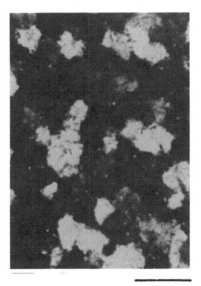

1 μm

Fig. 4
Backscattering random and aligned spectra for a Si(340nm)/BaF$_2$(300nm)/SiO$_2$(500nm)/Si(100) structure. Both the top Si and the underlying BaF$_2$ films were grown at 700°C.

Fig. 5
TEM dark-field micrograph of a Si film grown onto a BaF$_2$ film on SiO$_2$ at 700°C.

180

SUMMARY

Preferentially (111) oriented fluoride films can be grown on SiO_2 at the growth temperature of 700°C. The spread of the crystallite orientation in BaF_2 films was as small as 0.45°, while that in other fluoride films was larger. The ion channeling effect was observed in BaF_2 films, and the channeling minimum yield in the film grown at 700°C was as low as 9%. The size of crystallites in BaF_2 films was about 400 nm, and the crystallite orientation was randomly distributed in the planar direction. Si films can be grown epitaxially onto the preferentially (111) oriented BaF_2 films, though the lattice mismatch between Si/BaF_2 is large. Thus the use of fluorides as intermediate layers is useful to grow well-oriented, large-grain Si films on amorphous materials at relatively low temperatures.

ACKNOWLEDGMENTS

We gratefully acknowledge useful discussions with Prof. S. Furukawa. We are also thankful to the members of the steering committee and the staff of the Van de Graaff accelerator at Tokyo Institute of Technology, to Prof. M. Abe for the use of the x-ray diffractometer, and to Prof. T. Mori and Prof. A. Satoh for the use of the transmission electron microscope. This work is partially supported by the Hosobunka foundation.

REFERENCES

1. M. Matsui, Y. Shiraki, and E. Maruyama, J. Appl. Phys. 55, 1950(1984)
2. J. C. Sturm, M. D. Giles, and J. F. Gibbons, IEEE Electron Device Lett. EDL-5, 151(1984)
3. M. Matsui, Y. Shiraki, and E. Maruyama, J. Appl. Phys. 53, 995(1982)
4. T. Asano, H. Ishiwara, and N. Kaifu, Jpn. J. Appl. Phys. 22, 1474(1983)

MBE GROWTH OF LANTHANIDE TRIFLUORIDES ON SILICON (111)

R.F.C. FARROW, S. SINHAROY, R.A. HOFFMAN, J.H. RIEGER, W.J. TAKEI,
J.C. GREGGI, JR., S. WOOD, T.A. TEMOFONTE
Westinghouse Research and Development Center, 1310 Beulah Road,
Pittsburgh, PA 15235

ABSTRACT

In this paper we report the first epitaxial growth of films of rare earth lanthanide trifluorides by MBE. These materials (XF_3, where X = La, Ce, Nd) possess the hexagonal, tysonite crystal structure and grow epitaxially, in the growth temperature range 400-700°C, on atomically clean Si (111) 1x1 or 7x7 surfaces with the hexagonal basal units in orientational registry with hexagonal surface units of the silicon (111) surface and with XF_3 [COO1] ∥ Si [111]. The hard, water insoluble nature of these materials combined with significantly smaller thermal expansion coefficients than the cubic fluorite structure fluorides makes them attractive as exploratory films for semiconductor passivation, gate dielectrics and epitaxial interlayers. In situ and ex situ structural investigations and preliminary electrical properties of the films are reported.

INTRODUCTION

Since the first report [1] of growth of epitaxial fluoride films on semiconductors emphasis has been placed [2] on MBE growth of the cubic group II fluorides and their alloys because of the close similarity of their structure to that of the well-known semiconductors. This field of research has developed to the point where prototype FET's employing epitaxial CaF_2 have been reported [3] and exploratory semiconductor-insulator-semiconductor structures have been prepared for Si [4] and GaAs [5]. Nevertheless, the finite water solubility of the cubic fluorite structure fluorides and their large ($\geq$ 18 x 10^{-6} deg^{-1}) thermal expansion coefficients are factors which may complicate device processing. The hexagonal-structure lanthanide trifluoride family (XF_3 where X = La, Ce, Pr, Nd, Pm) includes members such as LaF_3, CeF_3 and NdF_3 which are known [6] to be water insoluble, have [7] significantly smaller thermal expansion coefficients and are [8] mechanically harder than the cubic fluorite structure fluorides. Table I summarizes a comparison between the physical properties of these materials. Clearly the lanthanide trifluorides have features which should make them attractive in exploratory applications as protective optical coatings, passivation layers, gate dielectrics or epitaxial interlayers on semiconductors. Indeed, LaF_3 is already used [8] as a protective overlayer on antireflection coatings inside an HF gas laser cavity.

Table I. Lanthanide Trifluorides (XF_3) vs. Group II Fluorides (XF_2):
Physical Properties

Crystal	Hardness (Knoop)	M. Pt. °C	Water Solubility gm/100 cc, R.T.	Thermal Expansion Coeff. α (10^{-6} K^{-1}) 300K	
				$\alpha_\perp$	α_{11}
MgF_2		1266	0.0076	9.0	14.0
LaF_3	400-600	1493	insoluble	17.0	10.7
CeF_3		1430	insoluble	12.9	16.5
NdF_3		1374	insoluble	17.4	14.7
CaF_2	120-160	1360	0.0016		19
SrF_2		1190	0.011		18
BaF_2	60-80	1280	0.17		18
NaF		988	4.22		34

Fortunately, the lanthanide trifluorides contain hexagonal symmetry units which are close in size to hexagonal symmetry sub-units of the Si (111), Ge (111) or GaAs {111} surfaces. Thus, epitaxy is to be expected, for example, with LaF_3 [0001] ‖ Si [111]. This relation and identical ones for CeF_3 and NdF_3 are indeed observed as we report in this paper.

On the basis of the most recent investigations of the structure of LaF_3 it may be concluded that LaF_3 and the other lanthanide trifluorides can exist in two different structural forms. Both are hexagonal and differ in basal plane parameter by a factor of $\sqrt{3}$. One has [9] the space group $P6_3$ cm with a unit cell containing six molecular units ($z = 6$). The other has [10] the space group $P6_3/mmc$ with a unit cell containing two molecular units ($z = 2$) as described by Wyckoff [11]. The $z = 2$ unit cell (sub cell) has a basal plane parameter related to that of the $z = 6$ unit cell (supercell) by: $a_{sub} \sqrt{3} = a_{sup}$. The $z = 2$ cell is a sub-cell of the $z = 6$ cell and it is the near coincidence of 2 a_{sub} with the hexagonal unit (or sub-unit in the case of Si (111) 7 x 7) lattice parameter
• $a_{Si} \sqrt{2}$ (or $a_{Ge} \sqrt{2}$) which favors epitaxy. Table II lists the lattice parameters of the lanthanide trifluorides together with the (290K) misfit between hexagonal basal units of the trifluorides and those of Si and Ge (111) surfaces. In this paper we report epitaxial growth of LaF_3, CeF_3 and NdF_3 on Si (111) 1x1 and 7x7 surfaces. It is evident from Table II, however, that this mode of epitaxy is likely to extend to all the lanthanide trifluorides on both Si and Ge (111) surfaces.

Table II. Lanthanide Trifluorides: Lattice Parameters (290K) and Misfit (290K) to Silicon and Germanium (111) 1x1 Surfaces

Crystal	a_{sup} (Å) Z=6, $P6_3$cm	a_{sub} (Å) z=2, $P6_3/mmc$	c (Å)	Basal Unit Misfit to Si (111) 1x1 $[2a_{sub}-a_{Si}\sqrt{2}]/2a_{sub}$	Basal Unit Misfit to Ge (111) 1x1 $[2a_{sub}-a_{Ge}\sqrt{2}]/2a_{sub}$
LaF_3	7.186	4.149	7.352	+7.4%	+3.6%
CeF_3	7.131	4.117	7.288	+6.7%	+2.8
PrF_3	7.078	4.086	7.240	+6.0%	+2.1
NdF_3	7.032	4.060	7.200	+5.4%	+1.5
PmF_3	6.970	4.024	7.190	+4.6%	+0.6

EXPERIMENTAL PROCEDURE

Film growth and in situ LEED/Auger analysis was carried out in a UHV system equipped with a 2-grid LEED unit, an Auger analyzer (PHI Model 10-150), sputter ion gun (PHI Model 04-161), and sample manipulator (PHI Model 10-501). Post-bakeout base pressures were below 10^{-10} Torr and pressures during growth were in the mid 10^{-9} Torr range. The LaF_3 molecular beam was generated by heating LaF_3 pellets in a helical coil of tungsten wire. The beam was collimated by a stainless steel tube and shuttered by a beam flag. Deposition rate (tyically 1-2 Å s^{-1}) was monitored by a quartz microbalance (Inficon XMS-1). Small silicon bar samples were heated resistively and held, relatively stress-free, by molybdenum spring clamps to avoid thermally-induced slip during sample heating at ≳ 800°C. Sample temperatures were measured using an infrared pyrometer (for T < 750°C) or an optical pyrometer. A clean, well ordered Si (111) surface was obtained by Ar$^+$ ion bombardment followed by annealing at 850-1000°C.

FILM GROWTH AND IN-SITU CHARACTERIZATION

Following sample cleaning, a well-defined (1x1) or (7x7) LEED pattern
was observed and Auger spectroscopy confirmed the absence of surface
impurities such as O and C. Film deposition was initiated by opening the
beam shutter when the sample temperature (usually in the range 400-700°C)
had stabilized. LEED was used to study the surface of the grown film.
Well-defined hexagonal symmetry (1x1) patterns were observed after
deposition of films $\gtrsim$ 1000Å thick. These patterns confirmed that the basal
plane of LaF$_3$, CeF$_3$ and NdF$_3$ was parallel to the (111) Si surface and that
the film lattice was rotationally aligned with the hexagonal cell of Si (111).
The LEED pattern   of the films exhibited the superlattice spots
characteristic [10] of the z = 6 unit cell. Measurement of the basal plane
lattice parameters of Si (111) and the films confirmed, from the LEED
patterns, within experimental error, that the surface of the films had a
parameter close to the bulk value, i.e., $\sim$ 6-8% greater than the Si value
(see Table II). Thus the misfit may be accommodated through an interfacial
dislocation network or an incommensurate interface. Attempts to examine
the film nucleation stage using LEED were not successful because only
faint, diffuse LEED patterns were observed in the initial stages of growth.
Films grown at room temperature were polycrystalline.

EX-SITU CHARACTERIZATION OF THE FILMS

The films were examined by RHEED (80 keV) and a variety of X-ray
diffraction techniques including texture analysis and single crystal
diffractometry. Figure 1 shows RHEED patterns recorded for a 1845Å thick
film of LaF$_3$. The patterns along two azimuths are shown:(a) along $\underline{a}^*$ and
(b) along $(\underline{a}^*+\underline{b}^*)$. These azimuths are referred to the supercell basis[10]
and are separated by an azimuthal rotation of 30°. For the $\underline{a}^*$ azimuth the
Ewald sphere intersects only the reciprocal lattice rods for the sublattice.
Rotation of the sample by 30° to the $(\underline{a}^*+\underline{b}^*)$ azimuth resulted in the Ewald
sphere intersecting both superlattice and sublattice rods giving an
apparent 3x pattern. The c/a ratio derived from the patterns is consistent
with the Z = 6, P6$_3$ cm structure for LaF$_3$. RHEED studies of NdF$_3$ films gave
similar patterns to those shown in Figure 1 consistent with the Z = 6,
P6$_3$ cm structure. On the other hand RHEED studies of CeF$_3$ films revealed
only sublattice rods for all azimuths indicative of the Z = 2, P6$_3$ mmc
structure. Transmission electron microscopy studies of the LaF$_3$ films
confirmed the Z = 6, P6$_3$ cm structure and in addition revealed, in some
early films, a network of microcracks. Optical microscope examination of
several early films also showed cracking related to underlying slip lines in
the substrate. These effects were later eliminated by improved sample
mounting to reduce mechanical stress at the sample ends. CeF$_3$ films showed
a greater tendency to cracking than either LaF$_3$ or NdF$_3$ films. Detailed
studies of film texture and structure as a function of growth temperature
and film thickness are still under investigation and will be reported
elsewhere [12].

Preliminary electrical (C-V) measurements of an 800Å thick LaF$_3$ film
grown on a p-type Si (111) substrate indicate a flat band shift, typically
to -3V. Net positive charge density in the film was typically $\sim$ 2.4$\cdot$10^{12}cm^{-2}.
AC (60 Hz) breakdown voltages, measured between Aℓ dots and the silicon
substrate, varied spatially between 2 x 10^5 and 6 x 10^6 V cm^{-1}. DC breakdown
voltages were $\sim$ 50% of the AC values. A dielectric constant of 12 was
derived from the C-V data.

184

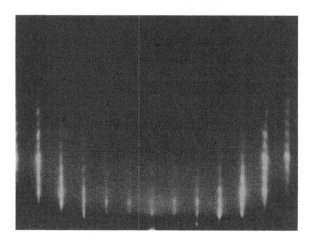

(a) azimuth along $\underline{a}^*$

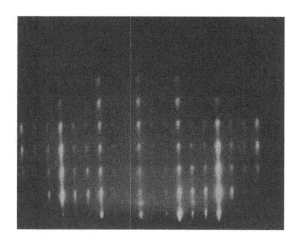

(b) azimuth along $\underline{a}^*+\underline{b}^*$

Figure 1. 80 keV RHEED patterns recorded from epitaxial LaF$_3$ film grown
on Si (111) substrate.

CONCLUSIONS

Epitaxial growth of LaF_3, CeF_3 and NdF_3 films on the (1x1) and (7x7) surfaces of Si (111) has been achieved at growth temperatures in the range 400-700°C. The c-axis of the trifluoride films is parallel to Si [111]. Preliminary electrical data for the LaF_3 films indicates that usefully high ($\geq 2 \times 10^6$ V cm^{-1}) breakdown voltages can be achieved for films of $\sim$ 800Å thick.

REFERENCES

1. R.F.C. Farrow, P. W. Sullivan, G. M. Williams, G. R. Jones, D. G. Cameron, J. Vac. Sci. Technol. 19 (3) 415 (1981).

2. Julia M. Phillips, L. C. Feldman, J. M. Gibson, M. L. McDonald, Thin Solid Films, 107 217 (1983).

3. T. P. Smith, III, J. M. Phillips, W. M. Augustyniak, P. J. Stiles, Appl. Phys. Lett. 45 (8) 907 (1984).

4. H. Ishiwara, T. Asano, S. Furukawa, J. Vac. Sci. Technol. B1 (2) 266 (1983).

5. P. W. Sullivan, G. M. Metze, J. E. Bower, In press; to appear in J. Vac. Sci. Technol., Jan. 1985.

6. Handbook of Chemistry and Physics, Ed. R. C. Weast, Published by Chemical Rubber Co., 45th Edition.

7. "Thermophysical Properties of Matter" Vol.13, Ed. Y. S. Touloukian, Plenum, N.Y. (1977).

8. J. Stanford, Personal communication (1984).

9. D. Gregson, C.R.A. Catlow, A. V. Chadwick. G. H. Lander, A. N. Cormack, B.E.F. Fender, Acta. Cryst. B39, 687 (1983).

10. O. Greis, D.J.M. Bevan, J. Solid State Chem. 24 113 (1978).

11. "Crystal Structures" Vol. 2, Second Edition, Ed. by R.W. Wyckoff, Interscience 1964, pp. 60-61.

12. S. Sinharoy, R. A. Hoffman, R.F.C. Farrow, Paper submitted for presentation at 12th International Conference on Metallurgical Coatings, April 15-19, 1985, Los Angeles, CA.

DIELECTRIC COATED FLUORINE PASSIVATED SILICON SURFACES

B. R. WEINBERGER, H. W. DECKMAN, E. YABLONOVITCH, T. GMITTER AND W. KOBASZ
Corporate Research Laboratory, Exxon Research and Engineering Company,
Annandale, New Jersey 08801

ABSTRACT

An alternative to oxide passivation of crystalline silicon surfaces is presented. By terminating the silicon surface bonds with fluorine we have achieved a highly passivated surface, $\leq 10^{10}$ electronically active centers per cm^2. Measurements of the surface recombination velocity and surface state density of such fluorinated surfaces using photoconductive lifetime measurements and C-V characteristics are also described.

INTRODUCTION

The termination of the ordered crystalline lattice at the surface of a semiconductor results in a surface electronic configuration drastically different from that found within the bulk of the semiconductor. Distorted bonds or unsatisfied dangling bonds at the surface create electronic states within the semiconductor's forbidden energy gap. Such states are parisitic to the proper operation of semiconductor devices, particularly surface junction devices like MISFET's. A crucial property of crystalline silicon which to a great measure has been responsible for its technological importance is the high quality of the interface it can form with its oxide. Thermally oxidized and annealed silicon surfaces can exhibit nearly perfect passivation, midgap densities of states as low as 1×10^{10} states/eV-cm^2 or ~1 surface state per 10^5 surface atoms [1].

Before silicon-oxide technology was firmly established, so-called "real" chemically treated and etched silicon surfaces were studied [2]. Some success in surface passivation was achieved. However, none achieved the stability in ambient atmospheres required for device fabrication. Recently, the use of epitaxial insulators such as CaF_2 have been explored as potential low defect heterolayers on silicon [3-6]. Alternate insulators such as Langmuir-Blodgett organic films deposited on top of a thin silicon native oxide layer have also been studied. [7,8] To date, these alternate surface passivations have not achieved the quality of the $Si:SiO_2$ interface.

WAFER PREPARATION AND MEASUREMENTS

Table 1

Silicon Surface Treatments

	STEP 1	STEP 2	SURFACE RECOMBINATION VELOCITY ACHIEVED
1	CP-4 etch	HF Acid dip	4 cm/s
2	Thermal Oxidation	HF Acid dip	4 cm/s
3	HF Acid Oxide Removal	XeF$_2$ Vapor	40 cm/s
4	KOH etch	HF Acid dip	>300 cm/s

Mat. Res. Soc. Symp. Proc. Vol. 37. ᶜ 1985 Materials Research Society

In our approach, the silicon surface is chemically treated with the final silicon surface terminated with fluorine bonds. Several silicon surface chemical treatments were employed with varying degrees of success at passivation. These are listed in Table 1. Of note is that technique 4 yielded poorly passivated surfaces despite the similarity to the highly successful technique 1. Preliminary XPS data suggests that the techniques employing a final hydrofluoric acid immersion step produce a silicon surface terminated with fluorine bonds similar to that obtained by procedure 3 [10]. We have measured, using a photoconductive decay technique, surface recombination velocities as low as 4 cm/s for the fluorine terminated silicon surface. Midgap interface state densities as low as 5×10^9/eV-cm^2 have been measured on MIS structures with fluorinated surfaces using standard capacitance-voltage (CV) techniques with non-lattice matched polycrystalline insulators. We conclude that fluorine termination produces a silicon surface of extraordinary electronic quality, a surface nearly free of any surface electronic states within the silicon energy gap.

The apparatus we have assembled for wafer processing is displayed in Figure 1. The data presented in this paper are for wafers prepared using techniques 1 and 2. Silicon wafers employed in this study were high bulk quality (bulk minority carrier lifetime $\tau_B > 3$ ms) float zone material obtained from Wacker Chemitronics. This high bulk quality enhanced the surface sensitivity of our characterization probes. N and P type wafers of both (100) and (111) crystallographic orientation were studied. The silicon wafer is mounted in a stainless steel cross on a teflon rod. The rod passes through an O-ring compression seal which permits both rotary and linear manipulation of the wafer. We have found that exposure of a fluorinated wafer to air results in a rapid deterioration of the surface passivation. The apparatus in Figure 1 permits all operations (etching, fluorination, subsequent coating depositions) to be performed either in an inert Ar atmosphere or under vacuum. Oxide removal and fluorination of the wafer surface can be accomplished by first purging the system with Ar, then immersing the wafer in an aqueous HF solution contained in a teflon vessel suspended below the cross and separated from it by a gate valve. The fluorinated wafer may then be withdrawn back into the cross without exposing the surface to air.

Figure 1. Wafer processing apparatus

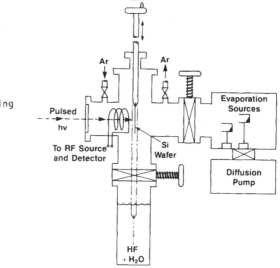

Air impermeable coatings are applied to the passivated surfaces by evacuating the cross through a second gate valve and depositing material from thermal evaporation sources through this same valve.

Also included in the cross is an r.f. coil inductively coupled to the silicon wafer. A transparent window is placed on one arm of the cross allowing illumination of the wafer. The condition of the wafer surface is monitored in-situ via the transient photo-induced eddy current losses following pulsed illumination of the wafer. From this photoconductive decay the minority carrier lifetime may be determined and the surface recombination velocity inferred [10]. The transient photoconductive decay technique is particularly suited to the study of well passivated silicon surfaces important in device technology. The surface sensitivity is limited only by the bulk quality of the material. In the work presented here, we have been able to detect electrically active surface states at a level of one per 10^6 surface atoms, a sensitivity far exceeding that of other surface analytical tools such as XPS.

The photoconductivity decay rate will be sensitive to surface recombination for thin wafers of high quality (long τ_B). The excess minority carrier density (holes for a n-type wafer) will be governed by a one dimensional diffusion equation.

$$D_p \frac{d^2 p}{dx^2} - \frac{p - p_0}{\tau_B} = \frac{dp}{dt} \tag{1}$$

The second term describes bulk recombination and the first term governs diffusion to the surface which when coupled with the boundary condition $D_p \frac{dp}{dx} = p \, S_p$, x = 0,L, includes the effect of surface recombination where D_p and S_p are the diffusion constant and surface recombination velocity for holes and L is the thickness of the wafer. In the absence of surface charging, S_p is time independent and the solution to (1) will have a time dependence of the form:

$$p(x,t) \propto \exp - (\frac{1}{\tau_B} + \frac{1}{\tau_s}) \, t = \exp - (\frac{t}{\tau_{ph}}) \tag{2}$$

where τ_s is the surface recombination lifetime and τ_{ph} is the measured photoconductive lifetime. For well passivated surfaces where S << D_p/L, τ_s = L/2S_p [11].

A representative photoconductive decay curve is shown in Figure 2. The data was obtained for a fluorine passivated (100) n-type wafer (L=275 μ, dopant density $N_D = 10^{15}$/cm^3) with the coating layers shown in the inset.

Figure 2. Photoconductive decay obtained for fluorinated silicon wafer (n-type, 5 Ω-cm, L = 275μ) coated with composite insulator.

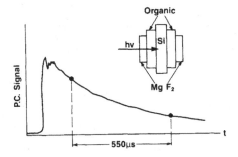

The measured τ_{ph} of 550 μs implies a surface recombination velocity S_p < 25 cm/sec. The upper bound assumes an infinite τ_B. If one assumes equal electron and hole capture cross sections and low level injection, for a density of surface states N_s at midgap, [12] $S_p = N_s v_t \sigma$, where v_t is the thermal velocity for electrons, 10^7 cm/s. Typical values for surface state hole capture cross sections obtained for oxidized surfaces are $\sigma \sim 5 \times 10^{-16}$ cm^2. [13] The measured τ_{ph} therefore implies a midgap surface state density of <5 x 10^9 cm^{-2}, or less than one electrically active interface trap per 10^5 surface atoms.

Measurements of τ_{ph} as large as 3 ms on fluorinated 275 μ thick wafers have been routinely obtained. Implied is a surface recombination velocity, assuming only surface recombination and no bulk recombination, S_p < 4 cm/s. Such results are independent of the crystallographic orientation of the wafer, (100) or (111). Significantly, such long surface photoconductive lifetimes have been obtained by fluorinating surfaces of wafers with widely varying conductivities. Surface recombination velocities less than 12 cm/s have been measured on n-type fluorinated wafers with dopant densities varying over the range 5 x 10^{13} - 1 x 10^{17} cm^{-3}. In other words, S_p was found to be independent of the position of the majority carrier Fermi level over the range 0.14 < E_c - E_F < 0.33 eV. The Shockley-Read-Hall treatment of surface recombination [12] demonstrates that under low level carrier injection, those surface states close to the band edges will have little effect on the surface recombination velocity since the residence time of carriers in these traps is too short to permit efficient recombination. In fact, for an n-type semiconductor it may be shown that only those traps with energy E_t in the range $E_v + (E_c - E_F) < E_t < E_F$ are effective. What this implies is that the surface photoconductive lifetime in more heavily doped wafers should be sensitive to surface states nearer the band edges. For thermally oxidized heavily doped wafers, we have been unable to obtain long lifetimes. We take this as a manifestation of the U-shaped density of surface states distribution of the Si:SiO$_2$ interface where the density of states rises precipitously away from midgap. [14] In contrast, the low S_p measured on fluorinated surfaces of similar wafers with E_F as close as 0.14 eV to the band edge, we take as preliminary evidence that the distribution of surface states associated with the fluorinated surfaces does not have the broad "band tails" the oxidized surface does.

The foregoing analysis is valid for the flat band situation for which surface charging is absent. For instance for a surface placed under accumulation S_p is diminished without a reduction of surface state density. [12] In effect, the surface potential acts as a minority carrier mirror repelling holes from the surface, reducing surface recombination because the arrival rate of minority carriers at recombination centers is the rate determining step in surface recombination. The effects of surface charging may be eliminated by high level injection of carriers which tends to flatten the surface band bending. The presence of appreciable surface charging will be quite apparent in the photoconductive lifetime measurements. The points along the decay, in fact, represent different levels of carrier injection. A strongly non-exponential decay indicates a surface recombination velocity which varies with injection level, or in other words a charged surface. The photoconductive decay displayed in Figure 2 exhibited no such serious charging effects so that the long photoconductive lifetime is indeed a firm indication of the low density of surface states at the fluorinated surface.

The density and properties of surface states may be probed by studying the capacitance of metal-insulator-semiconductor (MIS) structures as a function of bias voltage and frequency [1]. Several requirements, however, had to be satisfied by the gate electrode for our purposes. The finished device must have an air impermeable encapsulation as air exposure rapidly deteriorates the fluorinated silicon surface. The insulator preferably should only weakly interact with the surface so as not to perturb its near ideal electronic structure. The insulator should have high resistivity and a low density of trapped charge for optimal performance of the MIS device. Finally, the insulator should form pinhole free films on room temperature substrates by thermal evaporation so as to be compatible with our apparatus (Figure 1).

An MIS structure which satisfies the foregoing criteria is shown in the insets to Figures 2 and 3. The gate insulator employed is a two layer structure consisting of 1000Å of a polynuclear aromatic (PNA) organic film capped with 500Å of MgF_2. The PNA used to obtain the data in Figures 2-3 was rubrene (5, 6, 11, 12 tetraphenyl-napthacene). It forms high quality pinhole free films from thermal evaporation sources at ~300°C onto room temperature substrates. Like other PNA's rubrene is a high resistivity insulator [15]. The PNA's form molecular solids with a weak Van der Waals intermolecular interaction. Presumably the binding of the rubrene film to the fluorinated silicon surface is of a similar nature. This is consistent with our observation that the deposition has little effect on the measured τ_{ph} (Figure 2). The PNA's, such as rubrene, however, were found to form only poor diffusion barriers to air so that long term stability of the MIS devices necessitated an air impermeable overlayer. We found that a 500Å layer of evaporated MgF_2 served this purpose well. It should be noted that MgF_2 without the organic layer present was unsatisfactory as a gate insulator, severe degradation of τ_{ph} being observed during the deposition process.

Figure 3. CV data for fluorinated silicon MIS device. (Same wafer as Figure 2).

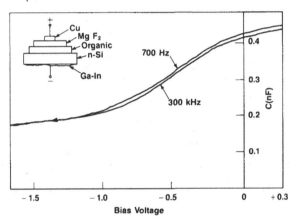

Shown in Figure 3 are high and low frequency capacitance data for the same fluorinated silicon wafer (n-type, 5 Ω-cm) whose photoconductive decay is shown in Figure 2. The area of the metallic gate electrode was 0.02 cm^2. The capacitance at positive bias (accumulation) should just be a measure of the insulator capacitance. (C_{ins}) We attribute the dispersion between high and low frequency capacitance observed in this bias range to the previously reported frequency dependent dielectric constant of poly-crystalline films of MgF_2 [16]. For negative biases (depletion) some of the dispersion between low and high frequency data may be attributed to the long response time of the surface state density (D_s) and hence the surface state capacitance (C_s) is given by [17]:

$$e\,D_s = C_s = \left(\frac{1}{C_{LF}} - \frac{1}{C_{ins}^{LF}}\right)^{-1} - \left(\frac{1}{C_{HF}} - \frac{1}{C_{ins}^{HF}}\right)^{-1} \tag{3}$$

where HF and LF refer to high and low frequency. From the measured dispersion of the capacitance at a bias corresponding to the majority carrier

Fermi level being near midgap, we obtain an interface density of states near midgap of D_s ~5 x 10^9 states/cm^2-eV, a value consistent with that inferred from the measurement of τ_{ph} (Figure 2).

CONCLUSIONS

To summarize, our measurements demonstrate that fluorine terminated silicon surfaces are, to an extraordinary degree, electrically inert. We have measured midgap surface state densities ~10^{10}/eV-cm^2. As a device technology, the air sensitivity of the fluorinated silicon surface is a potential drawback. However, the gate insulator we have described demonstrates that suitable device stability may be attained. The apparent lack of severe "band tailing" effects, the simple low temperature processing steps, and the use of non-epitaxial insulators alternative to SiO_2 suggest that a silicon interface technology based on fluorine chemistry may for many applications be superior to that based on oxide chemistry.

REFERENCES

1. E. H. Nicollian and J. R. Brews, MOS Physics and Technology (Wiley, New York, 1982).
2. T. M. Buck and F. S. McKim, J. Electrochem. Soc. 105, 709 (1958).
3. H. Ishiwara and T. Asano, Appl. Phys. Lett. 40, 66 (1982).
4. J. M. Phillips, L. C. Feldman, J. M. Gibson and M. L. McDonald, Thin. Sol. Films 107, 217 (1983).
5. T. Asano, H. Ishiwara and N. Kaifu, Jap. J. of Appl. Phys. 22, 1474 (1983).
6. T. P. Smith, J. M. Phillips, W. M.Augustyniak, and P. J. Stiles, Appl. Phys. Lett. (in press).
7. J. Tanguy, Thin Sol. Films 13, 33 (1972).
8. I. Lundstrom, Physica Scripta. 18, 424 (1978).
9. F. R. McFeely, J. F. Morar, N. D. Shinn, G. Landgren and F. J. Himpsel, Phys. Rev. B30, 764 (1984).
10. G. L. Miller, D. A. H. Robinson and S. D. Ferris, Proc. Electrochem. Soc. 78-3, 1 (1978).
11. T. Tiedje, J. I. Haberman, R. W. Francis and A. K. Ghosh, J. Appl. Phys. 54, 2499 (1983).
12. A. S. Grove, Physics and Technology of Semiconductor Devices (Wiley, New York, 1967), Chapter 5.
13. Ref. 1, p. 297
14. J. Singh and A. Madhukar, J. Vac. Sci. Technol. 19, 437 (1981).
15. D. C. Northrop and O. Simpson, Proc. Roy. Soc. A234, 124 (1956).
16. C. Weaver, Adv. Phys. 11, 83 (1962).
17. Ref. 1, p. 332

CORRELATION OF Si-SiO$_2$ INTERFACE ROUGHNESS
WITH MOSFET CARRIER MOBILITY

ZUZANNA LILIENTAL[*], O.L. KRIVANEK, S.M. GOODNICK[**], AND C.W. WILMSEN[**], Center for Solid State Science, Arizona State University, Tempe, AZ 85287.
[*]Now at Lawrence Berkeley Laboratory, University of California, Berkeley, CA 94720
[**]Department of Electrical Engineering, Colorado State University, Fort Collins, CO 80523.

Si-SiO$_2$ interface roughness in MOSFET (metal-oxide-semiconductor field-effect transistor) devices has long been recognized as a cause of carrier scattering[1], and hence a major limitation of carrier mobility in Si-SiO$_2$ inversion layers. Its role is particularly important at high carrier concentrations (high gate voltage V_g), where the charge carriers are confined close to the interface. It has also been suggested that the density of Coulomb centers at the interface increases with increased interface roughness[2]. The roughness may, therefore, also be indirectly limiting the mobility at low carrier densities, where scattering by Coulomb centers predominates.

Previous studies of the interface by high resolution electron microscopy[3] and HREM combined with Hall effect mobility measurements[4] have shown the interface roughness to range around 3Å. The aim of the present work is to attain a precise quantitative correlation of interface roughness and mobility by studying carrier mobility at intentionally roughened Si-SiO$_2$ interfaces. The gate oxides were therefore grown with Cl in the oxidizing ambient, since this is known to produce rough interfaces. Cross-sections for electron microscopy were prepared on the samples for which Hall mobility measurements were done earlier. The cross-sections were examined in a JEM 200CX electron microscope at what is nowadays frequently called "ultra-high resolution". Fig. 1 shows three different segments of the purposely roughened interface (a-c), which display roughness up to 10Å peak to peak. This is about two times larger than the smooth interfaces studied previously[4], one of which is shown in Fig. 1d.

To obtain a more quantitative measure of the roughness, the following procedure was followed[5]. Starting with an actual micrograph of an interface (Fig. 2a) the boundary between the outermost (and weakest) Si lattice protrusions and the amorphous SiO$_2$ was traced (Fig. 2b). The trace was then digitized, and corrections were made for possible shift and misorientation of the adopted system of coordinates so that the average height of the interface trace was exactly zero. Next, a Fourier transform was performed to obtain a power spectrum of the roughness (Fig. 3). The power spectrum was then fitted either with a Gaussian or an exponential curve (the latter generally gave a much better fit), and the measured r.m.s. roughness Δ_m was determined from the area under the fitted curve. The experimentally measured correlation length L_m, representing the average distance between "hills" at the interface, was determined as $L=4.7/$ (power spectrum full-width at half-maximum). For the interface of Fig. 2, this procedure gave $\Delta_m=2.8$Å and $L_m=12.6$Å.

The measured values Δ_m and L_m are, however, not the true values of interface roughness and mean correlation length.

194

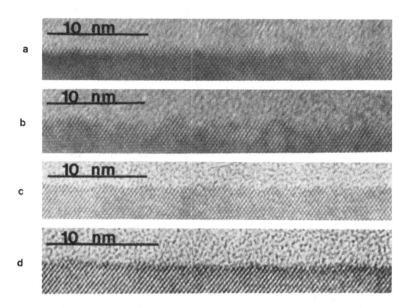

FIG. 1. (a-c) HREM images of intentionally roughened Si/SiO$_2$ interfaces.
(d) smoother Si/SiO$_2$ interface (not intentionally roughened).

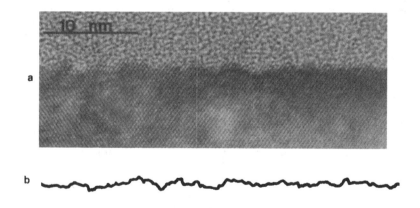

FIG. 2. a) HREM image of an Si-SiO$_2$ interface.
b) trace representing the interface profile.

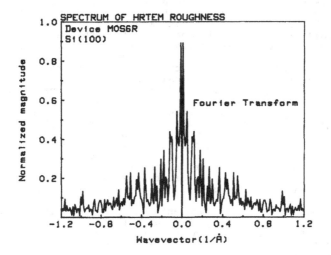

FIG. 3. The fast Fourier transform of the interface profile shown in Fig. 2b.

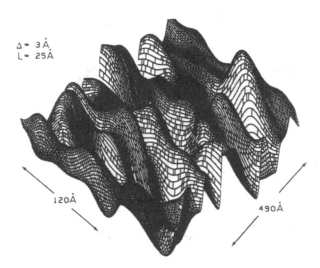

FIG. 4. A two-dimensional plot of a random surface of roughness equivalent to the Si-SiO$_2$ interfaces (vertical scale exaggerated).

This is because the EM micrographs do not show a true profile of the interface, but a projection of a typically 100Å to 200Å thick cross-section. The effect this has can best be appreciated by recalling the silhouette of a distant mountain range with just the peaks of many mountains showing, but none of the valleys revealed in their full depth. We simulated the effect by computer generating random surfaces with a range of Δ and L values, projecting these through thicknesses typical of the EM cross-sections, and determining the resultant Δ_m and L_m. A typical 2-D surface representative of the Si-SiO$_2$ interface used in these simulations is shown in Fig. 4 (with the vertical scale exaggerated).

The projection always results in diminished values of L_m and Δ_m relative to the true L and Δ. The magnitude of the effect depends on the ratio of t/L where t is the specimen thickness. It is negligible for t/L<1, but grows appreciably for the typical t/L values of 5-10. Assuming an average sample thickness of 120Å for the interface of Fig. 2 (judged from the position of thickness fringes in the image), the corrected values of roughness and correlation length become Δ=5Å and L=25Å.

The Hall effect measurements of mobility vs. carrier density were carried out at 52K, and the results were fitted by a theoretical model in which Δ and L are free parameters. The data for the chlorine-roughened sample is shown in Fig. 5, together with a fit for Δ=3.45Å and L=21Å. The roughness scattering is especially important at high carrier concentrations, whereas at low carrier concentrations the mobility is determined predominantly by Coulomb and phonon scattering.

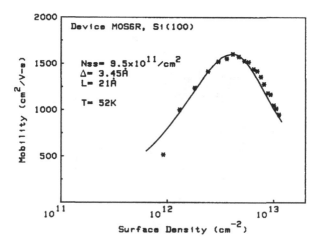

FIG. 5. Measured Hall mobility vs. carrier density (points) and the mobility predicted for roughness parameters Δ=3.45Å and L=21Å (solid line).

The slight difference in Δ determined by Hall measurements ($\Lambda = 3.45\text{Å}$) and by HREM ($\Delta = 5\text{Å}$) is probably due to the short length of the thin portion of the interface in a typical cross-section sample, and hence the limited statistical precision of the HREM result. Of the three rough interface segments shown in Fig. 1, the one actually used for the analysis (1a) is easily the roughest. Comparing the roughness previously determined by HREM for the "smooth" interface with the present rough one gives a factor of 2.1 difference. The mobility-determined roughnesses for the two interfaces give a factor of 1.7. This also shows that the interface segment shown in Fig. 1a and 2 was probably rougher than the average for the whole wafer. Another important effect is that the theory used to model the mobility data does not account for phonon scattering and is therefore exactly valid only at absolute zero.

In conclusion, the structures of the $Si\text{-}SiO_2$ interface as determined from electrical measurements, and from HREM observations, are in agreement within 30%, and the agreement is mainly limited by the statistical error of the HREM result.

We are grateful to Dr. D. Fathy for supplying the micrograph shown in Fig. 1d. This research is supported by NSF grants DMR-8117052, CHE-7916098, and the Office of Naval Research.

REFERENCES

1. Y.C. Cheng and E.A. Sullivan, Surface Science 34, 717 (1973).
2. P.O. Hahn and M. Henzler, J. Vac. Sci. Technology A2(2), 574 (1984).
3. O.L. Krivanek, D.C. Tsui, T.T. Sheng and A. Kamgar, in: the Physics of SiO and Its Interfaces (ed. S.T. Pantelides, Pergamon, NY), p. 356 (1978).
4. S.M. Goodnick, R.G. Gann, J.R. Sites, D.K. Ferry, C.W. Wilmsen, D. Fathy and O.L. Krivanek, J. Vac. Sci. Technol. B1, 803, 1983.
5. S.M. Goodnick, Ph.D. thesis, Colorado State University (1983).

ATOM PROBE ANALYSIS OF NATIVE OXIDES AND THE THERMAL OXIDE/SILICON INTERFACE.

C.R.M.GROVENOR,A.CEREZO and G.D.W.SMITH.
Department of Metallurgy and Science of Materials,Parks Road,OXFORD,UK.

ABSTRACT

The Pulsed Laser Atom Probe has been used to determine for the first time the
stoichiometry of both the native oxide on Si and of the SiO2/Si interface.

INTRODUCTION

The native and thermal oxides of silicon have important influences on the
satisfactory operation of many semiconductor devices. The native oxide is
usually sandwiched between any active silicon component and its metallisation ,
and is also necessarily already present during the initial stages of thermal
oxidation. The possible role of this oxide layer in retarding the chemical
reactions between silicon and metals , or oxygen , are not well understood.
Neither is the stoichiometry of the layer well characterised , although it is
often described as SiOx. Thermal oxides are important components of MOS
technology , and it is now clearly established that the properties of these
devices are controlled to a large extent by the details of charge trapping
close to the SiO2/Si interface [1]. The midgap trap states and the fixed
oxide charges close to the interface are possibly associated with structural
defects such as silicon dangling bonds and oxygen vacancies respectively [2,
8] and the density of these defects is governed by the chemistry of the
interface region. A very wide range of techniques has been used to to study
this interface including Auger Spectroscopy [3,4] , Photoelectron Spectroscopy
(XPS)[5] , Rutherford Backscattering (RBS)[6] , Transmission Electron
Microscopy (TEM)[7] , and Electron Spin Resonance [8]. Certain features of
the structure of the SiO2 interface have been clarified by these
experiments. There is general acceptance of the concept of a silicon rich
'transition layer' at the interface , although there is disagreement amongst
the techniques as to the thickness of the layer ; ranging from no observable
transition region [7] to 0.25nm by XPS [9] to > 1nm in RBS [6] and Auger
Spectroscopy [3]. This disagreement reflects the different depth resolutions
intrinsic to the various techniques and the lack of contrast from changes in
stoichiometry in TEM. Similarly the stoichiometry of this transition layer is
poorly characterised , although chemical bonding states distinct from those of
pure silicon and silicon in SiO2 have been reported [3,4,5,9]. Thus
although the silicon rich nature of the transition layer is clearly shown by
many of these experiments none of them have sufficient depth resolution to
distinguish between a gradual transition from stoichiometric bulk SiO2 to
pure silicon and a distinct interfacial layer. Litovchenko [10] has proposed·
a 3 layer model for the interface where crystalline silicon changes abruptly
to a crystalline oxide , of undefined stoichiometry , and thence to amorphous
SiO2 in less than a nanometre , while other authors propose a more gradual
transition where excess silicon atoms are somehow trapped in the amorphous
oxide [11]. Determination of the most nearly correct model for the interface
would greatly assist in the understanding of the character of the defects
responsible for its electrical character. This paper presents the first
observations of the stoichiometry of silicon native oxides and the SiO2/Si
interface using Atom Probe Microanalysis. This technique provides chemical
information with a depth resolution of one atomic plane spacing , and
excellent lateral resolution , and so is uniquely suited to the study of the
stoichiometry of thin layers and interfaces [12]. Recently the development of
the Pulsed Laser Atom Probe (PLAP)[13] has extended the applicability of this
technique to semiconducting and insulating materials[14].

EXPERIMENTAL DETAILS AND RESULTS.

The PLAP used in this work has been described elsewhere [14]. Silicon Field Ion specimens were prepared from Wacker single crystal ingots , P doped to $10^{18}/cm^3$, by chemical polishing in HNO3/HF/CH3COOH solutions. Native oxides were allowed to grow on both HF dipped samples and on atomically clean specimens , produced by field desorption of the contaminated surface layers in the PLAP , by air exposure at room temperature. The HF dipped specimens were air exposed for about 3 hours , while the field desorbed surfaces were allowed to react for a variety of times between 1 and 70 hours. Thin thermal oxides were grown on HF dipped silicon specimens by heating to 800°C in wet oxygen for 10 minutes. All the data presented in this paper was obtained from analysis close to the axis of randomly oriented specimens.

Native Oxides.

Figure 1 shows a typical PLAP profile through a native oxide layer on an HF cleaned silicon specimen. It can be seen that the stoichiometry of the oxide is very close to SiO at the surface , with indications of a very thin layer of approximate composition $Si_{0.7}O_{0.3}$ before the silicon substrate is reached. Estimation of layer thicknesses from such profiles are not straightforward , but an approximate depth scale is shown which indicates that the oxide has an approximate thickness of 1 – 1.5nm. (Each data point in the profile corresponds roughly to an atomic plane in the oxide.) This estimated thickness agrees quite well with that of the native oxide observed on HF cleaned silicon wafers after 29 days air exposure [15], but the oxide stoichiometry assumed in these experiments (SiO2) is shown here to be suspect. The development of the native oxide with time of air exposure has

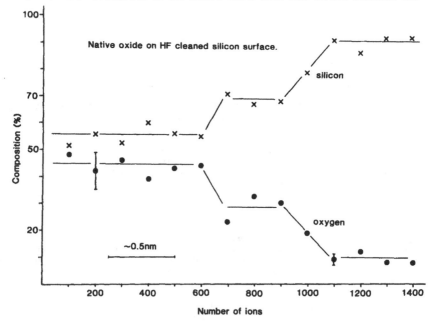

FIGURE 1. PLAP profile through a native oxide layer on silicon produced by 3 hours air exposure after an HF dip.

been studied on samples cleaned in the PLAP until only silicon ions are detected. Figure 2 shows profiles through two native oxides grown under identical conditions but for different times. 6 hours air exposure can be seen to produce an oxide layer with a surface stoichiometry of approximately $SiO_{.6}O_{.4}$. The total thickness of the layer is only 0.5nm. However a longer exposure to the oxidizing environment (70 hours) produces an oxide with surface composition SiO, and indications of a region before the silicon substrate with composition approximately $SiO_{.7}O_{.3}$, exactly as seen on the HF dipped sample. These profiles indicate that the development of a native oxide layer after both an HF dip or desorption cleaning leads to a

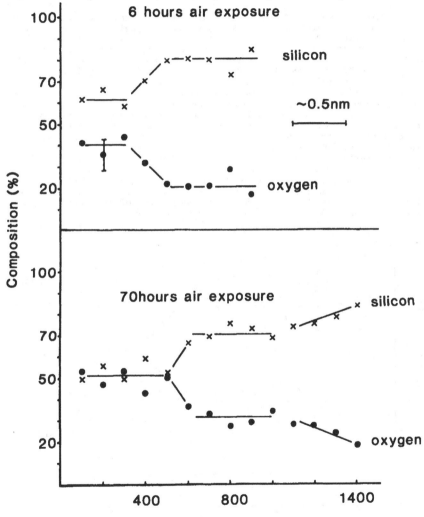

FIGURE 2. PLAP profiles through native oxides produced by air exposure of clean silicon surfaces for 6 and 70 hours.

stable stoichiometric SiO layer and a transition layer of lower oxygen content , but no evidence for any region of composition SiO2. The thickness of the oxide layer is seen to increase as the time of air exposure is extended , and the results may indicate that HF dipping enhances the rate of native oxide growth since only 3 hours air exposure after an HF dip produces an oxide layer as thick as was produced by 70 hours exposure of a clean silicon surface.

Thermal oxide.

Figure 3 shows a detail of the complete profile through a thermal oxide sample , (the remainder of the oxide layer had a composition very close to SiO2 as expected). As the SiO2/Si interface is approached a change in stoichiometry is seen , and it is tempting to interpret this region of the profile as showing a layer of stoichiometric SiO. However from such a profile it is difficult to be sure that the apparent SiO stoichiometry is not an artifact due to collection of ions from a pure silicon and an SiO2 plane at the same time. If the data is presented in the form of a "ladder diagram" where the individual ions are plotted in sequence of detection Figure 4 is obtained. Here the silicon species are shown vertically and the oxygen horizontally , and so the slope of the line gives directly the local stoichiometry of the oxide layer at any point. On such a diagram a gradual transition from SiO2 to silicon [11] , or a blurring of a sharp SiO2/Si interface by ion collection from two atomic planes , will appear as a smoothly curving transition between the oxide and silicon. By contrast a distinct transition phase will be a sharply defined region of slope intermediate between that of the oxide and the pure silicon. From Figure 4 it can be seen that the bulk oxide region has a slope of 1/2 , and that there is a very sharply defined region of unitary slope (SiO).

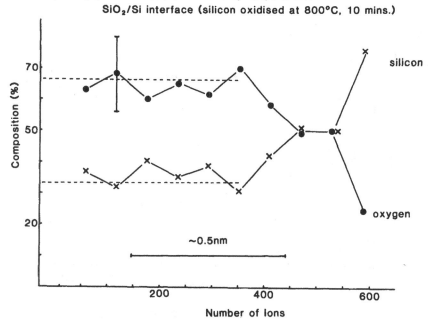

SiO₂/Si interface (silicon oxidised at 800°C, 10 mins.)

FIGURE 3. A detail of the PLAP profile through a thin thermal oxide on silicon grown at 800oC for 10 minutes in wet O2.

before the silicon substrate is reached. The thickness of this region can be
estimated to be no more than 2 atomic planes , or roughly 0.3nm , and its
composition is clearly characterised as SiO. This measured thickness of the
transition layer agrees well with that estimated from XPS experiments [9] ,
but this is the first time that the composition and the sharpness of the
edges of the transition layer have been measured for this interface.

DISCUSSION.

The results presented above demonstrate the unusual potential of PLAP
microanalysis in the determination of the chemistry of thin layers and
interfaces. It has been shown that native oxide layers on silicon develop
with time to give a stoichiometric layer of SiO separated from the silicon
substrate by an incompletely reacted layer with a lower oxygen content. It
has also been shown that the often proposed silicon rich transition layer
between thermal oxide and silicon is also stoichiometic SiO , and that the
thickness of this layer is about 0.3nm. These results lead us to propose that
the 3 layer model of Litovchenko [10] for the SiO2/Si interface is the
most accurate description currently available , and we are able to identify
both the composition and thickness of the thin transitional oxide layer in
this model. The crystalline nature of this layer proposed by Litovchenko
cannot be easily investigated in PLAP experiments , but RHEED experiments
give some evidence for a thin (unidentified) crystalline phase being

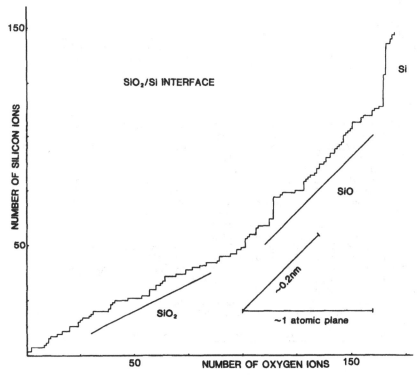

FIGURE 4. Ladder diagram of the interfacial region of Figure 3 showing the
sharp changes in composition that identify the discrete SiO layer.

localised at the interface[16], which may be the SiO interface layer analysed here. The fact that stoichiometric SiO is found in this work both as the major constituent of native oxides , and as the two dimensional phase at the SiO_2/Si interface, can be interpreted as an indication that SiO is a necessary transition phase between bulk SiO_2 and silicon. However whether kinetic or thermodynamic constraints are dominant in determining the composition and thickness of this interfacial phase remains to be determined. Some evidence of a high temperature SiO phase has been found[17] , but only in an amorphous form.

The knowledge of the stoichiometry and the thickness of the interfacial transition layer at the SiO_2/Si interface gained in this work will allow the construction of a detailed atomic model of the interfacial region. In such a model most of the silicon atoms at the interface would be bonded to oxygen atoms , in a manner similar to that illustrated by Litovchenko[10]. This is consistent with the rather low measured density of interfacial trap states relative to the density of interfacial silicon atoms[1]. Oxygen vacancies in the SiO layer could also contribute to fixed oxide charges close to the interface , as proposed for the same defect in SiO_2[2]. The influence of annealing treatments and impurity segregation could be included in this detailed model of the interface structure allowing calculation of the electrical properties of the 'real' SiO_2/Si interface.

ACKNOWLEDGEMENTS.

The authors would like to thank Professor P.B.Hirsch for the provision of laboratory facilities. CRMG gratefully acknowledges an IBM(World Trade) Fellowship , and AC thanks VG Scientific Ltd for a CASE award during this work. The PLAP was developed with the aid of a grant from the Paul Instrument Fund of the Royal Society , and additional support from the SERC.

REFERENCES.

1. Y.C.Cheng, Progress in Surface Science, 8, 181 (1977).
2. T.Sakurai and T.Sugano, J.Appl.Phys., 52, 2889 (1981).
3. C.R.Helms,Y.E.Strausser and W.E.Spicer, Appl.Phys.Lett., 33, 767 (1978).
4. J.F.Wager and C.W.Wilmsen, J.Appl.Phys.,50, 874 (1979).
5. S.I.Raider and R.Flitsch, J.Vac.Sci.Tech.,13, 58 (1976).
6. T.W.Sigmon,W.K.Chu,E.Lugujjo and J.W.Mayer, Appl.Phys.Lett.,24,105(1974).
7. O.L.Krivanek,T.T.Sheng and D.C.Tsui, Appl.Phys.Lett., 32, 407 (1978).
8. P.M.Lenahan and P.V.Dressendorfer, J.Appl.Phys., 54, 1457 (1983).
9. T.Hattori and T.Suzuki, Appl.Phys.Lett.,43, 470 (1983).
10. V.G.Litovchenko, Sov.Phys.-Semiconductors. 6, 696 (1972).
11. B.E.Deal,M.Sklar,A.S.Grove and E.H.Snow, J.Electrochem.Soc.,114,226(1967)
12. R.Wagner, Field Ion Microscopy, Springer-Verlag,Berlin (1982).
13. G.L.Kellogg and T.T.Tsong, J.Appl.Phys. 51, 1184 (1980)
14. C.R.M.Grovenor,A.Cerezo and G.D.W.Smith, Microscopy of Semiconducting Materials (Eds.A.G.Cullis,S.M.Davidson and G.R.Booker) Inst.Phys.,London(1983) p.109.
15. J.H.Mazur,R.Gronsky and J.Washburn, as [14] p.77.
16. B.Agius,S.Rigo,F.Rochet,M.Froment,C.Maillot,H.Roulet and G.Dufour, Appl. Phys. Lett. 44,48 (1984).
17. F.A.Shunk, Constitution of Binary Alloys Second Suppl. McGraw-Hill,New York 1969 p.572.

EPITAXIAL SILICON ON YTTRIA-STABILIZED CUBIC ZIRCONIA (YSZ) AND SUBSEQUENT OXIDATION OF THE Si/YSZ INTERFACE

L.M. MERCANDALLI, D. PRIBAT, M. DUPUY*, C. ARNODO, D. RONDI, D. DIEUMEGARD
Thomson-CSF/LCR,Domaine de Corbeville,B.P.10,91401 Orsay Cedex,France
*LETI/CRM, 85X, 38041 Grenoble Cedex,France.

ASTRACT

(100) single crystal silicon films have been deposited onto (100) oriented Yttria-Stabilized Zirconia (YSZ) substrates by pyrolysis of SiH_4 at $\sim 980°C$.

The as deposited epitaxial silicon films have been characterized by Reflexion High Energy Electron Diffraction and Transmission Electron Microscopy techniques.

The as deposited silicon films have also been oxidized by oxygen transport through the substrate, resulting in a Si(100)/amorphous SiO_2/YSZ(100) structure in which the most defective part of the epitaxial silicon deposit has been eliminated. The oxidized interfaces (with SiO_2 thicknesses in the 2000 Å range) have then been characterized by Transmission Electron Microscopy in order to assess the improvement in crystalline quality. Electrical measurements have also been performed on MOS-Hall bar structures.

1. INTRODUCTION

Within the last two years, extensive studies have been performed on the silicon/Yttria-Stabilized cubic Zirconia (YSZ) composite, in order to obtain high quality epitaxial silicon films on single crystalline insulating material [1-5]. The Si/YSZ structure is a potential alternative for Silicon On Sapphire (SOS) in applications requiring radiation-hard and high speed integrated devices.

When comparing the Silicon On Sapphire and silicon on cubic zirconia (YSZ) composites, the differences in lattice parameters between silicon and substrates are in the two cases of the same order of magnitude. However in the Si-YSZ system this difference (varying with the Y_2O_3 content between ~ 4.2 and 5.6%) is isotropic. As in the SOS structure the resulting strain in Si cannot be relaxed by interfacial misfit dislocations and the defects mainly consist in twins, their density being maximum at the Si/substrate interface. These defects are known to degrade the electrical properties of the submicron thick epitaxial silicon films, thus impeding the manufacture of bipolar devices in such heteroepitaxial systems [6].

Taking advantage of both the oxygen ion and electronic conductivities of YSZ at elevated temperatures (YSZ still being an excellent insulator at room temperature), we have recently shown [4,5] that thermal SiO_2 layers could be grown at the Si-YSZ interface, eliminating by silicon comsumption the most defective part of the epitaxial Si films.

In this paper, emphasis is put on the structural characterizations of the epitaxial silicon films, both after deposition and interface oxidation. The crystallographic investigations have been performed using Reflexion High Energy Electron Diffraction (RHEED) and Transmission Electron Microscopy (TEM). In complement preliminary electrical characterizations have been carried out on MOS-Hall bar structures.

2. RESULTS and DISCUSSION

We have been working with YSZ substrates doped with 18.0, or 21.0 mol percent (m/o) Y_2O_3.

The process for substrate preparation prior to silicon deposition has been described elsewhere [4]. The silicon deposits were performed in a "pan cake" industrial CVD reactor by pyrolysis of SiH_4 in Pd purified H_2 carrier gas. The deposition temperature was typically 980°C and the deposition rate $\sim$ 1.5 µm.mn^{-1} [4]. An in situ annealing stage at higher temperature prior to the deposition(typically 1175°C during 90 mn) was found to be necessary in order to obtain subsequent single crystal silicon growth at 980°C [1,3,4].

Fig. 1 shows the RHEED patterns obtained on a $\sim$ 6500 Å thick epitaxial silicon film deposited onto a 18.0 m/o doped YSZ substrate. These patterns have been obtained with a 100 keV primary electron beam falling in grazing incidence ($\sim$ 1°) onto the silicon surface. In these experimental conditions, the probed depth is $\sim$ 500 Å from the top surface of the silicon. The two RHEED patterns, taken after a 45° rotation in the silicon surface plane and indexed as (011) and (010), indicate that the silicon is of (100) surface orientation, with a rather high surface crystal quality, since Kikuchi lines are clearly evidenced. The punctual shape of the diffraction spots near the central spot indicates that the surface is grainy in agreement with SEM observations [4]. Note on Fig. 1b the extra spots at a distance R from the central one which are characteristic of microtwins of the Σ_3 type. This situation is quite similar to the one of silicon films on sapphire and particularly Σ_3 microtwins extend up to the surface.

TWIN SPOTS

R

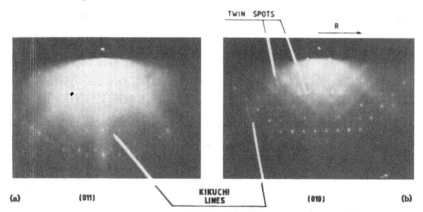

(a)   (011)   KIKUCHI LINES   (010)   (b)

Fig. 1 - RHEED patterns of a $\sim$ 6500 Å thick epitaxial Si film onto a 18 m/o Y_2O_3 doped YSZ substrate.

In order to gain more insights on the crystalline quality of the overall silicon film and especially at the Si-YSZ interface, we have performed TEM analysis on cross sections of the Si-YSZ structure. These cross sections have been prepared by mechanical thinning followed by ion milling with 6 keV Ar ions [7]. So far, we have not precisely determined the ion milling parameters (high voltage, tilt angle of the samples) leading to the same sputtering yield for both silicon and YSZ; it follows that the silicon films thin down more rapidly than YSZ. As a consequence the silicon layer and the YSZ substrate cannot be observed at the same time. Nevertheless, it should be pointed out that the YSZ substrates, when observed, appeared featureless. Fig. 2 displays a TEM cross section of a $\sim$ 7000 Å thick epitaxial silicon film deposited onto a 18 m/o doped YSZ substrate

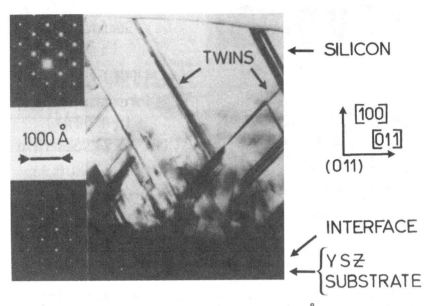

Fig. 2 - Transmission Electron Micrograph of a ∿ 7000 Å thick epitaxial Si
film onto a 18 m/o Y₂O₃ doped YSZ substrate.

with, in inserts, diffraction patterns for both silicon and YSZ. These
patterns are identical but the one relative to silicon shows that the film
is highly twinned; this observation is in agreement with the TEM micrograph which shows that the Σ_3 type twins extend up to the surface as already evidenced by RHEED. It is clear, from Fig. 2, that the defect density is much higher at the Si-YSZ interface, situation similar to the one of
silicon films on sapphire [8]. High resolution (∿ 3 Å) TEM micrographs
have been performed on the first 1000 Å of silicon from the interface [7];
in this region the Σ_3 type twins were of course observed, but in addition
Σ_9 twins were also found together with grain boundaries [7]. Moreover,
small crystallites (thickness ∿ 100 Å) of random orientations were also
found at the very interface. The randomly oriented crystallites together
with the Σ_9 twins rapidly disappear as the thickness increases probably
because they grow slowly due to the strong interfacial energy associated
with the boundaries either with the (100) silicon matrix or the Σ_3 twins
[9].

It is also clear according to these first observations that the overall crystalline quality of the epitaxial silicon films may greatly be enhanced if the ∿ 1000 Å thick interfacial silicon layer can be eliminated.

We have recently shown that the elimination of this interfacial layer
was indeed possible, using the peculiar high temperature oxygen ion and
electronic transport properties of YSZ material [4,5]. In other words, the
interfacial silicon layer can be thermally oxidized by oxygen transport
through the YSZ substrates, the result being a Si(100)/Amorphous SiO₂/YSZ
(100) structure. We have shown that in our conditions (T = 1150°C; 10^{-1}
atm < Po₂ < 2 atm; t < 40 h) the SiO₂ thickness (in the 1500 Å range) at
the Si-YSZ interface increases linearly as a function of both time and oxygen pressure [4]. For instance, Fig. 3 shows a Scanning Electron Micrograph obtained after thermal oxidation of a ∿5000 Å thick epitaxial silicon
film deposited onto a 21.0 m/o doped YSZ material. The oxidation has been

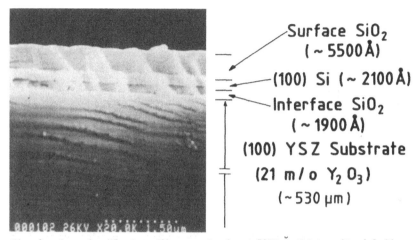

Surface SiO₂
(~ 5500Å)

(100) Si (~ 2100Å)

Interface SiO₂
(~ 1900Å)

(100) YSZ Substrate
(21 m/o Y₂O₃)
(~ 530 µm)

Fig. 3 - Scanning Electron Micrograph of a ∿5000 Å thick epitaxial Si film after thermal oxidation of the Si/YSZ interface (21 m/o Y₂O₃ doped YSZ substrate) (see text for details).

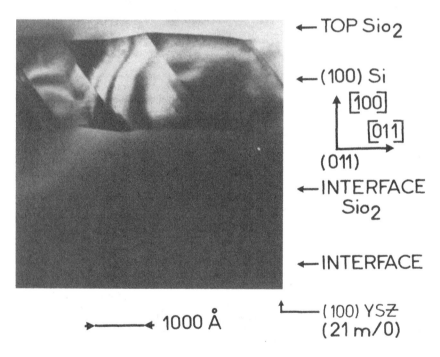

← TOP Sio₂

← (100) Si

↑ [Ī00]
[0Ī1]

(0Ī1)

← INTERFACE Sio₂

← INTERFACE

← (100) YSⱫ
(21 m/0)

•———• 1000 Å

Fig. 4 - Transmission Electron Micrograph of the sample of Fig. 3. Note that the only remaining defects after consumption of ∿1000 Å of Si are Σ₃ type twins.

performed at 1150°C in a ∿0.3 atm O_2 ambient for   40 h. The top surface
of the silicon deposit has also been oxidized, since it had not been encapsulated (pyrolytic Si_3N_4) prior to oxidation; we have grown ∿2000 Å of
SiO_2 at the interface corresponding to the consumption of ∿1000 Å of silicon.

Fig. 4 shows a TEM cross section obtained on the sample of Fig. 3.
YSZ is still too thick to be observed. Σ_3 twins are practically the only
remaining defects in the silicon film. In addition, it can be seen that
the oxidation velocity depends on the silicon orientation. (The twins
oxidize more rapidly than the matrix). Hence the resulting silicon film is
slightly undulating with an amplitude in the 100 Å range. The most significant feature is that the Si/SiO_2 interface is quite sharp and appears
of good quality. More information will soon be obtained by observing this
interface using the high resolution technique.

It is worth pointing out that despite the long annealing time and
elevated temperature involved here, SiO_2 does not crystallize, although
the interface SiO_2 layer is sandwiched between (100) Si and (100) YSZ.
Also note that apparently no compound is formed between SiO_2 and YSZ
substrate.

In addition   we have performed electrical measurements on MOS-Hall
bar structures (50 µm wide, 200 µm long). The test vehicle used in this
study is shown on Fig. 5. We started with epitaxial silicon films about
6000 Å thick. A 1000 Å thick thermal SiO_2
layer was first grown at the silicon surface
in order to protect the top surface of the
silicon films ; a 3000 Å thick pyrolytic
SiO_2 cap was subsequently deposited at 870°C
on the top of the thermal SiO_2 layer,
followed by a 1000 Å thick Si_3N_4 film also
deposited at 870°C. After these deposition
steps, the Si-YSZ interface was oxidized at
1150°C during 10h in a 1 atm oxygen ambient,
resulting in a ∿ 2000 Å thick interfacial
SiO_2 layer. The Si_3N_4 /pyrolytic SiO_2/thermal
SiO_2 cap was then removed by plasma etching
and the resulting $Si/SiO_2/YSZ$ structures were
processed according to a conventional MOS
fabrication sequence. We have been using
either n^+ or p^+ ohmic contacts (drain-
source), in order to determine both the type
and the average mobility of the silicon

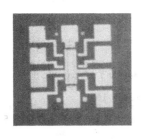

Fig. 5 - Test vehicule used
for MOS characterizations.

layers. From the $I_{DS}(V_{DS})$ characteristics (with either n^+ or p^+ ohmic
ohmic contacts) the silicon films were found to be n type. The breakdown
voltages of the p^+/n diodes were found to be in the 200-220 V range which
corresponds to a residual impurity concentration of 3-4 10^{15} cm^{-3}. The
average conductivity of the silicon layers has been obtained from resistance measurements performed on n^+/n configurations.

From these data the mean drift mobility of the n type silicon films is
evaluated to be at least 550 cm^2 V^{-1} s^{-1}.

It should be pointed out that although we have obtained saturation in
the MOSFETs channel, we have not presently been able to control the saturation current through the gate potential, probably because the gate oxides
(with thicknesses of ∿1000 Å) were leaky.

3. CONCLUSIONS

Heteroepitaxial silicon films on Yttria-Stabilized cubic Zirconia have
been characterized both after deposition and interface oxidation. The
Si/YSZ composite differs from all the other Si/bulk insulator systems
studied so far, because it allows oxidation of the Si/substrate interface

after silicon film deposition. Therefore the crystalline quality of silicon on YSZ after interface oxidation is potentially better that the one of SOS films since the most defective part of the silicon films may be removed by the oxidation process in the former case.

Our first electrical measurements indicate that the mean drift mobility of electrons in MOS channels is at least of 550 cm^2 V^{-1} s^{-1} after oxidation of the Si/YSZ interface.

ACKNOWLEDGEMENTS

The authors would like to thank M. Croset and J. Siejka for fruitfull discussions.

N. Nouailles is also acknowledged for the SEM work and V. Schwob for her technical assistance.

REFERENCES

1. I. Golecki, H.M. Manasevit, L.A. Moudy, J.J. Yang and J.E. Mee, Appl. Phys. Lett. 42 (1983) 501.
2. V.A. Loebs, T.W. Haas and J.S. Solomon, J. Vac. Sci. Technol. Al (1983) 596.
3. H.M. Manasevit, I. Golecki, L.A. Moudy, J.J. Yang and J.E. Mee, J. Electrochem. Soc. 130 (1983) 1752.
4. D. Pribat, L.M. Mercandalli, J. Siejka and J. Perrière, submitted to J. Appl. Phys.
5. D. Pribat, L.M. Mercandalli, M. Croset, D. Dieumegard and J. Siejka, Mat. Lett. 2 (A and B) (1984) 524.
6. I. Golecki in: Proceedings of the Symposium on Comparison of Thin Film Transistors and SOI Technologies, 1984 Spring Meeting of the MRS. Albuquerque, N.M., February 1984.
7. M. Dupuy, J. Microsc. Spectrosc. Electron. 9 (1984) 163.
8. M.S. Abrahams, J.L. Hutchison and G.R. Booker, Phys. Stat. Sol. (a) 63 (1981) K3; M.S. Abrahams and C.J. Buiocchi, Appl. Phys. Lett. 27 (1975) 325.
9. H.J. Möller, Philo. Mag. A, 43 (1981) 1045.

SILICON ON SAPPHIRE OF SINGLE CRYSTAL QUALITY OBTAINED BY DOUBLE SOLID
PHASE EPITAXIAL REGROWTH

M. A. PARKER*,+ R. SINCLAIR* AND T. W. SIGMON**
* Materials Science Dept., Stanford University, Stanford, CA 94305
+ Materials and Process Engineering Dept., IBM, Rochester, MN 55901
** Electrical Engineering Dept., Stanford University, Stanford, CA 94305

ABSTRACT

    A dual-implantation and annealing procedure has been refined which
results in silicon films of single crystal perfection on sapphire substrates. This double solid phase epitaxy (DPSE) procedure consists of
two self-implantation steps that serve to amorphize the defect structure
of the parent silicon film with subsequent annealing steps employed to
recrystallize the amorphized structure. High resolution cross-section
transmission electron microscopy (HRXTEM) and ion channeling have been
employed to study the microstructure of the silicon films, the sapphire
substrates, and the interface. The resulting perfection of the silicon
layer is shown to be a sensitive function of the first self-implant
energy, dose, and silicon film thickness. Also, a correlation has been
established between microtwin density and dechanneling profiles. HRXTEM
of microtwins in silicon on sapphire (SOS) starting material demonstrates
contrast features identified as rotational Moire' fringes between the
microtwins and the matrix by means of optical diffractometry. Best
results are achieved for the first implant energy of 170 keV ^{28}Si$^+$. Thus,
the DSPE process is shown to give optimal results in defect reduction.

INTRODUCTION

    Silicon on sapphire (SOS) films are useful for the fabrication of
very high speed integrated circuits (VHSIC) as well as radiation hardened
devices. The benefits for devices made from this material arise primarily
from the isolation provided by the insulating sapphire substrate. However, SOS films possess high defect densities due to the lattice mismatch
between sapphire and silicon. These defects have been shown to be microtwins, stacking faults and dislocations, are electrically active, and act
as scattering centers to reduce the mobility of the charge carriers in
the silicon. An associated reduction of carrier velocity results that
limits the application of SOS films for VHSICs [1]. Hence, the elimination
of these crystal defects through subsequent wafer processing is very
important. A double solid phase epitaxial (DSPE) regrowth process which
consists of two self-implant amorphization (SIA) and thermal annealing
treatments has significantly reduced defect densities as shown by ion
channeling experiments [2,3]. In this work, high resolution cross-section
transmission electron microscopy (HRXTEM) is used to give explicit evidence for the reduction of defects in these films after such treatments.
Furthermore, a correlation is established between dechanneling data and
the microtwin density which was estimated on the basis of a stereological
analysis of conventional cross-section transmission electron microscopy
(CXTEM). Thus further light has been shed on the optimization of processing parameters for the improvement of these films, as well as the interpretation of data that is obtained by these complimentary analytical
techniques.

EXPERIMENTAL DETAILS

The starting wafers were .3μ (001) chemically vapor deposited (CVD) silicon on (1̄102) sapphire substrates (Union Carbide). These films were first given a self-implant amorphization (SIA), i.e. $^{28}Si^+$ was implanted to amorphize the regions of high defect density in the silicon film using ion energies of 150, 170, or 190 keV at a dose of 1×10^{15} cm^{-2}. The subsequent annealing treatment was at 550°C for 3 hrs, followed by 1050°C for 1 hr which was followed by a second SIA (ion energy 100keV at a dose of 2×10^{15}cm^{-2}) with a further anneal (550°C for 3 hrs, followed by 1050°C for 72 hrs). The first SIA serves to amorphize the region of highest defect density close to the silicon-sapphire interface; the subsequent anneal results in the regrowth of this amorphous silicon layer from the surface portion of the film, which acts as a seed crystal of lower defect density and is not amorphized by the implant. The energy of this first implant determines the peak in the damage profile; it is desirable to choose an energy that amorphizes the silicon at the location where its defect density is the highest without unduly damaging the sapphire, since this can lead to significant doping of the silicon by aluminum. The second SIA amorphizes the surface layer of the silicon; the subsequent anneal results in the regrowth of the surface from a more nearly perfect underlying silicon layer that then acts as the seed crystal. Thus with two passes of SIAs, DSPE eliminates most of the defects in the silicon film and serves to homogenize the bulk distribution of any aluminum impurities that may have been introduced during the first higher energy implant [2,3].

For ion channeling, samples were oriented with the incident 2.2 MeV $^4He^+$ ion beam along the [001] zone axis (see Fig. 1a) [2]. Dechanneling spectra were obtained for four samples, viz., the starting material, and the three samples given the DSPE treatment for the initial SIA energies of 150, 170, or 190 keV. An advantage of the Rutherford backscattering (RBS) and ion channeling technique is the relative ease of sample preparation. If precautions are taken to assure the cleanliness of the sample surface by suitable decontamination and cleaning procedures, spectra can be obtained from samples with minimal further preparation.

Transmission electron microscopy (TEM) compliments ion channeling in that it provides further detailed information about the nature of the defects giving rise to dechanneling, but at the cost of extensive sample preparation. HRXTEM and CXTEM micrographs were obtained from the same four samples examined by ion channeling. A Philips EM400 microscope, equipped with the "super-twin" objective lens, was used at 120 keV with alignment of the electron beam along the silicon [110] zone axis (see Fig. 1b). It is worth mentioning that the orientation of the films for the cross-section TEM (XTEM) studies differs from that employed to obtain the dechanneling patterns. Cross-section samples were prepared by a standard procedure with some modifications for the specific case of SOS [4]. The wafers were either scribed and broken, or cleaved parallel to the wafer flat. This assured the correct crystal orientation of cross-sections for HRTEM experiments, i.e. parallel to the [110] zone axis of the silicon and the [2̄021] zone axis of the sapphire with the orientation relationship of (001) silicon on (1̄102) sapphire [6]. However, it is found experimentally that the microorientation at the interface differs slightly from the overall macroorientation of the SOS film [5]. So that it is difficult to find regions where the [110] zone axis of silicon and the [2̄021] zone axis of sapphire are exactly parallel. Further modifications to the cross-section sample preparation procedure typically employed for silicon, were the use of diamond abrasive for the various grinding, polishing, and dimpling operations. This was necessitated by the greater hardness of the sapphire substrate.

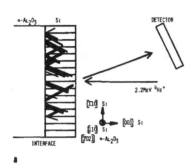

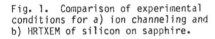

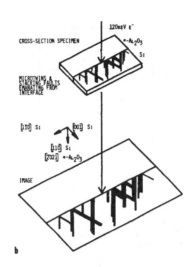

Fig. 1. Comparison of experimental
conditions for a) ion channeling and
b) HRTXEM of silicon on sapphire.

RESULTS OF ION CHANNELING AND CXTEM

Previous ion channeling work has demonstrated that the starting
material has a high density of defects that increases as the interface
between silicon and sapphire is approached (see Fig. 2) [2]. Ion channeling of CVD SOS samples that were given the dual implant and annealing
treatment demonstrates that an initial implant energy of 170keV was
successful in eliminating virtually all the defects within the film, i.e.
the dechanneling pattern is nearly identical to that of a single crystal
float zone (FZ) silicon wafer. (Note the low surface peak for 170keV and
190keV in Fig. 2.) However, at the lower implant energy of 150keV, there
was evidence of residual defect density near the sapphire interface; this
is seen as a rise in dechanneling counts near the silicon-sapphire interface. At the highest energy of 190keV, the dechanneling pattern is essentially identical to that for the 170keV implant. However, significant
out-diffusion of aluminum into the silicon as indicated by secondary ion
mass spectrometry (SIMS) data has been observed, since the higher energy
implant deposits more energy in the underlying sapphire substrate with a
concomitant increase in the number of displacements per atom (DPAs) and
aluminum interstitials.

Although the ion channeling data indicates the presence of defects
in the starting material with an increase of their density near the
interface, the nature of the defects giving rise to the dechanneling is
unclear (see Fig. 2). Hence, CXTEM was performed to gain further
information about these SOS films. CTEM of the starting materials shows
a high density of planar defects. Selected area diffraction (SAD) and
dark field TEM have revealed that these defects are primarily microtwins.
This is a well established observation [5,7-11]. The density of these
microtwins as a function of depth into the film was estimated by counting
the number, and measuring the length, of the regions of dark contrast on
bright field micrographs in ~30nm increments below the film surface.
Thus, a rough correlation has been established between the defect density
indicated by dechanneling and the density of microtwins indicated by
CXTEM (see Fig. 2). This gives only an estimate and improved measures of

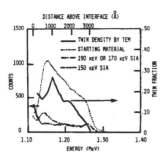

Fig. 2. Comparison of defect density obtained by CXTEM of starting material (solid line) and by ion channeling of starting material, as well as that given DSPE treatments (broken lines).

the defect density can be obtained by dark field TEM of the diffracted beams that correspond to the differently oriented microtwins parallel to the [100] zone axis. Furthermore, corrections for the taper of the foil thickness should be made by the convergent beam electron diffraction technique which is the most accurate method.

In the case of films that were given a DSPE treatment, the defect density was too sparse to make a rough estimate of the defect density as a function of depth within the film from CXTEM. Thus, a correlation with the ion channeling data was precluded in this case. However, we have reported elsewhere that CTEM of the 150keV SIA sample indicates the presence of some residual defects near the interface and within the silicon film [12]. The density of these defects is higher in the vicinity of the interface. CTEM of the 170keV SIA sample shows the absence of any defects in the silicon film. And, CTEM of the 190 keV implant shows linear defects with diffuse contrast and an orientation similar to the microtwins of the starting material. However, the density of such defects is exceedingly low. These observations support the conclusions drawn from the inspection of the dechanneling patterns. To obtain correlations of defect density between the two techniques in these cases will require the use of the more detailed analytical procedures that were mentioned earlier.

RESULTS OF HRXTEM

HRXTEM was performed to gain further information about the SOS films. HRTEM of the starting material reveals that the defects are primarily microtwins with some stacking faults also present. It is easy to confuse the image with that of arrays of stacking faults on every third {111} plane of the silicon structure because of the modulated contrast on every third {111} plane within these microtwinned regions (see Fig. 3a). However, optical diffractograms taken of these regions reveal diffraction patterns characteristic of those expected from the dynamical theory for Moiré fringes [13]. There are satellite spots associated with the double diffraction expected for rotational Moiré fringes (see Fig. 4). The spacing of $\Delta_g = 1/3 g_{111}$ gives rise to fringe contrast that is periodic on every third {111} plane. It is also noted from the image that the contrast of these Moiré fringes is modulated and reverses in moving away from the interface as the film thickness and the twin thickness change. The interface between the silicon and the sapphire is abrupt to within a few atomic planes. There is a region of altered contrast a few planes thick at the interface whose exact identity is unclear and may simply be a phase contrast effect associated with a slight misorientation

Fig. 3. HRXTEM micrographs of a) starting material and b) a representative SOS film given the DSPE treatment, i.e. the 150keV SIA treated film.

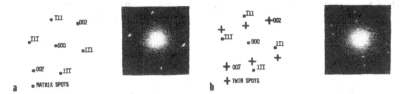

Fig. 4. Optical diffractograms and the expected electron diffraction patterns for a) the matrix (M) and b) the matrix overlaying a twin (T&M).

of the interface between the two crystals with respect to the electron beam,i.e. the beam is slightly non-parallel to the interface [14]. However, no thick intermediate compound phase appears to be present, as evidenced by the continuity of lattice fringes to the near vicinity of the interface, although this does not rule out the possibility of a layer with altered bonding between the silicon and sapphire that may be a few atomic layers thick (see Fig. 3a).

HRXTEM of the samples given the DSPE treatment indicates that they are relatively free of defects [12]. In the case of the 150keV SIA sample, some residual defects remain; these are dislocation loops and microtwin remnants. However, broad regions of the interface are found free of defects (see Fig. 3b). The silicon-sapphire interface of this sample again shows it is abrupt within a few atomic layers; there is also some evidence of ledges at the sapphire substrate. In the case of the 170keV and the 190keV SIA samples, HRTEM of the film shows the character of single crystal silicon without dislocations, stacking faults, and microtwins similar to that of the 150keV SIA sample. HRTEM of the interface reveals again that it is abrupt within a few atomic layers. However, HRTEM of the underlying sapphire substrate adjacent to the interface shows some disruption of the regular pattern of the sapphire structure image. This is probably due to some implantation damage of the substrate because of the increased range of the higher energy ions. In all cases, no amorphous layer was found at the interface as has been observed in the case of samples which are given an e-beam anneal, instead of the thermal anneal employed in this study [11].

CONCLUSIONS

In conclusion, it has been found that the defect density of SOS films given a DSPE treatment is low and approaches that of single crystal silicon in the case of samples that are given 170 and 190keV SIA treatments. The identity of the defects in the CVD SOS starting material is confirmed as primarily microtwins with some stacking faults and dislocations. A correlation has been made between microtwin density and dechanneling as measured by CXTEM and ion channeling, respectively. Optical diffractograms of contrast fringes appearing on every third {111} plane indicate that they are rotational Moiré fringes due to the microtwins. The nature of the SOS interface for both the starting material and the samples given the DSPE treatment is abrupt within a few atomic layers. There is no evidence for an amorphous layer at the interface as in the case of e-beam annealed samples. Some residual defects exist in the low energy, 150keV, SIA samples which are residual microtwins and dislocation loops and in the high energy, 190keV, samples, which are dislocations. Also, some sapphire substrate damage was evident in the higher energy, 170keV, SIA film, but not as severe as occurs with higher doses. Thus, an optimal method for eliminating defects in SOS has been found and characterized by the complimentary techniques of ion channeling and HRXTEM.

ACKNOWLEDGEMENTS

M. A. Parker wishes to thank IBM Corp. which made possible this research by providing an IBM fellowship, and also R. Reedy of Naval Ocean Systems Center, San Diego, who provided the material for analysis.

REFERENCES

1. G. W. Cullen, "The preparation and properties of chemically vapor deposited silicon on sapphire and spinel." Proc. 1st Intl. Conf. on Crystal Growth and Epitaxy from the Vapor Phase, Zurich, 107 (1970).
2. L. A. Christel, R. E. Reedy, and T. W. Sigmon, Appl. Phys. Lett., 42, 707 (1983).
3. T. Yoshii, S. Taguchi, T. Inoue and H. Tango, Japan. J. Appl. Phys., 21 (Supl. 21-1), 175 (1982).
4. J. Bravman and R. Sinclair, Jour. Elec. Micro. Techs., 1, 53 (1984).
5. K. A. W. Carey, Silicon on Sapphire: An Investigation of the Defect Structure, Ph.D. Thesis, Stanford (1981).
6. T. J. La Chapelle, A. Miller and F. L. Morritz, Prog. Solid State Chem., 3, 1 (1964).
7. M. S. Abrahams and C. J. Buiocchi, Appl. Phys. Lett., 27, 325 (1975).
8. F. A. Ponce, Defects in Semiconductors, edited by J. Narayan and T. Y. Tan (North-Holland, New York, 1980), pg. 285.
9. J. Amano, J. Aranovich, K. W. Carey and F. A. Ponce, J. Appl. Phys. 54, 4414 (1983).
10. J. L. Hutchinson, G. R. Booker and M. S. Abrahams, Microscopy of Semiconducting Materials 1981, edited by A. G. Cullis and D. C. Joy (Institute of Physics, London, 1981), p. 139.
11. D. J. Smith, L. A. Freeman, R. A. McMahon, H. Ahmed, M. G. Pitts and T. B. Peters, J. Appl. Phys., 56, 2207 (1984).
12. M. A. Parker, T. W. Sigmon and R. Sinclair, Appl. Phys. Lett., submitted.
13. P. Hirsch, A. Howie, R. B. Nicholson, D. W. Pashley and M. J. Whelan, Electron Microscopy of Thin Crystals, Krieger, New York 1977.
14. J. C. H. Spence, Experimental High-Resolution Electron Microscopy, Clarendon Press, Oxford 1981.

Strained-Layer Superlattices

STRAINED-LAYER SUPERLATTICES AND STRAIN-INDUCED LIGHT HOLES*

G. C. OSBOURN
Sandia National Laboratories, Albuquerque, New Mexico 87185

ABSTRACT

The capability of growing high quality strained-layer superlattices (SLS's) from lattice-mismatched semiconductors opens up new opportunities in both basic and applied material sciences. This flexibility in the choice of SLS layer materials allows the study of quantum well effects in a variety of new and interesting heterojunction systems. In addition, the large tetragonal distortions of the SLS layers allow these structures to exhibit unique features which are not found in bulk semiconductors. An exciting example is the possibility of tailoring certain SLS materials so that they exhibit small two dimensional hole effective masses and enhanced low field hole mobilities at low temperatures and low carrier concentrations. Band structure calculations indicate that the layer strains, the direct band gap magnitudes of the layers, the valence band offset, and the hole energies all can play a role in determining these small two dimensional hole effective mass values.

INTRODUCTION

Strained-layer superlattices (SLS's) are high quality multilayered structures grown from lattice-mismatched materials. The layers are kept sufficiently thin so that the mismatch is totally accommodated by elastic layer strains rather than by misfit dislocations at the SLS interfaces. A number of research groups have now studied the crystalline quality and material properties of various SLS systems [1-13]. Perhaps the most important aspect of SLS materials is the capability of independently tailoring a number of SLS material properties [14-15]. This tailorability results from the flexible choice of layer materials and thicknesses and from the effects of the zone folding, layer strains, and quantum size effects [15].

SLS tailorability allows a number of conventional quantum well materials and device studies to be extended to a variety of new mismatched material systems [16-19]. In addition, the large layer strains in these structures allow them to exhibit a number of unique features which have no analogue in bulk semiconductor materials. Examples of these unique features include the capability of independently varying the band gap and lattice constant of ternary SLS's [14] and the use of layer strains to reduce the band gaps of small gap III-V SLS's [20]. Another interesting example is the strain modification of the SLS valence band structure [2,21]. The (100) uniaxial component of the layer strain in (100) SLS's splits the bulk $m_J = \pm 3/2$, $m_J = \pm 1/2$ valence band degeneracy and causes a significant modification of the in-plane valence band dispersion. These valence band effects are discussed in detail in the next section.

STRAIN-INDUCED LIGHT HOLES

The "heavy hole" $m_J = \pm 3/2$ valence bands in biaxially strained III-V materials have small two-dimensional effective masses near the Brilloin zone center which are nearly independent of the strain magnitude. The "light

*This work performed at Sandia National Laboratories supported by the U.S. Department of Energy under contract number DE-AC04-76DP00789.

hole" $m_J = \pm 1/2$ bands in strained III-V materials on the other hand, have two distinct two dimensional effective masses near the zone center which depend on the strain magnitude. Layers which are compressed in the plane have the $m_J = \pm 3/2$ bands with the small in-plane masses as the uppermost valence bands. These bands can be preferentially populated by the holes in SLS's which have valence band offsets appropriate for hole confinement in the biaxially compressed SLS layers [21]. These SLS materials could exhibit improved low-field hole mobilities at low temperatures and might be useful for III-V complementary logic applications [22]. Values of the in-plane hole mass at the zone center for various bulk semiconductors under biaxial compressions from Ref. 21 are shown in Fig. 1. These masses are strong functions of the Γ gap of the material.

It is clear from the figure that these zone center hole masses are significantly less than the usual bulk density of states heavy hole masses ~ 0.4 m_e. Recent Shubnikov-deHaas studies of InGaAs/GaAs SLS's have indicated that the hole masses can be as small as 0.13 m_e [23]. Tight binding band structure calculations of the cyclotron hole masses in a variety of SLS structures as a function of hole energy have been carried out. The details of the tight binding model and the SLS material parameters will be described elsewhere. Results for several model SLS structures with identical layer materials and thicknesses with different values of layer strains are shown in Fig. 2. These model structures were chosen to have hole confinement in biaxially compressed layers and were also chosen to exhibit unstrained bulk cyclotron hole masses $m_h^*(m_J = \pm 3/2) \sim 0.34$ and $m_l^* = \pm 1/2)$ ~ 0.064 at the zone center of the hole-containing layers. Structures of this type were chosen in order to independently examine the effects of the $m_J = \pm 3/2$, $m_J = \pm 1/2$ valence splitting on the hole masses for SLS's with hole-containing layers that have unstrained bulk hole masses similar to those of GaAs. The zone center cyclotron masses for these particular SLS structures are all approximately $m_h^* \sim 0.09 m_e$ with a small structure-dependent contribution from the hydrostatic strain-induced shift in the band gap of the hole-containing laeyrs. At larger hole energies the hole masses associated with the $m_J = 3/2$ and $m_J = 3/2$ valence bands split and increase. The hole energy at which the masses become strongly energy dependent is roughly 1/4 of the $m_J = \pm 3/2$, $m_J = \pm 1/2$ zone center valence band splitting. It follows that the smallest hole masses will only be observed for hole concentrations which are low enough to keep the hole Fermi level below this transition energy. The transition energy can of course be increased by increasing the layer strain in the hole-containing SLS layers. Further results (not shown) for a variety of SLS's with different bulk band gap values and offset values for the SLS layers have also been obtained. They indicate that the small in-plane hole mass at the zone center can also be a function of the bulk band gap of the hole barrier layers (in addition to that of the well layers as shown in Fig. 1) if the valence band offset is sufficiently small. This is the regime in which the hole barrier height significantly influences the energies of the SLS valence bands.

CONCLUSIONS

The tailorability of SLS material properties allows these structures to exhibit a number of unique features such as small two dimensional hole masses. Such features make possible a number of interesting material science studies and may provide the basis for novel device applications.

BIAXIAL COMPRESSION

$m_{\ell H}$(BULK)/m_E

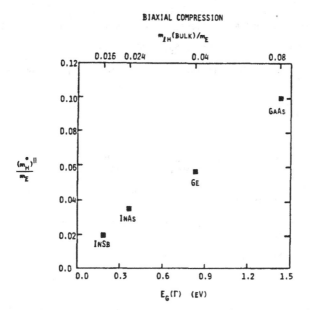

Figure 1. Calculated values of the in-plane zone center hole masses in (100) biaxially compressed materials as a function of the energy gap value at the zone center. The corresponding bulk light hole masses for the un-strained materials are shown at the top of the figure.

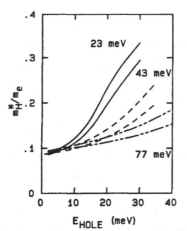

Figure 2. Calculated cyclotron hole masses vs. hole energy for the $m_J = \pm 3/2$ valence bands of several model SLS structures with hole confinement in biaxially compressed layers. The hole energy is defined to be the posi-tive difference between the valence band energy of interest and the valence band maxima. The different pairs of curves correspond to different strain magnitudes in the hole-containing layers and the values of the $m_J = \pm 3/2$, $m_J = \pm 1/2$ zone center valence band splitting are given for each pair.

REFERENCES

1. J. W. Matthews and A. E. Blakeslee, J. Cryst. Growth 32, 265 (1976).

2. G. C. Osbourn, J. Appl. Phys. 53, 1586 (1982).

3. G. C. Osbourn, R. M. Beifeld, and P. L. Gourley, Appl. Phys. Lett. 41, L72 (1982).

4. I. J. Fritz, L. R. Dawson, and T. E. Zipperian, Appl. Phys. Lett. 43, 846 (1983).

5. J. Y. Marzin and E. V. K. Rao, Appl. Phys. Lett. 43, 560 (1983).

6. M. J. Ludowise, W. T. Dietze, C. R. Lewis, M. D. Comras, N. Holonyak, B. K. Fuller and M. A. Nixon, Appl. Phys. Lett. 42, 487 (1983).

7. W. D. Laidag, P. J. Campbell, Y. F. Lin, and C. K. Peng, Appl. Phys. Lett. 44, 653 (1984).

8. S. M. Bedair, T. Katsuyama, M. Timmons, and M. A. Tischler, IEEE Electron. Dev. Lett. EDL-5, 45 (1984).

9. H. Temkin and W. T. Tsang, J. Appl. Phys. 55, 1413 (1984).

10. J. C. Bean, T. T. Sheng, L. C. Feldman, A. T. Fiory, and R. T. Lynch, Appl. Phys. Lett. 44, 102 (1984).

11. P. Voisin, C. Delalande, M. Voos, L. L. Chang, A. Segmuller, C. A. Chang, and L. Esaki, Phys. Rev. B30, 2276 (1984).

12. J. Klem, R. Fischer, W. T. Masselink, W. Kropp, and H. Morkoc, J. Appl. Phys. 55, 3843 (1984).

13. F. J. Grunthaner, Proc. of 1st Intl. Conf. on Superlattices, Micro-structures, and Microdevices, Champaign, IL, 8/12-16/84.

14. G. C. Osbourn, J. Vac. Sci. Technol. B1, 379 (1983).

15. G. C. Osbourn, Mat. Res. Soc. Symp. Proc. Vol. 25 (1984).

16. J. E. Schirber, I. J. Fritz, L. R. Dawson, and G. C. Osbourn, Phys. Rev. B28, 2229 (1983).

17. T. E. Zipperian, L. R. Dawson, G. C. Osbourn, and I. J. Fritz, Proc. of IEEE Intl. Electron Devices Meetings, 696 (1983).

18. P. L. Gourley and R. M. Biefeld, Appl. Phys. Lett. 45, 749 (1984).

19. R. M. Biefeld, P. L. Gourley, I. J. Fritz, and G. C. Osbourn, Appl. Phys. Lett. 43, 759 (1983).

20. G. C. Osbourn, J. Vac. Sci. Tehnol. B2, 176 (1984).

21. G. C. Osbourn, Proc. of 1st Intl. Conf. on Superlattices, Micro-structures, and Microdevices, Champaign, IL 8/12-16/84.

22. R. J. Chaffin, G. C. Osbourn and T. E. Zipperian, unpublished.

23. J. E. Schirber, L. R. Dawson, and I. J. Fritz, unpublished.

LUMINESCENCE PROPERTIES OF $In_xGa_{1-x}As$ - GaAs STRAINED-LAYER SUPERLATTICES

N. G. Anderson, W. D. Laidig, G. Lee, Y. Lo, and M. Ozturk
Department of Electrical and Computer Engineering
North Carolina State University
Raleigh, NC 27695-7911

Abstract

The low-temperature (20K) photoluminescence of $In_xGa_{1-x}As$ and $In_xGa_{1-x}As$ - GaAs strained-layer superlattices (SLS's) grown by molecular beam epitaxy (MBE) is investigated. Data are presented for thick (bulk) epitaxial layers grown directly on GaAs and for relatively-thin (~600Å) $In_xGa_{1-x}As$ layers under biaxial compression. Data are also presented for two series of SLS's. In the two series of SLS's, the $In_xGa_{1-x}As$ layer thickness (L_Z) is held constant while only the GaAs layer thickness (L_B) is varied. The photoluminescence (PL) spectra of the crystals are useful in analyzing the effects of biaxial strain, carrier confinement, and barrier layer thicknesses in SLS's. Results are compared with calculations based upon a modified Kronig-Penney model which incorporates the appropriate deformation potentials for SLS analysis. This type of analysis, in agreement with experimental data, suggests that the electron-to-light-hole transition can be lower in energy than the electron-to-heavy-hole transition in SLS's, depending upon layer thickness and crystal composition.

Introduction

Semiconducting strained-layer superlattices (SLS's) are structures consisting of thin alternating layers of different semiconductors which in their bulk crystal form would have unequal lattice constants. In a SLS, however, if the layer thicknesses are kept below strain-dependent critical values (h_c), the lattice mismatch will be accommodated entirely by elastic strain instead of misfit generation [1,5]. Thus, it is possible to have high quality heterointerfaces in spite of significant mismatches (several percent) in the bulk lattice constants of the component materials. This is evidenced by the demonstration of photoexcited strained-layer lasers [6] and low threshold strained-layer diode lasers [7,8].

It should be noted that neither the concept nor the realization of SLS's is entirely new. As early as 1971, Blakeslee [9] reported the growth of GaAs - $GaAs_{1-x}P_x$ SLS's by organometallic chemical vapor deposition (OMCVD). At that time, however, epitaxial growth technology was relatively immature by present standards. The development of improved growth systems and epitaxial techniques, along with the availability of high-purity source materials, has resulted in the capability to grow epitaxial III-V compounds with electrical and optical characteristics far superior to those of the previous decade. These improvements, coupled with the slow growth rates and layer uniformity made possible with molecular beam epitaxy (MBE) and OMCVD, have helped to stimulate renewed interest in SLS's.

Although SLS's warrant investigation from purely physical considerations, there are important practical implications as well. The primary advantage offered by SLS's arises from a loosening of the lattice-matching constraints normally imposed on high-quality devices (especially optical emitters). This extra degree of freedom provides added flexibility [10] in both device design and materials growth. In addition, because of the unrelieved internal strain [10] and quantum size effects [11,12] (due to increased carrier confinement), many of the crystal properties will be altered from those characteristic of bulk materials.

In this work, the luminescence properties of SLS's utilizing the $In_xGa_{1-x}As$ - GaAs crystal system are studied. Efforts are concentrated on examining the low temperature (20K) photoluminescence spectra in order to determine the SLS energy gap and the associated electron, heavy hole and light hole energies. The dependence of these energies on layer thickness, strain, and crystal composition are investigated.

Crystal Growth and Characterization

All of the samples described here are grown by MBE (Varian 360 system) on <100>-oriented GaAs substrates (Si doped, $4\times10^{18}cm^{-3}$) at a substrate temperature of 520°C. Furnace temperatures are ~1020°C, 860°C, and 320°C for Ga, In, and As, respectively. This results in beam equivalent pressures of ~5.8×10^{-7} T, 2×10^{-7} T, and 8×10^{-6} T for Ga, In, and As, respectively. Growth rates are ~3Å/s for GaAs and somewhat faster for $In_xGa_{1-x}As$ depending on the crystal composition. The In furnace temperature and the beam equivalent pressure stated above represent typical values for $In_xGa_{1-x}As$ with x~0.2 and are modified appropriately for other compositions.

Several thick layers (referred to as bulk) of undoped $In_xGa_{1-x}As$ have been grown with various compositions x. These layers are typically 2 μm thick and are grown on a 0.5 μm GaAs buffer layer. The crystal composition is determined by x-ray diffraction. In addition, four undoped strained double heterostructures (DH) have been grown for comparison with the bulk (unstrained thick layer) samples described above. Each of the four pairs of samples (strained DH and bulk) were grown consecutively under identical conditions. The strained DH's consist of a 600Å $In_xGa_{1-x}As$ layer (0.03 < x < 0.06) grown on a 0.5 μm GaAs buffer layer. Following $In_xGa_{1-x}As$ growth, a GaAs cap layer (400Å thick) is deposited. Since the mole fraction of In is so small in these samples, the 600Å layer thickness is below the critical thickness, resulting in elastically strained $In_xGa_{1-x}As$.

Two series of SLS's, each consisting of four undoped consecutively-grown samples, have been grown with a structure similar to that shown schematically in Fig. 1. The primary difference in the two series is in the In composition. In one series x ~ 0.14, while in the other series x ~ 0.20. Within each series, the $In_xGa_{1-x}As$ quantum well thickness (L_z) remains constant, as does the total SLS thickness (0.4 μm). Only the GaAs barrier layer thickness (L_B) is allowed to vary. For the x ~ 0.14 series, L_z ~ 30Å and L_B ~ 25-76Å. For the x ~ 0.20 series L_z ~ 38Å and L_B ~ 30-90Å. Both SLS's have been grown on 0.5 μm GaAs buffer layers.

The structural quality, the crystal composition, and the layer thicknesses are determined by x-ray diffraction [5,13]. A comparison of the layer thicknesses determined from x-ray diffraction, from direct cross-section measurements, and from fixed growth-time ratios for consecutively grown samples substantiate the linear growth rate reported earlier [5,14]. PL of the SLS's is obtained with cavity-dumped Ar-ion laser light focussed onto the SLS sample surface [14]. The samples are cooled to 20K as measured by a thermocouple directly in contact with the sample mount. The peak power density is approximately 10^4 W/cm^2.

$In_xGa_{1-x}As$ – GaAs SLS

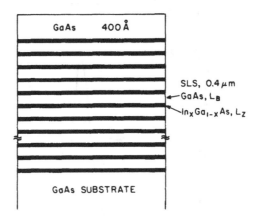

Fig. 1 Schematic cross section of strained layer superlattices (SLS's). The $In_xGa_{1-x}As$ quantum wells (L_z) and the GaAs barriers (L_B) are grown by molecular beam epitaxy (MBE) on a GaAs substrate and a GaAs buffer layer (not shown).

$In_xGa_{1-x}As$: Bulk

The term "bulk" is used throughout this discussion to refer to relatively thick layers which demonstrate bulk-crystal properties. In this work, a number of ~ 2μm thick layers of $In_xGa_{1-x}As$ were grown as a baseline for MBE calibration and PL analysis. Since these layers are sufficiently thick (>>h_c), virtually all of the lattice mismatch present between the GaAs substrate and the $In_xGa_{1-x}As$ epitaxial layer is accomodated by formation of dislocations near the substrate-epilayer interface. The strain relaxation near the interface results in an epitaxial layer with the properties of bulk $In_xGa_{1-x}As$ of the same composition. X-ray diffraction and PL measurements on these samples allow determination of the lattice constant (thus crystal composition) and band gap of epitaxial $In_xGa_{1-x}As$ without the strain or quantum-size effects present in thinner layers grown under identical conditions.

The 20K PL spectra for a thick $In_{0.16}Ga_{0.84}As$ epitaxial layer grown on GaAs is shown in Fig. 2. A PL halfwidth (full width at half maximum) of ~7.5meV for this sample indicates that the material is of quite high quality, in spite of the 1.1% lattice mismatch present between the substrate and the epilayer. Luminescence from this undoped film is attributed primarily to band-to-band rather than excitonic transitions, given the fairly high excitation level used in these experiments.

The PL peak energy of the x ~ 0.16 sample described above along with several similar samples of various compositions are plotted as a function of crystal composition in Fig. 3. The solid curve is a plot of the expression

$$E_g(x) = 1.515 - 1.24x + 0.18x^2 \text{ (eV)}. \tag{1}$$

This expression fits the experimental data within ~10meV and terminates at x=1 near Eg = 0.46eV for InAs. These data are fairly linear, in agreement with recent published results [15].

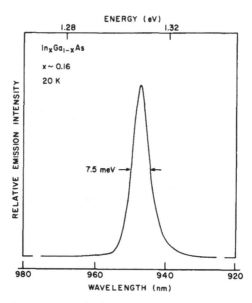

Fig. 2 20K photoluminescence (PL) spectra of an undoped 2 μm $In_xGa_{1-x}As$ (x~0.16) sample. The excitation source is a Ar-ion laser (λ~5145Å) focussed to obtain a pump power density of ~10^4 W/cm².

$In_xGa_{1-x}As$: Strained

The effects of externally applied uniaxial and hydrostatic stresses upon the band structures of several elemental and compound semiconductors have been examined by several researchers [16,20]. Epitaxial layers in strained-layer superlattices are, however, under biaxial compression and tension, since the lattice mismatch betwen the SLS constituent materials is accommodated by biaxial elastic strains in the planes of the heterointerfaces. Only recently have the effects of biaxial compression and tension upon the band structure of III-V compounds been investigated [21,22].

The expected behavior of the upper valence bands and lower conduction band with biaxial strain in a zinc blend semiconductor is shown in Fig. 4. Note that with biaxial strain, the degeneracy of the valence bands at k=0 is split, since the light hole band shifts more than the heavy hole band. Thus, the band-gap decreases (increases) with biaxial tension (compression) as the valence bands move toward (away from) the conduction band. Note also that, in the case of biaxial tension (compression), the lowest energy optical transition occurs between the conduction band and the light hole (heavy hole) band. The relative curvatures of the two valence bands, hence the assignment of "light" and "heavy" hole band designations, actually depends upon crystallographic direction in the strained (tetragonally distorted) lattice. Thus, it is important to specify that the crystallo-graphic direction perpendicular to the biaxial stress plane is to be assumed in all energy band diagrams and references to light and heavy holes in this work. Both Olsen et. al. [21] and Asai and Oe [22] have studied the shift in the lowest energy optical transition with biaxial strain in epitaxial $Ga_xIn_{1-x}P$ grown on GaAs and have obtained similar results which agree well

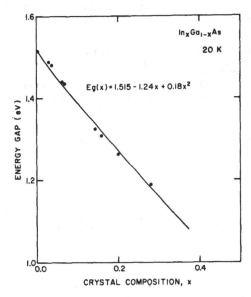

Fig. 3. Energy gap (PL peak energy) at 20K for several compositions of $In_xGa_{1-x}As$. The temperature measurement is made with a thermocouple in direct contact with the sample mount. Crystal composition is determined by x-ray diffraction. The solid line corresponds to the best quadratic fit to the data points and is accurate to ~10 meV.

BIAXIAL STRAIN

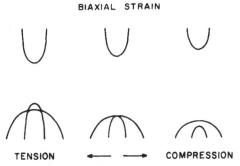

Fig. 4. Schematic representaions of the shift in conduction and valence bands due to biaxial strain in the crystallographic direction perpendicular to the biaxial stress plane. The bands move apart under compression and together under tension, with the light-hole band shifting more than the heavy hole band in each case.

with a simple calculation carried out by Asai and Oe [22]. Although SLS's consisting of alternating layers of $In_xGa_{1-x}As$ and GaAs have attracted considerable attention in recent years, experimental information regarding the effects of biaxial strain upon the band structure of $In_xGa_{1-x}As$ is not currently available. In this section, results of PL experiments designed to determine the shift in the lowest energy optical transition for epitaxial $In_xGa_{1-x}As$ (on GaAs) under biaxial compression are presented and compared with results of a simple calculation.

To determine the band-gap shift with biaxial strain in epitaxial $In_xGa_{1-x}As$, several pairs of samples were grown as described previously. The thickness of the $In_xGa_{1-x}As$ layer in the strained DH is ~600Å. This is below the estimated critical layer thickness [1,23] above which relaxation of strain occurs by formation of dislocations, but sufficiently thick such that quantum-size effects due to carrier confinement in the $In_xGa_{1-x}As$ layer are neglegible. Comparison of the PL peak energies for a pair of samples (bulk and strained DH) with composition x directly gives the band gap shift of $In_xGa_{1-x}As$ strained to GaAs for that composition.

20K PL spectra for bulk and strained DH $In_xGa_{1-x}As$ layers with x~0.04 (as determined by x-ray diffraction) are shown in Fig. 5. The PL spectra of the strained DH (labeled "STRAINED") occurs ~8meV higher in energy than that of the thick layer (labeled "BULK"). The small peak at ~1.515 eV is emission from the GaAs cap layer. Note that the PL peak for the strained DH is very narrow (~1.3 meV full width at half maximum), indicating that the material is of high quality (i.e., relatively dislocation free). The PL

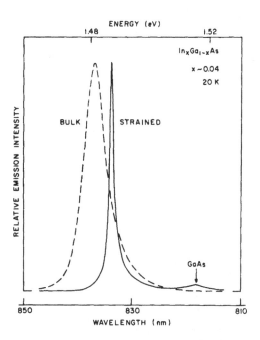

Fig. 5. Comparison of a 2 μm thick x ~0.04 $In_xGa_{1-x}As$ layer (labeled "BULK") with a strained DH sample consisting of ~600Å of $In_xGa_{1-x}As$ (x~0.04) sandwiched between two GaAs layers (labeled "STRAINED"). This layer is thin enough to be elastically strained to the GaAs, but too thick to exhibit quantum size effects. A shift of ~8 meV is observed in the 20K PL spectra. Note also the extremely narrow spectrum (full width half maximum ~1.3 meV) for the strained DH.

spectra for both bulk and strained $In_xGa_{1-x}As$ layers with x~0.06 are shown in Fig. 6. The results are similar to those obtained for the samples with x ~ 0.04, however, the PL peak energy of the strained DH is higher in energy than that of the bulk layer by ~27 meV. The PL half width for this strained DH is also quite narrow (~2.5 meV).

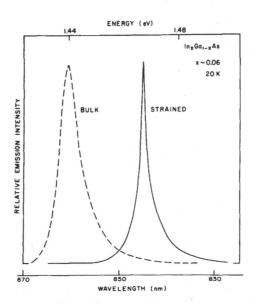

Fig. 6. 20K PL spectra of unstrained (BULK) and strained $In_xGa_{1-x}As$. The samples are similar to those described in Fig. 5 but with x ~0.06 in this case. The strain results in a transition energy shift of ~27meV.

Similar results were obtained for two other pairs of samples (not shown). The increase in energy gap is plotted as a function of strain (or crystal composition) in Fig. 7. The four experimental data points correspond to the shift in the four sample pairs described above. The solid line corresponds to the expected energy gap shift based on a calculation similar to that of Asai and Oe [22]. This calculation will be described further in the following section.

SLS Model

A fairly simple model is used to aid in understanding and predicting the properties of SLS's. The model uses a Kronig-Penney type of analysis similar to that used for unstrained superlattices. In the case of SLS's, however, the band gaps of the wells and barriers are altered due to biaxial compression or tension (see Fig. 4). The change in the energy gap is given by

$$\Delta E = -\alpha \, \varepsilon, \tag{2}$$

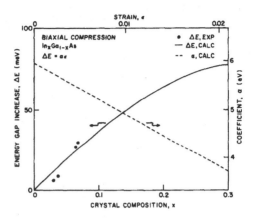

Fig. 7. Energy gap increase for $In_xGa_{1-x}As$ biaxially compressed to lattice match GaAs. The four data points are experimental results. The solid line is a calculation based upon the coefficient α for heavy holes (see text).

where ε is the strain (negative for compression) and α is given by [22]

$$\alpha = -2a(C_{11} - C_{12})/C_{11} \pm b(C_{11} + 2C_{12})/C_{11}. \tag{3}$$

In equation (3) the "+" sign corresponds to the electron-heavy-hole energy gap while the "-" sign corresponds to the electron-light-hole energy gap. The parameters a and b are the hydrostatic and shear deformation potentials, respectively. C_{11} and C_{12} are the elastic constants. Table I lists the values of these material parameters for both InAs and GaAs. For $In_xGa_{1-x}As$, a linear interpolation between the two binary compounds is assumed.

Table I. Material Parameters for GaAs and InAs

	a	b	C_{11}	C_{12}	$\partial E_0/\partial P$
	eV	eV	10^{12} dyn/cm²		10^{-12} eV/dyn-cm²
GaAs	-8.67*	-1.7(27)	12.25(24)	5.70(24)	11(26)
InAs	-3.01*	-1.8(18)	8.64(25)	4.85(25)	5(26)

* derived using $a = (-1/3) (C_{11} + 2C_{12}) \partial E_0/\partial P$ (See Ref. 22)

Following the procedure described above, α can be calculated as a function of x for $In_xGa_{1-x}As$ biaxially strained to lattice-match GaAs. This is shown by the dashed curve in Fig. 7, corresponding to the lowest energy band gap (conduction band to heavy hole band for $In_xGa_{1-x}As$ under compression). Using the relation given in Eq. (2) the increase in the $In_xGa_{1-x}As$ energy gap is estimated, and shown by the solid line in Fig. 7. These calculations are in the range expected based on the experimental data described in the previous section.

For the $In_xGa_{1-x}As$ -GaAs SLS's described here the $In_xGa_{1-x}As$ is under compression (Eg increase with respect to bulk) and the GaAs is under tension (Eg decrease with respect to bulk). The altered band gaps for both wells and barriers are calculated as described above, with the strain in each layer determined using ratios of layer thicknesses and shear moduli for each layer as described by Osbourn [10]. Fig. 8 shows schematically the results of these calculations and the corresponding SLS square-well approximation used for the Kronig-Penney model. Note that because of the light-hole band (dashed) shifts more than the heavy-hole band (solid) the light-hole barrier height is substantially lower than that of the heavy hole. In fact, for layers that are highly strained, it is expected that the light-hole bands in the GaAs and $In_xGa_{1-x}As$ will cross each other, resulting in light-hole collection in the GaAs and spatial separation of the light and heavy holes.

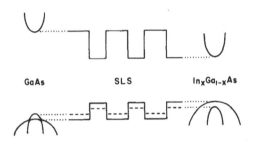

GaAs   SLS $In_xGa_{1-x}As$

Fig. 8. Simplified diagram of the band structure of GaAs under tension and of $In_xGa_{1-x}As$ under compression. The modified band gaps are used to establish the well depths and barrier heights necessary for a Kronig-Penney analysis. Note that the splitting between the light and heavy hole bands results in two quite different valence-band square wells.

Figures 9 and 10 show the calculated transition energies for SLS's with x ~0.14 $In_xGa_{1-x}As$ (L_z ~ 30Å) and with x ~ 0.20 $In_xGa_{1-x}As$ (L_z ~ 38Å), respectively. The electron-to-light-hole transitions are shown by the dashed curves and the electron-to-heavy-hole transitions are shown by the solid curves. Two sets of curves are shown in each figure, corresponding to valence band discontinuities of 15% and 25% of the difference in the strained band gaps (using the conduction-band to heavy-hole-band energy gap). This range of discontinuities seems the most appropriate for this materials system based on the GaAs-AlAs system and what limited data is available for InAs-GaAs heterojunctions [28]. Both Figs. 9 and 10 indicate that the transition energies are fairly insensitive to the choice of band discontinuities.

The most significant feature of these curves is that the lowest energy transitions are electron-to-light-hole transitions for small L_B and electron-to-heavy-hole transitions for larger L_B. The crossover point is marked by an arrow in the figures. This is a consequence of the low-energy

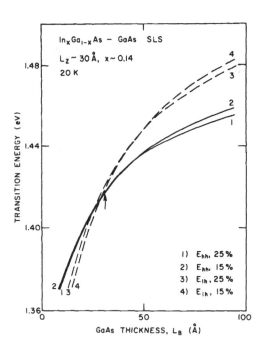

Fig. 9. Calculation of the lowest transition energies (20K) for a SLS with x~0.14 $In_xGa_{1-x}As$ wells of thickness $L_z \sim 30\text{Å}$. The dashed curves (electron-to light-hole transitions) and the solid curves (electron-to-heavy-hole transitions) cross at a point near $L_z \sim L_B$. The calculation is fairly insensitive to changes in the choice of energy band discontinuities. Curves are shown assuming both a 25% and 15% valence band discontinunity (see text).

light hole states in the highly strained GaAs layers. These results are compared with experimental data in the following section.

SLS Photoluminescence

The PL spectra (20K) of the x ~0.14 series of SLS's are shown in Fig. 11. In all four samples $L_z \sim 30\text{Å}$, however, the GaAs barrier layer thickness varies from $L_B \sim 25\text{Å}$ to $L_B \sim 76\text{Å}$. The PL spectra narrow and shift to higher energy as the barrier layer thickness increases.

A similar set of spectra is obtained for the x ~ 0.20 series of SLS's (Fig. 12). In this case $L_z \sim 38\text{Å}$ and L_B varies from 30Å to 90Å. The same narrowing and shift to higher energy with increasing L_B are observed for these samples. It should be noted that the intensities of the spectra are shown in relative order, but with different scaling factors. Actually the samples with largest L_B were nearly an order of magnitude stronger in intensity than the samples with smaller L_B.

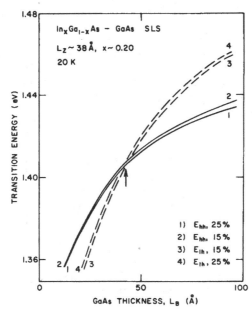

Fig. 10. Transition energy calculation for a SLS with x ~0.20 and L_z ~38Å. These calculations (as in Fig. 9) indicate that for small L_B the electron-to-light-hole transition is lower in energy than the electron-to-heavy-hole transition.

It is interesting to observe the variation of the PL full width at half maximum for these samples. This data for both series is plotted in Fig. 13. The SLS's, with the exception of those with the thinnest barrier layers, exhibit PL half widths in the range of 8-12 meV, with the narrowest spectra occuring for thickest barriers. The width of PL spectra for superlattices with thin layers is usually determined by slight deviations in layer thicknesses (on the order of a monolayer) which result in significant shifts in the transition energy (due to the quantum size effect). This is likely to account for much of the increase from ~2 meV in the strained DH (single thick layer) to ~8 meV for the SLS's with larger L_B. Within each series, the change in PL halfwidth is attributed to the decreasing sensitivity of the transition energy to variations in L_B with increasing L_B, and to narrowing of the n=1 electron and hole subbands with increasing separation between (decoupling of) the quantum wells.

The peak energies of the PL spectra for each series is plotted in Fig. 14. The large change in the transition energy for the two SLS's in each series with smallest L_B values is apparent. The transition energy change for the SLS's with larger L_B values is much more gradual. This behavior is to be expected based on the calculations described in the previous section. For thin barriers, the electron-to-light-hole transition is of lowest energy. The energy of these transitions increases quickly with increasing L_B, until the crossover point (marked by arrows) occurs. The lowest energy transition then becomes the electron-to-heavy-hole transition, which is much less sensitive to changes in L_B. The calculations shown in Figs. 9 and 10 do in fact predict the observed behavior. The absolute transition energies

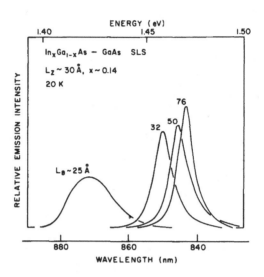

Fig. 11 PL spectra (20K) for a series of SLS's with $L_z \sim 30\text{Å}$ and $x \sim 0.14$. The GaAs layer thickness L_B is different for each sample.

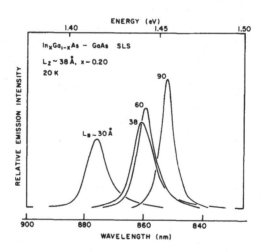

Fig. 12 Set of PL spectra similar to those of Fig. 11 but for a series of SLS's with $L_z \sim 38\text{Å}$ and $x \sim 0.20$. The PL spectra narrow and shift to higher energy with increasing L_B.

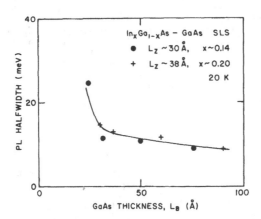

Fig. 13 Full width at half maximum of the PL spectra for the two series of
SLS's of Figs. 11 and 12. The decrease of the halfwidth (to ~ 8 meV) is attributed to added confinment as a result of the thicker barrier layers, and
to decreased sensitivity to small deviations in L_B with increasing L_B.

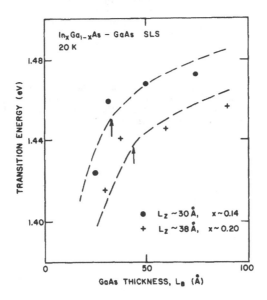

Fig. 14   Plot of peak energies from the PL spectra of Figs. 11 and 12.   In
both SLS series, with increasing L_B a sharp increase in transition energy is
first observed, followed by a substantially more gradual increase.   This is
to be expected based upon calculations (dashed lines, see text) which indicate electron-to-light-hole transitions for values of L_B below the arrows
and electron-to-heavy-hole transitions for larger L_B.

are slightly lower than observed ($\sim$30 meV) corresponding to a $\sim$2% error in the calculated transition energy. The dashed lines in Fig. 14 represent the lowest transition energies (from Figs. 9 and 10) using a 15% valence band discontinuity, and offset by 30 meV for comparsion purposes. This relatively small offset could be easily accounted for by using a slightly larger $In_xGa_{1-x}As$ band gap, perhaps indicating that the band gap increase due to biaxial compression is slightly larger than calculated. This is certainly plausible since the data (Fig. 7) are obtained only for small strains and even then (for x $\sim$0.06) the measured ΔEg is slightly greater than the ΔEg used in calculations.

Summary

The low temperature (20K) PL of $In_xGa_{1-x}As$-GaAs structures grown by MBE has been studied. PL from relatively thick layers has been used to determine the variation in transition energy with composition. PL half-widths of $\sim$8 meV are observed for $In_xGa_{1-x}As$ with x $\sim$0.16. A comparison between PL from thick-layers (unstrained $In_xGa_{1-x}As$) and strained DH's (600 A of $In_xGa_{1-x}As$ biaxially strained to lattice-match GaAs) allows a direct measurement of band gap increase in biaxially compressed $In_xGa_{1-x}As$. These data correspond well with calculated results.

A modified Kronig-Penney analysis is described, which accounts for band-gap alterations due to both tension and compression in the SLS wells and barriers. For the two series of SLS's described here, this model predicts that for small L_B the electron-to-light-hole transition is lower in energy than the electron-to-heavy-hole transition. PL spectra on these SLS's are in support of this conclusion, based upon the observed rate of change in transition energy with variation in the GaAs layer thickness.

Acknowledgements

The authors are grateful to Drs. S. M. Bedair, J. R. Hauser, M. A. Littlejohn and J. J. Wortman for helpful discussions, and to P. J. Caldwell, Y. F. Lin, V. Y. Stone and D. M. Banasz for their assistance. This work was supported by the National Science Foundation.

References

1. J. W. Matthews and A. E. Blakeslee, J. Cryst. Growth 27, 118 (1974).

2. J. W. Matthews and A. E. Blakeslee, J. Cryst. Growth 32, 265 (1976).

3. G. C. Osbourn, J. Appl. Phys. 53, 1586 (1982).

4. G. C. Osbourn, R. M. Biefeld, and P. L. Gourley, Appl. Phys. Lett. 41, 172 (1982).

5. W. D. Laidig, C. K. Peng, and Y. F. Lin, J. Vac. Sci. Technol. B2, 181 (1984).

6. M. D. Camras, J. M. Brown, N. Holonyak, Jr., M. A. Nixon, R. W. Kaliski, M. J. Ludowise, W. T. Dietze, and C. R. Lewis, J. Appl. Phys. 54, 6183 (1983).

7. W. D. Laidig, P. J. Caldwell, Y. F. Lin, and C. K. Peng, Appl. Phys. Lett. 44, 653 (1984).

8. W. D. Laidig, Y. F. Lin, and P. J. Caldwell, J. Appl. Phys. 56 (Dec. 1, 1984).

9. A. E. Blakeslee, J. Electrochem. Soc. 118, 1459 (1971).

10. G. C. Osbourn, J. Vac. Sci. Technol. B1, 379 (1983).

11. R. Dingle, in Festkorper Probleme XV (Advances in Solid State Physics), H. J. Queisser, Ed. New York: Pergamon, 1975, pp. 21-48.

12. N. Holonyak, Jr., R. M. Kolbas, R. D. Dupuis, and P. D. Dapkus, IEEE J. Quan. Electron QE-16, 170 (1980).

13. M. Quillec, L. Goldstein, G. LeRoux, J. Burgeat, and J. Primot, J. Appl. Phys. 55, 2904 (1984).

14. N. G. Anderson, W. D. Laidig, and Y. F. Lin, J. Electron. Mater. (to be published, Mar. 1986).

15. B. J. Baliga, R. Bhat, and S. K. Ghandhi, J. Appl. Phys. 46, 4608 (1975).

16. M. Chandrasekhar and F. H. Pollak, Phys. Rev. B 15, 2127 (1977).

17. C. W. Higginbotham, M. Cardona, and F. H. Pollak, Phys. Rev., 184, 821 (1969).

18. P. Y. Yu, M. Cardona, and F. H. Pollak, Phys. Rev. B3, 340-346 (1971).

19. F. H. Pollak and M. Cardona, Phys. Rev. 172, 816 (1968).

20. W. Paul, J. Appl. Phys. 32, 2082, (1961).

21. G. H. Olsen, C. J. Nuese, and R. T. Smith, J. Appl. Phys. 49, 5523 (1978).

22. H. Asai and K. Oe, J. Appl. Phys. 54, 2052 (1983).

23. I. J. Fritz, S. T. Picraux, L. R. Dawson, and W. R. Allen, Materials Research Society Meeting, Paper D4.4 (Boston, Nov. 1984).

24. Extrapolated to 20K from values given by F. M. Gashimzada, V. E. Khartsiev, Fiz. Tv. Tela $\underline{3}$, 1453.

25. Extrapolated to 20K from values given by D. Gerlich, J. Appl. Phys. $\underline{34}$, 2915 (1963).

26. J. I. Pankove, Optical Processes In Semiconductors, New York: Dover, 1971, pp. 412-413.

27. I. Balslev, Solid State Commun. $\underline{5}$, 315 (1967).

28. S. P. Kowalozyk, W. J. Schaffer, E. A. Kraut, and R. W. Grant, J. Vac. Sci. Technol. $\underline{20}$, 705 (1982).

TEM STUDY OF GaSb/InAs STRAINED LAYER SUPERLATTICES

B. C. DE COOMAN AND C. B. CARTER
Dept. of Materials Science and Engineering, Cornell University
G. W. WICKS AND T. TANOUE
School of Electrical Engineering, Cornell University, Ithaca, NY 14853

Cross-sectional TEM of GaSb/InAs superlattices grown by MBE on (100) GaAs and (100) GaSb substrates shows an unusual defect structure within the strained layers. Dislocations are present within the layers and at the interface. High-resolution TEM analysis of the structure of the InAs layers suggests that these layers grow by an island mechanism. A crystal structure different from the zinc blende, is found to be present within the GaSb layers.

INTRODUCTION

The electronic characteristics of GaSb/InAs superlattices has been studied theoretically by Sai-Holasz et al. [1] and the structure of similar superlattices has been characterized by Saris et al. [2] and Chu et al. [3] using Rutherford ion-backscattering (RBS). In both RBS studies, oscillations in the random direction RBS spectra were attributed to the periodic mass interference of Ga, Sb, In, and As atoms in the layers; these oscillations were considered to be proof of the good lattice match and epilayer quality. The RBS analysis did not indicate the presence of dislocations which might be expected to influence the energy dependence of the dechanneling yield. No previous studies of similar superlattices grown on (100) GaAs substrates are available, although it has been reported [3] that such layers could not be grown successfully.

For superlattices grown on (100) GaSb substrates, the dechanneling along <110> directions is large even though no dechanneling occurs along the [100] growth direction. High dechanneling along <110> directions has also been observed for GaSb/AlSb superlattices [4]. Saris [2,3] proposed a bond-relaxation model to explain the RBS results, but subsequent calculations [5,6] of the dechanneling rate using this model showed that bond relaxation alone cannot account for the observed dechanneling. Barrett [5,6] has proposed an elastic strain model in which the InAs and GaSb layers have a tetragonal crystal structure due to the presence of the strain. In the present study, (002) dark-field TEM and cross-sectional HREM of the layers in the [110] orientation show that bond-relaxation and the formation of a crystal structure different to the zinc blende structure for GaSb actually combine to produce the observed RBS dechanneling.

EXPERIMENTAL

GaSb/InAs superlattices were grown in a Varian MBE system by periodically opening and closing the Ga, Sb, In, and As effusion cells. During the growth the substrate temperature was 350°C, the As-cell temperature was 328°C and the As-partial pressure was 1.5×10^{-5} torr. The superlattice growth was preceded by the deposition of a 250 nm thick buffer layer. The substrate materials used were (100) GaSb or (100) GaAs. The epilayer structure nominally consisted of 100 layers each of alternately 10 nm GaSb and 10 nm InAs. The TEM analysis was performed using a Siemens 102 electron microscope operating at 125 keV. TEM samples were prepared by mechanical polishing and subsequent ion milling.

RESULTS

The results of different experiments are presented in a systematic fashion in the following subsections. The (002) dark-field electron microscopy technique allows for the chemical identification of the layers since the (002) reflection has a structure factor which is given by $F_{200} = 4(f_{III} - f_V)$ where f_{III} and f_V are the atomic scattering amplitudes of the group III and V elements, respectively [7]. Electron diffraction makes it possible to study both the strain between the layers and the possible existence of interfacial layers, the latter is prediced to occur in the model proposed by Saris et al. [2]. High-resolution electron microscopy and computer image simulation provide direct information on the local crystal structure [8].

Dark-Field Electron Microscopy

a. GaSb/InAs on GaSb (Fig. 1)

The layers grow in the expected manner, but the thickness of the InAs layers can vary by up to 40%. The InAs-GaSb interface appears as a black line in such (002) dark-field images. This observation can be understood by a simple kinematical calculation which shows that if the intensity I_{GaSb} of the (002) beam for GaSb is taken to be unity, i.e. $I_{GaSb} = 1$, then $I_{InAs} = 0.64$, $I_{GaAs} = 0.01$, and $I_{InSb} = 0.007$. It is, therefore, suggested that the black lines at the interface are due to thin layers of GaAs and InSb between the InAs and GaSb layers. These new thin layers are most likely only 1 or 2 unit cells in thickness ($\sim$1 nm) and the structure is thus very similar to the model of Saris et al. [2,3,9]. Finally, it should be noted that there are very few dislocations in these epilayers.

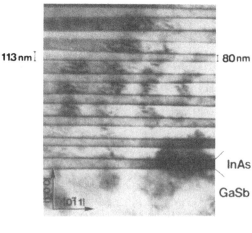

113 nm 80 nm

InAs

GaSb

Fig. 1. (002) Dark-field image of a GaSb/InAs SLS grown on (100) GaSb. Note the thickness inhomogeneity in the InAs layer and the characteristic black contrast for the interface.

b. GaSb/InAs on GaAs (Fig. 2)

The layers generally grow in a periodic manner, but do not form a well-defined superlattice because the InAs shows a preference for island growth; the GaSb grows in layers. Dislocations are present throughout the layers and there is a dislocation array present in the plane of the GaAs/InAs heterojunction, i.e. between the GaAs substrate and the initial InAs layer. The average dislocation spacing within this array is 6.6 nm.

Electron Diffraction

The results presented concern only diffraction patterns for which the electron beam is parallel to the [110] zone axis.

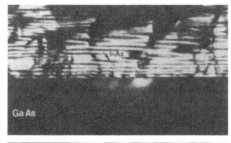
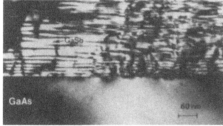

Fig. 2. (002) Dark-field image of a GaSb/InAs SLS grown on (100) GaAs. The InAs island in (a) and GaSb "channels" in (b) are clearly visible. The large number of dislocations present in the SLS (e.g. on the right of Fig. 2b) causes changes in the diffraction conditions.

a. GaSb/InAs on (100) GaSb substrates (Fig. 3)

All the spots in the diffraction pattern are elongated in a direction parallel to the growth direction. However, the width of the streaks cannot be related to the strain between GaSb and InSb even in interfacial layers of GaAs and InSb are present since this strain would produce much shorter streaks. It is, therefore, proposed that the spot elongation is a shape effect due to the thin phase between each layer. This thin boundary phase may be GaAs or InSb, as suggested by the (002) dark-field results. Both the (001) and (1$\bar{1}$0) reflections are clearly present in these diffraction patterns; these reflections have a zero structure factor according to the selection rules for the zinc blende structure and are never observed in, for example, diffraction patterns from GaAs.

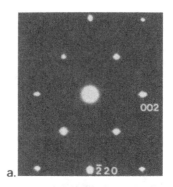

Fig. 3. (a) SAD pattern with the electron beam parallel to [110]. (b) Detail of (a) showing weak (001) reflection.

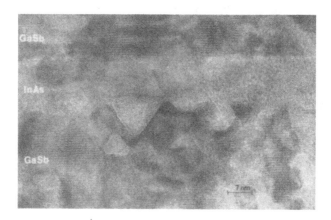

Fig. 4. Lattice-fringe image of the GaSb/InAs SLS grown on (100) GaSb. The GaSb has a clear lattice period in the [100] growth direction.

b. GaSb/InAs in (100) GaAs substrates

In the diffraction pattern the GaSb/InAs spots are again elongated although the GaAs spots from the buffer layer are clearly not elongated. The forbidden (001) and (1$\bar{1}$0) reflections also occur, but they are generally weaker than in the case of GaSb/InAs grown on GaSb substrates. The streaks in the diffraction pattern do not run parallel to g_{200} which indicates that they are associated with an interface or thin layer which is not normal to the [100] growth directions.

High Resolution Microscopy

Since the Siemens 102 has a coefficient of spherical aberration C_s of 2.1 mm, it is not possible to obtain a true structure image of any of the III-V compounds studied here. It is, however, possible to obtain clear lattice-fringe images which contain valuable information on the crystal structure and the local orientation of the lattice planes.

a. GaSb/InAs on (100) GaSb (Fig. 4)

The GaSb layers show a very pronounced periodicity in the [100] growth direction which is a factor of two larger than that expected and found elsewhere for the zinc-blende structure. There are regions where the image shows a clear transition from that expected for the zinc-blende structure into that of the GaSb defect structure. This transition does not involve the presence of dislocations.

b. GaSb/InAs on (100) GaAs (Fig. 5)

The new 0.6-0.7 nm periodicity normal to the growth direction is again observed on the lattice-fringe images. The regions over which the new GaSb structure extends are smaller in this case and an interfacial dislocation array is clearly detectable at the InAs/GaAs interface.

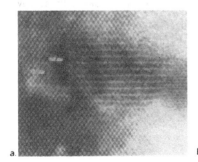 

a.   b.

Fig. 5. Lattice-fringe images of the GaSb defect structure in SLS grown on (100) GaAs.

DISCUSSION AND CONCLUSIONS

High-resolution TEM has now revealed an unusual crystal structure present in MBE-grown epilayers of GaSb. This structure is such that the (001) and (1$\bar{1}$0) spots in the diffraction patterns which are forbidden for the zinc-blende structure are clearly visible. The lattice-fringe images of this new structure do not show the usual pronounced {1$\bar{1}$1} type fringes, but instead show a 0.6-0.7 nm periodicity normal to (001) and a clear (1$\bar{1}$0) periodicity. In order to compare the observed new structures with those suggested by Barrett, computer simulated, multislice images have been generated for a series of tetragonal unit cells with different c/a ratios. The results, which will be reported elsewhere, show that the observed images cannot be explained by the model of a simple tetragonal distortion.

The large dechanneling effect observed for the <110> direction in RBS experiments can be explained by the occurrence of a crystal structure for the GaSb which is not the zinc blende structure. In addition, the (002) dark-field contrast also strongly suggests that the interface between the GaSb and InAs layers does have composition close to GaAs or InSb: this result is then in agreement with the original model suggested by Saris [2]. It is thus proposed that both models [2,6] which have been advanced to explain the observed dechanneling are correct in that they each explain part of the observed effect.

ACKNOWLEDGMENTS

The authors thank Dr. Frans Saris for informative discussions and Mr. Ray Coles for maintaining the microscopes which are supported by the NSF through the Materials Science Center at Cornell. This research is supported in part by the U.S. Army Research Office under contract No. DAAG-29-82-K-0148.

REFERENCES

1. G. A. Sai-Halasz, R. Tsu, L. Esaki, Appl. Phys. Lett. 10(12) (1977), 651.
2. F. W. Saris, W. K. Chu, C. A. Chang, R. Ludeke, L. Esaki, Appl. Phys. Lett. 37(10) (1980), 931.
3. W. K. Chu, F. W. Saris, C. A. Chang, R. Ludeke, L. Esaki, Phys. Rev. B26(4) (1982), 1999.
4. W. K. Chu, C. K. Pan, C. A. Chang, Phys. Rev. B28(7) (1983), 4033.
5. J. H. Barrett, Appl. Phys. Letts. 40 (1982), 482.
6. J. H. Barrett, Phys. Rev. B28 (1983), 2328.
7. P. M. Petroff, A. Y. Cho, F. K. Reinhart, A. C. Gossard, W. Wiegmann, Phys. Rev. Lett. 48(3) (1982), 170.
8. J.C.M. Spence, Experimental High Resolution Microscopy, Clarendon Press, Oxford (1978), Chapt. 6.
9. C. A. Chang, J. Vac. Sci. Technol. B1(2) (1983), 346.
10. B. C. De Cooman, N.-H. Cho, Z. Elgat, C. B. Carter, to be published in Ultramicroscopy (1985).

Molecular Beam Epitaxy of Ge_xSi_{1-x}/(Si, Ge) Strained-Layer Heterostructures and Superlattices

J. C. Bean

AT&T Bell Laboratories
Murray Hill, New Jersey 07974

ABSTRACT

This paper reviews recent work on Ge_xSi_{1-x}/Si(100) strained-layer epitaxy and reports new findings on Ge_xSi_{1-x}/Si(111) and Ge_xSi_{1-x}/Ge(100) growth as well as results on modulation doping. Layer synthesis and evaluation techniques are described along with tabulations of strain and critical layer thickness data. Evaluation techniques include Low Energy Electron Diffraction, Rutherford backscattering, X-ray diffraction, Raman scattering and cross-sectional Transmission Electron Microscopy. Synthesized structures range from simple heterojunctions to 100 period superlattices.

Until recently, work with silicon-based heterojunction structures was frustrated by the absence of compatible semiconducting materials. Lattice-matched semiconductors such as GaP[1] proved chemically incompatible. Other column IV semiconductors such as Ge present large lattice-mismatches. Although mismatch can be accommodated by lattice deformation in thin layers, calculations indicated that defects could be avoided only in layers on the order 10-100Å thick.[2] Early Ge_xSi_{1-x}/Si experiments supported these calculations and further indicated that smooth films could be grown only for $x \leq 0.15$.[3-5] Low-temperature, high-quality molecular beam growth techniques[6] have radically altered this situation. With low starting defect densities and inhibition of defect migration, strained-layer Ge_xSi_{1-x} growth can be maintained on silicon to thicknesses of $\sim 1~\mu$m and smooth films of all compositions can be grown.[7-9] This has yielded both simple heterostructures and complex superlattices virtually free of dislocations. This paper will review recent published work on Ge_xSi_{1-x}/Si(100) growth and present new findings on Ge_xSi_{1-x}/Si(111) and Ge_xSi_{1-x}/Ge(100) structures as well as results on modulation doping experiments.[10-11]

The bandstructures of pure Si and Ge are shown in Fig. 1. Despite fundamental similarities the materials have indirect bandgaps with conduction band minima in different k-space directions[12] (a (100) minimum $\sim 3/4$ of the way to the X zone boundary in Si versus a (111) minimum at the L zone boundary in Ge). In the alloy, the cross-over of these minima yields the nonlinear variation of energy gap[13] shown in

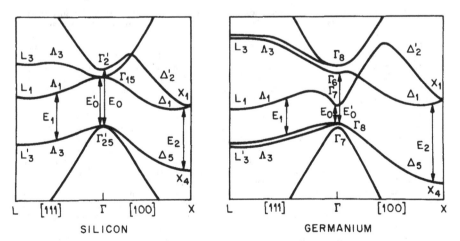

Fig. 1. Band structures of pure, unstrained Si and Ge based on electroreflectance measurements and pseudopotential calculations (after Ref. 12).

Fig. 2. These data are for pure elements or bulk unstrained alloys. As of yet little theory or data exist for strained, two-dimensional Ge_xSi_{1-x} heterostructures. Differences may be significant and at least two theorists have predicted a possible direct superlattice bandgap.[14-15] Because many proposed heterostructures require bandstructure changes ≥ 100 meV, composition changes (Δx) of 20-30% will be necessary if Si substrates are employed. Given the 4.2% lattice mismatch between Ge and Si this means that defect-free heteroepitaxy is required for mismatches $\geq 1\%$.

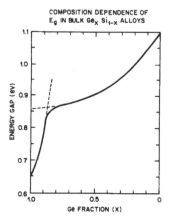

COMPOSITION DEPENDENCE OF
E_g IN BULK $Ge_x Si_{1-x}$ ALLOYS

Fig. 2. Variation of minimum bandgap in bulk, unstrained GeSi alloys as determined by optical absorption measurements (after Ref. 13).

Germanium Silicide films discussed below were grown in a UHV dual e-gun MBE system described previously.[16] While most experiments to date used 75 mm Si(100) Wacker substrates, limited work has also been done with 50 mm Si(111) AT&T Technologies and Ge(100) Eagle-Pitcher wafers. Prior to loading, all wafers were degreased and Si wafers were subjected to a hot acidic oxidizing bath. In situ cleaning consisted of a 1.5 min., 1.0 kV argon ion bombardment followed by a moderate temperature 10 min. anneal (850°C for Si, 650°C for Ge). In the past this sputter cleaning procedure has yielded high quality silicon on silicon epitaxy[17,18] and in this work it permitted smooth, high-quality growth of Ge on Ge at temperatures below 500°C (i.e., 400°C cooler than previously reported[19]). For deposition, Si and Ge fluxes were produced by e-beam evaporation sources, each of which was separately monitored and controlled at deposition rates of 0.1-10Å/sec.

Immediately after growth, films were evaluated by an in situ rear-view LEED system. Ge_xSi_{1-x} growth on both Si(100) and Ge(100) substrates yielded (2×1)(100) diffraction patterns. Reconstructions on Si(111) were considerably more complex depending on both alloy composition and thickness. For thin commensurate (i.e., strained but largely defect free) Ge_xSi_{1-x} layers, the (111) reconstruction changed from (7×7) to (5×5) to (7×7) as x increased from 0.0 to 0.5 to 1.0.[20] The anomalous reconstructions are apparently induced by strain. This is borne out by experiments with growth of 100-5000Å Ge films on Si(111). As film thickness increases above ~1000Å the Ge(7×7) pattern reverts to the more familiar c(2×8) reconstruction (subsequent X-ray measurements indicate that the transformation occurs when the Ge in-plane lattice spacing relaxes to within 0.5% of the bulk Ge spacing).[21]

After removal from the MBE system a variety of techniques were used to evaluate film quality and to determine criteria for strained-layer epitaxy. As indicated in Fig. 3, Rutherford backscattering (RBS) and channeling were used to evaluate both layer quality and strain. In normal incidence backscattering (Fig. 3a) an abrupt increase in the ratio of channeling to random yields ($\chi_{min.}$) signalled the transition from commensurate to incommensurate growth as alloy layer thickness increased. In off-normal incidence (Fig. 3b) the shift in channeling directions between substrate and epitaxial layer provided direct information on the actual film distortion. While these techniques worked well for Ge_xSi_{1-x} on Si films (where substrate and Epi backscattering peaks are well separated) other techniques are required for Ge_xSi_{1-x} on Ge or for more critical evaluation of layer quality in commensurate films where defect densities are well below RBS sensitivities. Both X-ray diffraction and transmission-electron-microscopy (TEM) were used extensively although they required substantial time for sample preparation and measurement.

A particularly rapid and versatile tool for film evaluation was found in Raman spectroscopy.[22-24] As illustrated in Fig. 4, the wavelength shift in Raman lines gives information on strain while increases in

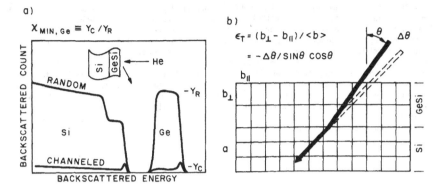

Fig. 3. Schematic diagrams on use of Rutherford Backscattering to determine commensurate-incommensurate transition in Ge_xSi_{1-x}/Si growth. Figure 3a: measurement of He channeling in Epi layer to detect disorder. Figure 3b: measurement of off-normal channel shift to quantify epitaxial distortion. Note, for Ge_xSi_{1-x}/Ge growth, superposition of epitaxial Si peak with substrate Ge counts compromises both measurements.

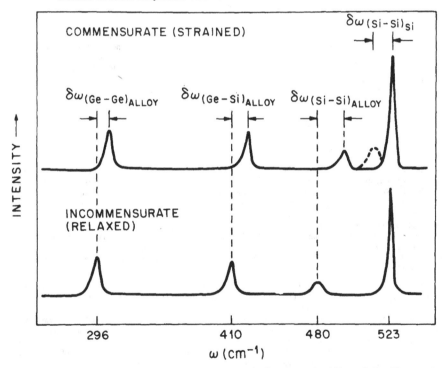

Fig. 4. Simplified Raman scattering spectra for strained (top) and unstrained (bottom) $Ge_{.65}Si_{.35}$ growth on Si. Compressive strain in alloy displaces peaks to higher energy. Absence of displaced (Si-Si)$_{substrate}$ satellite peak at lower energy (dashed line) indicates absence of significant tensile strain in cladding Si layers. Spectra substantially unchanged for superlattice vs. single layer samples (after Refs. 22 and 23).

linewidth indicate lattice disorder. The bottom panel of Fig. 4 shows the spectrum of a thick (incommensurate) $Ge_{.65}Si_{.35}$ layer on Si(100). Four peaks are evident due to the Ge-Ge, Ge-Si, and Si-Si bonds in the alloy and to the Si-Si bond in the substrate (the Si-Si alloy and substrate peaks are at different energies because of the different bandgap of these layers). The top panel shows the spectrum from a thin (commensurate) layer of the same composition. The three alloy peaks are shifted to higher energy due to compressive stress but, significantly, a lower energy satellite of the Si-Si substrate peak is not observed as would be expected for tensile relaxation. This satellite is not observed in either two layer structures or in superlattices of up to 100 periods. Its absence indicates that in Ge_xSi_{1-x}/Si commensurate heterostructures virtually all of the strain (i.e., lattice accommodation) occurs in the alloy layer. With this simple asymmetric strain accommodation Raman peaks should shift linearly with layer strain providing a simple nondestructive measure of this quantity.[23] As shown in Fig. 5, peaks shift as expected and have been correlated with other direct measures of strain.

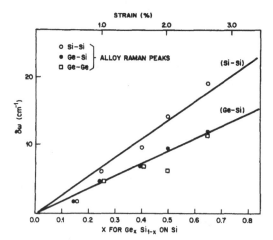

Fig. 5.

Compilation of data on Raman alloy peak shifts in commensurate layers as a function of alloy composition. Shift is linear as predicted by models and provides simple means for quantifying strain (top axis) (after Ref. 23).

Figure 6 summarizes data on strain in Ge_xSi_{1-x} on Si(100) films as determined by RBS, X-ray and Raman experiments. Measurements were typically made on samples where a shutter was stepped across a growing Ge_xSi_{1-x} layer to produce stripes of different thickness but identical composition. Absolute calibration of composition was based on RBS. It is evident that strain in the alloy layers increases linearly with composition up to some peak value and then gradually decreases. This peak value corresponds to the onset of defect formation as growth shifts form commensurate to incommensurate growth. The transformation point is a strong function of alloy layer thickness with thinner layers accommodating increasingly Ge-rich commensurate growth. This interpretation of strain data is supported by direct TEM imaging of defect structures as typified by Fig. 7, which shows the onset of incommensurate growth in buried $Ge_{.50}Si_{.50}$ films.

The critical thicknesses for Ge_xSi_{1-x} strained-layer growth are plotted in Fig. 8. Critical thickness varies from 3/4 μm for $Ge_{.20}Si_{.80}/Si(100)$ to 20Å for $Ge_{.70}Si_{.30}/Si(100)$. Although original data were only for growth on Si(100), recent experiments indicate similar critical thicknesses for Si(111) substrates. More significant, experiments with Ge_xSi_{1-x} on Ge(100) growth (x = 0.7-1.0) show a strongly complementary behavior (as indicated by the (Δ) points referred to the upper horizontal axis). These data suggest that Fig. 8 can be generalized to a plot of critical thickness vs. x for A_xB_{1-x}/B where A, B represent either Ge or Si of (100) or (111) orientation. The critical thicknesses determined are more than an order of magnitude larger than those expected from a comparison of the integrated energy involved in strained vs. misfit relieved growth,[2,3] suggesting a metastable growth mechanism. Apparently, defect generation and migration are inhibited by a combination of low temperature growth and careful substrate preparation which yields a low starting defect density. This inhibition is also evident in the very gradual relaxation of strain displayed in Fig., 6 (e.g., although the critical thickness for $Ge_{.50}Si_{.50}$ growth is 100Å, only 2/3 of the strain has relaxed in a 500Å film).

STRAIN IN Ge$_x$Si$_{1-x}$ LAYERS ON Si AND Ge

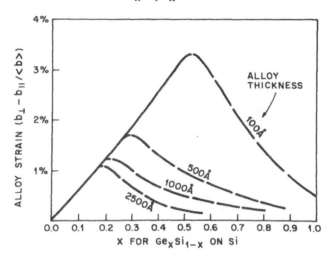

Fig. 6.   Data on strain in Ge$_x$Si$_{1-x}$/(Si, Ge) films as determined by backscattering, X-ray and Raman measurements with layer composition and film thickness as parameters. Strain peaks at commensurate to incommensurate transition which is a function of thickness. Bulk of data is for Ge$_x$Si$_{1-x}$/Si growth but data on Ge$_x$Si$_{1-x}$/Ge ($x = .70$-1.0) is in agreement if Si fraction is plotted on horizontal axis. This agreement indicates complementary strain accommodation mechanism.

Fig. 7.   TEM cross sections of buried Ge$_{.50}$Si$_{.50}$/Si(100) films showing transition from commensurate to incommensurate growth between 100 and 500Å. Transition emphasized by propagation of dislocation up into Si capping layer.

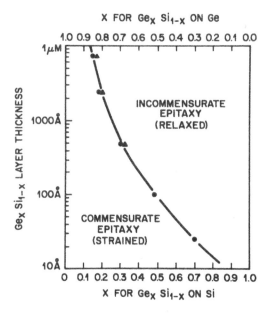

Fig. 8.

Summary of data on critical Ge_xSi_{1-x} layer thickness for commensurate growth. Bottom axis for growth on Si (●). Top axis for growth on Ge (△).

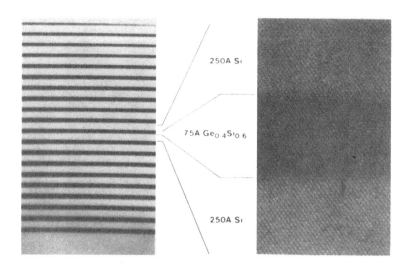

Fig. 9. Low and high magnification cross-sectional transmission electron micrographs of 20 period Ge_xSi_{1-x}/Si strained-layer superlattice.

The critical thicknesses plotted in Fig. 8 apply only to single Ge_xSi_{1-x} layers. As indicated in the Raman data of Fig. 4, silicon layers cladding Ge_xSi_{1-x} experience little deformation. As such, a second Ge_xSi_{1-x} layer grown on such a cladding layer behaves essentially as if it were grown on a semi-infinite Si substrate (provided that the Si is 2-3 times the alloy thickness). Integrated Ge_xSi_{1-x} layer thickness may thus greatly exceed single layer critical thicknesses. This is shown, for example, in the TEM cross-sections of Fig. 9 where a defect free twenty period strained-layer-superlattice[18,25] is shown. In this structure the net $Ge_{.40}Si_{.60}$ thickness is 15 times the single layer critical value. In a similar 100 period structure[26] net alloy thickness was 75 times the single layer value and growth was still fundamentally commensurate.

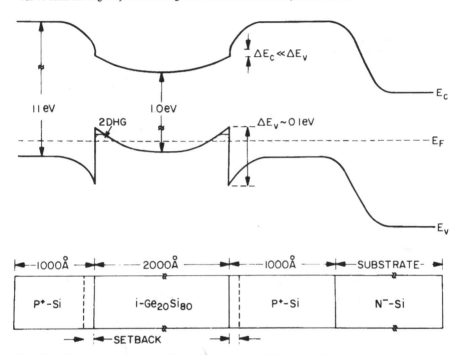

Fig. 10. Schematic cross-section and band diagram of p-modulation doped heterostructure (Refs. 10 and 11).

The above data demonstrate that Ge_xSi_{1-x} strained-layer growth provides an effective means of adding heterojunction capabilities to silicon based structures. Although these structures are metastable, preliminary annealing experiments[27] indicate little strain relaxation at subsequent processing temperatures as high as 900°C. These results suggest a large number of device applications; some of these are unique, many are simply analogs of existing compound semiconductor structures. A dramatic early success was provided by the modulation doped structure[10] indicated in Fig. 10. A 2000Å thick undoped $Ge_{.20}Si_{.80}$ layer is clad by doped Si layers. As demonstrated in III-V devices, one expects carriers generated in the Si to fall into the lower energy Ge_xSi_{1-x} well. If this occurs carrier freezeout should be avoided and mobilities will be enhanced by the reduction of ionized impurity scattering. This indeed occurs as shown in the Hall hole mobility data of Fig. 11. In even these early structures hole mobilities meet or exceed values measured in p-channel MOS devices. The two-dimensional nature of the hole gas is also demonstrated by the magneto-resistance data of Fig. 12 which displays Shubnikov-de-Haas oscillations indicating bound particle-in-a-box states. Further experiments[11] on P-type modulation doped structures, varying parameters such as well width, doping setback distance from the well, and doping level, display systematic variations consistent with first order models of modulation doping (and with far less scatter than observed in the better developed III-V modulation doped structures[28]). Preliminary experiments with N-type doping indicate premature carrier freezeout, suggesting that most of the $Ge_{.20}Si_{.80}/Si$ bandgap discontinuity occurs at the valence band edge as shown at the top of Fig. 10. More Ge-rich alloys will therefore be necessary to achieve a complementary N-type modulation doped structures. Modulation doped transistors, photodetectors, heterojunction bipolar transistors and a variety of other structures and will be reported in the near future.

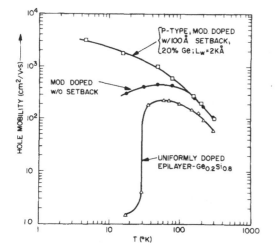

Fig. 11.

Temperature vs. Hall hole mobility for thick uniformly doped $Ge_{0.2}Si_{0.8}$ and modulation doped heterostructures with and without setback of dopant from 2000Å $Ge_{.20}Si_{.80}$ well. Note absence of freezeout and beneficial effect of setback in modulation doped films.

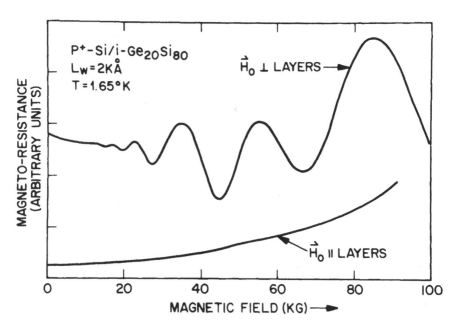

Fig. 12. Magneto-resistance data for structure of Fig. 10 showing Shubnikov-de-Haas oscillation for perpendicular H indicating existence of two dimensional hole gas.

To summarize, molecular beam deposition techniques now allow synthesis of $Ge_xSi_{1-x}/(Si, Ge)$ heterostructures of useful quality and thickness. This occurs by a metastable strain accommodated growth mechanism wherein lattice mismatch is accommodated by plastic deformation of alloy layers. Critical thicknesses are from one to two orders of magnitude larger than predicted by equilibrium theory in single and multiple heterointerface structures respectively. This impressive degree of metastability depends on both low-temperature growth and low-defect-density substrate surfaces to inhibit nucleation and migration of strain relieving dislocations. Quality Ge_xSi_{1-x}/Si growth opens the door to a wide range of silicon-based heterojunction structures as typified by the silicon modulation doped transistor.

The results of this study were critically dependent upon the sustained, enthusiastic collaboration of a large number of co-workers, some (but not all) of whom are cited in the references of this paper.

REFERENCES

[1] T. de Jong, W. A. S. Douma, J. F. van der Veen, F. W. Saris, and H. Haisma, Appl. Phys. Lett. 42, 1037 (1983).

[2] J. H. Van der Merwe and C. A. B. Ball in Epitaxial Growth, edited by J. Matthews (Academic, New York, 1975), Part b.

[3] E. Kasper, H. J. Herzog and H. Kibbel, Appl. Phys. 8, 199 (1975).

[4] E. Kasper and W. Pabst, Thin Solid Films, 37, L5 (1976).

[5] E. Kasper and H. J. Herzog, Thin Solid Films, 44, 357 (1977).

[6] J. C. Bean, Chap. 4, Impurity Doping Processes in Silicon, F. F. Y. Wang Ed., North Holland, Amsterdam (1981).

[7] J. C. Bean, T. T. Sheng, L. C. Feldman, A. T. Fiory and R. T. Lynch, Appl. Phys. Lett. 44, 102 (1984).

[8] J. C. Bean, L. C. Feldman, A. T. Fiory, S. Nakahara and I. K. Robinson, J. Vac. Sci. Technol. A 2(2), 436 (1984).

[9] A. T. Fiory, J. C. Bean, L. C. Feldman and I. K. Robinson, J. Appl. Phys. 56(4), 1227 (1984).

[10] R. People, J. C. Bean, D. V. Lang, A. M. Sergent, H. L. Störmer, K. W. Wecht, R. T. Lynch and K. Baldwin, to be published Appl. Phys. Lett., Dec. 1984.

[11] R. People, J. C. Bean and D. V. Lang, submitted to J. Vac. Sci. Technol. A.

[12] J. S. Kline, F. H. Pollack and M. Cardona, Helv-Phys. Acta 41, 968 (1968).

[13] R. Braunstein, A. R. Moore and R. Herman, Phys. Rev. 109, 695 (1958).

[14] J. A. Moriarity and S. Krishnamurthy, J. Appl. Phys. 54, 1892 (1983).

[15] S. Krishnamurthy, Ph.D. Thesis Univ. of Cincinnati, Dept. Physics (1984).

[16] J. C. Bean and E. A. Sadowski, J. Vac. Sci. Technol. 20, 137 (1982).

[17] J. C. Bean, G. E. Becker, P. M. Petroff and T. E. Seidel, J. Appl. Phys. 48, 907 (1977).

[18] J. C. Bean, to be published J. Cryst. Growth, early 1985.

[19] Y. Ota, J. Cryst-Growth 61, 431 (1983).

[20] H. J. Gossman, J. C. Bean and L. C. Feldman, Surf. Sci. 138, L175 (1984).

[21] H. J. Gossman, J. C. Bean, L. C. Feldman, E. G. McRae and I. K. Robinson, submitted to J. Vac. Sci. Technol. A.

[22] F. Cerdeira, A. Pinczuk, J. C. Bean, B. Batlogg and B. A. Wilson, Proc. 3rd Int. Conf. on MBE, San Francisco (Aug. 1984).

[23] F. Cerdeira, A. Pinczuk, J. C. Bean, B. Batlogg and B. A. Wilson, to be published Appl. Phys. Lett.

[24] F. Cerdeira, A. Pinczuk and J. C. Bean, submitted to Phys. Rev. Rapid. Comm.

[25] R. Hull, J. M. Gibson and J. C. Bean, to be published Appl. Phys. Lett.

[26] R. Hull, A. T. Fiory, J. C. Bean, J. M. Gibson, L. Scott, J. L. Benton and S. Nakahara, Proc. 13th Int. Conf. on Defects in Semiconductors, Coronado, CA (Aug. 1984).

[27] A. T. Fiory, J. C. Bean, R. Hull and S. Nakahara, submitted to Phys. Rev. Rapid Comm.

[28] J. C. M. Hwang, A. Kastalsky, H. L. Störmer and V. G. Keramides, Appl. Phys. Lett. 44, 802 (1984).

EPITAXIAL GROWTH OF Si_xGe_{1-x} FILMS ON Si BY SOLID PHASE EPITAXY.

C. S. PAI AND S. S. LAU
Department of Electrical Engineering and Computer Sciences, Mail Code C-014
University of California, San Diego, La Jolla, CA 92093

ABSTRACT

It has been demonstrated in the literature that amorphous Si (or Ge) can be transported across a metal layer and grown epitaxially on Si(Ge) single crystal substrates in the solid phase. The objective of this study is to investigate if amorphous Si_xGe_{1-x} mixtures can be transported uniformly across a medium and grown epitaxially on single crystal substrates without phase separation. The samples were prepared by e-beam evaporation of thin Pd films onto Si<100> substrates, followed by co-evaporation of Si_xGe_{1-x} alloyed films (0<x<1) without breaking vacuum. The samples were annealed in vacuum at 300°C to form a Pd silicide-germanide layer at the interface, then at 500°C for transport of the alloyed layer across the Pd silicide-germanide layer and subsequent epitaxial growth on Si substrate. The samples were investigated by x-ray diffraction and by MeV ion backscattering and channeling. The results show the alloyed film transports uniformly with no phase separation detected. The channeling result shows the grown alloyed layer is epitaxial with some Pd trapped in the layer. This simple technique is potentially useful for forming lattice-matched non-alloyed ohmic contacts on III-V ternary and quaternary compounds.

INTRODUCTION

Homostructures and heterostructures have many important applications in microelectronics. One of the interesting applications is the utilization of heterojunction to form a non-alloyed ohmic contact on GaAs. It has been demonstrated that contact resistivities of the order of 10^{-7} Ω cm^2 can be formed by the growth of n^+Ge layers using molecular beam epitaxy on n^+ GaAs substrates[1]. Recently solid phase epitaxy with a transport medium has been used to form non-alloyed ohmic contacts on GaAs with reasonable success[2]. In this case, a sample with an initial configuration of Ge (amorphous)/Pd/GaAs(100) is annealed at ~ 350°C to yield a final sample configuration of PdGe/Ge (epitaxial layer)/GaAs(100). The amorphous Ge layer migrates across the transport medium (the PdGe layer) and grows epitaxially on the GaAs substrate. This solid phase epitaxial (SPE) technique relaxes the high vacuum requirement for sample preparation. The objective of this work is to investigate the growth of Si_xGe_{1-x} layers on semiconductor substrates by means of the SPE process. The epitaxial growth of an alloy film has the potential advantage of lattice matching with the underlying substrate and band gap adjustment. In this investigation, Si substrate with (100) orientation was chosen. Epitaxial growth of Si_xGe_{1-x} on Si, over the entire compositional range, by means of molecular beam epitaxy has been reported in the literature[3]. Our particular interest in this experiment is to investigate if an amorphous Si_xGe_{1-x} alloy can be transported uniformly across a medium without phase separation. In other words, the point of interest is to find out if the amorphous alloy migrates as an integral unit or one of the components, say Ge, migrates first followed by the other component during the SPE process.

Experimental

Si substrates with (100) orientation were first degreased using acetone, isopropyl alcohol and D.I. water. After the final etching in 10% HF for 30 seconds and rinse in D.I. water, samples were blown dry using nitrogen gas and were loaded immediately into an e-gun evaporator equipped with ion pumps. For each sample Pd (∼ 300 Å) was first evaporated onto Si substrate followed by an evaporation of a near constant composition Si_xGe_{1-x} alloy film (∼ 3000 Å) on top of the Pd layer without breaking vacuum. Alloy layers with composition x = 0.90, 0.80, 0.64, 0.30 and 0.20 were deposited. The pressure before and during evaporation was 1×10^{-8} torr and 1×10^{-7} torr, respectively. The evaporation rates of Pd and the alloy film were 10 Å/sec and 25 Å/sec, respectively. These samples were then loaded into a vacuum annealing furnace (∼ 1×10^{-7} torr) for a two step annealing cycle. Samples were first annealed at 300°C to form a Pd silicide-germanide layer at the interface, then at 500°C for the transport of the alloy layer across the Pd silicide-germanide layer and subsequent epitaxial growth on the Si substrate. X-ray diffraction (Read camera) technique was used for determining the phases of the reaction products and MeV ion backscattering and channeling techniques were used for depth profiling and monitoring the epitaxial growth. Before channeling measurements, the Pd-Si-Ge ternary compound on the sample surface was removed as much as possible by etching in a hot aqua regia solution.

Experimental Results and Dicussion

For Si_xGe_{1-x}/Pd/Si(100) samples with x>0.5 (Si rich), x-ray diffraction experiments showed the possible formation of a $Pd_2(Si_xGe_{1-x})$ ternary compound after annealing at 300°C for one hour (figure 1a). For samples with x<0.5 (Ge rich), the presence of both $Pd_2(Si_xGe_{1-x})$ and PdGe compounds were detected after the same annealing cycle (figure 2a). Since Pd_2Si and Pd_2Ge are isomorphic with very similar lattice constants[4] (Pd_2Si, hexagonal, a = 6.493 Å, c = 3.429 Å; Pd_2Ge, hexagonal, a = 6.67 Å, c ≅ 3.39 Å) and with the limited resolution of the Read camera, we could not confirm directly that a $Pd_2(Si_xGe_{1-x})$ ternary compound was formed at 300°C by x-ray measurements. However, it has been argued that when two isomorphic compounds are mixed together, very likely a ternary compound would be formed such as in the case of $TbSi_{1.7}$ mixed with $TbGe_{1.7}$ resulting in a Tb-Ge-Si ternary compound[5]. We believe that the formation of a $Pd_2(Si_xGe_{1-x})$ ternary compound prevails in our case.

The x-ray diffraction patterns obtained from samples either with x>0.5 or x<0.5 after additional annealing at 500°C for 10 hours, (where transport and growth took place) were almost identical. Only one set of diffraction rings pertaining to the $Pd_2(Si_xGe_{1-x})$ compound and a few diffraction spots pertaining to an epitaxial Si_xGe_{1-x} layer were observed (figures 1b and 2b). It is interesting to note that the PdGe compound which co-existed with the $Pd_2(Si_xGe_{1-x})$ phase before the 500°C annealing was converted into the $Pd_2(Si_xGe_{1-x})$ phase after the annealing for samples with x<0.5 (Ge rich, see figure 2b). For samples with x>0.5, no detectable change of the Pd-Si-Ge compound was observed. The diffraction spots pertaining to the epitaxial layer were asterisk in shape and in some cases a faint diffraction ring connected these asterisk spots (see figures 1b and 2b). This observation indicates that the crystalline quality of the epitaxial layers were not perfect, also shown by channeling measurements presented later. The spots did not appear to be split, suggesting that a homogeneous Si_xGe_{1-x} layer gave rise to these diffraction spots and not from a phase-separated Si and Ge mixture. (The resolution limit of our Read camera is good enough to distinguish diffraction lines between Si and Ge). The 2θ position of these asterisk spots is clearly related to the composition of the epitaxial layer, indicating again that the epitaxial layer has not phase separated.

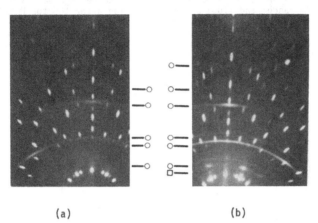

(a)   (b)

Figure 1 X-ray diffraction patterns of a sample with a configuration of $Si_{64}Ge_{36}$/Pd/Si(100) after annealing at (a) 300°C for 1 hr and (b) additional 500°C for 10 hrs. The diffraction rings from the $Pd_2(Si_xGe_{1-x})$ compound are marked with o and the diffraction spots from grown $Si_{64}Ge_{36}$ layer are marked with □ . The diffraction angle 2θ measured from these diffraction spots is 27.9.

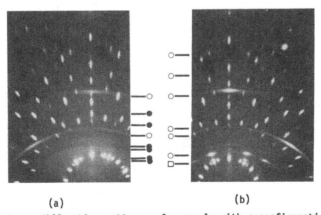

(a)   (b)

Figure 2 X-ray diffraction patterns of a sample with a configuration of $Si_{30}Ge_{70}$/Pd/Si(100) after annealing at (a) 300°C for 1 hr and (b) additional 500°C for 10 hrs. The diffraction rings from the $Pd_2(Si_xGe_{1-x})$ compound and the PdGe compound are labled o and ● respectively. The diffraction spots from the grown $Si_{30}Ge_{70}$ layer are marked with □ . The diffraction angle 2θ measured from these diffraction spots is 27.5.

Figure 3a shows the backscattering spectra for a sample with a configuration of $Si_{64}Ge_{36}$ (a, ~ 3000 Å)/Pd (~ 300 Å)/Si(100) before and after annealing at 300°C for 1 hour and 500°C for 10 hours. It can be seen that after the 500°C annealing, the Pd signal appears on the sample surface, indicating that the amorphous $Si_{64}Ge_{36}$ mixture has transported across the $Pd_2(Si_xGe_{1-x})$ layer. Figure 3b shows the backscattering spectra of the same sample in the random and (100) aligned direction after the top $Pd_2(Si_xGe_{1-x})$ has been removed by hot aqua regia. It can be seen that the grown layer (~ 2800 Å thick) has the same composition as that before transport i.e., $Si_{64}Ge_{36}$. Channeling effect was also evident although relatively poor. This observation is consistent with the x-ray result in that the grown layer

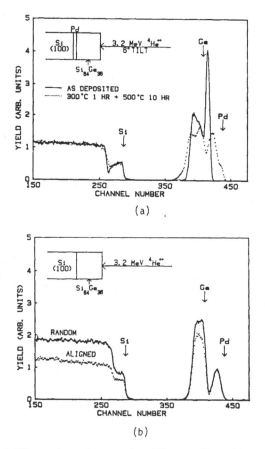

(a)

(b)

Figure 3 (a) RBS spectra of a sample with a configuration of $Si_{64}Ge_{36}$/Pd/Si(100), before and after annealing at 300°C for 1 hr and 500°C for 10 hrs, (b) (100) aligned channeling and random spectra of the same sample after the top $Pd_2(Si_xGe_{1-x})$ compound layer was removed. The thickness and the x_{min} of the grown layer are ~ 2800 Å and ~ 80%, respectively.

is of poor crystalline quality. It is also evident from the spectra that the grown layer is uniform in composition in depth; this observation in conjunction with un-split x-ray diffraction asterisks shown in figure 1b lead us to believe that there was no phase separation between Si and Ge during transport and that the grown layer is a homogeneous alloy of the composition of $Si_{64}Ge_{36}$. The Pd signal centered around channel #430 represents the Pd compound trapped inside the grown layer. The trapping of the transport compound is a common phenomenon in solid phase epitaxy with a transport medium[6,7], and the amount of trapping generally decreases with slower growth rate[8].

Figure 4a shows the backscattering spectra for a sample with a configuration of $Si_{30}Ge_{70}$ (a, ~ 3000 Å)/Pd (~ 300 Å)/Si(100) before and

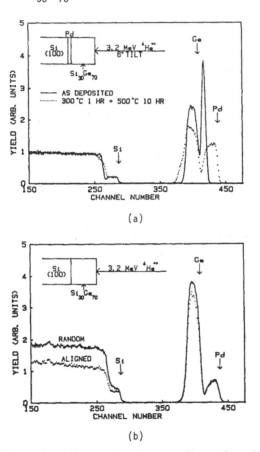

(a)

(b)

Figure 4(a) RBS spectra of a sample with a configuration of $Si_{30}Ge_{70}$/Pd/Si(100), before and after annealing at 300°C for 1 hr and 500°C for 10 hrs, (b) (100) aligned channeling and random spectra of the same sample after the top $Pd_2(Si_xGe_{1-x})$ compound layer was removed. The thickness and the x_{min} of the grown layer are ~2800 Å and ~90%, respectively.

after annealing at 300°C for 1 hour and 500°C for 10 hours. The amorphous mixture with a composition of $Si_{30}Ge_{70}$(Ge rich in this case) can be seen to have transported across the Pd-Si-Ge medium. Figure 4b shows the backscattering spectra of the same sample after the removal of the top Pd-Si-Ge compound in the random and the (100) aligned direction. Channeling effect was again observed on the grown layer which also appears to be a homogeneous alloy.

In summary, we have demonstrated that an amorphous Si_xGe_{1-x} mixture can be transported across a metallic compound medium and grown epitaxially onto a Si substrate without phase separation. The crystalline quality of the grown layers is relatively poor, however, it may be good enough for ohmic contact purposes.

The financial support from NSF (Grant No. DMR-8106843, L. Toth) is greatly appreciated.

References

[1] R. Stall, C. E. C. Wood, K. Board and L. F. Eastman, Elect. Lett. 15, 800 (1979).
[2] E. D. Marshall, C. S. Wu, D. M. Scott, S. s. Lau and T. F. Kuech, in Thin Films and Interfaces II, edited by J. E. E. Baglin, D. R. Campbell and W. K. Chu, MRS Symposia Proceedings, volume 25, North-Holland (1984, p. 63).
[3] J. C. Bean, T. T. Sheng, L. C. Feldman, A. T. Fiory and R. T. Lynch, Appl. Phys. Lett. 44, 104 (1984).
[4] See ASTM cards, 6-558 and 6-559.
[5] J. E. E. Baglin, F. M. d'Heurle and C. S. Petersson, J. Appl. Phys. 52 2841 (1981).
[6] S. S. Lau. Z. L. Liau and M.-A. Nicolet, Thin Solid Films 47, 313 (1977).
[7] G. Majni, G. Ferrari, R. Ferrai, C. Canali, F. Catellani G. Ottaviani and G. Della Mea, Thin Solid Films, 44, 193 (1977).
[8] W. Tseng, Z. L. Liau, S. S. Lau, M.-A. Nicolet and J. W. Mayer, Thin Solid Films, 46, 99 (1977).

INTERFACIAL STRUCTURE AND STABILITY IN Ge$_x$Si$_{1-x}$/Si STRAINED LAYERS.

R.HULL, J.C.BEAN, J.M.GIBSON, K.J.MARCANTONIO, A.T.FIORY and S.NAKAHARA
AT&T Bell Laboratories, 600 Mountain Avenue, Murray Hill, NJ 07974

ABSTRACT

High resolution electron microscopy is used to probe the atomic scale structure of interfaces and defects in the Ge$_x$Si$_{1-x}$/Si system. By careful quantification of lattice images, it is shown that molecular beam epitaxy may be used to grow Ge$_x$Si$_{1-x}$/Si (100) and (111) interfaces which are sharp on the scale of the unit cell and flat to within a few atomic planes when about 5000 Å^2 of the interface are sampled. Interfacial quality is retained in single and multiple quantum well structures. Conditions for superlattice stability against misfit dislocations are discussed. It is shown that Ge$_x$Si$_{1-x}$/Si interfaces produced by molecular beam epitaxy at 550°C can exist in a metastable state which relaxes upon thermal annealing.

INTRODUCTION

The ultra-high resolution technique of direct lattice imaging from electron diffraction and the complementary techniques of diffraction contrast electron microscopy and ion channeling which sample larger (and thus more representative) areas of the crystal, are a powerful combination of tools in the study of epitaxial semiconductor interfaces [1]. In this paper, we correlate the atomic-scale information provided by high resolution electron microscopy (HREM) with the average interfacial properties obtained from the other two techniques. In particular, we show how HREM may provide accurate quantitative information about interfacial profiles via the quantification of lattice images. The detailed analysis of such information is discussed.

SAMPLE GROWTH

The optimum conditions for molecular beam epitaxial growth of Ge$_x$Si$_{1-x}$ alloys upon silicon are discussed elsewhere in these proceedings [2]. Here we emphasise only those results which have a direct bearing on this paper. Previous work on the Ge$_x$Si$_{1-x}$/Si system [3,4,5] has established that Ge$_x$Si$_{1-x}$ alloys may be grown epitaxially upon Si (100) surfaces if the deposition parameters are carefully controlled. In this work, we use molecular beam epitaxy (MBE) to deposit Ge$_x$Si$_{1-x}$ layers on both Si (100) and (111) surfaces. It is found that the alloy layers may be deposited in an epitaxial and commensurate (i.e. dislocation-free) fashion, up to a critical thickness which increases rapidly as the germanium concentration decreases [2,3,4,5]. Below the critical thickness, the alloy exists as a strained layer, with the same lattice parameter parallel to the interface as the Si substrate (in its relaxed state, Ge$_x$Si$_{1-x}$ has a larger lattice parameter than Si, the mismatch increasing with higher Ge concentration). As the critical thickness is exceeded, misfit dislocations relieve the strain in the film.

EXPERIMENTAL DETAILS

Lattice imaging was performed on a JEOL 200CX high resolution electron microscope, which has a point-to-point resolution of 2.5 Å. With the electron beam along the relevant <011> axis of the crystal which is parallel to the interface, two sets of {111} planes may be imaged. Under the correct combination of crystal thickness and objective lens defocus, the image intensity is approximately proportional to the projected crystal potential [6]. This relationship holds only for extremely thin crystals (so-called "weak phase objects" where the only effect of the crystal upon the incident electrons is assumed to be a weak modification of their relative phases) at a specific value of the objective lens defocus, called Scherzer defocus. Images of Si or Ge$_x$Si$_{1-x}$ in the JEOL 200CX then consist of a two-dimensional array of white dots (see Fig.1, for example) which correspond to tunnels in the structure. Intervening darker areas correspond to columns of atom pairs aligned along the beam direction. Images which do not correspond to these ideal conditions may be interpreted by careful comparison with simulations produced by numerical calculation, usually using the "multislice" method [7].

Ion channeling experiments were performed using 1.8 MeV He ions, whilst conventional diffraction contrast transmission electron microscopy was carried out on a Phillips 420 TEM.

INTERFACIAL QUALITY

Figure 1 shows lattice images of commensurate interfaces between Ge_xSi_{1-x} alloys and (a) a Si(100) substrate and (b) a Si(111) substrate. Detailed analysis of interfacial contrast and subsequent deductions about interfacial sharpness and flatness require considerable computation. It is found that contrast between alloy layer and substrate depends critically upon a variety of experimental parameters, primarily crystal thickness, objective lens defocus and alloy composition. In the ideal imaging limit of extremely thin crystals at the Scherzer defocus for the objective lens, interfacial contrast should be proportional to the Ge content of the alloy. This is demonstrated in Figure 2, where the solid line indicates the projected potential for low-angle electron scattering [8]. Also shown in the same figure are multislice calculations of the ratio of mean image intensities between Si and Ge_xSi_{1-x} as a function of composition and crystal thickness. Note that for extremely thin crystals (see e.g. curve for 31 Å in Fig. 2), the contrast corresponds closely to the variation in projected potential. For thicker crystals, multiple scattering of diffracted electrons becomes significant and the interfacial contrast bears no obvious relationship to the change in projected potential across the interface. For certain combinations of crystal thickness and alloy composition interfacial contrast is significantly enhanced, whilst under certain conditions the Ge_xSi_{1-x} layer appears lighter than the silicon substrate. It is also found that interfacial contrast has a periodic dependence on objective lens defocus [9] which arises essentially from phase differences arising from the different extinction distances for Ge_xSi_{1-x} and Si. Finally, the effects of near-surface elastic relaxation may be significant [10]. The extreme sensitivity of contrast between alloy layer and substrate to experimental conditions means that HREM is not well-suited to determining absolute Ge concentrations in the alloy (this can be determined more accurately and over a more representative area using ion channeling). By quantifying (i.e. making image intensity measurements of) interfacial images, however, we can obtain relatively good sensitivity to changes in Ge concentration. In particular, we can determine whether there is diffusion of Ge from the alloy into the substrate during the deposition process leading to a broadening of the interface. Calculations such as those represented in Figure 2 lead us to expect that under optimum conditions we should be sensitive to a variation in the absolute Ge concentration of about 5% [9].

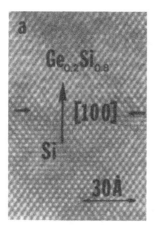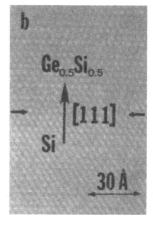

Figure 1: Lattice images of commensurate interfaces between (a) 1000 Å of $Ge_{0.2}Si_{0.8}$ and Si (100) and (b) 100 Å of $Ge_{0.5}Si_{0.5}$ and Si(111).

Two further parameters to consider in such measurements are the effect of alloy disorder and interfacial roughness. The latter will ensure that any quantification which is averaged over a significant area of the interface will only yield information about the combined effects of interfacial roughness and interdiffusion. To differentiate between these two effects, the interface area scanned should be as small as possible. For a microdensitometer trace, this corresponds to a slit which ideally samples an area of only about one square lattice parameter projected through an extremely thin crystal. As the slit area is reduced, however, the effects of noise in the image become more severe: each point in the image has superimposed upon it a significant amount of random intensity due primarily to the effects of contamination and damage arising from specimen preparation. Thus reduced slit areas lead to noisier data and a subsequent loss in sensitivity in detecting variations in Ge concentration. A final factor to consider is that the random nature of the alloy will produce apparent interfacial roughness. In the extremely thin specimens used for lattice imaging, the number of atoms in each column is relatively small (for a 50 Å crystal in Si, there will be 26 atoms in each unresolved column pair) and statistical fluctuations in the number of Ge atoms in each column will be significant. This will produce further variations in the image intensity in the alloy, the effect of which at the interface will be to make it appear less abrupt.

These effects merit detailed quantitative investigation, as they will have to be taken into account if accurate atomic-scale measurements of interfaces are to be attempted. We have undertaken preliminary quantitative measurements of interfacial geometry in the Ge_xSi_{1-x}/Si system. Figure 3 shows a microdensitometer scan across a $Ge_{0.2}Si_{0.8}/Si$ (100) interface. In this case, the combined effects of interdiffusion, alloy disorder and interfacial roughness are such as to cause the interface to extend over about 3 (200) planes within the area (about 5000 Å^2) of interface sampled. Figure 4 shows (a) an interface in a 50 Å $Ge_{0.4}Si_{0.6}/$ 50 Å Si(100) superlattice and (b) to (d) microdensitometer traces across the interface with decreasing slit areas. Note that as the sampling area of interface, and hence the sampling volume within the crystal, decreases the apparent interface breadth decreases, until in (d) where an area of less than one square lattice parameter projected

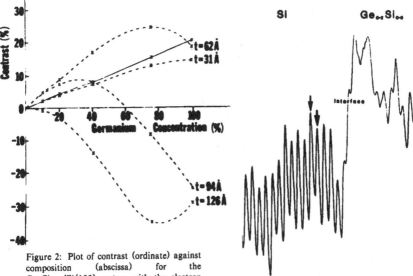

Figure 2: Plot of contrast (ordinate) against composition (abscissa) for the $Ge_xSi_{1-x}/Si(100)$ system with the electron beam along <011>. Positive contrast is defined by the ratio of the mean image intensity in Si to that in Ge_xSi_{1-x}. The solid line shows the projected potential for low angle electron scattering, whilst broken lines show multislice calculations for the relevant crystal thicknesses.

Figure 3: Image intensity (microdensitometer) trace across a $Ge_{0.2}Si_{0.8}$ /Si (100) interface. The interface area sampled by the microdensitometer slit is approximately 5000 Å^2. Arrows indicate (200) planes 2.7 Å apart in Si.

through the crystal thickness (approximately 70 Å) is sampled at a time, the interface appears atomically sharp. This is in agreement with expected bulk diffusion lengths of Ge in Si at the deposition temperature (550°C) used for this sample. It is worth noting, however, that such diffusion lengths may not be applicable to the atomic-scale diffusion encountered here. Indeed, lattice image quantification is probably able to detect smaller interdiffusion lengths than any other experimental technique. We conclude that the interfaces we have grown in the Ge_xSi_{1-x}/Si system at 550°C have interdiffusion lengths less than the unit cell dimension and for a typical sampling volume of about 15000 Å^3 (corresponding to about 5000 Å^2 of interface area) are constrained to about 3 (200) planes - about 8 Å.

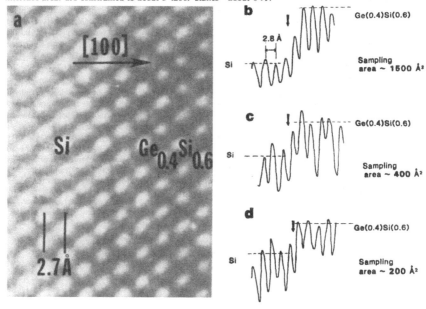

Figure 4: (a) Lattice image of a $Ge_{0.4}Si_{0.6}/Si(100)$ interface and (b)-(d) microdensitometer traces with decreasing slit area scanned across the interface in (a).

MULTIPLE AND SINGLE QUANTUM WELLS

We have been able to grow commensurate single and multiple (i.e. superlattice) quantum well structures in the $Ge_xSi_{1-x}/Si(100)$ system which have interfacial quality comparable with single $Ge_xSi_{1-x}/Si(100)$ interfaces. Figure 5 shows a 20-period 75 Å $Ge_{0.4}Si_{0.6}/250$ Å Si(100) superlattice which appears to be commensurate throughout its extent. Although the total thickness of $Ge_{0.4}Si_{0.6}$ alloy is significantly greater than the critical thickness for this composition, all interfaces remain commensurate as each individual layer is well below that critical thickness. As the number of periods increases, however, misfit dislocations start to relieve the strain. The preferred location for these dislocations is at the interface between the substrate and the first alloy layer, as in the electron microscope image of a 100-period 75 Å $Ge_{0.4}Si_{0.6}/250$ Å Si(100) superlattice in Figure 6. This preferred site for dislocations in the larger superlattice may be understood in terms of strain energy relief: misfit dislocations which relieve strain at the substrate/alloy interface serve to expand the $\overline{average}$ in-plane lattice parameter of the succeeding superlattice. Although the strain is not fully relieved, it will be shared more equally between the Ge_xSi_{1-x} alloy and silicon layers in the superlattice, the former being under compression and the latter under tension. In the dislocation-free superlattice, all the strain is contained in the Ge_xSi_{1-x} layers [11]. The strain energy in the system is equal to the sum of the strain energies in each individual superlattice layer. The strain due to the lattice mismatch, $\phi(x)$, between Ge_xSi_{1-x} and Si will be proportional to x and will be denoted here by $\epsilon(\phi)$. Misfit dislocations at the substrate/alloy interface will expand the average lattice parameter of the superlattice by an amount β. For alloy and Si layers of thicknesses d,D respectively, the elastic strain energy in the superlattice will be approximately proportional to $[d\epsilon^2(\phi-\beta)+D\epsilon^2(\beta)]$. For the dislocation-free case, $\beta = 0$, and the case where all the strain at the substrate/alloy interface is relieved, $\beta = \phi$, this reduces simply to $d\epsilon^2(\phi)$ and $D\epsilon^2(\phi)$ respectively. For $\frac{|d-D|}{(d+D)}\phi$ (x) $< \beta < \frac{2d}{(d+D)}\phi$ (x), the elastic strain energy will be less than either of these values. Note that misfit dislocations which reside at interfaces within the superlattice do not relieve as much strain energy as those which lie at the substrate/alloy interface, as the former will relieve strain energy only at interfaces between themselves and the surface. Dislocations lying at the latter location also have the additional advantage of not interfering with electronic conduction in the bulk of the superlattice. A condition for maximum strain energy relief due to dislocations at the substrate/alloy interface may be obtained by setting to zero the derivative of E with respect to β, yielding $\beta \simeq d\frac{\phi}{(d+D)}$. This value for β corresponds to the weighted (over d,D) average lattice parameter in the superlattice. This analysis would, of course, be modified by including the dislocation energy, E_d, but assuming dislocation formation to be energetically favourable, the tendency for $0 < \beta < \phi$ to be the minimum energy configuration would still be true.

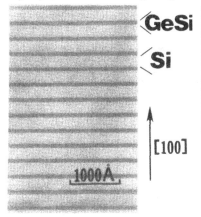

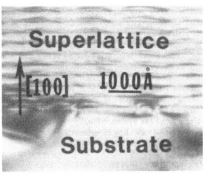

Figure 5: Part of a 20-period 75 Å Ge$_{0.4}$Si$_{0.6}$/250 Å Si(100) superlattice.

Figure 6: Dislocations at the interface between a Si (100) substrate and the first alloy layer in a 100-period 75 Å Ge$_{0.4}$Si$_{0.6}$/250 Å Si superlattice. Intensity modulations perpendicular to the interfaces are thought to be due to buckling of the highly-strained thin specimen prepared for electron microscopy.

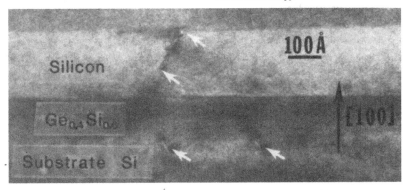

Figure 7: Dislocations (arrowed) in a 20-period 75 Å Ge$_{0.4}$Si$_{0.6}$/250 Å Si(100) superlattice annealed for 30 minutes at 750°C.

METASTABILITY

Measured critical thicknesses [2,5] in the Ge_xSi_{1-x}/Si system are as much as an order of magnitude larger than predicted by calculations balancing strain and dislocation energies [3,12], indicating that these films may exist in a metastable state. This possibility has been investigated by comparing dislocation densities before and after thermal annealing. Detailed results will be published elsewhere, but briefly HREM and ion channeling indicate that strain relief occurs by two mechanisms: dislocation formation and interdiffusion, demonstrating that the unannealed films are indeed in a metastable state. Figure 7 shows the superlatice imaged in Figure 5 after 30 minutes annealing at $750^{\circ}C$. Dislocations have clearly formed during annealing. The preferred site is again the substrate/alloy interface, but some dislocations are now present at interfaces within the superlattice. This probably indicates that the previously commensurate 75 Å $Ge_{0.4}Si_{0.6}$ layers are now relaxing away from a metastable state.

CONCLUSIONS

It is shown how high resolution electron microscopy can provide atomic-scale information about epitaxial semiconductor-semiconductor interfaces in the Ge_xSi_{1-x}/Si system. Quantification of lattice images yields accurate information about interfacial profiles and geometries, and it is shown how HREM can be used as a sensitive tool for detecting and measuring diffusion lengths which are comparable to the unit cell dimensions. Interfacial quality is retained in single and multiple quantum well structures, making the growth of well-defined extremely thin layers possible. Although strained layer superlattices may be grown with total alloy thicknesses significantly greater than the critical thickness for a single interface of the relevant alloy composition, it is shown that a superlattice with a sufficiently large number of periods relaxes via the introduction of misfit dislocations mainly at the interface between the substrate and the first alloy layer. It is shown how formation of dislocations at this location is most effective in lowering the strain energy of the system. Finally, the existence of a metastable state is demonstrated in as-deposited films via thermal annealing experiments.

ACKNOWLEDGEMENTS

The authors gratefully acknowledge the technical assistance of T.Boone, R.T.Lynch and M.L.McDonald and useful discussions with L.C.Feldman. Figure 3 was produced by L.Scott. of Southern University, Louisiana.

REFERENCES

[1] R.Hull, A.T.Fiory, J.C.Bean, J.M.Gibson, L.Scott, J.L.Benton and S.Nakahara, to be published in J.Elec.Mat.

[2] J.C.Bean, these proceedings

[3] E.Kasper, H.J.Herzog and H.Kibbel, Appl. Phys. 8, 199 (1975).

[4] H.M.Manasevit, I.S.Gergis and A.B.Jones, J.Elec. Mat. 12, 637 (1983).

[5] J.C.Bean, L.C.Feldman, A.T.Fiory, S.Nakahara and I.K.Robinson, J. Vac. Sci. Technol. A2, 436 (1984).

[6] See, for example, High Resolution Electron Microscopy, J.C.H.Spence (Clarendon Press, Oxford, 1981).

[7] P.Goodman and A.F.Moodie, Acta. Cryst. A30, 280 (1974).

[8] P.A.Doyle and P.S.Turner, Acta. Cryst. A24, 390 (1968).

[9] R.Hull, J.M.Gibson and J.C.Bean, to be published in Appl. Phys. Lett.

[10] J.M.Gibson, M.M.J.Treacy, R.Hull and J.C.Bean, these proceedings.

[11] F.Cerdeira, A.Pinczuk, J.C.Bean, B.Batlogg, and B.A.Wilson, Appl. Phys. Lett. 45,1138 (1984).

[12] J.H. Van der Merwe and C.A.B.Ball in Epitaxial Growth, Part b, edited by J.W.Matthews (Academic Press, New York, 1975).

TRANSMISSION ELECTRON MICROSCOPY OF STRAINED-LAYER SUPERLATTICES

J. M. GIBSON*, M. M. J. TREACY**, R. HULL* and J. C. BEAN*
* AT&T Bell Laboratories, 600 Mountain Avenue, Murray Hill, NJ 07974
** Exxon Research and Engineering Company, Annandale, NJ 08801

INTRODUCTION

Transmission electron microscopy provides a powerful means of studying compositionally modulated materials. In such materials there is usually a local variation in electron scattering power along with a lattice dilatation wave which both accompany the local composition. The most revealing geometry for studying such materials has the lattice modulation direction lying within the plane of the thin foil. However, shear stresses accompanying the dilatation wave can be significantly relaxed by the presence of the thin foil surfaces, modifying the local atomic displacement field such that it is representative of neither the bulk, nor the free unstressed material. Two pertinent semiconductor examples which we have studied are spinodally decomposed quaternary III-V layers and strained-layer superlattices of Si/Si_xGe_{1-x}. We provide experimental evidence demonstrating relaxation in these cases and a simple elasticity model to describe it. Our data and model show a thickness dependence to relaxation and can explain previously reported 'anomalous' lattice parameter measurements from a strained-layer superlattice [1]. In this paper we concentrate on the effects of dilatation and relaxation on imaging and diffraction from a strained-layer superlattice.

RELAXATION IN LATTICE-MODULATED SEMICONDUCTORS

Transmission electron microscope contrast from the quaternary semiconductor $In_xGa_{1-x}As_yP_{1-y}$ has been reported in detail elsewhere [2]. In this study it was found that a long period (~2000A) strong diffraction contrast seen in certain thin foils of liquid phase epitaxially grown material arose from elastic relaxation of a lattice dilatation associated with spinodal decomposition. From this point, consideration was given to the more general case of elastic relaxation in other lattice-modulated thin films. Our original model for the spinodal decomposition was a one imensional sinusoidal unit cell parameter modulation (amplitude ε_0, wavelength Λ) in the plane of a thin foil of thickness t. By simple linear elasticity theory, it was derived [2] that the local value of strain along the modulation direction (ε_{xx}) is distorted from the unstressed value of ε_0 to

$$\varepsilon_{xx} = 2\varepsilon_0 \frac{(1+\sigma)}{(1-\sigma)} \left[\frac{(\alpha t/2)\cosh(\alpha t/2)\cosh(\alpha z)-(\alpha z)\sinh(\alpha z)\sinh(\alpha t/2)}{-(1-2\sigma)\sinh(\alpha t/2)\cosh(\alpha z)} - 1/2 \right] \cos(\alpha x)$$

(1)

Where the film surfaces are at $z = \pm t/2$, and $\alpha = 2\pi/\Lambda$, and σ is Poisson's ratio.

For the bulk case when $t/\Lambda \to \infty$, $\varepsilon_{xx} = (1+\sigma)/(1-\sigma)\varepsilon_0$. For the very thin film case when $t/\Lambda \to 0$, $\varepsilon_{xx} = (1+\sigma)\varepsilon_0$. For intermediate values of t/Λ, ε_{xx} varies with depth, giving rise to lattice plane bending, whose form can be derived by differentiation of Equation 1 with respect to z. The case of a strained-layer superlattice, studied in cross-section, is very much analogous [3]. The commensurate strain arises from the difference in lattice parameter $2\varepsilon_0$ between the layers. For a cubic material this strain causes a tetragonal distortion modulation in the bulk of amplitude $\varepsilon_0(1+\sigma)/(1-\sigma)$. By adding a suitable Fourier series of sinusoidal modulations, the strained layer superlattice can readily be synthesized [3]. Each sinusoidal component of strain amplitude ε_n and wavelength $\Lambda_n = 2\pi/\omega_n$ can be independently treated by linear elasticity theory so that the strain modulation is a series of terms like Equation (1). In this way relaxation can be calculated for the superlattice. Additional constant strain is sometimes present due to a mismatch between the superlattice mean spacing and the substrate spacing. This remains unrelaxed and has no major effect.

By analogy with the discussion of a single sinusoidal component, there are two extreme cases for the lattice dilatation. For $t/\Lambda_n \to \infty$ the strain amplitude is $(1+\sigma)/(1-\sigma)$ which is the amplitude of the tetragonal distortion which has been confirmed for the Si/Si_xGe_{1-x} system by X-ray diffraction [4]. (Note that for Si/Si_xGe_{1-x} as grown on a Si substrate, the Si_xGe_{1-x} layers contain all the tetragonal distortion due to the addition of a mean strain of $\varepsilon_0(1+\sigma)/(1-\sigma)$ [4]). In the very thin film case the strain amplitude becomes $(1+\sigma)$, which is typically 33% smaller than the bulk value. Figure 1 shows the calculated mean value of lattice parameter as a function of position across a half period of a strained-layer superlattice for various values of the parameter t/Λ. Fifty Fourier components were considered. The two extreme cases $t/\Lambda = 10$ and 0 are as predicted. For intermediate cases the lattice parameter relaxes for the relatively long-period components to $\varepsilon(1+\sigma)$ but remains unrelaxed for the short-period components whose effect is localized near the interface. It is worth noting that this treatment can be easily extended to the case of a single interface by making Λ_1 very large. Then it will be seen that strain is constant (and relaxed) except very near an interface, which may be important in interpreting images from such systems. Auret et al. [5] have previously demonstrated such an effect for a single strained interface. Our model allows ready treatment of any nonperiodic strained system by Fourier synthesis. A useful rule of thumb to emerge from this is that such strain near a single interface relaxes to within 10% of $\varepsilon(1+\sigma)$ (its 'fully' relaxed value as a thin uniform film) beyond a distance of about $t/6$ from the interface.

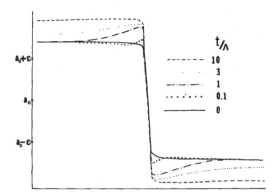

Figure 1 Calculated mean value of lattice parameter as a function of position across half a period of a strained layer superlattice for various values of t/Λ.

Measuring such small strains in TEM samples is nontrivial. We have used two techniques for the strained-layer superlattice $Si/Si_{0.6}Ge_{0.4}$ grown by molecular beam epitaxy as described by Bean [4]. Selected-area diffraction is limited to the study of areas ~1$_{l}$m in size, over which thickness is rarely constant. Brown et al. [1] have used this technique to examine $In_{0.27}Ga_{0.73}As/GaAs$ superlattices which have ε_0 = 0.0096 and a relative structure factor amplitude 0.0127. The latter authors reported anomalous values of strain in superlattices of very short period ($\Lambda_1 <$ 200Å) which apparently had no lattice parameter modulation.

The kinematically diffracted intensity in the zero order Laue zone from a single sinusoidal component of a superlattice, with wavelength Λ_n and modulation amplitude ε_n is a series of satellites at positions $1/a + m/\Lambda_n$ with intensities proportional to [6] ($\varepsilon_n\Lambda_n/a$). For $\varepsilon_n \ll 1$ two limits should be noted for diffraction from the above formula. When $\varepsilon_n\Lambda_n/a \gg 1$, the diffraction pattern resembles two well-defined Bragg spots separated by $2\varepsilon_n/a$ as would be expected. However, when $\varepsilon_n\Lambda_n/a \lesssim 1$, the superlattice satellites remained fixed in positions $1/\Lambda_n$, independent of ε_n. Only their intensities depend on ε_n. Physically, the criteria for discrete spots ($\varepsilon_n\Lambda_n/a \gg 1$) corresponds to the need for a sufficiently large wavelength ($\Lambda_n/2$) in which to define the diffraction vector $\underline{g}$ to accuracy $2\varepsilon_n/a$

Because of limitations on layer width ($\Lambda_1/2$) due to elastic strain energy, it is difficult to exceed values of $\varepsilon_n\Lambda_n/a \approx 1$ for a strained layer superlattice without introducing distortions. Figure 2 is a diffraction pattern of a 100A period $Si/Si_{0.6}Ge_{0.4}$ superlattice where ε_0 = 0.0084. This falls into the latter category of diffraction patterns where only spot intensities can be used to deduce ε_0. The local thickness is probably ~500A ± 200A. In the data of Brown et al. [1], the two superlattices of period 70A and 170A have such diffraction patterns (e.g. Figure 1(a) in [1]) from which the lattice parameter modulation has been mistakenly assumed to be zero. Apart from these values, the data of Brown et al. [1] shows little significant variation in strain amplitude with period. However, the mean strain amplitude found for ~400A period superlattices was 0.008 ± .002 which is a factor of two less than the bulk tetragonal distortion ($\varepsilon_0(1+\sigma)/(1-\sigma) \approx 0.019$). This is consistent with relaxation, although the value is a little smaller than quantitatively predicted from our simple model ($\varepsilon_0(1+\sigma) \approx 0.013$).

Figure 2 Electron diffracted intensity around the (400) position for a (100) $Ge_{0.4}Si_{0.6}/Si$ superlattice of 100A period.

To calculate the detailed diffraction pattern from a superlattice, one must sum each appropriate relaxed sinusoidal component and then correlate the result with diffraction from the 'square-wave' structure factor modulation. Our calculations assume kinematical diffraction and looked only in the zero order Laue zone so they would not be expected to give very good quantitative agreement with experimental diffraction data. However, we have calculated such diffraction patterns and find the satellite intensities in Figure 2 consistent with relaxation for a thickness of ~500Å ± 200Å. They definitely correspond to a lower modulation amplitude than the bulk, which would give rise to four, rather than the observed three, bright satellites.

The limitation of averaging over large areas of varying thickness can be overcome by using high-resolution electron microscopy for lattice parameter measurements. Figure 3 shows an image from a thin area of the same $Si/Si_{0.6}Ge_{0.4}$ superlattice taken near the Scherzer Focus with a JEOL 200 CX high-resolution microscope having point-to-point resolution 2.5Å. Lattice parameter measurements at the 1% level are beset with problems due to misalignments and image distortion [7]. However, we do find an experimentally significant variation from the thinnest ~50Å thick areas, where measured $\varepsilon = 0.007 \pm .004$, to thicker areas (~300Å) where measured $\varepsilon = 0.013 \pm .004$. These are qualitatively consistent with our relaxation model although it is again interesting to note that they are a little smaller (i.e., more relaxed) than expected.

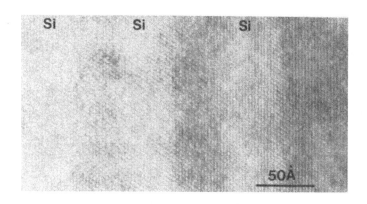

Figure 3 High resolution bright field image taken in the [110] direction of a (001) $Si/Si_{0.6}Ge_{0.4}$ superlattice. Thickness increases from left to right.

It is difficult to measure with sufficient accuracy relaxation effects in such thin films by lattice parameter measurement. The fact that all our observations of strain amplitude are on the slightly lower side compared with our model may be attributable to the fact that anisotropic and inhomogeneous effects were not included in the model, and also the possibility of additional modes of relaxation due to bending, buckling or thickness variations in the y (or third) dimension. However, the experimental accuracy is not yet sufficient to make much of these points.

The bending which occurs for intermediate values of $t/\Lambda n$ can be a strong source of diffraction contrast [2]. Figure 4 reveals that diffraction contrast from relaxation and dilatation effects is sufficiently large to enable visualization of a weak modulation in a $Si/Si_{0.95}Ge_{0.05}$, 500Å period superlattice with $\varepsilon_0 = 0.001$. Figure 4(a) is taken with a (220) reflection in the plane of the substrate surface which

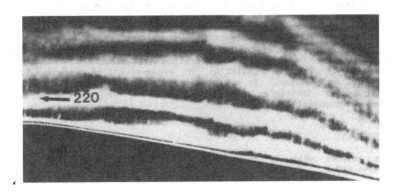

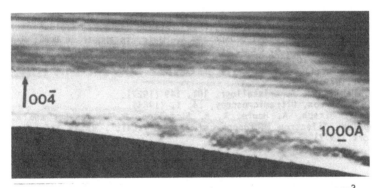

Figure 4   Si/Si$_{0.95}$Ge$_{0.05}$ superlattice of 500Å period, and $\varepsilon = 10^{-3}$.
a) $2\bar{2}0$ dark field, b) $00\bar{4}$ dark field.

shows only weak contrast due to variations in extinction distance between
the layers.   However, Figure 4(b) reveals much stronger contrast in
a $(00\bar{4})$ reflection (parallel to the modulation direction) from which the
maximum effect on image contrast would be expected from relaxation-induced
bending and dilatation.   Extinction distance contrast would be very much
the same in both images.   We have some difficulty in obtaining accurate
agreement with image calculations using the column approximation and the
Howie-Whelan equations [8] for Figure 4(b).   Since $\varepsilon_0 \Lambda_1/a = 0.38$ the column
approximation would not be expected to apply in studying contrast due to
'dilatation' and related effects, since the diffraction effects from one
whole superlattice period and not an individual layer are necessary to
simulate zero order Laue' zone scattering.   In any case, this mode of
imaging allows visualization of rather weak, periodic strain fields and
perhaps detailed study of elastic properties on a very small scale through
careful image simulation and full elasticity theory.

CONCLUSIONS

In conclusion we have shown evidence for elastic relaxation in thin
films of lattice-modulated semiconductors, such as strained-layer superlattices, which are in qualitative agreement with simple linear isotropic

272

elasticity theory. Detailed agreement may require the use of inhomogeneous anisotropic theory and more sophisticated image and diffraction calculations. In any case, experimental data is not sufficiently accurate as yet. We can explain previously reported anomalies in microscopy of strained-layer superlattices and provide a framework for dealing with this important effect. Only by taking relaxation into account can meaningful TEM studies of lattice modulation in cross-sectioned strained-layer superlattices be effected. It may also provide a method for the study of weak, otherwise invisible, modulations and of microscopic elastic properties of superlattices.

ACKNOWLEDGMENTS

The authors wish to thank J.M. Brown and A. Howie for useful discussions and M.L. McDonald for help with specimen preparation.

REFERENCES

(1)   J. M. Brown, N. Holonyak, Jr., R. W. Kaliski, M. J. Ludowise, W. T. Dietze and C. R. Lewis, Appl. Phys. Letts., 44, 1158, (1984).
(2)   M. M. J. Treacy, J. M. Gibson and A. Howie, to appear in Phil. Mag.
(3)   J. M. Gibson and M. M. J. Treacy, to appear in Ultramicroscopy.
(4)   J. C. Bean, this volume.
(5)   F. D. Auret, C. A. B. Ball and H. C. Snyman, Thin Solid Films, 61, 289 (1979).
(6)   U. Dehlinger, Z. Kristallogr, 101, 149 (1927).
(7)   J. M. Gibson, Ultramicroscopy, 14, 1, (1984).
(8)   P. B. Hirsch, A. Howie, R. B. Nicholson, D. W. Pashley and M. J. Whelan, "Electron Microscopy of Thin Crystals', R. E. Krieger, New York (1977).

Metastable Structures

THE EPITAXIAL GROWTH OF METASTABLE PHASES

ROBIN F. C. FARROW
Westinghouse Research and Development Center
1310 Beulah Road, Pittsburgh, PA 15235

ABSTRACT

Pseudomorphism, the epitaxial growth of metastable phases, is reviewed. Emphasis is given to recent developments in the growth and investigation of α-Sn films on InSb and CdTe since this is a case where there is a very large strain energy barrier to the metastable to stable phase transformation. As a result, pseudomorphic growth of α-Sn can be sustained to film thicknesses of $\sim$1 μm.

Progress towards stabilization of direct gap group IV semiconductors is discussed and possible applications of pseudomorphism considered.

INTRODUCTION AND HISTORICAL BACKGROUND

The discovery of the phenomenon of epitaxial growth of metastable phases of materials came early in the development of epitaxy when in 1950 L. G. Schulz [1] discovered that ionic crystals such as CsCl and CsI assumed a rock-salt structure when nucleated on rock-salt structure substrates such as NaCl and LiF. This structure is not the equilibrium form for bulk CsCl and CsI at room temperature but is of lower density and has been observed only at temperatures near to the melting point of bulk CsCl. Its existence in the form of 100-200Å edge islands in parallel epitaxial orientation on NaCl substrates constitutes the first example of stabilization of a metastable phase. The stabilization process was not observed for deposition on amorphous substrates and was clearly of limited range since beyond the island coalescence stage, the deposit assumed the normal CsCl structure. The phenomenon was later named "pseudomorphism" by W. A. Jesser [2] who reported a variety of metal on metal systems in which thin (≲20Å) films of metals assumed a structure, appropriate to a higher temperature bulk phase of the overlayer metal, but coherent with the substrate. One example was stabilization of the fcc phase of Fe (γ-Fe) by epitaxial growth, at room temperature, onto a Cu (001) surface in ultra-high vacuum. This is a striking example of pseudomorphism since γ-Fe is stable in bulk form only in the limited temperature range $\sim$900-1300°C.

Recently, a spectacular example of pseudomorphism has been discovered [3] in which α-Sn films up to $\sim$0.5 μm thick grow epitaxially on clean InSb (001) or CdTe (001) substrates at temperatures well above the bulk transformation temperature for the α→β phase of Sn. This is a case of stabilization of a phase which, in bulk form, is stable only below room temperature and which is extremely difficult to prepare in the form of high perfection, high purity single crystals. In this example, the exceptionally great range ($\sim$0.5 μm) of pseudomorphic growth was sufficient to permit the first quantitative determination of strain anisotropy in the α-Sn films. This study [3] confirmed that the interface was commensurate and that the lattice misfit between overlayer and substrate was accommodated by a uniform elastic strain in the film. As a result of this discovery, the first studies [4] of optical properties of α-Sn surfaces have now been made possible and the α→β phase transformation in thin films has been monitored [5] by Raman spectroscopy. In addition, the growth and investigation of single quantum well structures of α-Sn has resulted [6] in the first evidence for direct transitions between sub-band levels of confined electrons and holes in a symmetry-induced zero-gap semiconductor.

In this review, emphasis is given to these recent developments in the growth and investigation of α-Sn films. The possible ways in which pseudomorphism can be utilized to engineer direct-gap character into the α-Sn band structure are considered and possible useful applications of the concept of pseudomorphism are speculated on.

EPITAXIAL GROWTH OF α-Sn AND α-Sn:Ge -- CASES OF VERY LONG RANGE PSEUDOMORPHIC GROWTH

In the early examples of pseudomorphism of ionic crystals and metals, the range of pseudomorphic growth was very limited. For example, Schulz [1] observed pseudomorphism only for discontinuous films of CsCl and CsI. On the other hand, in the cases of metals on metals, coherent layer by layer growth of the metastable phase at room temperature was observed as for fcc Co on Ni [7] and γ-Fe on Cu [8]. The range of pseudomorphism in these and other early examples remained limited to ≲20Å beyond which preferential nucleation and growth of the stable phase occurred. This limitation may be due to a variety of factors which include the extremely high (>4 x 10^8 cm^{-2}) dislocation density of the substrates [9] and small difference in density of metastable and stable phases. The former factor is believed [7] to provide a mechanism for conversion of the metastable to stable phase while the latter factor precludes a strain energy barrier tending to stabilize the metastable phase. The recently discovered case of pseudomorphism of α-Sn on InSb and CdTe substrates is, in contrast to these early examples, maintained to very considerable ranges (∼1 μm). It is likely that the reason for this exceptional behavior is the strain energy barrier hindering the α→β metastable to phase transformation in the growing film. This barrier arises from the very large density change [$\rho(\alpha)$ = 0.79 $\rho(\beta)$] in the transformation which causes tensile stress in the α-Sn along the interphase boundary. This stress opposes the lattice collapse required for the transformation.

The close lattice match (Δa/a = 0.14% for α-Sn/InSb) between α-Sn and InSb (or CdTe) is responsible for the onset of pseudomorphic growth, but the continuation of such growth is aided by the strain energy barrier. The available experimental evidence [3],[10] suggests that there may be no well defined upper limit to pseudomorphic growth on InSb (001) surfaces, near room temperature, since the α→β transformation nucleates preferentially at defect sites at the film-substrate interface and that more perfect interfaces lead to thicker pseudomorphic films with improved stability. That lattice match has a controlling influence in initiation of pseudomorphic growth of α-Sn is demonstrated by the fact [11] that on the atomically clean GaAs (001) surface (at room temperature) only β-Sn nucleates. Furthermore, Williams et al. [12] have shown that β-Sn nucleates in three dimensional islands on atomically clean InP (011) cleaved surfaces. In these cases, tetragonal β-Sn (a = b = 5.8313Å) has a closer geometric fit to the substrates than α-Sn and grows in preference to the metastable phase.

Optical Investigations of α-Sn Films and the α→β Phase Transformation

The technique of pseudomorphic growth of α-Sn on InSb has recently been used by workers from the Max Planck Institute fur Festkorperforshung, Stuttgart to prepare α-Sn films on (001), (111) A and (111) B InSb surfaces. These studies are significant since they reveal, for the first time, substrate orientation dependent differences in the growth behavior and stability of the metastable α-Sn films. Earlier work by Mattern and Luth [13] had revealed growth of structurally disordered α-Sn films on UHV cleaved (011) InSb surfaces with significant In outdiffusion into the films.

Menéndez and Höchst used the technique of pseudomorphic growth of α-Sn onto ion bombarded and annealed surfaces of InSb to prepare samples for optical investigations by Raman [5] and reflectance [4] techniques. The Raman technique was used to probe the α→β transformation in films in the thickness range 250-2000Å, by observing the temperature dependence of the shift of the $\vec{q}{\sim}0$ optical Raman phonon. Early [14] Raman investigations of bulk α-Sn samples had showed that the $\vec{q}{\sim}0$ optical phonon had a Raman frequency of 196.7 cm^{-1}. Menéndez and Höchst observed a value of 197.4(1) cm^{-1} for the epitaxial films, consistent with the expected frequency shift due to the measured [3] uniaxial compression of the α-Sn films on InSb. At the α→β transformation, the intensity of the $\vec{q}{\sim}0$ phonon fell abruptly to zero. Figure 1 shows the temperature dependence of the integrated Raman signal of the α-Sn films measured for various film thicknesses. The vertical scale

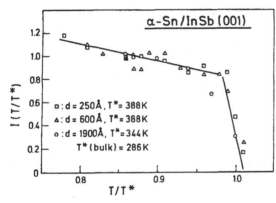

FIGURE 1

Temperature dependence of integrated Raman signal of α-Sn films grown on InSb (001).

After Menendez and Hochst [5]

(see text)

has been chosen to normalize $I(T/T^*)$, where T^* is the transformation temperature, to unity for $T/T^* = 0.85$. This permits a comparison between the shapes of the transformation curve for different thickness films. Several significant features are evident from the data. Firstly, the transformation temperature for the films is in all cases greatly enhanced over the accepted bulk transformation temperature of ∿13°C (286°K). The transition temperature (71°C) for a 1900Å film is in excellent agreement with the value of 70°C reported by the author and coworkers for films of similar thickness. The second feature of the data is that thinner films exhibit a greater stability against the α→β transformation. One possible explanation for this may be the influence of interfacial coherency on short wavelength shear in thin films. Clearly, the α-Sn film is laterally constrained by interfacial coherency. The α→β transition, which requires short wavelength shear [15] will be inhibited at the interface by interfacial bonding. It also seems possible that the α-Sn lattice may be laterally "stiffened" by coherency. Measurements of the TA phonon frequency in the film as a function of film thickness would provide a test of this idea and a probe of the range of interfacial stiffening.

In the case of Sn deposition onto (011), (111) and (1̄1̄1̄) InSb surfaces, the Raman studies did not reveal a single temperature for the α→β transformation but rather a continuous increase in the β component of the film with increasing temperature. This suggests mixed nucleation of α and β phase components; a point supported by the observation of discontinuities in 300Å thick films grown at room temperature. At this time, it is not clear whether nucleation of the β phase crystallites is an intrinsic feature of {111} and (011) surfaces or whether these surfaces retain impurities following the in-situ cleaning procedures. Nucleation and growth of β-Sn is known [3] to occur on C and O contaminated InSb (001) surfaces. A further uncertainty which needs to be resolved in the work of Menéndez and Höchst is that of sample heating by the laser beam. Despite defocussing of the beam by cylindrical lenses to reduce the optical power density at the samples to $\sim$10W cm^{-2}, a possibility remains [16] that the samples are destabilized and that the true $\alpha \rightarrow \beta$ transformation temperatures are somewhat higher.

A significant finding of Menéndez and Höchst was that the Raman spectra of (001) α-Sn films showed no evidence of the heavy p-type doping of the films reported by the author and coworkers [3]. Such high ($>10^{18}$ cm^{-3}) p-type doping would have resulted [17] in significant changes in frequency, peak width and peak shape of the $\tilde{q}=0$ Raman peak. Since Menéndez and Höchst used [16] a graphite container for Sn in their effusion source, this finding supports the view [3] that the p-type doping of α-Sn films reported by the author and coworkers [3] was due to extrinsic boron from the pyrolytic boron nitride container used to contain the Sn.

Vina et al. [4] have recorded dielectric function spectra for α-Sn films using the technique of automated ellipsometry to probe 950Å thick α-Sn films grown by MBE on (001) InSb substrates. Critical point parameters and interband transitions were mapped as a function of temperature. The temperature dependencies were in agreement with existing theoretical models.

Angle Resolved Photoemission Spectroscopy Studies of α-Sn Films

Angle resolved photoemission spectroscopy studies of (001) [18] and {111} [19] surfaces of α-Sn films have been carried out by H. Höchst and I. Hernández-Calderon. The measured energy difference between the Fermi energy and the extrapolated top of the valence band was found to be 0 ± 0.05 eV for (001) oriented α-Sn at 300K. This result is consistent with the Groves-Paul [20] semimetallic band structure of α-Sn. From an analysis of the photoemission valence band spectra of (001) and {111} surfaces, the dispersion E(k) of initial state bands was derived in order to compare the dispersion data with the various band structure calculations of α-Sn. The E(k) data were derived by assuming that the photoemission transitions were to parabolic free electron-like final state bands. The best agreement was with the nonlocal pseudopotential band structure calculation by Chelikowsky and Cohen [21].

STABILIZATION OF DIRECT-GAP GROUP IV SEMICONDUCTORS

The demonstration that α-Sn and α-Sn:Ge can be stabilized by epitaxy has opened up the prospect of growth of new thin film structures in which the indirect-gap band structure of group IV semiconductors is modified to a direct-gap band structure. In this section, progress towards growth of such structures is reviewed.

There is general agreement that α-Sn is a symmetry-induced, zero-gap semiconductor with the Groves-Paul [20] band structure. The difference in band structure between Ge and Sn is illustrated in Figure 2. Ge is indirect gap because the indirect transition Γ_8^+-L_6^+ requires less energy than the direct transition Γ_8^+-Γ_7^-. As one moves in composition from Ge to Sn, the Γ_7 and Γ_8 bands approach each other and eventually cross over. The original valence bands then become the new valence and conduction bands. This structure has a degeneracy of Γ_8 bands at the zero center imposed by the cubic lattice symmetry; hence, the term "symmetry-induced, zero-gap semiconductor." Here we consider three ways in which this degeneracy can be lifted and the zero-gap state modified. These are (a) alloying α-Sn with Ge to form a diamond structure group IV alloy $Ge_{1-x}Sn_x$; (b) imposition of elastic strain through pseudomorphic growth; and (c) confinement of carriers in an α-Sn quantum well.

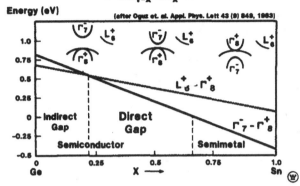

BAND STRUCTURE OF
Ge$_{1-x}$ Sn$_x$

Energy (eV)

(after Oguz et. al. Appl. Phys. Lett 43 (9) 848, 1983)

FIGURE 2

Ge$_{1-x}$Sn$_x$ -- A Direct-Gap Group IV Semiconducting Alloy?

As has already been pointed out [10],[22] a linear interpolation between the band structures of Ge and Sn predicts a composition range over which Ge$_{1-x}$Sn$_x$, if it exists, should exhibit direct-gap character. This interpolation indicates (see Figure 2) a direct-gap regime from x∼0.23 to x∼0.7. Uncertainty in the band structure parameters for α-Sn precludes an exact prediction of the end points of the regime. Growth of 1 μm thick films of the alloy with x = 0.99 has been achieved by MBE [3],[10]. No other attempts to achieve epitaxy of the alloy have been reported to date. Indeed, the choice of substrate for pseudomorphic growth of Ge$_{1-x}$Sn$_x$ is a crucial issue. InSb is probably only suitable as a substrate for compositions near x=1. For example, the misfit (Δa/a) between Ge$_{1-x}$Sn$_x$ and InSb exceeds 0.5% for x<0.95. Even if pseudomorphic growth, with misfit accommodation by elastic strain, occurred for such a large misfit, the films would be subjected to considerable uniaxial compression with significant modifications to band structure from strain effects alone (see next section). Assuming Vegard's law to apply for Ge$_{1-x}$Sn$_x$, then InP, InAs, or GaSb would be suitable substrates for exploratory studies of epitaxy of Ge$_{1-x}$Sn$_x$ (see Figure 4) in the direct-gap regime. The cubic fluorite structure, mixed fluoride system -- Ba$_x$Sr$_{1-x}$F$_2$ -- can be grown by MBE [23] across the entire composition range and lattice-matched to InAs [24]. This would allow growth of a lattice-matched buffer layer between the substrate and Ge$_{1-x}$Sn$_x$ overlayer to avoid possible problems of interdiffusion at elevated growth temperatures. It would also permit growth of a lattice-matched protective film over the Ge$_{1-x}$Sn$_x$.

Until recently, the existence of diamond structure Ge$_{1-x}$Sn$_x$ for x<0.99 was pure speculation. However, Oguz et al. [22] recently synthesized Ge$_{1-x}$Sn$_x$ as a diamond structure phase with x∼0.22 (a=5.838Å) by pulsed laser annealing of sputter deposited amorphous Ge$_{0.7}$Sn$_{0.3}$ films 0.1-1 μm thick. This technique minimized phase separation of the Ge-Sn by allowing insufficient time for formation of equilibrium phases such as β-Sn. However, the films were microcrystalline and contained some β-Sn precipitates. Nevertheless, electroreflectance spectra for the films showed features characteristic of a crystalline semiconductor. Clearly, in view of the prospect of a new, high lattice mobility semiconductor in which carrier scattering by polar optic phonons is absent [25], the work of Oguz et al. provides added impetus to attempts to prepare diamond structure Ge$_{1-x}$Sn$_x$ by pseudomorphic growth.

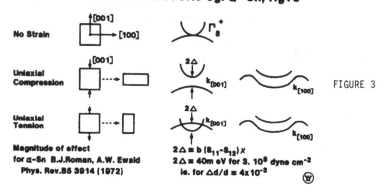

[after M. Cardona Sol.St. Comm. 5 233 (1967) & L. Liu, W.Leung Phys. Rev. B12, 2336 (1975)]

EFFECT OF STRAIN ON BAND STRUCTURE OF ZERO GAP SEMICONDUCTORS eg. α -Sn, HgTe

FIGURE 3

No Strain

Uniaxial Compression

Uniaxial Tension

Magnitude of effect for α-Sn B.J.Roman, A.W. Ewald Phys. Rev.B5 3914 (1972)

$2\Delta \cong b\,(S_{11}-S_{12})\,X$

$2\Delta \cong 40m\ eV$ for $3.\ 10^9$ dyne cm^{-2}

ie. for $\Delta d/d \cong 4\times10^{-3}$

Band Structure Modifications Induced by Coherency Strain in α-Sn and α-Sn:Ge

In the cases of epitaxy of thin (0.2 μm) films of α-Sn and $Ge_{0.01}Sn_{0.99}$ on InSb (and CdTe) quantitative measurements [3] of strain anisotropy in the films confirmed that each film was in a state of well defined uniform uniaxial strain imposed by coherency. The misfit between substrate and overlayer was accommodated entirely by elastic strain and no strain relaxation through the film thickness was evident. This is a symmetry-breaking effect which converts the film from cubic to tetragonal symmetry. Since the degeneracy of Γ_8 bands in α-Sn is imposed by the cubic lattice symmetry, the departure ·from cubic symmetry causes significant modifications to the band structure of α-Sn. A similar effect would also occur for HgTe films under uniaxial strain since HgTe is also a symmetry-induced zero gap semiconductor. There is an extensive literature on this subject. Cardona first calculated the shape of the energy bands for α-Sn under uniaxial [001] and [111] compression. This is the type of strain to be expected if the films were in lateral tension in the film plane, as for a film with smaller bulk (relaxed) lattice constant than the substrate. Such strain would occur [3] in practice for $Ge_{1-x}Sn_x$ films on InSb with $x \lesssim 0.98$. Cardona's predictions are illustrated in Figure 3. A direct gap is opened up for carriers moving along the strain axis. For carriers moving transverse to the axis, an indirect gap is formed and the upward curvature of the valence band for small k implies that holes moving transverse to the strain axis can have negative effective mass. Under an applied field, such holes can move to the inflexion points of the band and revert to normal hole behavior. Goodman [26] recently proposed this effect as the basis for a negative differential resistance oscillator in α-Sn. However, the original idea for using valence band distortions to provide regions of negative effective mass for holes and hence negative differential resistance was due to Kroemer [27]. As discussed by Liu and Leung [28], the reason why such devices were not realized in bulk Ge or Si was that the population of negative effective mass carriers at equilibrium was too small. However, in pseudomorphic films of $Ge_{1-x}Sn_x$, hole injection from a p-type InSb substrate can overcome this problem.

In the case of uniaxial tension of α-Sn (or HgTe), Liu and Leung [28] showed that the Γ_8 bands cross for k along the strain axis (see Figure 3) leading to a region of negative effective mass for both electrons and holes moving parallel to the strain axis. A region (centered around $k=0$) of

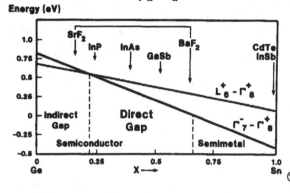

FIGURE 4

negative effective mass for holes was predicted for all directions of holes.
Liu and Leung carried out carrier transport calculations to show that a
weak negative differential resistance effect occurred for hole transport
in α-Sn or HgTe, under an applied field, at low (≲77K) temperatures. The
magnitude of this effect would be enhanced by increasing the population of
negative mass holes by injection from a p-type InSb substrate.

   For these concepts of band structure control through coherency strain
to be useful, the magnitude (2Δ) of band overlap or band splitting must be
significant (>>kT) at the levels of coherency strain which can be sustained
in pseudomorphic growth. In the case of epitaxy of α-Sn on InSb (001), the
[001] tensile strain has been measured [3] as 2.7×10^{-3}. The strain
induced overlap (2Δ) in this case is estimated at 13 meV (>2 kT at 77K) for
a deformation potential of 2.3 eV [29] so the band structure modifications
should indeed be significant. Values of $2\Delta > 40$ meV have been measured [29]
for bulk crystals of α-Sn subjected to uniaxial compression with stress
values of ∿3×10^9 dyne cm^{-2}. Similar levels of stress may be accessible
for thin (∿0.2 μm) pseudomorphic films of Ge$_{1-x}$Sn$_x$ (x∿0.05) grown on InSb.

Carrier Confinement in α-Sn Quantum Wells

   In 1980, Broerman [30] introduced the concept of carrier confinement
in a quantum well of a zero gap semiconductor, such as α-Sn or HgTe, to
generate a series of sub-band energy states with energies set by quantum
well thickness and barrier height. Hence, by controlling well thickness
for α-Sn (or HgTe) sandwiched between two wide gap semiconductors, or one
semiconductor and vacuum, the transition energies and, hence, optical band
edge could be engineered to the spectral region of interest. Broerman's
theory did not take into account the overlap of Γ_8 bands at points away
from the zone center for unstrained HgTe. In addition, the effects of
misfit strain on band structure of the zero-gap semiconductor was neglected
in the theory. Despite these uncertainties, single quantum well structures
of α-Sn have recently been realized [6] by MBE growth and exhibit band to
band transitions in qualitative agreement with the simple theory.
Takatani and Chung [6] prepared these structures by pseudomorphic growth of
α-Sn onto CdTe {111} surfaces in ultrahigh vacuum. Transitions between
confined particle sub-band states in the well were probed in situ by high
resolution electron energy loss spectroscopy (HREELS). HREELS spectra
were recorded for samples at ∿0°C with film thicknesses in the range 47-76Å.

These spectra revealed a peak which shifted to lower energy with increasing film thickness. The thickness dependence of onset energy of the peak was in qualitative agreement with the direct transition energy calculated for n=1 valence → n=1 conduction band states for carriers confined in the well. In the highly simplified calculations of Takatani and Chung, a square quantum well was assumed with infinite confining potential at the α-Sn/vacuum interface. The conduction band offset was treated as an adjustable parameter in fitting the data. The best fit was obtained for a conduction band offset of 0.5 eV, i.e., a valence band offset of 0.94 eV. The latter value compares reasonably well with Harrison's LCAO calculation [31] of 1.28 eV.

At this stage, the results obtained by Takatani and Chung can be considered as only tentative evidence for direct gap sub-band transitions. The observed transitions are rather weak, and the simple theory does not include the significant effect of misfit strain on the α-Sn band structure. However, the qualitative agreement with theory supports the case for further investigation in this area.

POSSIBLE APPLICATIONS OF PSEUDOMORPHISM

In the widest sense, pseudomorphism is of value because it provides a route to the synthesis and stabilization of a wide variety of phases of metals, semiconductors, and insulators. Some of these phases have potentially useful properties. For example, heteroepitaxy of α-Sn on InSb has opened up the prospect of growth of p-type InSb:n-type $Ge_{1-x}Sn_x$ heterojunctions, which, as Goodman [26] has speculated, might lead to a new type of negative differential resistance oscillator   .
Furthermore, if pseudomorphic growth of $Ge_{1-x}Sn_x$ alloy films in the direct-gap composition regime proves to be possible, then the potential applications multiply. For example, in addition to cooled FET's utilizing the high electron mobility of the alloy [25], there is a possibility that $Ge_{1-x}Sn_x$ (x∿0.5), lattice-matched to GaSb, could have a direct gap in the 0.1→0.2 eV range. Depending on the exact value of the gap at 77K, it may be attractive as a materials system for far infrared detection. The present leading contender for this spectral range is the ternary alloy (Hg, Cd)Te which has continuing problems in device production because of its chemical instability.

In the area of pseudomorphic metals, continuous films are in most cases limited to thicknesses below ∿20Å. However, this is thick enough to permit the first in situ examination of the physical properties of such phases. For example, the magnetic properties of γ-Fe, fcc Co, fcc Cr could all be probed at room temperature by techniques such as spin-polarized electron diffraction. These properties may reflect the strain anisotropy in the films. In addition, compositionally modulated structures involving metastable phases may provide a method of stabilizing larger volumes of the metastable phase.

Perhaps the most exciting use of pseudomorphism is that of preparation of new phases of materials; phases which are not accessible by bulk crystal growth techniques or which are not known to exist, a priori. Recent examples include the discovery [32] of a new structure of the element Mn (stabilized by epitaxy on Ru (001) up to at least 30 atomic layers) and the growth [10] of 2000Å-thick, epitaxial dielectric films of $Ba_xCa_{1-x}F_2$ compositions (0.02<x<0.92) not accessible by bulk crystal growth techniques. Speculation on applications of such new materials must await a full exploration of their properties.

CONCLUSIONS

The epitaxial growth of metastable phases (pseudomorphism) is an area of research which is generating renewed interest since it provides a route to growth and stabilization of phases of metals, semiconductors, and insulators with novel and potentially useful properties. This is a field in which experiment is far in advance of theory. For example, existing theories of pseudomorphism apply [2] only to particular modes of film growth or exclude [33] the possibility of growth of thick pseudomorphic films of soft materials (such as α-Sn) at temperatures above 4.2K. Furthermore, no attempts to model pseudomorphic growth of compounds have yet been reported.

REFERENCES

1. L. G. Schulz, Acta. Cryst. 4, 487 (1951).
2. W. A. Jesser, Mater. Sci. Eng. 4, 279 (1969).
3. R. F. C. Farrow, D. S. Robertson, G. M. Williams, A. G. Cullis, G. R. Jones, I. M. Young, and P. N. J. Dennis, J. Cryst. Growth 54, 507 (1981).
4. L. Vina, S. Logothetidis, H. Höchst, to be published in the proceedings of the 17th International Conference on the Physics of Semiconductors, San Francisco, August 6-10, 1984.
5. J. Menéndez, H. Höchst, Thin Solid Films 111, 375 (1984).
6. S. Takatani, Y. W. Chung, to be published in the Proceedings of the Fall 1983 TMS Symposium on Submicron Device Structures.
7. W. A. Jesser, J. W. Matthews, Acta Met. 16, 1307 (1968).
8. W. A. Jesser, J. W. Matthews, Phil Mag 15, 1097 (1967).
9. J. W. Matthews, J. Vac. Sci. Technol. 3 (3), 133, 1968.
10. R. F. C. Farrow, J. Vac. Sci. Technol. B1(2), 222 (1983).
11. A. J. Noreika, R. F. C. Farrow, unpublished work.
12. R. H. Williams, A. McKinley, G. J. Hughes, T. P. Humphreys, J. Vac. Sci. Technol. B2(3), 561 (1984).
13. M. Mattern, H. Luth, Surface Science 126, 502 (1983).
14. C. J. Buchenauer, M. Cardona, F. H. Pollak, Phys. Rev. B3, 1243 (1971).
15. J. C. Phillips in "Bonds and Bands in Semiconductors," Materials Science and Technology Series, Academic Press, New York, pp. 93-94.
16. H. Höchst, personal communication.
17. F. Cerdeira, M. Cardona, Phys. Rev. B5, 1440 (1972).
18. H. Höchst, I. Hernández-Calderon, Surface Science 126, 25 (1983).
19. I. Hernández-Calderon, H. Höchst, in press, Surface Science (1984).
20. S. Groves, W. Paul, Phys. Rev. Lett. 11, 194 (1963).
21. J. R. Chelikowsky, M. L. Cohen, Phys. Rev. B14, 556 (1976).
22. S. Oguz, W. Paul, T. F. Deutsch, B. Y. Tsaur, D. V. Murphy, Appl. Phys. Lett. 43, 850 (1983).
23. P. W. Sullivan, Appl. Phys. Lett. 44, 190 (1984).
24. K. Sugiyama, J. Appl. Phys. 56, 1733 (1984).
25. C. H. L. Goodman, IEE Proc. 129, P1-1(5), 189 (1982).
26. C. H. L. Goodman, IEE Proc. in press, 1984.
27. H. Kroemer, Phys. Rev. 109, 1856 (1958).
28. L. Liu, W. Leung, Phys. Rev. B12, 2336 (1975).
29. B. J. Roman, A. W. Ewald, Phys. Rev. B5, 3914 (1972).
30. J. G. Broerman, Phys. Rev. Lett. 45, 747 (1980).
31. W. A. Harrison, J. Vac. Sci. Technol. 14, 1016 (1977).
32. B. Heinrich, C. Liu, A. S. Arrott, Paper presented at International Conference on Molecular Beam Epitaxy, August 1-3, San Francisco, in press J. Vac. Sci. Technol.
33. E. S. Machlin, P. Chaudhari, pages 11-29 in "Synthesis and Properties of Metastable Phases," Ed. by E. S. Machlin, T. J. Rowland, Pub. by The Metallurgical Society of AIME 1980.

A REVIEW OF RECENT RESULTS ON SINGLE CRYSTAL METASTABLE SEMICONDUCTORS: CRYSTAL GROWTH, PHASE STABILITY, AND PHYSICAL PROPERTIES

S.A. BARNETT, B. KRAMER, L.T. ROMANO, S.I. SHAH, M.A. RAY, S. FANG, AND J.E. GREENE
Department of Metallurgy, the Coordinated Science, Laboratory, and the Materials Research Laboratory, University of Illinois, Urbana, Illinois, 61801

ABSTRACT

Recent results on metastable semiconducting alloys, concerning in particular the growth of new Sn-based alloys $(GaSb)_{1-x}(Sn_2)_x$ and $Ge_{1-x}Sn_x$ and the physical properties of $(GaAs)_{1-x}(Ge_2)_x$ and $(GaSb)_{1-x}(Ge_2)_x$, are discussed. $(GaSb)_{1-x}(Sn_2)_x$ and $Ge_{1-x}Sn_x$ alloy films were grown with x-values as high as 0.20 and 0.15, respectively, well in excess of equilibrium Sn solid solubility limits ($\leq$ 1%) while epitaxial $(GaAs)_{1-x}(Ge_2)_x$ and $(GaSb)_{1-x}(Ge_2)_x$ alloys were obtained on (100) GaAs at compositions ranging across the pseudobinary phase diagram. Low energy ion bombardment induced collisional mixing and preferential sputtering during film growth played a critical role in obtaining single phase alloys. An optimal ion energy, which depended on the ion flux and the alloy composition, was determined, allowing in most cases growth at temperatures T_s sufficient for obtaining single crystal alloys on (100) GaAs and (100) Ge substrates. Decomposition of the Sn-based alloys occurred above a critical T_s-value via α-Sn-rich precipitates which were stable above the β-Sn melting point. X-ray diffraction, STEM, EXAFS, and Raman spectroscopy measurements, performed on single crystal $(GaAs)_{1-x}(Ge_2)_x$ and $(GaSb)_{1-x}(Ge_2)_x$ alloys, indicate that there is a transition in the long-range order from zincblende to diamond with increasing x while the short-range order remains perfect at all compositions, i.e. no V-V or III-III bonds are observed. These results are discussed in light of recent models which relate $(GaAs)_{1-x}(Ge_2)_x$ atomic structure to its band structure and optical properties.

I. INTRODUCTION

A metastable state exists at a local free energy minimum and exhibits an activation barrier to phase transitions to the (lower energy) equilibrium state as indicated schematically in Figure 1. The transition from the metastable to the equilibrium state may involve additional intermediate states when access to such phases requires a smaller activation energy than that necessary for the direct transition to the equilibrium state. Some metastable phases such as amorphous glasses form readily in nature. However, it was the development of splat cooling by Duwez and co-workers[1] in 1960, and the subsequent use of this technique to produce a variety of metastable amorphous and crystalline phases,[2] that first focused scientific interest on metastable materials. It is only much more recently that vapor phase techniques have been developed for the growth of phases which are not only thermodynamically metastable but are, in addition, good quality single crystals. The best known examples are superlattice structures which are discussed in some detail elsewhere in this volume. Work on pseudomorphically constrained structures, such as the α-Sn/InSb system, are also covered in this volume.[3]

It is the purpose of the present paper to review recent results concerning the growth, phase stability, and physical properties of a new class of semiconductors -- epitaxial metastable alloys. Examples include alloys involving non-isovalent substitution such as $(GaAs)_{1-x}(Si_2)_x$,[4] $(GaAs)_{1-x}(Ge_2)_x$,[5-8] $(GaAs)_{1-x}(Sn_2)_x$,[9] $(GaSb)_{1-x}(Ge_2)_x$,[10-16] and $(GaSb)_{1-x}(Sn_2)_x$[17] as well as isovalent but non-isostructural substitutions such as $Pb_{1-x}Cd_xS$,[18,19] $InSb_{1-x}Bi_x$,[20-22] and $Ge_{1-x}Sn_x$[23,24] (PbS crystallizes in the NaCl structure, CdS is either wurtzite or zincblende, InSb is zincblende, InBi is tetragonal, and Sn is tetragonal above 13.2°C). Note that the Ge and Si-substituted $(III-V)_{1-x}(IV_2)_x$ alloys are essentially isostructural in the sense that both end members have a fcc space lattice with a two-atom basis while the Sn-substituted $(III-V)_{1-x}(IV_2)_x$ alloys are both non-isovalent and non-isostructural.

Figure 2 shows the equilibrium pseudobinary phase diagrams for $(GaSb)_{1-x}(Ge_2)_x$[12] (typical of Ge and Si-substituted $(III-V)_{1-x}(IV_2)_x$ alloys) and $(GaSb)_{1-x}(Sn_2)_x$[25] (typical of Sn-substituted $(III-V)_{1-x}(IV_2)_x$ alloys). Both are eutectic systems (the eutectic in the $(GaSb)_{1-x}(Sn_2)_x$ system occurs very near the melting point of pure β-Sn) exhibiting very little solid solubility. Nevertheless, single crystal metastable $(GaSb)_{1-x}(Ge_2)_x$ alloys have been grown at compositions across the entire range while $(GaSb)_{1-x}(Sn_2)_x$ alloys have so far been grown with x up to 0.2. Initial work in this field was reviewed by Greene[26] and the present paper will concentration on recent results, primarily in the systems $(GaAs)_{1-x}(Ge_2)_x$, $(GaSb)_{1-x}(Ge_2)_x$, $(GaSb)_{1-x}(Sn_2)_x$, and $Ge_{1-x}Sn_x$.

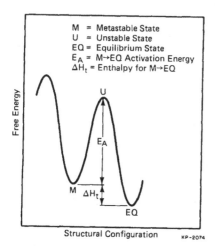

Figure 1. Schematic energy level diagram showing metastable, unstable, and equilibrium states.

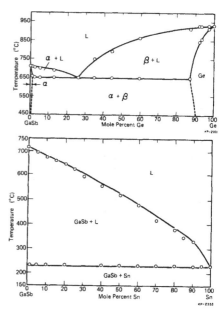

Figure 2. Pseudobinary phase diagrams for GaSb-Ge (from reference 12) and GaSb-Sn (from reference 25).

II. CRYSTAL GROWTH AND PHASE STABILITY

A key feature in the growth of most of the alloy systems listed above is the use of low energy (typically 10 to ~ 200 eV) self-ion and/or inert-ion bombardment of the growing film. "Low energy" in this context refers to energies with a corresponding ion range of the order of one or two monolayers. As has been discussed in recent review articles by Greene and Barnett[27] and Greene,[28] such low energy ion bombardment of the growing film results in large changes in film nucleation kinetics, near-surface diffusivities, elemental incorporation probabilities, and film growth kinetics. However, in the work to be discussed here, the primary role of the low-energy ion irradiation is to provide dynamic collisional mixing of the upper one or two layers of the growing film during deposition.

Inhibition of phase separation from the mixed to the equilibrium state places limits on both the materials systems chosen for study as well as the ion energy. These constraints are especially severe if the object of the experiment is to grow a single crystal metastable alloy. The self-diffusion coefficients of the alloy constituents must be low enough at the epitaxial temperature that atoms quenched-in to bulk lattice positions as the ion-mixed layer is covered up make few or no jumps during deposition of the remaining film. The ion energies used during deposition must be low so that the corresponding ion ranges are short enough to prevent enhanced diffusion in the "bulk" frozen-in lattice.

III-V based systems are reasonable candidates for metastable alloys since they are known to have low self-diffusion coefficients. Thermodynamic measurements carried out in the $(GaSb)_{1-x}(Ge_2)_x$ system, for example, show that the driving force for the transition from the single phase metastable to the equilibrium two-phase state, ΔH_t in Figure 1, is 27 meV for $(GaSb)_{0.36}(Ge_2)_{0.64}$ while the kinetic activation barrier, E_a in Figure 1, to the transition is 3.1 eV, more than two orders of magnitude higher.[16] Thus if one succeeds in fabricating the metastable alloy, it will be thermally and temporally stable to quite high temperatures.[10,11]

A. Role of Low—Energy Ion Irradiation During Crystal Growth

With the possible exception of $(GaAs)_{1-x}(Ge_2)_x$ alloys,[6,7] for which ΔH_t is expected to be small due to the low value of the strain energy component of the total free energy, low energy ion-mixing during deposition was found to be essential for the growth of these metastable alloys. For example, $Ge_{1-x}Sn_x$ films grown by dc glow discharge sputter deposition on (100) GaAs substrates with no applied negative substrate bias V_a were always found to be amorphous or two phase, depending on the substrate temperature T_s. The negative induced substrate bias V_i, additive to V_a, with respect to the positive plasma potential was estimated to be ~ 5 V during deposition. The average incident ion energy (the predominant ion species was Ar^+) was thus ~ 5 eV, too low to yield significant ion-mixing in the growing film.

However, when V_a was increased above ~ 50 V giving a total bias $V_s = V_i + V_a \sim 55$, it was possible to obtain crystalline single phase $Ge_{1-x}Sn_x$ alloys with x up to 0.11 (the solid solubility of Sn in Ge is $\leq 1\%$ and the alloy eutectic temperature is $231°C$[29]). In this case, the average incident ion energy is above the threshold energy for atomic displacements in the solid and ion-surface interactions such as ion mixing and preferential sputtering become important. Both of these effects play a crucial role in the growth of metastable alloys. Increasing V_s results in an increase in the maximum allowable growth temperature T_s at which single phase alloys could be obtained until, at each value of x, T_s itself reached a maximum value at V_a. Increasing $V_a > V_a$ resulted in a decrease in T_s.

Figure 3 shows a three dimensional plot of the transformation temperature T_s as a function of the substrate bias and the alloy composition. The T_s (x,V_a) surface was obtained from X-ray and electron diffraction analyses of the structure of more than 50 films. V_a decreased from 180 to 110 V and the maximum T_s decreased from 150 to 90°C as x increased from 0.02 to 0.11. In general, the growth parameter range for crystalline metastable alloy growth increased with decreasing x. This trend was observed previously for $(GaSb)_{1-x}(Ge_2)_x$ and is due to the decrease in the energy difference between the metastable and equilibrium states as x approaches the solid solubility limits.[3] The maximum T_s was found to be $> 140°C$, high enough to allow the growth of single crystal metastable $Ge_{1-x}Sn_x$ on (100) GaAs substrates, for $x \leq 0.05$.

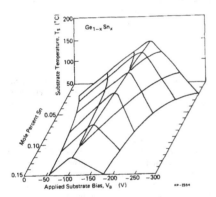

Figure 3. A growth phase map for dc sputter deposited $Ge_{1-x}Sn$ alloys. The three dimensional surface corresponds to the transition, as a function of the Sn fraction x, the substrate temperature T_s, and the applied substrate bias V_a, from a single phase metastable alloy to a three phase state. (Data from reference 24).

A similar effect of substrate bias on the growth of $(GaSb)_{1-x}(Ge_2)_x$ alloys by rf sputter deposition has recently been observed. Figure 4 shows a constant composition (x=0.10) section of a $(GaSb)_{1-x}(Ge_2)_x$ growth phase-map plotted as a function of T_s and the total substrate bias V_s. The result illustrates the same basic trend as shown in Figure 3, but with several important differences. First, the induced substrate bias during rf sputtering is typically much larger than in dc sputtering. Hence the results in Figure 4 were plotted as a function of $V_s = V_a + V_i$, where V_i was estimated from previous measurements.[30] Second, T_s was much higher for $(GaSb)_{1-x}(Ge_2)_x$ than for $Ge_{1-x}Sn_x$. This was due largely to the relatively low melting point of Sn which severely limited the T_s-range for the growth of $Ge_{1-x}Sn_x$. The larger T_s values in the $(GaSb)_{1-x}(Ge_2)_x$ system made it possible to grow single crystals on (100) GaAs substrates at all x-values.[10,11] Third, the V_s-value at which T_s is a maximum was considerably lower for the rf-sputtered $(GaSb)_{1-x}(Ge_2)_x$ than the dc-sputtered $Ge_{1-x}Sn_x$. This is due to the fact that substrate ion currents are much larger during rf sputtering than dc sputtering. This latter difference explains the recent result that single phase crystalline $Ge_{1-x}Sn_x$ alloys have been obtained with larger Sn fractions (x = 0.15) by rf sputtering than by dc sputtering (x = 0.11). The ratio of ions to atoms impinging on the growing film during glow discharge sputtering is estimated to be ~ 0.1 to 1.

The effect of ion bombardment on the growth of metastable alloys, as shown in Figures 3 and 4, can be explained based upon ion-surface interactions. Ar^+ ion energies of ~ 20 to 200 eV, corresponding to an ion range of the order of one monolayer, provided collisional mixing of the surface layer without affecting the "bulk" lattice. Once a well-mixed layer was buried by subsequent film growth it was frozen into the metastable state due to relatively low bulk self-diffusion rates (provided that T_s was sufficiently low). T_s' decreased rapidly when V_s was increased above 100 to 200 V, however. The upper limit on V_s is due to the ion range becoming sufficiently large to promote enhanced diffusion and hence phase separation in the bulk film.

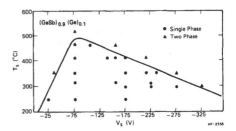

Figure 4. Phase map, plotted as a function of the growth temperature T_s and total substrate bias V_s, for $(GaSb)_{0.9}(Ge_2)_{0.1}$ films grown on (100) GaAs.

Cadien et al.[16] have recently shown that low energy ion irradiation during the deposition of amorphous GaSb/Ge mixtures also affects the local structure and hence the subsequent reaction paths upon post-annealing. That is, ion-mixed films deposited at $T_s = 60°C$ on Corning 7059 substrates were observed to transform to single-phase polycrystalline $(GaSb)_{1-x}(Ge_2)_x$ alloys upon annealing at 500°C. Films grown with no, or very little ion bombardment, transformed immediately to the two-phase equilibrium state. Transformation enthalpies were determined by differential scanning calorimetry. Preliminary EXAFS measurements showed that films grown with little or no ion bombardment contained incipient Ge and GaSb-rich clusters. Such clusters acted as natural nucleation sites for second phase formation upon annealing. Collisional mixing in the ion irradiated films inhibited cluster formation. For a well-mixed amorphous alloy, transformation to a metastable single phase upon crystallization is the reaction path with the lowest activation barrier since it involves only slight atomic rearrangement over distances of the order of an atomic diameter or less.

B. Thermal Effects During Crystal Growth

As shown in the previous section, low energy ion bombardment plays a crucial role in the vapor phase growth of metastable alloys. However, thermal effects are also important since they determine the resulting film crystal structure (i.e., amorphous, polycrystal, or single crystal) and the kinetics of phase transitions. In this section, results concerning the effect of the substrate temperature on metastable alloy growth are presented.

Thermal effects on the growth of $(GaSb)_{1-x}(Ge_2)_x$ alloys have been discussed previously.[11] The general findings can be illustrated with results for $(GaSb)_{0.37}(Ge_2)_{0.63}$ alloys grown on (100) GaAs. At 470°C, the films were single crystals as indicated by electron channeling and transmission electron microscopy. X-ray diffraction spectra showed only a (400) peak. However, for films grown at 509°C, the matrix alloy, represented by the position of the (400) diffraction peak, became more Ge-rich as side peaks, corresponding to GaSb-rich and Ge-rich second phases appeared. The precipitates were also single crystals and coherent with the matrix. At higher T_s values, the side peaks increased in intensity and shifted toward their equilibrium positions for GaSb saturated with Ge and Ge saturated with GaSb. Thus, in agreement with post-deposition annealing studies, the transformation from the single phase metastable state in this alloy to the two phase equilibrium state proceeds through a continuous series of metastable GaSb-rich and Ge-rich states. This is somewhat different behavior than observed for $(GaSb)_{1-x}(Sn_2)_x$ as discussed below where there is the additional structural constraint of maintaining Sn in the cubic phase.

Single-crystal zincblende structure $(GaSb)_{1-x}(Sn_2)_x$ alloys with compositions ranging from $x=0$ to $x=0.14$ were grown by rf sputter deposition on (100) GaAs substrates even though the maximum equilibrium solid solubility of Sn in GaSb is $\leq 1\%$.[25] The films, typically 1.5 μm thick, were grown by co-sputtering from a two-phase GaSb-Sn target as well as by multi-target sputtering[31,32] from pure GaSb and Sn targets. In both cases, the Sn and GaSb targets were undoped with purities of $\geq$ 99.999%. Deposition temperatures were typically low enough to allow the growth of stoichiometric GaSb without the use of an excess Sb flux. rf sputter deposition was carried out with $V_a = 0$ and $P_{Ar} = 15$ mTorr corresponding to $V_s = V_i \sim 75V$, near the optimal ion accelerating potential shown in Figure 4. The sputter deposition was the same one used by Barnett et al.[33] to grow high resistivity and Mn doped p-type GaAs with carrier mobilities equivalent to those obtained by molecular beam epitaxy and liquid phase epitaxy.

Figure 5 shows the X-ray diffraction pattern taken from a x=0.06 alloy, $\sim$ 1 μm thick, grown on (100) GaAs at a substrate temperature $T_s = 212°C$. The only reflections observed were (200) and (400) peaks from both the alloy and the substrate. (These appear as doublets due to reflection of both $K\alpha_1$ and $K\alpha_2$ X-rays.) The peaks are symmetric giving no indication of second-phase shoulder peaks. Electron channeling indicated that the film was a single crystal. The alloy peaks are broader than the substrate peaks due largely to misfit dislocations produced by the $\sim$ 8% film-substrate lattice mismatch. The (200) alloy peak intensity relative to the (400) peak is close to that for pure GaSb indicating that the alloy had a zincblende structure. (Zincblende-to-diamond structure transitions in $(III-V)_{1-x}(IV_2)_x$ alloys are discussed in Section III.)

Figure 6 shows X-ray diffraction results as a function of T_s for polycrystalline $(GaSb)_{0.93}(Sn_2)_{0.07}$. At $T_s = 160°C$, the film was a single-phase polycrystalline alloy with a strong (220) preferred orientation and a grain size of the order of 20-40 nm. The first evidence of second phase formation was observed for T_s between 180 and 200°C. As T_s was increased to 230°C, the intensity of the (220) and (311) peaks increased due to an increase in grain size. However, the alloy peak positions shifted slightly towards that expected for pure GaSb as small (111) GaSb-rich and α-Sn-rich (diamond phase) peaks appeared. The new phases are Sn-rich relative to the alloy resulting in a decrease in the alloy Sn content. This trend continues at $T_s = 410°C$, where the (111) GaSb-rich and Sn-rich peaks have grown at the expense of the (220) alloy peak which in turn has shifted further towards pure GaSb. The X-ray spectrum for a film grown at $T_s = 430°C$ indicates an essentially equilibrium two phase mixture with (111), (220) and (311) peaks corresponding to GaSb saturated with Sn, and (200) and (101) reflections from pure β-Sn (the tetragonal phase).

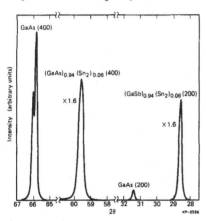

Figure 5. X-ray diffraction pattern from a single crystal $(GaSb)_{0.93}(Sn_2)_{0.07}$ alloy film grown on a (100) GaAs substrate at a substrate temperature of 212°C. (Data from reference 17).

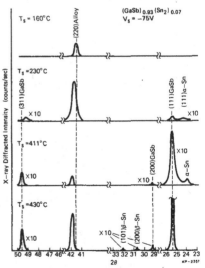

Figure 6. X-ray diffraction patterns from polycrystalline $(GaSb)_{0.93}(Sn_2)_{0.07}$ alloy films grown on glass at the temperatures T_s indicated.

Figure 7 summarizes the diffraction results showing the (220) alloy peak position, the (111) α-Sn-rich peak intensity, and the (200) β-Sn peak intensity as a function of T_s for $x = 0.07$. The alloy peak position shifts continuously with increasing T_s until it reaches a value corresponding to GaSb saturated with Sn. The Sn lost from the alloy appears in the α-Sn phase, which becomes increasingly Sn-rich as T_s increases, until $T_s = 420°C$ where an abrupt transition to essentially pure β-Sn occurs. The narrow temperature range ($\leq 5°C$) over which the phase transition from diamond to tetragonal structure Sn occurs suggests that Sn melts. After film growth when the sample is slowly cooled the Sn re-freezes and rejects most of the dissolved GaSb resulting in the observed essentially pure β-Sn and GaSb phases. The surface morphology of samples with β-Sn were quite rough as observed by SEM, in contrast to the featureless surfaces obtained at $T_s < 420°C$.

The Sn-rich diamond phase is stable to film growth temperatures as high as 420°C, considerably above the melting point of Sn, 232°C. This result is similar to that obtained for metastable zincblende $InSb_{1-x}Bi_x$ films where single-phase alloys were obtained well above the melting point of tetragonal InBi.[21] The zincblende matrix acts to stabilize the second phase in the cubic structure.

The interrelationship between thermal and ion bombardment effects during the growth of metastable alloys is clearly illustrated by the results shown in Figure 8 which is a plot of the Sn concentration in $(GaSb)_{1-x}(Sn_2)_x$ films grown with the same incident fluxes as a function of T_s for $V_s = 175$ V. The Sn concentration was measured by wavelength-dispersive analysis in an electron microprobe and, on the scale of the several μm diameter probe area, found to be uniform over the entire sample. For $T_s \leq 150°C$, $x = 0.14$. However, for $T_s \geq 150°C$, the maximum growth temperature for obtaining single phase $(GaSb)_{0.93}(Sn_2)_{0.07}$ alloys at $V_s = 175$ V, the x-value decreased rapidly becoming 0.04 at $T_s = 230°C$. The explanation for these results is that Sn is preferentially sputtered from incipient Sn-rich precipitates at a much higher rate than from the $(GaSb)_{1-x}(Sn_2)_x$ matrix due to a decrease in the Sn surface-binding energy. This is similar to an effect observed previously in the growth of InSb at low temperatures where, as the In/Sb flux ratio was varied, preferential sputtering always acted to inhibit second phase formation by preferentially removing the excess species.[34]

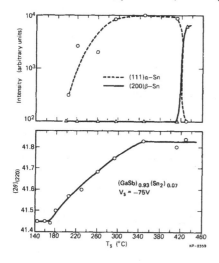

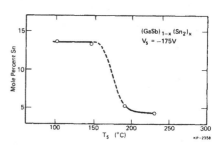

Figure 8. The Sn fraction x as a function of growth temperature T_s for $(GaSb)_{1-x}(Sn_2)_x$ films grown with the same incident flux composition and an applied substrate bias $V_s = 100$ V.

Figure 7. A summary of X-ray diffraction data from $(GaSb)_{0.93}(Sn_2)_{0.07}$ alloys showing (a) the angular position of the (220) alloy peak and (b) the intensities of the (111) α-Sn-rich and the (200) β-Sn peaks as a function of film growth temperature T_s.

III. PHYSICAL PROPERTIES OF SINGLE CRYSTAL METASTABLE ALLOYS

Detailed structural and optoelectronic measurements have been carried out on two metastable systems: $(GaAs)_{1-x}(Ge_2)_x$ and $(GaSb)_{1-x}(Ge_2)_x$. A variety of experimental techniques including optical absorption, spectroscopic ellipsometry, Raman spectroscopy, Hall measurements, X-ray diffraction, analytical electron microscopy, and extended X-ray absorption fine structure (EXAFS) analysis, were employed.

Single crystals of both $(GaAs)_{1-x}(Ge_2)_x$ and $(GaSb)_{1-x}(Ge_2)_x$ have been grown with $0 \leq x \leq 1$ on (100) GaAs. The $(GaSb)_{1-x}(Ge_2)_x$ alloys were deposited by multitarget sputtering as described earlier using a two-phase $Ga_{0.3}Sb_{0.7}$ target as the GaSb source in order to provide an excess Sb overpressure during growth. $(GaAs)_{1-x}(Ge_2)_x$ alloys, on the other hand, were grown in an ultra-high vacuum ion beam sputtering system.[5] Two modified 2.5 cm-diameter Kaufman ion guns with all metal seals were used to simultaneously Ar$^+$ ion sputter 5-cm diameter high purity, undoped, water-cooled, single crystal GaAs and Ge wafers. An effusion cell charged with 99.9999% pure As was used to provide an As_4 overpressure during film growth at T_s between 550°C (for pure GaAs) to 450°C (for pure Ge). The alloys were grown with no intentional ion bombardment although there was an incident flux of energetic Ar atoms backscattered from the targets and sputter deposited species had average incident kinetic energies of the order of 10-15 eV. No trapped Ar was detected in the $(GaAs)_{1-x}(Ge_2)_x$ (or the $(GaSb)_{1-x}(Ge_2)_x$) alloys by either electron microprobe or SIMS analysis due to the low ion range and high growth temperatures.

Preliminary transport measurements showed that the $(GaSb)_{1-x}(Ge_2)_x$ alloys were p-type with room temperature carrier concentrations of ~ 1.5 μm thick films ranging from 10^{16} to 10^{19} cm^{-3} with corresponding hole mobilities form 10 to 720 cm^2/V-s. $(GaAs)_{1-x}(Ge_2)_x$ films grown at ~ 1 μm h^{-1} and an As_4 flux of $\sim 2 \times 10^{15}$ cm^{-2}s^{-1} were n-type for GaAs-rich alloys and p-type for Ge-rich alloys. Increasing the As_4 flux decreased the hole concentration in p-type films and increased the electron concentration in n-type films. Electron channeling and X-ray diffraction analyses using a double crystal spectrometer showed that the $(GaAs)_{1-x}(Ge_2)_x$ alloys are highly perfect single crystals with X-ray diffraction line-widths of ~ 19 s of arc, comparable to that of the best MBE GaAs films. Analytical scanning transmission electron microscopy analysis gave no evidence of phase separation down to the limits of resolution, ~ 3 nm.

Optical absorption measurements (Figure 9) of the direct Γ-point bandgap E_o of $(GaAs)_{1-x}(Ge_2)_x$ shows a strong negative, but non-parabolic, bowing as a function of x. E_o decreases rapidly with increasing Ge concentration until $x \sim 0.3$. Further increases in x result in a gradual increase in E_o to the value for pure Ge. Spectroscopic ellipsometry data for the x-dependence of the higher band. E_1 vs x show qualitatively similar behavior. Furthermore, photoluminescence measurements of E_o vs x in $(GaAs)_{1-x}(Ge_2)_x$ grown by OM—CVD[7] agrees well with the results shown in Figure 8, although optical absorption measurements of the OM-CVD grown alloys show less deviation from parabolicity and place the minimum near $x \sim 0.7$.[6] In any case, the large non-parabolic bowing in E_o vs x is not well described by conventional virtual crystal models which would treat $(GaAs)_{1-x}(Ge_2)_x$ as being composed of averaged $Ga_{1-x}Ge_x$ cations and $As_{1-x}Ge_x$ anions.

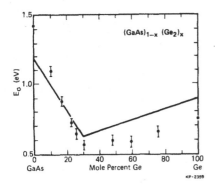

Figure 9. The direct Γ-point energy gap E_o as a function of x for $(GaAs)_{1-x}(Ge_2)_x$ alloys. The solid line is the theoretical prediction. (Data from reference 8).

The abrupt change in slope of E_o occurring near x = 0.30 in Figure 8 was interpreted by Newman et al.[8,37] as being due to a second order phase transition from a zincblende to a diamond structure. That is, at x $\leq$ 0.3 there are well defined sublattices. In the "disordered" diamond phase there are comparable numbers of Ga and As atoms on both cation and anion sites. In either structure, no restrictions were placed on the formation of Ga-Ga or As-As bonds. Using these assumptions about the atomic structure, an order parameter was obtained and used within the virtual crystal approximation and an sp^3s tight binding model to derive a Hamiltonian from which the band structure was calculated. The results provided a reasonable description of the measured data as shown in Figure 8. However, Holloway and Davis[38] have recently described the experimental results with a model which assumes perfect short-range order and in which the bowing is only weakly dependent upon the existence of a long-range order-disorder transition at high Ge concentrations. Perfect short-range order in this context implies no Ga-Ga or As-As nearest-neighbor pairs.

The fact that both models provide reasonable fits to the data while starting from very different assumptions concerning the atomic arrangement of the alloys is due at least in part to the fact that the energy gap, an opto-electronic property, is related to the atomic structure in a complicated and indirect manner. However, recent X-ray diffraction, EXAFS, and Raman results on $(III-V)_{1-x}(IV_2)_x$ alloys have begun to provide a much more detailed understanding of atomic ordering in these materials. Most of the experiments have been carried out on $(GaSb)_{1-x}(Ge_2)_x$, which also exhibits a large and non-parabolic bowing in E_o vs x, since the atomic numbers of constituent species in $(GaAs)_{1-x}(Ge_2)_x$ are so close. It is difficult, in the latter case, to separate the alloy diffraction peak from that of the GaAs substrate or to distinguish among backscattered EXAFS signals from Ga, As, and Ge.

Raman spectra from $(GaSb)_{1-x}(Ge_2)_x$ exhibited GaSb-like and Ge-like optical modes with a "one-two" type mode behavior.[15,35] The mode pattern was due to the lack of possible localized modes on the Ge-rich side. Figure 10 shows the x-dependence of the observed optical phonon frequencies. The frequency of the GaSb-like longitudinal optical phonon mode did not shift appreciably with x, but more importantly, there is evidence of a small peak due to a transverse optical phonon, indicating a zincblende structure, out to at least x $\sim$ 0.1. The fact that it was not observed between x $\simeq$ 0.1 and 0.3 provides no information since the TO mode is not Raman-active for the backscattering geometry used. In addition, the Raman spectra shows no evidence for Sb-Sb pairs (the Raman peak due to Ga-Ga pairs would have been obscured by the alloy peak).

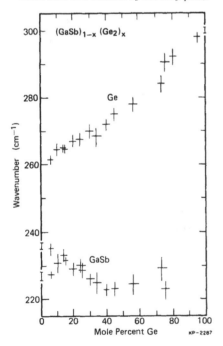

Recent EXAFS results are in agreement with the Raman data and indicate no Sb-Sb, or Ga-Ga, nearest neighbors.[36] That is, the alloy system exhibits perfect short range order, to within the sensitivity of the technique, throughout the entire composition range. Long range order in the alloys was also investigated using X-ray diffractometry to determine the intensity ratio r_I of superlattice (200) to fundamental (400) reflections as a function of x. The results are shown in Figure 11. r_I, which is related to the long range order parameter, decreases with increasing x to become essentially zero for x $\gtrsim$ 0.3 indicating the existence of a zincblende to diamond structural transition. Initial results from diffraction experiments carried out on $(GaAs)_{1-x}(Ge_2)_x$ alloys on (100) GaP (GaP was used in order to eliminate interference between substrate and film reflections) show a similar behavior.

Figure 10. Wave number of optical phonons as a function of x in $(GaSb)_{1-x}(Ge_2)_x$ alloys. (Data from reference 35).

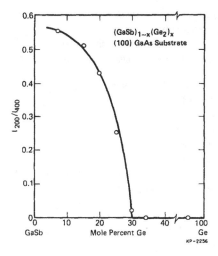

Figure 11. Ratio of the intensities of the (200) superlattice X-ray reflection to the (400) fundamental reflection as a function of Ge fraction from single crystal $(GaSb)_{1-x}(Ge_2)_x$ alloy films grown on (100) GaAs substrates. (Data from reference 24).

Thus, while the composition of the zinc blende to diamond transition is close to that predicted by Newman et al.[8,37] in $(III-V)_{1-x}(IV_2)_x$ alloys, the assumption of a corresponding change in short-range order was not observed. More work is clearly required in order to fully understand the relationship between the band structure and the atomic structure in these materials.

IV. CONCLUSIONS

The Sn-based metastable alloys $(GaSb)_{1-x}(Sn_2)_x$ and $Ge_{1-x}Sn_x$ have been grown as single crystals, adding to the range of single crystal metastable alloy films investigated, which includes $(GaSb)_{1-x}(Ge_2)_x$, $(GaAs)_{1-x}(Ge_2)_x$, $(GaAs)_{1-x}(Si_2)_x$, $Pb_{1-x}Cd_xS$, $InSb_{1-x}Bi_x$, $GaSb_{1-x}As_x$, and α-Sn. The Sn-based alloys are complicated by the low melting point and polymorphic nature of Sn, and as such the results on the growth and phase stability of these materials help to extend the general understanding of the growth of metastable alloys. The recent experimental and theoretical results on the atomic structure and optical properties of $(III-V)_{1-x}(IV_2)_x$ alloys provide a much clearer understanding of the physical properties of these materials.

ACKNOWLEDGMENTS

The authors greatfully acknowledge the financial support of the Joint Services Electronics Program under contract #N00014-84-C-0149 and the Materials Sciences Division of the Department of Energy during the course of this work.

REFERENCES

1. P. Duwez, R.H. Willens, and W. Klement, Jr., J. Appl. Phys. 31, 1136 (1960).

2. H. Jones, Rep. Prog. Phys. 36, 1425 (1973).

3. See the paper by R.F.C. Farrow in this volume.

4. A.J. Noreika and M.H. Francombe, J. Appl. Phys. 45, 3690 (1974).

5. S.A. Barnett, M.A. Ray, A. Lastras, B. Kramer, J.E. Greene, and P.M. Raccah, Electron. Lett. 18, 891 (1982).

6. Zh.I. Alferov, M.Z. Zhingarev, S.G. Konnikov, I.I. Mokin, V.P. Ulin, V.E. Umanskii, and B.S. Yavich, Sov. Phys. Semicond. 16, 532 (1982).

7. Zh.I. Alferov, Sov. Phys. Semicond. 16, 567 (1982).

8. K.E. Newman, A. Lastras, B. Krammer, S.A. Barnett, M.A. Ray, J.D. Dow, J.E. Greene, and P.M. Raccah, Phys. Rev. Lett. 50, 1466 (1983).

9. S. Fang and J.E. Greene, unpublished.

10. K.C. Cadien, A.H. Eltoukhy, and J.E. Greene, Appl. Phys. Lett 38, 773 (1981).

11. K.C. Cadien, A.H. Eltoukhy, and J.E. Greene, Vacuum 31, 253 (1981).

12. S.I. Shah, K.C. Cadien, and J.E. Greene, J. Electron. Mater. *11*, 53 (1982).

13. K.C. Cadien and J.E. Greene, Appl. Phys. Letters *40*, 329 (1982).

14. J.E. Greene, S.A. Barnett, K.C. Cadien, and M.A. Ray, J. Cryst. Growth *56*, 389 (1982).

15. T.N. Krabach, N. Wada, M.V. Klein, K.C. Cadien, and J.E. Greene, Solid State Commun. *45*, 895 (1983).

16. K.C. Cadien, B.C. Muddle, and J.E. Greene, J. Appl. Phys. *55*, 4177 (1984).

17. L.T. Romano, S.A. Barnett, and J.E. Greene, unpublished.

18. A.K. Sood, K. Wu and J.N. Zemel, Thin Solid Films *48*, 76 (1978).

19. A.K. Sood, K. Wu and J.N. Zemel, Thin Solid Films *48*, 87 (1978).

20. J.L. Zilko and J.E. Greene, J. Appl. Phys. *51*, 1549 (1980).

21. J.L. Zilko and J.E. Greene, J. Appl. Phys. *51*, 1560 (1980).

22. A.J. Noreika, W.J. Takei, M.H. Francombe, and C.E.C. Wood, J. Appl. Phys. *53*, 4932 (1982).

23. R.F.C. Farrow, D.S. Robertson, G.M. Williams, A.G. Cullis, G.R. Jones, I.M. Young, and P.N.J. Dennis, J. Cryst. Growth *54*, 507 (1981).

24. S.I. Shah and J.E. Greene, unpublished.

25. F. Gerdes and B. Predel, J. Less-Common Metals *79*, 281 (1981).

26. J.E. Greene, J. Vac. Sci. Technol. *B1*, 229 (1983).

27. J.E. Greene and S.A. Barnett, J. Vac. Sci. Technol. *21*, 285 (1982).

28. J.E. Greene, CRC Critical Reviews of Solid State and Materials Science *11*, 47 (1983) and *11*, 189 (1984).

29. W. Klemm and H. Stohr, Z. Anorg. Chem *241*, 305 (1939).

30. S.A. Barnett and J.E. Greene, Surf. Sci. *128*, 401 (1983).

31. C.E. Wickersham and J.E. Greene, J. Appl. Phys. *47*, 4734 (1976).

32. C.E. Wickersham and J.E. Greene, J. Appl. Phys. *47*, 2289 (1976).

33. S.A. Barnett, G. Bajor, and J.E. Greene, Appl. Phys. Letters *37*, 734 (1980).

34. J.L. Zilko, S.A. Barnett, A.H. Eltoukhy, and J.E. Greene, J. Vac. Sci. Technol. *14*, 595 (1980).

35. R. Besserman, J.E. Greene, M.V. Klein, T.N. Krabach, T.C. McGlinn, L.T. Romano, and S.I. Shah, Proc. 17th Internat. Conf. Phys. Semicond., San Francisco, August, 1984.

36. E.A. Stern, F. Ellis, K. Kim, L. Romano, S.I. Shah, and J.E. Greene, unpublished.

37. K.E. Newman and J.D. Dow, Phys. Rev. *B27*, 7495 (1983).

38. H. Hollowy and L.C. Davis, Phys. Rev. Letters *53*, 830 (1984).

INTERDIFFUSION AND STABILITY OF COMPOSITIONALLY MODULATED FILMS

Frans SPAEPEN[+]
Division of Applied Sciences, Harvard University, Cambridge, MA 02138

ABSTRACT

The thermodynamics and kinetics governing the stability of compositionally modulated materials are reviewed. Typical results of interdiffusion experiments on fcc and amorphous metals are discussed to demonstrate the sensitivity and special features of the technique. Two coarsening mechanisms, and the stability of orthogonal waves in phase separating systems are analyzed.

INTRODUCTION

Compositionally modulated materials are produced by multiple alternate deposition of very thin layers of different composition. The deposition methods include vapor deposition [1,2], molecular beam epitaxy [3], sputtering [4,5], and chemical vapor deposition [6]. If a structure has a reasonably low free energy over an extended continuous composition range, and if there is sufficient intermixing during the deposition process, the multiple layering process can result in the formation of a compositionally modulated material, i.e. one in which the composition profile is closely approximated by a simple harmonic function :

$$c = c_0 + A \cos \beta x \qquad (1)$$

where c_0 is the average composition, x the distance coordinate normal to the film plane, A the modulation amplitude and β the modulation wavenumber ($\beta = 2\pi/\lambda$; λ is the modulation wavelength). Of interest here are materials with λ between 0.2 and 10 nm. This composition modulation can be measured by X-ray diffraction [7,8]. It gives rise to a peak at $\bar{k} = \beta\hat{x}$, with intensity proportional to A^2. This peak can be considered as a satellite of the (000) forward scattering peak, and is independent of the atomic scale structure (crystalline or amorphous) of the material. In crystalline materials, satellites also appear at $\bar{k}_{hkl} + \beta\hat{x}$ around the higher order Bragg peaks (hkl); their intensity, however, is not just dependent on A^2, but also on the variation in lattice spacing. Deviations from a simple harmonic composition modulation (eq. (1)) can be detected from the higher order peaks at $\bar{k} = n\beta\hat{x}$, (n = 2, 3, ...).

Crystalline compositionally modulated metallic films have been made most successfully in binary fcc systems with full solid solubility. For example : Ag-Cu [2], Cu-Pd [9], Au-Ni [10], Au-Cu [11] and Cu-Ni [12,13]. Usually a mica substrate was used, which produced films with a strong (111) texture that were fully coherent for small enough wavelength.

Amorphous compositionally modulated metallic films have so far only been reported for metal-metalloid alloys, such as $(Pd_{85}Si_{15})$ - $(Fe_{85}B_{15})$ [5, 14-16]. Modulated binary amorphous systems have only very recently been obtained; for example Cu-Zr [17-18] and Ni-Zr [18]. These amorphous modulated films differ in important ways from the crystalline ones :

[+] Visiting Professor, Metallurgy Department, University of Leuven, Belgium, Fall 1984-1985.

(i) They can be produced at any value of λ, without interference between the periodicity of the atomic structure, as is the case at short modulation wavelengths in crystals. Because the spatial distribution of atom centers in an amorphous system is continuous and the composition c in equation (1) is the average one over the yz-plane at position x, λ can, in principle, be made arbitrarily small. In an X-ray diffractometer experiment with the scattering vector $\vec{k}$ along $\hat{x}$, the same averaging is done, and therefore a peak at an arbitrarily large value of β could be observed. On a _local_ scale, however, the composition gradient can not be much greater than 1/a (a : average interatomic distance), even if dc/dx in the averaged equation (1) is very large at ·small λ. This is important in view of the local composition gradient contributions to the free energy, discussed below. In crystalline materials, the continuum description must be corrected when λ approaches 2a; a similar correction seems required for amorphous materials.
(ii) In many systems, amorphous alloys can be produced over a much larger continuous composition range than their crystalline counterparts. Producing a composition modulation in such a system may therefore only be possible in the amorphous state.
(iii) Coherency strains, which can contribute considerably to the free energy of crystalline modulated films and must be taken into account in the analysis of interdiffusion [9], are absent, presumably at least initially, in amorphous metallic modulated films. They may arise from chemical changes during the interdiffusion process [19].
(iv) Short circuit diffusion paths, especially grain boundaries (see below), are not found in the amorphous films. This simplifies the analysis of the early stages of the interdiffusion, and, from a practical point of view, enhances the stability and useful life of the films.

Artificial compositionally modulated materials described above are thermodynamically unstable. They have an unusually high density of interfaces, which can be eliminated in two ways : homogenization by interdiffusion, or sharpening of the composition profile followed by coarsening. The thermodynamic conditions determining the evolution of a modulated structure will be discussed in Section 2. Monitoring the intensity of the modulation satellites as a function of annealing time is the most sensitive method for measuring interdiffusivity, which is especially useful for the study of amorphous metals; this will be discussed in Section 3. Some mechanisms of coarsening are identified and analyzed in Section 4. A more detailed survey of some of the theory and experiments discussed here can be found in a forthcoming review paper [20].

THERMODYNAMICS

The theory of the thermodynamics and stability of systems, such as the modulated films, that are inhomogeneous on a very fine scale was first fully formulated by Cahn and Hilliard [21-23]. A detailed exposition can be found in the excellent review paper by Hilliard [24]. A review of the historical development was given by Cahn in his 1967 Institute of Metals Lecture [25]. In this section, the most important results are outlined.

In an inhomogeneous system, the free energy of a volume element is not only dependent on the local composition, but also on that of the neighboring volume elements through the formation of interfaces with a composition gradient. For an inhomogeneity in one dimension, $\hat{x}$, the total free energy of the system can then be written as :

$$F = \alpha \int [f(c) + \kappa(\frac{\partial c}{\partial x})^2] \, dx \tag{2}$$

$c\ell$ is the area perpendicular to $\hat{x}$; $f(c)$ is the free energy per unit volume of a homogeneous system of composition c; κ is the gradient energy coefficient. If F is minimized, under the constraint of conservation of mass the following stability condition is obtained :

$$f' - 2 \kappa \frac{\partial^2 c}{\partial x^2} = \alpha \tag{3}$$

where α is a constant if the system is in equilibrium. Note that $f' = (\mu_A - \mu_B)/\bar{V} = $ constant (μ_i : chemical potential of element i, $\bar{V}$: molar volume) is the equilibrium condition for a system with negligible gradient energy. If such a system is not in equilibrium, the gradient of f' becomes the driving force for the interdiffusion flux, $\bar{J}$, towards equilibrium :

$$\bar{J} = - M \frac{\partial f'}{\partial x} \tag{4}$$

where M is the mobility. In the presence of gradient energy contributions, the driving potential for interdiffusion becomes α in equation (3) :

$$\tilde{J} = - M \frac{\partial \alpha}{\partial x} \tag{5}$$

Combination of equations (3) and (5) with the conservation of mass, $\partial c/\partial t= -\bar{V}$. div $\tilde{J}$, gives the diffusion equation :

$$\frac{\partial c}{\partial t} = \tilde{D}[\frac{\partial^2 c}{\partial x^2} - \frac{2\kappa}{f''} \frac{\partial^4 c}{\partial x^4}] \tag{6}$$

where M, f'' and κ are assumed constant to obtain a linear equation. The bulk interdiffusion coefficient is :

$$\tilde{D} = M f'' \bar{V} \tag{7}$$

The evolution of a composition modulation wave is obtained by inserting equation (1) in the diffusion equation, which yields :

$$A = exp [- \tilde{D}\beta^2 (1 + \frac{2\kappa\beta^2}{f''}) t] \tag{8}$$

The relative rate of change in amplitude is defined as an amplification factor :

$$R = \frac{1}{A} \frac{\partial A}{\partial t} = - \tilde{D}_\lambda \beta^2, \tag{9}$$

where

$$\tilde{D}_\lambda = \tilde{D} (1 + \frac{2\kappa\beta^2}{f''}) \tag{10}$$

is the wavelength-dependent interdiffusion coefficient.

The growth or decay of a modulation wave now depends on the sign of R. Three different cases, illustrated on Figure 1, can be distinguished. To facilitate the interpretation it is useful to consider the predictions of the regular solution model for f'' and κ. For an equiatomic binary mixture, with a molar energy of mixture ΔU :

$$f'' = \frac{4}{\bar{V}} [RT - 2\Delta U] \tag{11}$$

The gradient energy coefficient is [21] : $\kappa = \frac{2 \Delta U a^2}{\bar{V}}$ \hfill (12)

298

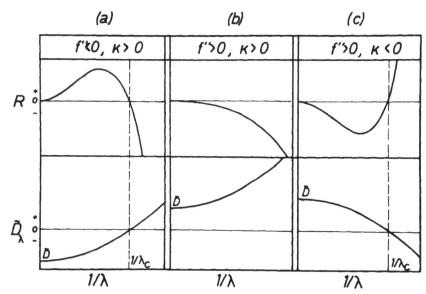

Figure 1 : Dependence of the amplification factor, R, and the interdiffusivity, $\tilde{D}_\lambda$, on the modulation wavelength for a phase separating
system inside the spinodal (a), and outside the spinodal (b),
and for an ordering system (c).

Figure 1a ($f'' < 0$, $\kappa > 0$) corresponds to a phase separating system ($\Delta U > 0$)
inside the spinodal. Modulations with wavelengths greater than $\lambda_c = 2\pi/\beta$
$= 2\pi(-2\kappa/f'')^{1/2}$ grow; this is the classical case of spinodal decomposition, corresponding to uphill diffusion ($\tilde{D} < 0$, eq. (7)). Modulations with
$\lambda < \lambda_c$ decay, due to their high density of interfaces which make a positive
gradient energy contribution.
Figure 1b ($f'' > 0$, $\kappa > 0$) corresponds to a phase separating system ($\Delta U > 0$)
outside the spinodal. The diffusion coefficient is positive for all wavelengths, and increases with decreasing wavelength due to the increasing
driving force for homogenization from the interfacial energy. Figure 1c
($f'' < 0$, $\kappa < 0$) corresponds to an ordering system ($\Delta U < 0$), and is the exact
negative of Figure 1a. At long wavelengths ($\lambda > \lambda_c$), the system homogenizes
($\tilde{D} > 0$, eq. (7)). Modulations with short wavelength ($\lambda < \lambda_c$) are predicted
to grow; since λ_c in crystalline systems is usually on the order of the atomic spacing, the development of very short wavelength "phase separation"
can be interpreted as an ordering process. It should also be kept in mind
that the continuum approximation used above must break down at small λ.
Corrections for the discreteness of the crystalline lattice have been made
by Cook et al. [26]. The fourth possibility ($f'' < 0$, $\kappa < 0$) is an unlikely
one. In fact, it can not occur in the regular solution approximation (equations (11) and (12)).

Since the lattice parameter of an alloy depends on its composition
($\eta = d(\ln a)/dc$), matching of the lattice planes along a modulation wave
introduces coherency strains. The free energy of eq. (2) now acquires an
additional term [22,23] :

$$F = \alpha \int [f(c) + \kappa (\frac{\partial c}{\partial x})^2 + \eta^2 Y (c - c_0)^2] dx \qquad (13)$$

where Y is a modulus that depends on the crystal symmetry and the direction of the modulation wave. For an isotropic continuum [22] :

$$Y = E/(1 - \nu) \tag{14}$$

E : Young's modulus; ν : Poisson's ratio.
As a result, the interdiffusion coefficient now contains an additional positive (i.e. homogenizing) term :

$$\tilde{D}_\lambda = \tilde{D} \left[1 + \frac{2\kappa\beta^2}{f''} \right] + 2Mn^2 Y \tag{15}$$

Note that f" now must be more negative to cause spinodal decomposition than in the absence of coherency strains. The condition $f'' < - 2n^2 Y$ defines the "coherent spinodal".

INTERDIFFUSION

Since the intensity, I, of the (000) satellite is proportional to A^2, (as are the (hkl) satellites in the absence of coherency strains) is the slope of a plot of ln I vs t, according to eq. (9), equal to $- 2 D_\lambda \beta^2$. Such a plot usually shows an initial transient with a decreasing slope before settling down to a "terminal slope" from which the interdiffusivity is taken. Cook and Hilliard [2] have attributed these transients to increased homogenization in moving grain boundaries as a result of texture-driven grain growth (see Figure 2).

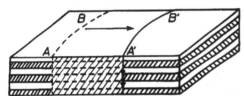

Figure 2 :
Schematic diagram of the homogenization by diffusion in a grain boundary moving from AB to A'B'.

In interdiffusion experiments on fcc-metallic compositionally modulated materials, the wavelength dependence of D_λ was found to obey equations (10) or (15), and the first direct determinations of the gradient energy coefficient, κ, were made. Cook and Hilliard [2] studied the Ag-Au system, in which the coherency strain effects are negligible ($n = 0.15$ %). A fit of their data to eq. (10) gave $\kappa = - 2.6 \times 10^{-11}$ J.m^{-1}, a negative value as expected for an ordering system (Figure 1c). In the temperature range 200-250°C they measured bulk interdiffusivities, $\tilde{D}$, as low as 10^{-24} m^2.s^{-1}, which were in perfect agreement with an extrapolation of high temperature (> 900°C) data ($\tilde{D} > 10^{-14}$ m^2.s^{-1}), which clearly represent lattice diffusion. The modulated film technique is therefore not only the most sensitive one, but it also allows determination of the lattice diffusivity in a temperature regime (T < $2T_M/3$) where other techniques would only detect short-circuit diffusion along grain boundaries or dislocations. The reason for this is, of course, that the diffusion distance for lattice diffusion ($\lambda/2$) is much lesser than either the grain size or the distance between dislocations, so that the fraction of the material homogenized by short circuit diffusion is negligible to that homogenized by lattice diffusion. The only short circuit mechanism that can homogenize an appreciable amount of material is that involving moving grain boundaries, as discussed above (see Figure 2). This mechanism also explains the very high effective diffusivities observed in the earliest compositionally modulated materials [1], which had a fine-grained polycrystalline structure.

The effect of the coherency strain on the diffusivity was confirmed by Philofsky and Hilliard [9] for the Cu-Pd system (η = 7 %). By comparing the intensities of the satellites at $\bar{k}_{111} + \beta$ and $\bar{k}_{111} - \beta$, they established that for full coherence the modulation wavelength had to be less than 2.8 nm; beyond that, a gradual loss of coherence was observed with full incoherence beyond 3.8 mm. The wavelength dependence of $\tilde{D}_\lambda$ was found to obey equation (15) for $\lambda < 2.8$ mm; beyond that a gradual decrease of $\tilde{D}_\lambda$ with λ was observed, corresponding with the decrease in the positive strain term in equation (15).

The modulated film technique has special advantages in the study of diffusion in metallic glasses.
(i) Since metallic glasses crystallize rapidly if their diffusivity is greater than about 10^{-20} m^2.s^{-1}, sensitive detection techniques are required. The solution of all other methods for measuring the diffusivity is determined by the spatial resolution of the "slicing" technique used to analyze a single diffusion junction. Specialized techniques, such as sputter profiling, Rutherford backscattering or nuclear reactions must be used on the metallic glasses, and they are limited to only a few orders of magnitude of D. Below 10^{-23} m^2.s^{-1}, only the modulated film technique is available. Its sensitivity is a result of its being composed of several hundred diffusion junctions, so that very small changes in the composition profile, corresponding to a diffusion distance of no more than 10 pm, can be detected by the X-ray technique.
(ii) Since glasses are thermodynamically unstable, they undergo a continuous series of transformations to states of lower free energy. This process of structural relaxation affects all physical properties, but most strongly the atomic transport properties such as viscosity [27] and diffusivity [15]. Since the modulated film technique is non-destructive it can be used conveniently to measure the time dependence of the diffusivity. Metallic glasses show a very large transient in their ln I vs t curve, which can only be attributed to structural relaxation [15]; indeed, for a (Pd$_{85}$Si$_{15}$)$_{50}$/ (Fe$_{85}$B$_{15}$)$_{50}$ film, the quantity $\tilde{D}_\lambda^{-1}(t)$ was observed to increase linearly with time, exactly as one would expect from a scaling relation with the viscosity, which has also been observed to increase linearly with annealing time [27]. The technique also allows iso-configurational measurements of the temperature dependence of $\tilde{D}$. After sufficient annealing, structural relaxation becomes slow enough so that upon cycling of the temperature, D can be reproduced. The iso-configurational activation energy for diffusion measured this way was very close to that of the viscosity in a similar system [15,27].
(iii) Since interdiffusion is driven by a chemical driving force, the modulated film technique allows insight into the thermodynamics of metallic glasses, which are difficult to obtain otherwise. Cammerata and Greer [16] have studied the wavelength dependence of $\tilde{D}_\lambda$ in the (Pd$_{85}$Si$_{15}$)$_{50}$/(Fe$_{85}$B$_{15}$)$_{50}$ system. Their results are redrawn on Figure 3, together with the fit to equation (10). The data points were taken from the terminal slopes of ln I vs t curves. Since the films had identical fabrication and thermal histories, and exhibited similar relaxation kinetics, the data were assumed to correspond to the same degree of relaxation. Nevertheless, part of the scatter in the data may still be due to the difficulty in reproducing an identical structural state. The system is clearly a homogenizing one. Since the bulk interdiffusivity, $\tilde{D}$, is positive, f" > 0 according to equation (7). Since the slope of $\tilde{D}_\lambda$ vs β is negative, κ must be negative, which, according to the regular solution approximation (equation (12)), corresponds to an ordering system. The equiatomic heat of mixing, ΔU, for Pd and Fe, which are the heaviest elements whose interdiffusion is mainly being observed by the X-ray technique, has indeed been calculated to be

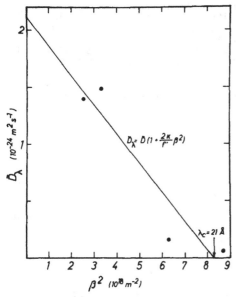

$D_\lambda = D(1 + \frac{2\kappa}{f''}\beta^2)$

$\lambda_c = 21\ \text{Å}$

Figure 3 :
Wavelength dependence of the
interdiffusivity in an amor-
phous $(Pd_{85}Si_{15})_{50}/(Fe_{85}B_{15})_{50}$
modulated film annealed at
250°C. After Cammarata and
Green [16].

- 6 kJ . mole^{-1} [28], and ordered phases occur in the crystalline state.
A self-consistent calculation of ΔU from Figure 3 and equations (11) and
(12), however, is not possible. It is probably necessary to extend the calculation of κ beyond the nearest neighbor approximation [29].

The scaling relation between diffusivity, D, and viscosity, η, [30]is:

$$\eta D = \frac{kT}{L} \tag{16}$$

where L is a characteristic length. In the liquid state, and around the
glass transition temperature, $L = 3\pi a$, corresponding to the Stokes-Einstein
relation. To use the value of Figure 4, it is necessary to correct for the
chemical driving force [24] :

$$\tilde{D} = D\,\frac{c(1-c)\ f''\bar{V}}{RT} \tag{17}$$

or, with c = 0.5 and the regular solution approximation of equation (11)
$\tilde{D} = D (1 - 2\Delta U/RT)$. Using the calculated value $\Delta U = -6$ kJ.mole [28], this
gives : $\tilde{D} = 3.8\ D$. Combined with earlier results on the viscosity [15,16,
27], this gives in equation (16) : L = 0.06 a. This is still 150 times
smaller than the Stokes-Einstein value, but closer to it than an earlier
estimate without the driving force correction [16]. It should, of course,
be kept in mind that this analysis is only approximate, since there is
insufficient information to take possible chemical effects of the metalloids,
B and Si, on the interdiffusion of Fe and Pd into account.

COARSENING

In a phase separating system inside the spinodal (see Figure 1a), a
composition modulation of sufficiently large wavelength ($\lambda > \lambda_c$) grows.
This has been observed in modulated films of the Au-Ni [10] and Cu-Ni [12,
13] systems. The modulation amplitude continuous to increase until it

reaches the equilibrium composition at the annealing temperature. At the same time, "squaring" of the composition profile occurs, and continues until each point is at one of the two equilibrium compositions, except for the interfaces, which must have a finite width due to the gradient energy [21]. This final state in the evolution of a growing modulation is called a "stationary state". It is clear that the evolution towards this state can no longer be described by the linear diffusion equations (6), since κ, f" and M can only be assumed constant if the amplitude, A, is small. For larger amplitudes, the composition dependence of f can be taken into account with a composition dependent diffusion coefficient, which is usually done by a Taylor expansion in $\Delta c = c - c_0$:

$$\tilde{D} = \tilde{D}^0 + \tilde{D}' \, (\Delta c) + \tfrac{1}{2} \tilde{D}'' \, (\Delta c)^2 \tag{18}$$

The coefficients are proportional, respectively, to the second, third and fourth derivatives of f, and are therefore related to the spinodal and equilibrium compositions [13]. Analyses based on the resulting non-linear diffusion equation have been made by de Fontaine [31], Cahn [32], Langer [33] and Tsakalakos [12,13,33]. In the non-linear regime, the Fourier components of a composition profile are coupled; as a result, higher harmonics, leading to "squaring" of an initially purely sinusoidal wave and to saturation of growth at a stationary state, are introduced. By fitting the experimentally observed evolution, as a function of annealing time, of the fundamental and the harmonics to the theory, the coefficients in equation (18) can be determined [12,13,35].

The stationary state is still thermodynamically unstable due to its density of interfaces. The free energy can therefore be lowered further by a continuous process of coarsening. Since this process is driven by the surface tension, σ, it is useful to recall Cahn and Hilliard's [21] result for a diffuse interface :

$$\sigma = 2 \, \frac{f''^2}{f^{1V}} \, \xi, \tag{19}$$

where

$$\xi = \frac{1}{2} \sqrt{\frac{-\kappa}{f''}} \tag{20}$$

The width of the interface is 2ξ. In the classical Lifshitz-Slyozov-Wagner theory of Ostwald ripening the coarsening rate of a population of spherical precipitates in the diffusion-limited regime is given by [36] :

$$R^3 - R_0^3 = C_1 \bar{V} M \sigma t \tag{21}$$

where R is the average particle size, R_0 the initial average particle size, and, C_1 a proportionality constant. Langer [33] has developed a theory for identifying some of the stationary states in one- and three-dimensional spinodally decomposing systems, and has demonstrated this instability. From the rate of decay of these states, he was able·to predict the coarsening rate. For a three-dimensional system of spherical particles, he predicts:

$$R^3 - R_0^3 = C_2 f'' \, \bar{V} M \xi t \tag{22}$$

where C_2 is again a proportionality constant. Taking into account eq. (19), the functional dependence on time and surface tension in equations (21) and (22) are seen to be the same.

For a one-dimensional system, such as the modulated structures considered here, he has identified as "uniform" coarsening process, illustrated in Figure 4, in which alternating layers of one component thicken and thin down, leading to an eventual doubling of the wavelength. The coarsening rate, expressed as a continuous process, is then :

$$\lambda - \lambda_0 = \frac{\xi}{2} \ln (1 - \frac{8f''\bar{V}Mt}{2} e^{- 2\lambda_0/\xi})$$ (23)

where λ_0 is the initial wavelength.

A second mechanism for coarsening in modulated films involves the motion of fault lines (i.e. edges of missing planes), as observed in the coarsening of lamellar eutectics. As illustrated on Figure 5, the half plane recedes by diffusion of atoms from its tip to the thickening adjacent lamellae. A simple analysis of the diffusional flow driven by the difference in curvature between the tip (positive) and the adjacent lamellae (negative), gives for the coarsening rate [37] :

$$\lambda - \lambda_0 = C_3 \rho_f \bar{V} M \sigma t$$ (24)

where C_3 is a geometrical constant, estimated at about 300; ρ_f is the fault line density (i.e. the number of configurations of Figure 5 per unit area).

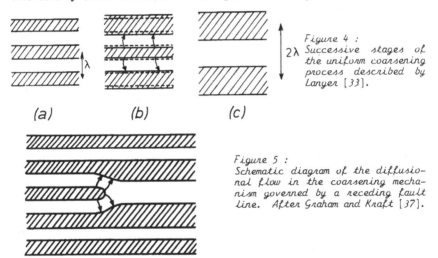

(a)   (b)   (c)

Figure 4 :
2λ Successive stages of the uniform coarsening process described by Langer [33].

Figure 5 :
Schematic diagram of the diffusional flow in the coarsening mechanism governed by a receding fault line. After Graham and Kraft [37].

Although this fault mechanism is likely to dominate the uniform one, due to the higher driving force, it is still interesting to compare equations (23) and (24). The linear dependence on time of the fault mechanism is found only approximately in the uniform mechanism of equation (23), and only if $\lambda \gg \xi$, which is the macroscopic limit. The linear proportionality of the coarsening rate, $d\lambda/dt$, with the surface tension found in the fault mechanism, however, does not have a simple equivalent in the uniform mechanism. From equation (23) it is seen that $d\lambda/dt$ indeed increases with ξ (proportional to σ ; equation (19)) in the macroscopic limit ($\lambda \gg \xi$), but not linearly. For short wavelengths ($\lambda < \xi/2$), the coarsening rate, even decreases with increasing ξ, presumably due to increasing overlap between the diffuse interfaces of adjacent layers. It seems likely that the fault mechanism also needs to be corrected in the short wavelength limit ($\lambda \leq \xi$).

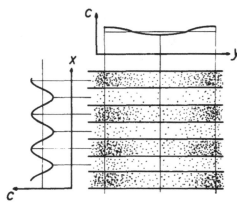

Figure 6 :
Schematic diagram of the com-
positional profile resulting
from two orthogonal modula-
tion waves.

In modulated materials of phase separating systems inside the spinodal, one also must consider the possibility of additional phase separation in directions other than these of the original modulation, as illustrated on Figure 6. This situation is different from that usually encountered in spinodal decomposition, where the two orthogonal waves usually have similar amplitudes. In a modulated film, the pre-existence of a large amplitude modulation affects the stability of small waves orthogonal to it. This will be demonstrated below on a simple example.

Consider a two-dimensional superposition, as in Figure 6 :

$$\Delta c = c - c_0 = C_x \cos \beta_x x + C_y \cos \beta_y y \tag{25}$$

The total free energy of the system is given by the two-dimensional equivalent of equation (2) :

$$F = \mathscr{L} \iint [f(c) + \kappa(\vec{\nabla} c)^2] \, dxdy \tag{26}$$

The bulk free energy must have at least one non-parabolic term, for example:

$$f(c) = -\frac{1}{2} A (\Delta c)^2 + \frac{1}{4} B (\Delta c)^4 \tag{27}$$

The total free energy per unit volume is then :

$$F = -\frac{A}{4} (C_x^2 + C_y^2) + \frac{3B}{32} (C_x^4 + C_y^4 + 4C_x^2 C_y^2) + \frac{\kappa}{2} (C_x^2 \beta_x^2 + C_y^2 \beta_y^2) \tag{28}$$

From this equation it is immediately clear that if the amplitudes are small ($C^4 \ll C^2$), or if the free energy is parabolic (B = 0, linear diffusion equation), the two waves are independent and have the same stability criterion ($\partial F/\partial C_i \leqslant 0$), $\beta^2 \leqslant A/2\kappa$, which is identical to the expression derived from equation (10). The case of interest here is : B $\neq$ 0, and $C_y \ll C_x$, so that higher powers of C_y can be neglected. The stability criteria for the two waves are now different :

$$\beta_x < (A - \frac{3}{4} BC_x^2)/2\kappa \tag{29}$$

$$\beta_y < (A - \frac{3}{2} BC_x^2)/2\kappa \tag{30}$$

Comparison of equations (29) and (30) shows that the spectrum of new small amplitude orthogonal waves, C_y, is not only more restricted than that of waves forming from a homogeneous solution, but even than that of the large amplitude pre-existing modulations.

A similar qualitative conclusion can be reached from Cahn's non-linear analysis of the later stages of spinodal decomposition [32], in which he derives expressions for the growth rates of coupled orthogonal waves of a particular wavenumber, β, by successive approximation In the problem of interest here, the spectrum of the first approximation is dominated by a single wave of amplitude C_x. The amplification factors, at t = 0, for waves in the $\hat{x}$ and $\hat{y}$ directions in the second approximation are then :

$$R_x = \frac{1}{C_x} \frac{dC_x'}{dt} = R\ [1 + \frac{3}{4} \frac{\tilde{D}''}{\tilde{D}_\lambda}\ C_x^2\ ] \tag{31}$$

$$R_y = \frac{1}{C_y} \frac{dC_y'}{dt} = R\ [1 + \frac{3}{2} \frac{\tilde{D}''}{\tilde{D}_\lambda}\ C_x^2\ ] \tag{32}$$

where C_x' and C_y' are the amplitudes of the waves in Cahn's second approximation, R is the amplification factor in the linear regime (equation (9)), $\tilde{D}_\lambda$ is the interdiffusion coefficient of equation (10) (negative), D" is the quadratic coefficient in equation (18), which is proportional to $f^{iv} = 6B$. Again, the growth rate of the orthogonal wave is only affected in the non-linear regime (B ≠ 0), and if the original modulation is large. In that case, its growth rate is not only less than in the linear regime, but also less than that of the original modulation.

CONCLUSIONS

Artificial compositionally modulated materials are inherently unstable. The modulation disappears, either by homogenization (in long wavelength ordering systems, in phase separating systems outside the spinodal or for short wavelengths inside the spinodal), or by initial sharpening of the profile to a stationary state (in long wavelength phase separating systems or short wavelength ordering systems) followed by coarsening.

Measurements of the rate of homogenization are a very sensitive method for determining interdiffusion coefficients. For crystalline materials, the method allows determination of the lattice diffusion coefficient at very low homologous temperatures. For metallic glasses, it permits time dependent measurements of $\tilde{D}$, which are essential to take into account structural relaxation.

Two coarsening mechanisms, a uniform one, and one governed by the motion of fault lines, have been discussed. Further investigation into the coarsening rate of the fault mechanism for wavelengths on the order of the interface thickness seems desirable. The presence of an initial, large amplitude modulation wave stabilizes the structure, through the non-linear effects in the diffusion equation, against the growth of certain waves orthogonal to the original one. A full analytic or numerical treatment of this question, especially taking into account crystal anisotropy, would be of interest.

ACKNOWLEDGEMENTS

It is a pleasure to acknowledge much collaboration and discussion on this subject with Lindsay Greer, especially during the writing of ref. 20. I am also grateful to the Metallurgy Department of the University of Leuven for their hospitality during the writing of the paper. Our research in this area is currently supported by the Office of Naval Research, under contract

306

number N00014-85-K-0023.

REFERENCES

1. J. DuMond and J.P. Youtz, J. Appl. Phys., 11, 357 (1940).
2. H.E. Cook and J.E. Hilliard, J. Appl. Phys., 40, 2191 (1969).
3. L.L. Chang, L. Esaki, W.E. Howard, R. Ludeke and G. Schul, J. Vac. Sci. Technol., 10, 655 (1973).
4. A.H. Eltoukhy and J.E. Greene, Appl. Phys. Lett., 33, 343 (1978).
5. M.P. Rosenblum, F. Spaepen and D. Turnbull, Appl. Phys. Lett., 37, 184 (1980).
6. B. Abeles and T. Tiedje, Phys. Rev. Lett., 51, 2003 (1983).
7. A. Guinier, "X-ray diffraction", Freeman, San Francisco, (1963), p. 169, 209-212.
8. D. de Fontaine, in "Local atomic arrangements studied by X-ray diffraction", ed. by J.B. Cohen and J.E. Hilliard, Gordon Briach, N.Y. (1966).
9. E.M. Philofsky and J.E. Hilliard, J. Appl. Phys., 40, 2198 (1969).
10. W.M.-C. Yang, Ph.D. Thesis, Northwestern Univ., (1979).
11. W.M. Paulson and J.E. Hilliard, J. Appl. Phys., 48, 2117 (1977).
12. T. Tsakalakos, Thin Sol. Films, 86, 79 (1981).
13. T. Tsakalakos, Scripta Met., 15, 225 (1981).
14. M.P. Rosenblum, Ph.D. Thesis, Harvard University (1979).
15. A.L. Greer, C.-J. Lin and F. Spaepen, Proc. 4th Int. Conf. on Rapidly Quenched Metals, ed. by T. Masumoto and K. Suzuki, Jap. Inst. Metals, Sendai, (1982), p. 567.
16. R.C. Cammarata and A.L. Greer, J. Non-Cryst. Solids, 61/62, 889 (1984).
17. R.C. Cammarata, unpublished results.
18. R.E. Somekh and A.L. Greer, unpublished results.
19. G.B. Stephenson, J. Non Cryst. Solids, 66, 393 (1984).
20. A.L. Greer and F. Spaepen, in "Synthetic Modulated Structure Materials", ed. by L.L. Chang and B.C. Giessen, Academic, New York, in press.
21. J.W. Cahn and J.E. Hilliard, J. Chem. Phys., 28, 258 (1958).
22. J.W. Cahn, Acta Met., 9, 795 (1961).
23. J.W. Cahn, Acta Met., 10, 179 (1962).
24. J.E. Hilliard, in "Phase Transformations", ed. by H.I. Aaronson, Am. Soc. Metals, Metals Park, Ohio, (1970), p. 497-560.
25. J.W. Cahn, Trans. Met. Soc. AIME, 242, 166 (1968).
26. H.E. Cook, D. de Fontaine, and J.E. Hilliard, Acta Met., 17, 765 (1969).
27. A.I. Taub and F. Spaepen, Acta Met., 28, 1781 (1980).
28. A.R. Miedema, Philips Tech. Rev., 36, 217 (1976).
29. A.L. Greer, private communication.
30. F. Spaepen, in "Physics of Defects", Les Houches Lectures XXXV, ed. by Balian et al., North-Holland, Amsterdam, (1981), p. 133.
31. D. de Fontaine, Ph.D. Thesis, Northwestern University (1967); some results also in ref. 24.
32. J.W. Cahn, Acta Met., 14, 1685 (1966).
33. J.S. Langer, Ann. Physics, 65, 53 (1971).
34. T. Tsakalakos, Ph.D. Thesis, Northwestern University (1977).
35. R.M. Fleming, D.B. McWhan, A.C. Gossard, W. Wiegmann and R.A. Logan, J. Appl. Phys., 51, 357 (1980).
36. C. Wagner, Z. Elektrochem., 65, 581 (1961).
37. L.D. Graham and R.W. Kraft, Trans. Met. Soc. AIME, 236, 94 (1966).

ION IMPLANTATION DISORDER IN STRAINED-LAYER SUPERLATTICES*

G. W. ARNOLD, S. T. PICRAUX, P. S. PEERCY, D. R. MYERS, R. M. BIEFELD AND
L. R. DAWSON
Sandia National Laboratories, Albuquerque, New Mexico 87185

ABSTRACT

Cantilever-beam bending and RBS channeling measurements have been used
to examine implantation-induced disorder and stress buildup in $In_{0.2}Ga_{0.8}As/$
GaAs SLS structures. Implantation fluences from 10^{11} to $10^{15}/cm^2$ were used
for 150 keV Si, 320 keV Kr, and 250 keV Zn in SLS and GaAs bulk materials.
The critical fluence for saturation of compressive stress occurs prior to
amorphous layer formation and is followed by stress relief. For all the
ions the maximum ion induced stress scales with energy density into atomic
processes and stress relief occurs above $\sim 1 \times 10^{20}$ keV/cm^3. Stress relief
is more pronounced for the SLSs than for bulk GaAs. We suggest that stress-
relief may lead to slip or other forms of inelastic material flow in SLSs,
which would be undesirable for active regions in device applications. Such
material flow may be avoided by limiting maximum fluences or by multiple-
step implantation and annealing cycles (or hot implants) at high fluences.

INTRODUCTION

Strained-layer superlattices (SLSs) can be made by epitaxial growth of
materials with differing lattice constants (~ 0.5 to $\sim 5\%$). The wide range
of material combinations which can be used for SLSs give these structures an
inherent advantage over the much more restricted material choices available
for forming lattice-matched superlattices. The sensitivity of the SLSs to
the additional strain induced by ion implantation, however, has been a
question of interest. In earlier work, the structural integrity of the
$In_{0.2}Ga_{0.8}As/GaAs$ SLS was found to be maintained for light ion implantations
(e.g., 2×10^{15} 75 keV N/cm^2) and for heavier ions (Si, Zn) at lower fluences
[1]. A more comprehensive review of implantation-induced disorder production, tion, damage annealing, implanted dopant activation, and alloying by
ion beam mixing (heavy ions and high fluence) for $In_{0.2}Ga_{0.8}As/$ GaAs and
$GaAs_{0.2}P_{0.8}/GaP$ SLSs has recently been given [2].
In the present paper we report studies using the cantilever-beam
technique, as in [1], and Rutherford backscattering (RBS) to measure lattice
disorder induced by Si, Kr, and Zn implantations into $In_{0.2}Ga_{0.8}As/GaAs$ SLSs.
Specifically, we have examined the dependence of disorder on ion fluence and
have shown how it scales with energy into atomic collisions. The earlier
work [1,2] established that ion beam mixing occurs at sufficiently high
fluences and ion masses (e.g., 2×10^{15} 150 keV Si/cm^2 in $In_{0.2}Ga_{0.8}As/GaAs$
SLSs). In this paper we explore the possible fluence limit due to stress
relief of the implantation disorder-induced compressive stress buildup.

EXPERIMENTAL

The SLS samples used in these experiments were all taken from the same
molecular beam epitaxial growth structure (#M556). This 18 period alternating layer structure of $In_xGa_{1-x}As/GaAs$ was determined by x-ray diffraction

---

*This work performed at Sandia National Laboratories and supported by the
U.S. Department of Energy under contract number DE-AC04-76DP00789.

to have an x = 0.228 with a thickness per period of ~ 24 nm. From nominal growth conditions the individual layers were assumed to be of equal thickness; this was in agreement with RBS observations. The SLS was grown on a buffer (400 nm) of $In_{0.1}Ga_{0.9}As$ whose in-plane lattice constant approximately matched that of the SLS. The buffer and SLS were grown on a semi-insulating (100) GaAs wafer.

The cantilever-beam technique [3] used samples (15 x 3 x 0.4 mm) which were metallized on the back surface with 100 nm of Cr. RBS measurements were made using 3 MeV ^{4}He with channeling analysis along the [100] growth direction. All implantations were carried out at room temperature in a vacuum of $< 10^{-6}$ Torr using electrostatically swept beams of 150 keV Si, 250 keV Zn$^-$, or 320 keV Kr.

RESULTS AND DISCUSSION

The cantilever-beam technique [1,3] measures the bending of samples induced by lattice expansion [$\Delta V/V$] when implanted on the lateral surfaces. Using the elastic constants of the substrate and the beam dimensions, the bending can be quantified to give the induced lateral stress (in dynes/cm) integrated over the implanted depth. These values can be converted to dynes/cm^2, an average stress over the region of damage, by dividing by the depth over which damage-produced stress occurs. The depths used (corresponding to the calculated depth over which approximately 95% of the damage is induced) for the Si, Kr, and Zn implants were 2150 Å, 1910 Å, and 1560 Å, respectively. In Fig. 1 are shown the values of implanted stress in dynes/cm^2 for the (100) oriented InGaAs/GaAs SLS and for bulk (100) GaAs as a function of ion fluence for 150 keV Si, 320 keV Kr, and 250 keV Zn. These data show that the implantation-induced stress in the SLS reaches a maximum value at a fluence which is dependent on ion mass. The maximum in stress is followed by a rather steep decrease due to a stress-relief mechanism. Further studies of the state of the material after stress relief will be reported later. The stress relief for bulk GaAs is seen in Fig. 1 to be less pronounced than for the SLS. Figure 1 also shows that the maximum implantation-induced stress varies from 2 - 4 x 10^9 dynes/cm^2. This is comparable to the calculated value of ~ 1 x 10^{10} dynes/cm^2 for the alternating compressive and tensile stresses in the $In_{0.2}Ga_{0.8}As$ and GaAs layers of the as-grown SLS structure.

The apparent mass dependence revealed in Fig. 1 suggests that the implantation stress depends on the collisional damage component of ion energy. In Fig. 2, the stress for the SLS is plotted as a function of the ion energy density (keV/cm^3) deposited into atomic processes for the various implantations. These plots show that the maxima in implantation stress occur at very nearly the same energy density, i.e., ~ 1.5-3.0 x 10^{20} keV/cm^3. The data for the Zn implant show a nearly linear buildup of stress, as is expected for non-interacting isolated damage tracks. For Si, however, the initial stress buildup is clearly less than linear. This behavior suggests that more dilute damage in the lighter ion tracks allows track interaction through migration and recombination of defects even at fluences as low as 10^{12}/cm^2. This result is consistent with the known long-range migration and annealing of defects at room-temperature in GaAs [4]. Lighter ions will produce larger numbers of point defects and this damage will be susceptible to annealing. The mobility of these defects will also extend the damage depth beyond that calculated and used to obtain the average implantation stress values plotted in Figs. 1 and 2. The difference between the actual and calculated damage depth should be greatest for the lighter ion implantations and a correction would tend to bring the maximum stress levels for the Si, Kr, and Zn implants into closer agreement.

Implantation-induced damage in SLSs has also been monitored by 3 MeV ^{4}He RBS channeling measurements. Typical spectra are shown in Fig. 3 for 150 keV

Si implanted at various fluence levels into (100) $In_{0.2}Ga_{0.8}As$/GaAs SLSs. The disorder is seen to extend to depths of beyond 2000-2500 Å, which is consistent with the calculated damage depth. The disorder level near the peak has been measured as a function of ion fluence for 150 keV Si and 250 keV Zn. The disorder relative to the random level is plotted in Fig. 4 as a function of energy density into atomic processes. When compared with Fig. 2, it is clear that the maxima in yield stress occur for RBS disorder fractions well below 0.5, for which crystalline order is still dominant. The fluence at which the channeling results show complete conversion of the layer to the amorphous state for these ions occurs at energy deposition levels for which stress relief has already occurred.

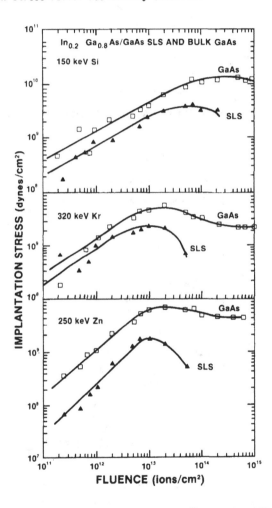

Fig. 1.   Implantation-induced stress versus fluence for 150 keV Si, 320 keV Kr, and 250 keV Zn incident on (100) $In_{0.2}Ga_{0.8}As$/GaAs SLS (▲) and (100) bulk GaAs (□).

310

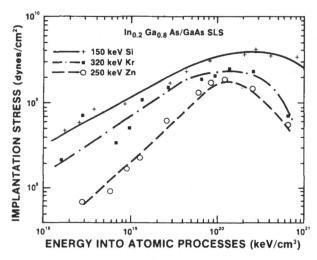

Fig. 2. Implantation-induced stress versus energy density into atomic processes for 150 keV Si (+), 320 keV Kr (■), and 150 keV Zn (○) implanted into $In_{0.2}Ga_{0.8}As/GaAs$ SLS.

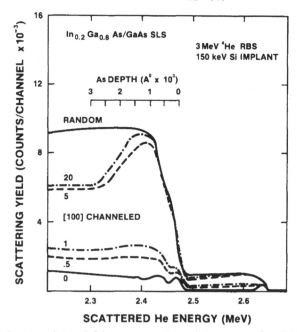

Fig. 3. Scattered He yield versus scattered He energy for 3 MeV ^{4}He channeled in (100) $In_{0.2}Ga_{0.8}As/GaAs$ implanted with 150 keV Si. The numbers associated with the various spectra are in units of 10^{14} Si ions/cm^2.

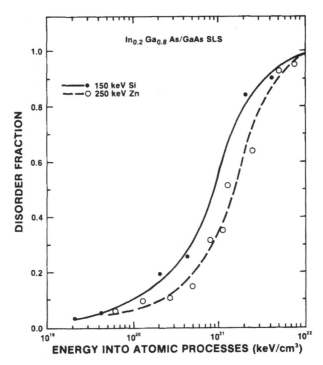

Fig. 4. RBS disorder fraction versus energy density into atomic
processes for 150 keV Si (●) and 250 keV Zn (○) incident
on $In_{0.2}Ga_{0.8}As/GaAs$ SLS.

CONCLUSIONS

The data presented in this paper show that ion implantation for the
doping of SLS structures used as devices can be a viable procedure subject
to certain conditions. Figures 2 and 4 suggest that in certain cases it
may be desirable to keep the implanted dopant energy deposition densities at
levels below ~ 2 x 10^{20} keV/cm^3. The stress relief mechanism above this
energy deposition level probably generates slip and this may not be desirable
in active regions for certain device applications.

The evidence for in situ implant damage annealing shown in Figs. 1
and 2 suggest that hot implants, or annealing cycles at intermediate implantation fluences, may be useful in preventing stress buildup and relief at
high implantation fluences.

REFERENCES

1. G. W. Arnold, S. T. Picraux, P. S. Peercy, D. R. Myers and L. R. Dawson
    Appl. Phys. Lett. 45, 382 (1984).
2. S. T. Picraux, G. W. Arnold, D. R. Myers, L. R. Dawson, R. M. Biefeld,
    I. J. Fritz and T. E. Zipperian, Nucl. Instr. Meth. B (to be published).
3. E. P. EerNisse, J. Appl. Phys. 45, 167 (1974).
4. See, e.g., F. L. Vook and S. T. Picraux, Ion Implantation in Semiconduc-
    tors, edited by I. Ruge and J. Graul (Springer-Verlag, 1971) p. 141.

AN EVALUATION OF IMPLANTATION-DISORDERING OF (InGa)As/GaAs STRAINED-LAYER SUPERLATTICES**

D. R. MYERS, C. E. BARNES, G. W. ARNOLD, L. R. DAWSON, R. M. BIEFELD, T. E. ZIPPERIAN, P. L. GOURLEY, and I. J. FRITZ
Sandia National Laboratories
Albuquerque, NM 87185

ABSTRACT

We have examined the optical and transport properties of $In_{.2}Ga_{.8}As/GaAs$ strained-layer superlattices (SLS's) which have been implanted either with $5 \times 10^{15}/cm^2$, 250keV Zn^+ or with $5 \times 10^{14}/cm^2$, 70keV Be^+ and annealed under an arsenic overpressure at 600°C. For both cases, electrical activation in the implantation-doped regions equalled that of similar implants and anneals in bulk GaAs, even though the Be implant retained the SLS structure, while the Zn implant intermixed the SLS layers to produce an alloy semiconductor of the average SLS composition. Photoluminescence intensities in the annealed implanted regions were significantly reduced from that of virgin material, apparently due to residual implant damage. Diodes formed from both the Be- and the Zn-implanted SLS's produced electroluminescence intensity comparable to that of grown-junction SLS diodes in the same chemical system, despite the implantation processing and the potential for vertical lattice mismatch in the Zn-disordered SLS device. These results indicate that Zn-disordering can be as useful in strained-layer superlattices as in lattice-matched systems.

INTRODUCTION

Strained-layer superlattices(SLS's) are man-made materials whose unique properties [1] have been exploited in prototype electronic and optoelectronic devices [2]. Localized doping of SLS's has been accomplished by implantation of light ion species (typically Be) followed by controlled-atmosphere annealing [3]. These implantation technologies have proved useful for the fabrication of planar homojunction SLS photodetectors [4,5,6] which retain the useful properties of the virgin SLS in completed device structures. Intentional compositional disordering of superlattice structures has been achieved by the appropriate implantation of heavy ions into both the "lattice-matched" [7] superlattices and into strained-layer superlattices [8,9]. The localized loss of superlattice ordering is of great technological interest, since the alloy formed by disordering the superlattice in these chemical systems has a larger energy bandgap than the original superlattice. Thus, this localized disordering process allows the controlled formation of heterojunction contacts between the virgin and the disordered superlattice regions. These heterojunction contacts are useful in optical emitters, where such compositionally disordered regions have been shown to produce both optical and electrical confinement in lattice-matched systems [10,11,12].

Unlike the lattice-matched (AlGa)As/GaAs superlattice system, in which the material in each layer essentially retains the cubic symmetry of its bulk form, the material in each layer of a strained-layer superlattice is tetragonally distorted from its bulk form due to the strain fields inherent in the SLS. Thus, in forming the essentially cubic alloy material by disordering the tetragonally distorted SLS, a possibility exists for vertical lattice mismatch at the lateral boundary between the SLS and the implant-disordered region (Fig. 1). This vertical mismatch could in principle lead to the formation of defects which would degrade diode properties and reduce electro-

**This work was performed at Sandia National Laboratories and supported by the U. S. Department of Energy under contract number DE-AC04-76DP00789.

314

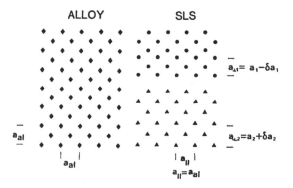

SCHEMATIC: POSSIBLE
VERTICAL LATTICE MISMATCH IN
IMPLANTATION–ALLOYED SLSs

Fig. 1   Schematic illustration of the potential for vertical lattice mis-match between locally disordered and undamaged regions of strained-layer superlattices. Layers in the undamaged SLS are distorted along the growth direction to values either larger or smaller than the in-plane lattice constant; while the compositionally disordered alloy has an isotropic lattice spacing whose value is given by the in-plane SLS lattice constant.

luminescence intensity. In contrast, this possibility does not arise in the lattice-matched systems, where high quality junctions have been demonstrated in compositionally disordered superlattices. In this study, we have examined the consequences of implant disordering in the (InGa)As/GaAs SLS system by comparing the properties of regions that have been doped by implantation of Be (which retains the SLS structure) with those of SLS regions that have been disordered by high-dose implantation of Zn to empirically determine the prac-tical utility of implant disordering of strained-layer superlattices.

EXPERIMENTAL PROCEDURE

A variety of $In_xGa_{1-x}As/GaAs$ SLS's were studied, with the In content in the ternary layers chosen near x=0.2. Typically, 25 periods of layers with equal thickness (~110Å/layer) were grown over a buffer layer $In_yGa_{1-y}As$ with y~x/2. Details of the growth process are presented elsewhere [13]. Either $^9Be^+$ was implanted at 70 keV to a dose of $5x10^{14}/$ cm^2 or $^{64}Zn^+$ was implanted at 250 keV to a dose of $5x10^{15}/cm^2$ at room temperature into SLS's which had been partially masked by patterned photoresist over silox. Follow-ing the implants, the masking materials were removed and the wafers annealed at 600°C for 10 minutes (the Be implant) or at 680°C for 30 minutes (the Zn implant) under an appropriate As overpressure. The anneal for the Zn implant was chosen to increase the depth of the SLS that was disordered by the in-diffusion of Zn [8]. Ohmic contacts to the SLS were achieved by patterned alloyed contacts of Au-Be/Ni/Au for the implanted p-type regions and by Au-Ge/Au for the n-type ($\sim10^{15}/cm^3$) unimplanted SLS.

The samples were examined by a variety of techniques. X-ray diffraction [14] was used to examine the structure of the implanted and annealed SLS's, as the location and spacing of the satellite peaks in the diffraction pattern can be analyzed to determine the composition and structure of the SLS, while the sharpness of the satellite peaks is related to the uniformity of the SLS throughout its depth. High-resolution Rutherford backscattering was employed to determine the compositional modulation of the SLS and to probe the residual disorder in the implanted regions after anneal. A conventional Hall-effect system was used for measurements of electrical activation of the implanted species at room temperature. Photoluminescence was taken at 4 K using 10mW of 750 nm-wavelength light focussed to a spot of 50μ m diameter for excitation and analyzed with a SPEX 1404 double monochrometer and a PbS detector. Junction electroluminescence was taken at 77K using 100μs-wide pulses of 20mA at 4.0 kHz and was analyzed using a SPEX 1801 1m monochromator and a photomultiplier tube with a S1 response.

RESULTS

X-ray diffraction indicated survival of the control SLS (unimplanted, 680°C, 30 min. anneal) and the 5×10^{14}/cm^2, Be-implanted SLS (600°C, 10 min. anneal) with no detectable structural degradation, similar to other studies of light-ion implanted SLS's [3,5]. In contrast, x-ray diffraction patterns from the 5×10^{15}/cm^2, 250keV Zn-implanted SLS exhibited significantly reduced intensity from the satellite peaks with a corresponding broadening of their lineshapes when compared to those of the virgin SLS, thereby indicating an intermixing of the top layers of the SLS to form an alloy of the average SLS composition for this high-dose Zn implant.

Examination of the In compositional modulation in the SLS as a function of ion dose for 250keV Zn indicated that a fluence of 6×10^{14}/cm^2 was sufficient to detectably intermix the SLS layers; while the SLS compositional modulation was retained for a dose of 4.0×10^{14}/cm^2 250keV Zn. The RBS measurements not only reinforce the X-ray diffraction measurements reported above, but in addition provide an estimate of the effective Zn-ion dose required to intermix the SLS layers.

Hall-effect measurements of the sheet carrier densities for the implanted annealed regions indicate 68% electrical activation with a sheet mobility of 73 cm^2/V-s for the Be implant; while the Zn-implanted regions exhibit a 32% electical activation and a sheet carrier mobility of 62cm^2/ V-s. Carrier densities in unimplanted SLS regions remained below 10^{15}/cm^3 after annealing.

The values obtained for the activation and mobilities in the implanted and annealed SLS's correspond well to the values of 70% electrical activation and 80 cm^2/V-s obtained for similar Be-implants and anneals into bulk GaAs [15], and to the values of 17% electrical activation and 80 cm^2/V-s for similar Zn implants and anneals of GaAs [16]. Thus, the fact that implants of either ion species into the SLS produces electrical activation consistent with that observed for a similar implant and anneal of the nearest binary semiconductor (irrespective of the retention or loss of the SLS structure) suggests that the incorporation of the implanted species into electrically active sites and the removal of compensating defect centers by thermal annealing does not stongly depend on the long-range ordering (i.e., SLS layering) of the host semiconductor. These results are also consistent with the results of an examination of radiation effects on Ga(AsP)/GaP SLS's [17].

The compositional disordering of superlattices and quantum well structures has found its primary application in the fabrication of solid-state LASERs [10,11,12]. We have chosen to examine photon output from both implantation-doped (SLS retained) and implantation-disordered SLS's in simple diode structures, where photon output is dependent on fewer additional variables than in LASER structures.

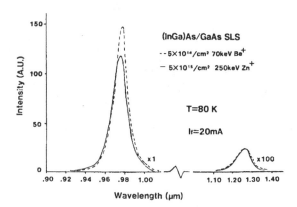

Fig. 2 Wavelength dependence of the junction electroluminescence (EL) for the Be implanted and the Zn-implanted SLS diodes. The variation in band-edge intensity (near 0.98μm) is within the sample-to-sample variation seen for each type, and exceeds the longer wavelength EL by a factor of roughly 600.

In Fig. 2, we compare the wavelength dependence of the junction electroluminescence (EL) intensities for representative Be-implanted and Zn-implanted $In_{.2}Ga_{.8}As/GaAs$ SLS's at the same drive current. The difference in the magnitude of the peak EL intensity between the two different devices is within the sample-to-sample variation of EL intensities seen for diodes of the same type. Additionally, both diodes produce EL intensities at a given drive current equal to that obtained from mesa-type grown junction SLS diodes in the same growth series (not shown). For both the Be-implanted and the Zn-implanted SLS's, the the emission at wavelengths corresponding to the band-edge transitions (~0.98μm) is roughly 600 times more intense than the defect-related luminescence at wavelengths centered near 1.25μm. This ratio of band-edge EL intensity to defect-related EL intensity is also representative of grown-junction SLS light-emitting diodes (LED's) in this chemical system. Thus, implantation and annealing has not detectably degraded the luminescence output from (InGa)As/GaAs SLS LED's, and also has not produced an increase in detectable defect-related luminescence centers. It should be noted that the average free carrier density in the implantation-doped regions exceeds that of the virgin SLS by roughly a factor of 10,000. Thus, most of the minority carrier injection (which is responsible for the EL emission) is into the lightly doped, unimplanted regions. Thus, junction electroluminescence is not sensitive to any residual defects that may remain in the implanted regions. Sample-to-sample variations in chemical composition of these SLS's make it impossible to determine whether or not the slight shift in the peak emission wavelength is due to band offset effects from the wider-bandgap disordered region in the Zn-implanted samples.

In contrast to the EL measurements, the photoluminescence (PL) intensities from the Be-implanted regions are reduced by a factor of 50 from that of the virgin SLS; while those from the Zn-implanted regions are similarly reduced by a factor of at least 1000. The reduction in PL intensities is probably due to recombination centers in the implanted regions related to residual ion damage. This result reinforces the interpretation that most of the EL intensity arises from injection into the unimplanted SLS regions. Such a lack of

PL intensity is not surprising in view of the low annealing temperatures used in this study, which were chosen to provide electrical activation of the implanted species and to minimize layer interdiffusion in the undamaged regions. A similar effect is seen in bulk GaAs, where the annealing temperatures required to obtain electrical activation of ion-implanted Be (600°C) are considerably lower than those (800°C) required to activate Be-related low-temperature photoluminescence [18].

DISCUSSION

The above measurements have demonstrated that localized disordering of SLS's by Zn-implantation does not detectably degrade the EL efficiency of SLS diodes compared to that of either Be-implanted SLS diodes or to grown-junction SLS diodes. A priori, the EL output of the Zn-implanted devices would have been thought to be degraded by two factors: (1) the damage-induced stresses for the Zn implant exceed the yield stress of the ion-damaged region while those for the Be-implant do not [10]; and (2) the potential for vertical lattice mismatch between the adjacent undamaged SLS and the compositionally disordered SLS (Fig. 1). Several additional considerations arise, however, to resolve these apparent difficulties.

Raman measurements of ion damage in compound semiconductor suggest that the yield of III-V semiconductors from implantation damage occurs within the ion damaged region itself [19]. Thus, any extended defects arising from the yield of the Zn-implanted region would occur in the heavily doped implanted region. Since few minority carriers are injected into the Zn-implanted region, those extended defects would not directly affect the EL intensity.

Three factors mitigate against the possibility for diode degradation from vertical lattice mismatch. Since lattice spacings in each SLS layer along the growth direction are alternately expanded or reduced from that of the uniform in-plane lattice spacing of the SLS, lattice planes between the adjacent undamaged and implant-alloyed regions align at least once every period. Thus there is no net macroscopic strain on the SLS from the alloy region. Secondly, the mismatch of 1.4% between the $In_{.2}Ga_{.8}As$ layers and the GaAs layers for our SLS is divided nearly equally between the two layers. Therefore, the vertical distortion in each layer is 0.7%. This distortion is small compared to the 7.2% distortions of the column-V sublattice in bulk (InGa)As alloys [20]. The alloy's sublattice is distorted because the In-As and Ga-As bonds retain their spacing throughout the entire composition range of the (InGa)As system. Thus, minimal microscopic strain is induced by localized disordering. Finally, the transition between the disordered regions and the undamaged SLS is not abrupt, but rather is graded over a finite distance. We have employed the ion range code of Winterbon [21], using a parameterization [22] of the WHB ion-atom potential [23], to approximate the Zn-implant studied here. This formulation assumes a homogeneous $In_{.1}Ga_{.9}As$ alloy as an approximation to the SLS, and allows for the transport of energy by recoiling target atoms. The simulations indicate that the ion density distribution parallel to the surface (which determines the electrical junction) decreases from the opening in implant mask with a standard deviation of 303Å . In contrast, the damage distributions (which are responsible for the compositional disordering) both lie within the p-type region: the lateral standard for the atomic damage distribution is 207Å , while that for the ionization energy distribution is 150Å . Thus, the junction occurs within SLS material free of compositional disordering. These results thus suggest that the optical benefits of localized compositional disordering of an SLS can be achieved without additional degradation of the electrical properties of an ion-implanted SLS diode.

318

ACKNOWLEDGEMENTS

The authors would like to acknowledge useful discussions with S. T. Picraux, G. C. Osbourn, and D. K. Brice of Sandia, as well as with J. H. Albers of the National Bureau of Standards. The expert technical assistance of J. Snelling, T. Plut, J. Clifton, D. Wrobel, L. Hansen, K. Baucom, R. Hibray, A. McDonald and N. Wing is also appreciated.

REFERENCES

1. G. C. Osbourn, Mat. Res. Symp. Proc. 25, 455 (1984).
2. T. E. Zipperian, L. R. Dawson, C. E. Barnes, J. J. Wiczer and G. C. Osbourn, IEDM Tech. Dig., 1984, to be published.
3. D. R. Myers, R. M. Biefeld, I. J. Fritz, S. T. Picraux and T. E. Zipperian, Appl. Phys. Lett. 44, 1052 (1984).
4. D. R. Myers, T. E. Zipperian, R. M. Biefeld and J. J. Wiczer, IEDM Tech. Dig., 1983, 700 (1983).
5. D. R. Myers, J. J. Wiczer, T. E. Zipperian and R. M. Biefeld, IEEE Electron Device Letters, EDL-5, 326 (1984).
6. G. E. Bulman, D. R. Myers, J. J. Wiczer, L. R. Dawson, R. M. Biefeld and T. E. Zipperian, IEDM Tech. Dig., 1984, to be published.
7. M. D. Camras, J. J. Coleman, N. Holonyak, Jr., K. Hess, P. D. Dapkus and C. G. Kirkpatrick, Inst. Phys. Conf. Series 65, 233 (1983).
8. S. T. Picraux, G. W. Arnold, D. R. Myers, L. R. Dawson, R. M. Biefeld, I. J. Fritz and T. E. Zipperian, Proc. IBMM'84 to be published in Nucl. Instrum. and Methods B.
9. G. W. Arnold, S. T. Picraux, P. S. Peercy, D. R. Myers and L. R. Dawson, this session.
10. T. Fukuzawa, S. Semura, T. Ohta, Y. Uchida, K.L.I. Kobayashi and H. Nakashima, IEDM Tech. Dig. 1983. p.746.
11. K. Meeham, J. M. Brown, N. Holonyak, Jr., R. D. Burnham, T. L. Paoli and W. Streifer, Appl. Phys. Lett. 44, 700 (1984).
12. W. D. Laidig, J. W. Lee and P. J. Caldwell, Appl. Phys. Lett. 45, 485 (1984).
13. L. R. Dawson, T. E. Zipperian, C. E. Barnes, J. J. Wiczer and G. C. Osbourn, Proc. Int. Conf. on GaAs and Related Compounds (Biarritz, France, 1984) to be published.
14. R. M. Biefeld, G. C. Osbourn, P. L. Gourley and I. J. Fritz, J. Electron. Mater. 12, 903 (1983).
15. W. V. McTevidge, M. J. Helix, K. V. Vaidyanathan and B. G. Streetman, J. Appl. Phys. 48, 3342 (1977).
16. S. S. Kular, B. J. Sealey and K. G. Stevens, Electon. Lett. 14, 2 (1978).
17. C. E. Barnes, G. A. Samara, R. M. Biefeld, T. E. Zipperian and G. C. Osbourn, Proc. 13th Int. Conf. on the Physics of Defects in Semiconductors (San Diego, CA Aug. 12-17, 1984) to be published.
18. J. P. Donnelly, Nucl. Instrum. Methods, 182, 553 (1981).
19. D. R. Myers, P. L. Gourley, and P. S. Peercy, J. Appl. Phys. 54, 5032 (1983).
20. J. C. Mikkelsen, Jr. and J. B. Boyce, Phys. Rev. B28, 7130 (1983).
21. K. B. Winterbon, Atomic Energy of Canada, Limited report AECL 5536 (1976).
22. W. K. Chu, R. H. Kastl, and P. C. Murley, Rad. Eff. 47, 1 (1980).
23. W. D. Wilson, L. G. Haggmark, and J. P. Biersack, Phys. Rev. B15, 2458 (1977).

STRAIN AND DAMAGE MEASUREMENTS IN ION IMPLANTED $Al_xGa_{1-x}As$/GaAs SUPERLATTICES

A. H. HAMDI*, J. L. TANDON**, T. VREELAND, JR.*, AND M.-A. NICOLET*
*California Institute of Technology, Pasadena, CA 91125
**Applied Solar Energy Corporation, City of Industry, CA 91744

ABSTRACT

Strain measurements in $Al_xGa_{1-x}As$/GaAs superlattices have been carried out before and after Si ion implantation. For doses up to 5×10^{15} cm^{-2}, no atomic intermixing of the sublayers is observed by backscattering spectrometry. However, with x-ray rocking curve measurements, significant changes in the strain profiles are detected for implantations with doses as low as 7×10^{12} cm^{-2}. Interpretation of the rocking curves suggests that low-dose implantations release strain in the $Al_xGa_{1-x}As$ sublayers. The strain profile recovery of the implanted samples, upon annealing at $\sim 420°C$, implies that the damage caused by implantation is largely reversible.

INTRODUCTION

Superlattices composed of thin alternating layers of different compound semiconductors offer unique possibilities in the conception and fabrication of novel optical and electronic devices [1,2]. To exploit these structures, localized and controlled doping by ion implantation constitutes a desirable processing method. Recently, a few studies have explored ion implantation in AlAs/GaAs, InGaAs/GaAs and GaAsP/GaP superlattices [3-6]. The primary objective of these studies is to demonstrate that superlattices can be doped by implantation without materially degrading the superlattice. Underlying these studies are the fundamental questions of the nature of the damage created by an ion irradiation in a superlattice, and its evolution upon subsequent thermal annealing.

Almost all superlattices possess strain in one or both sublayers because of the lattice mismatch of the constituents. By measuring this strain as a function of depth, and monitoring its evolution after implantation and annealing, accurate information on the cumulative effects of defects on the superlattice can be obtained. In this paper, the technique of x-ray rocking curves is used to make such measurements. The changes in the rocking curves of $Al_xGa_{1-x}As$/GaAs superlattices are measured as a function of the dose after Si ion implantations, and also upon thermal annealing. The curves are interpreted in terms of changes in the strain after using the kinematic model of x-ray diffraction [7]. The results help in understanding the onset of damage created by ion implantation.

EXPERIMENTAL

$Al_xGa_{1-x}As/GaAs$ (x = .88) strained-layer-superlattice (SLS) structures were prepared by metal organic chemical vapor deposition. Alternating layers (10 each) of $Al_xGa_{1-x}As$ and GaAs were grown with thicknesses of 140 and 270 Å respectively, on (100) GaAs substrates. Si ions were implanted at room temperature with an energy of 200 keV. To minimize channeling during implantation, the incoming beam was oriented $\sim 7°$ with respect to the sample's surface normal. The range of the ions is about 2000 Å with a range straggling of ~ 900 Å which places the Si and the damage profiles fully within the superlattice. The implantation doses ranged from 7×10^{12} to 5×10^{15} Si/cm^2. Thermal annealing was carried out, on a few samples, in a forming gas atmosphere (85% N_2 + 15% H_2) at $\sim 420°C$ for one hour.

X-ray rocking curve measurements were made in the (400) reflection, using a double-crystal diffractometer with $FeK\alpha_1$. The x-ray beam was collimated and rendered nearly monochromatic by a (400) reflection from a high quality (100) GaAs crystal (the first crystal). The beam spot was adjusted to ~ 0.5 mm x 1 mm by slits placed between the first crystal and the sample. The measured rocking curves were fitted with calculated curves using Speriosu's model of kinematical theory of x-ray diffraction in thin epitaxial layers [7]. In this model, the diffraction by the substrate is treated dynamically. For fitting, the structure factors of $Al_{.88}Ga_{.12}As$ (= 117.4) and GaAs (= 157.0) were calculated using the tabulated atomic scattering factors [8].

Backscattering spectrometry measurements were also carried out using a 2 MeV He beam. All data refer to room temperature unless stated otherwise.

RESULTS AND DISCUSSION

Effect of Si Ion Implantation

Figure 1 shows the measured (dashed line) and the calculated (solid line) rocking curves for the as-grown $Al_xGa_{1-x}As/GaAs$ SLS structure. The angle $\Delta\theta$ is plotted relative to the Bragg angle (θ_B) of the substrate peak, P_{sub}, at $\Delta\theta = 0$. The reflecting power, plotted on the vertical axis, is normalized with respect to the intensity of the incoming x-ray beam. The periodicity in the structure factors and in the strain of the layers in the SLS generates the subsidiary peaks observed in the rocking curves. The displacement of the peak P_o from P_{sub} measures the average strain in the SLS. For symmetric reflection, as in this case, the equal separation between the subsidiary peaks measures the average thickness of one period in the SLS. The calculated curve in Fig. 1a was obtained using the strain distribution in Fig. 1b. The details of the calculations are given elsewhere [7,9]. From Fig. 1b, the number

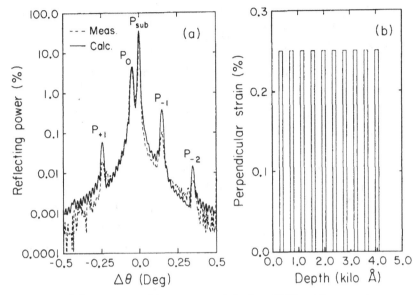

Figure 1 (a). X-ray rocking curves of Fe Kα$_1$ (400) reflection for an
as-grown Al$_x$Ga$_{1-x}$As/GaAs SLS structure with 10 periods. The calculated
curve (solid line) was obtained using the depth-strain distribution
shown in (b).

of periods in the SLS are verified to be equal to ten. The average thicknesses of the GaAs and the Al$_x$Ga$_{1-x}$As sublayers are 270 and 140 Å respectively.
The average perpendicular strains relative to the substrate are 0.0%
and 0.25% respectively. The remaining discrepancy between the measured and
the calculated curves as observed in Fig. 1a has been attributed to thickness
variations in the periods of the SLS and to the nonabrupt interfaces of the
sublayers [9]. Thus, from the analyses of the rocking curve measurements,
the strain profiles in the SLS are accurately determined.

Rocking curves on the Al$_x$Ga$_{1-x}$As/GaAs SLS samples were obtained after Si
implantations with doses ranging from 7 x 10^{12} to 5 x 10^{15} cm^{-2} (see Fig. 2).
The intensity of subsidiary peaks, except P$_0$, diminishes. Also P$_0$ shifts to
an increased angular position P$'_0$.

Since the separation between P$_{sub}$ and P$'_0$ measures the average strain
in the damaged SLS, the average strain clearly increases after this implantation. Further insight into the change in strain upon implantation is provided
by fitting the measured rocking curve with a calculated (solid line) one
(see Fig. 2a). To this end, the strains of the sublayers in the model of
Fig. 1a were changed and calculations were made iteratively to obtain a best

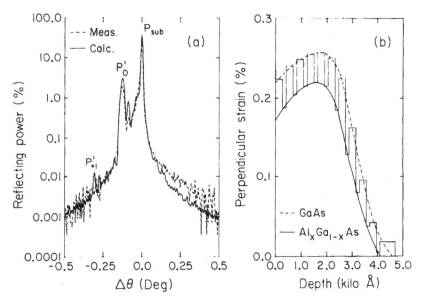

Figure 2. Fe $K\alpha_1$ (400) rocking curves of an $Al_xGa_{1-x}As/GaAs$ SLS structure implanted with 200 keV Si ions to a dose of 7×10^{12} cm^{-2}. The calculated curve was obtained using the sublayers strain distribution shown in (b).

fit. The strain profile used in this fit of Fig. 2a is shown in Fig. 2b. It is important to note that the damage induced by the low dose implantation considered here reduces the strain in the $Al_xGa_{1-x}As$ sublayers. On the other hand, the strain increases in the GaAs sublayers, so that the net effect is an overall increase in the average strain of the SLS. A similar increase in the strain of the bulk GaAs after implantation of a corresponding dose has been previously reported [10]; a reduction of strain in a single $Al_xGa_{1-x}As$ layer has also been measured after implantation with similar doses [11]. The current data are thus consistent with those measurements. It should be pointed out that an equally good fit in Fig. 2a could be obtained by interchanging the strains in the $Al_xGa_{1-x}As$ and GaAs sublayers in Fig. 2b. This is because of a small difference between the strains in the two sublayers for this particular case. From measurements performed on single $Al_xGa_{1-x}As$ layers, the strain profile in Fig. 2b appears more probable.

The average strain in the SLS structures increases as the implantation dose rises above 7×10^{12} Si/cm^2. The measured average strain as a function of dose is shown in Fig. 3. The average strain initially increases with dose and tends to saturate, to a level of $\sim 0.44\%$ beyond a dose of $\sim 5 \times 10^{14}$

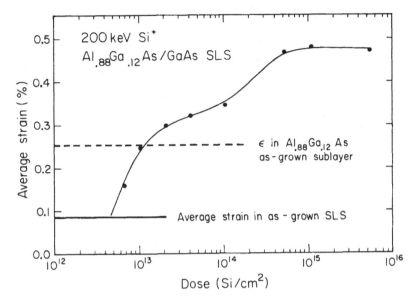

Figure 3. The average strain in the damaged SLS structures as a function of Si ion dose. The average strain was measured by the angular separation of P_o^1 from P_{sub}, using the relation

$$\Delta\theta = - <\varepsilon^{\pm}>\tan\theta_B$$

Si/cm^2. A nonlinear behavior of strain has also been observed in bulk GaAs. Strain measurements carried out on single $Al_xGa_{1-x}As$ layers on GaAs show that the strain in these layers decreases with doses up to 1×10^{13} Si/cm^2, and then increases [11]. It would be constructive to pursue such investigations in bulk material to understand the nonlinear behavior of strain with dose in SLS structures.

Thermal Annealing

The recovery of the strain profiles of the implanted SLS structures upon annealing is demonstrated in Fig. 4. X-ray rocking curves were obtained before and after annealing at 420°C for one hour of a sample implanted with 1×10^{14} Si/cm^2. In the curve of the as-implanted sample (Fig. 4a), the angular separation between the peaks P_{sub} and P_o' is about twice that in Fig. 2a because of the higher dose (see Fig. 3). Another striking feature of Fig. 4a is that the peak P_{+1}' is more pronounced than in Fig. 2a. The intensity of this peak is related to the difference in the strain between the sublayers of the SLS [12]. The low intensity of P_{+1}' in Fig. 2a thus is due to the small difference in the strains of the sublayers, as shown in Fig. 2b.

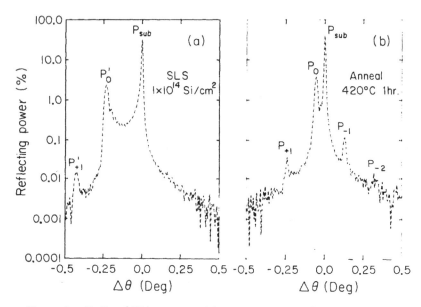

Figure 4. Fe $K\alpha_1$ (400) x-ray rocking curves of an $Al_xGa_{1-x}As/GaAs$ sample implanted with 1×10^{14} Si/cm^2, (a) before annealing (b) after annealing, at 420°C for one hour.

After annealing, the rocking curve (Fig. 4b) reverts very nearly to that of the as-grown sample (Fig. 1a), which implies an almost complete recovery of the strain profile of the implanted SLS structure.

This investigation establishes that within the resolution of the rocking curves, thermal annealing restores the original state of the SLS. This conclusion was tested by conducting backscattering spectrometry measurements[12]. Insignificant atomic intermixing in depth was observed up to 5×10^{15} Si/cm^2 after irradiation and subsequent heat treatment. We conclude that the reversible strain alteration produced by the implantation is due to damage generated within the sublayers.

CONCLUSIONS

Ion implanted $Al_xGa_{1-x}As/GaAs$ superlattices have been investigated before and after annealing. Strain profile measurements in these strained-layered-structures have been carried out by x-ray rocking curves. The technique is sensitive enough to detect and measure changes in the strain below 0.1% induced by implantations with doses as low as 7×10^{12} Si/cm^2.

The analyses of the rocking curves show that in these SLS structures the initial stages of implantation reduces the strain in the $Al_xGa_{1-x}As$ sub-layers. In addition, backscattering analyses establish that up to doses of 5×10^{15} Si/cm^2, long-range displacements of atoms are not dtectable. This observation is further substantiated by the recovery of the SLS structure upon annealing at a relatively low temperature of 420°C.

The study of implanted SLS structures, as carried out in this paper, opens up new avenues in investigating the fundamental aspects of ion-solid interactions. In particular, details of damage creation and ion mixing could be clarified by conducting studies similar to this one on highly strained SLS structures with atomically sharp interfaces.

ACKNOWLEDGMENTS

The authors thank Y. C. M. Yeh, D. A. Smith, and A. Mehta at Applied Solar Energy Corporation for supplying the as-grown samples. Partial financial support by the Office of Naval Research under contract N00014-84-C-0736 through Rockwell International is acknowledged as well.

REFERENCES

1. Y.-H. Wu, M. Werner, and S. Wang, Appl. Phys. Lett., 45, 606 (1984).
2. D. Arnold, J. Klem, T. Henderson, M. Morkoc, and L. P. Erickson, Appl. Phys. Lett., 45, 764 (1984).
3. J. J. Coleman, P. D. Dapkus, C. G. Kirkpatrick, M. D. Camras, and N. Holonyak, Jr., Appl. Phys. Lett., 40, 904 (1982).
4. D. R. Myers, R. M. Biefeld, I. J. Fritz, S. T. Picraux, and T. E. Zipperian, Appl. Phys. Lett., 44, 1052 (1984).
5. G. W. Arnold, S. T. Picraux, P. S. Peercy, D. R. Myers, and L. R. Dawson, Appl. Phys. Lett., 45, 382 (1984).
6. S. T. Picraux, G. W. Arnold, D. R. Myers, L. R. Dawson, R. M. Biefeld, I. J. Fritz, and T. E. Zipperian, IBMM'84, Ithaca, New York (July 16-20, 1984); proceedings in Nucl. Instr. Meth. B (in press).
7. V. S. Speriosu and T. Vreeland, Jr., J. Appl. Phys. 56, 1591 (1984).
8. J. A. Ibers and W. C. Hamilton, eds., International Tables X-Ray Crystallography, Vol. IV, (Kymoch, Birmingham, 1974).
9. A. H. Hamdi, V. S. Speriosu, J. L. Tandon, and M-A. Nicolet, Phys. Rev. B (in press).
10. V. S. Speriosu, B. M. Paine, M-A. Nicolet, and H. L. Glass, Appl. Phys. Lett. 40, 604 (1982).
11. A. H. Hamdi, J. L. Tandon, T. Vreeland, Jr., and M-A. Nicolet, (to be published).
12. A. H. Hamdi, J. L. Tandon, and M-A. Nicolet, 8th Intl. Conf. on the Application of Accelerators in Research and Industry, Denton, Texas, (November 12-14, 1984); proceedings in Nucl. Instr. Meth. B.

STABILITY AND GROWTH OF LENNARD-JONES STRAINED LAYER SUPERLATTICE INTERFACES

BRIAN W. DODSON
Division 1131, Sandia National Laboratories, Albuquerque, NM 87185

ABSTRACT

In the context of a model system whose atoms interact via Lennard-Jones (LJ) interatomic potentials, we have studied the stability of an initially perfect strained layer superlattice interface (SLS) and the process of growth of a mismatched layer on a substrate using Monte Carlo techniques. An initially perfect SLS interface is found to be metastable up to 12% mismatch, a much higher value than is found on real SLS systems. In contrast, we find that, within the limitations of the calculations, a perfect SLS interface cannot be grown in an LJ system. The implications of these results for understanding SLS growth processes is discussed.

INTRODUCTION

The discovery and characterization of semiconductor strained layer superlattice (SLS) systems is of great signifigance to device applications because of the extremely fine control possible over electronic properties by adjusting the degree of mismatch and layer thickness [1]. It is therefore of great interest to understand the conditions under which SLS interfaces can grow in and maintain registry. To date, however, despite much empirical information on stability limits, there is little understanding of what causes the observed limitations on registry.

The present work is an early attempt to address some of these issues. First, we examine the stability of an initially perfect SLS interface, using a Monte Carlo technique to determine the limits of metastability as a function of material mismatch. Following this is a description of a Monte Carlo technique to simulate crystal growth, which is then used to examine the growth of an interface between a substrate material and a mismatched depositing material. These results suggest that the process of growth is likely the principal limiting factor determining production of registered SLS structures.

MODEL SYSTEM

The same model system is used for both calculations. It consists of atoms which interact via a Lennard-Jones, or 6-12 potential, such as is characteristic of spherically symmetric systems which interact by the van der Walls potential, which in normalized units is

$$\phi(r) = (r^{-12} - \alpha r^{-6})/\alpha^2 \tag{1}$$

This simple pairwise potential allows us to utilize rather complicated numerical simulation techniques and remain within practical computational times. As a further simulation, we are only concerned in this paper with a one-dimensional interface between two-dimensional solids. Previous microscopic simulations of lattice stability have demonstrated that results in two- and in three-dimensional systems are qualitatively very similar, and that the quantitative results are not grossly different [2].

METASTABILITY OF A PERFECT SLS INTERFACE

In the case of an initially perfect interface between two mismatched materials, a question of interest concerns the degree of mismatch at which the interface will become unstable to damage. Similar questions arise in the theory of ultimate strength of perfect crystals, and lead to the conclusion that the interface will remain stable until the mismatch strain energy is sufficently large that nucleation of dislocations is an energetically favorable event. We have previously applied a Monte Carlo technique to such a stability analysis [3], and adapt that method for analysis of the SLS interface stability question.

The initial system is a lattice 24 atoms high by 120 atoms long where the bottom 12 rows of atoms have $\alpha=1$ in the LJ potential (Eq.1), and the top 12 rows use $\alpha<1$, and thus are larger in size. Both layers initially are isotropically compressed or expanded so that the unit cells have the average dimensions of the two isolated materials. Owing to the size of the model system (2880 atoms), the interaction potential has been restricted to action only upon nearest neighbors.

To determine stability of the above system, free boundary conditions are imposed, and the entire system is equilibrated using a conventional Monte Carlo process at zero temperatureso that only mechanical mismatch effects are included. (Thermal instability of an initially perfect SLS interface will be considered in a later publication.) We find that a total of 2000 Monte Carlo steps per atom is sufficent to approximate closely the mismatch threshold for instability. This threshold is approximately 12%, or the top atoms are 12% bigger than are the lower atoms before the atomically perfect interface becomes unstable to damage.

To find that SLS registry is stable to such a large mismatch parameter is very surprising in view of the experimental work which indicates that realistic systems with more than 1-2% mismatch will not form high-quality interfaces. However, stability limits for LJ lattices undergoing uniaxial strain are quite consistent with the 12% SLS result. We have previously determined that a 2-d perfect LJ lattice with nearest-neighbor potentials will have a damage threshold of 8.5% uniaxial strain. In the current SLS interface model, the mismatch strain is almost equally shared by the two layers. Given that the microscopic equivalent of the Poisson effect is also observed, the equivalent uniaxial strain felt by the SLS layers is very near 8.5%. The mechanism for interfacial instability in the initially perfect SLS modelled here is therefore strain-induced nucleation of dislocations.

SIMULATION OF CRYSTAL GROWTH

In order to examine the question of registry resulting from SLS growth, a method to accurately simulate crystal growth for a long-range LJ potential had to be developed. Such a simulation is much harder to accomplish than the more conventional solid-on-solid simulations, which are isomorphic with the Ising model [4]. By contrast, the technique described below will allow use of a potential with a range of several interatomic spacings, and will include such finite-temperature effects as phonon displacements which cannot be treated in the simpler models.

The basis of the technique is the following. First, we set up a 4x80 lattice of the substrate material with periodic boundary conditions so that the substrate is 4 layers high and 80 atoms in circumference. Then a random angle around the substrate hoop is generated, and an atom of the material to be deposited is positioned at that angle and roughly 2 interatomic spacings above the substrate. The atom is then lowered until the binding energy reaches a maximum. The system then equilibrates for 10000 Monte Carlo steps, and another atom is positioned. The system is held at a finite temperature, and because all of the atoms (including substrate) are included in the Monte Carlo equilibration, all but the last few deposited atoms have settled into reasonable locations. A total of 220 atoms are deposited on top of the substrate, which represents not quite three layers of new material. These calculations were carried out initially on a Cray-1, taking about 1 hour per run, but were later transfered to an HP-9000 Model 520 desktop computer, which resulted in much faster (overnight) turnaround time.

This method was tested by simulating the growth of a matched crystal. The LJ potential was used, with a cutoff at 3 normalized distance units to reduce the computational time. The substrate temperature was 0.03 normalized energy units. The product was a defect-free single crystal, having no failures of local order. The deposition process gave a rather rough surface, consistent with earlier simulations of LJ crystal growth.

SIMULATED GROWTH OF A MISMATCHED LATTICE SYSTEM

The method described in the last section was used to determine the range of mismatch parameters over which an SLS interface can be grown in our model system. The deposited atoms are bigger than the substrate atoms, but have identical binding energies. The substrate temperature is maintained at 0.03 normalized energy units, which is about 25% of the melting temperature.

In a range of mismatch parameters from 5 to 0.5%, imperfect interfaces result. In the systems with larger mismatch, the interface defects are of the form shown in Figure 1, where the defect is a pentagonally coordinated hole in a hexagonally coordinated lattice. Such a defect represents a region where two zones of new material growth interfere with each other, so that the stable location for the next deposited atom is straddling the two rather than between them. (This is determined by observing the sequential growth of the deposited layers.)

In systems with mismatches near or below 1%, the finite size of the substrate limits the sensitivity of the simulation to growth non-registry. Since the substrate is only 80 atoms around, if the materials had 0.1% mismatch, the accumulated length difference of the equilibrium materials would only be 0.08 interatomic spacings, which would be easily accomodated. However, in the 0.5% mismatch system, the smallest simulated, one can see a lattice disturbance which is similar to, although less prominant than, the defect shown in Figure 1. Since such configurations do not appear in the simulated growth of the matched lattice, we suggest that, in a larger model system, the same type of interfacial defects would appear associated with very small mismatch parameters.

DISCUSSION

When considering the significance of these results, one must remember the limitations of the model system. The model system is two dimensional, and has interatomic interactions characterized by a weak binding force and a soft core, both of which are rotationally symmetric. This is very far from the ternary covalently bonded three-dimensional materials which form most practical SLS systems. Therefore, these efforts must be viewed first as developmental steps toward treating more realistic models, and also as model systems, however unrealistic, from which insight concerning the processes governing real systems might be obtained.

Keeping the above caveats in mind, one can quickly infer that the growth process seems to be a more significant limitation on registry in SLS systems than any intrinsic instability in an initially perfect interface. Thus, maintaining an SLS structure is much easier, at least in the model system, than growing it.

330

ACKNOWLEDGEMENTS

I thank Paul Taylor for many invaluable discussions. This work was performed at Sandia National Laboratories supported by the U.S. Dept. of Energy under contract # DE-AC04-76DP00789.

REFERENCES

1. See, for example, G.C. Osbourn, J.Appl.Phys. 53,1586 (1982).
2. B.W. Dodson and P.A. Taylor, Phys.Rev.B 30,1679 (1984).
3. P.A. Taylor and B.W. Dodson (in preparation).
4. H. Muller-Krumbhaar, in Monte Carlo Methods in Statistical Physics, K. Binder, ed., Springer-Verlag, New York (1979).

SINGLE CRYSTAL GROWTH

GROWTH OF A MISMATCHED SYSTEM

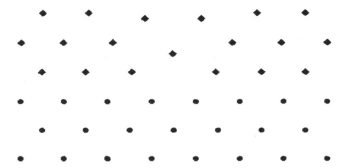

MISFIT DISLOCATIONS IN EPITAXIAL LAYERS OF SI ON GaP (001) SUBSTRATES

M.P.A. VIEGERS*, C.W.T. BULLE LIEUWMA*, P.C. ZALM* and P.M.J. MAREE**

* Philips Research Laboratories, P.O. Box 80.000, 5600 JA Eindhoven, The Netherlands.
** FOM Institute for Atomic and Molecular Physics, Kruislaan 407, 1098 SJ Amsterdam, The Netherlands

ABSTRACT

Misfit dislocations in epitaxial layers of Si grown by MBE at 570°C on GaP(001) substrates have been studied by TEM. It is found that layers as thick as 500 Å at least reside coherently on the substrate without misfit dislocations. In 1000 Å layers of Si the misfit strain is accommodated in part by 60-degree type dislocations with their Burgers vector inclined with respect to the interface, and by stacking faults intersecting the Si layer. The dislocations are dissociated into 30- and 90-degree Shockley partial dislocations. It is shown that in the case of a biaxial strain field, which is tensile in a (001)-plane, the 90-degree partial must be nucleated first. Only then can the 30-degree partial follow on the same glide plane. This geometrical effect explains the presence of dislocations as well as stacking faults in the Si layer.

INTRODUCTION

Recent studies of Molecular Beam Epitaxy (MBE) of Si overlayers on (001)-oriented GaP substrates (1) have demonstrated that epitaxial growth can provide good crystal quality in spite of an appreciable lattice mismatch. At room temperature the bulk lattice parameter of Si is a = 5.43105 Å, while for GaP it is a = 5.45140 Å. Consequently there is a mismatch of 0.37%. From this we might expect, that only relatively thin layers can grow without misfit dislocations. The strain accumulated in thicker films can provide a driving force for dislocation formation.

The present study is concerned with the nature of the strain relaxation. Our interest has grown because of the observation that pseudomorphic growth of epitaxial layers , i.e. without misfit dislocations, is possible up to at least 500 Å of Si on (001)-GaP. Similar results were observed in layers of Si(Ge) grown on Si (2)(3). There it was found that pseudomorphic growth is possible until critical thicknesses are achieved an order of magnitude larger than predicted by Van de Merwe's theory for the formation of misfit dislocations (4). This theory assumes thermodynamic equilibrium between an array of misfit dislocations along the interface and the strained film. Apparently, therefore, there exists a barrier to dislocation formation.

In this context little attention has been paid to the geometry of the formation of misfit dislocations. In the hope of shedding some light on the large critical thicknesses, we undertook a TEM study of lattice defects in MBE:Si on GaP .

EXPERIMENTAL ASPECTS

Details of substrate preparation and MBE procedure have been given elsewhere (1). TEM specimens were made in plane- as well as in cross-section. The plane-view specimens were prepared by ultrasonically drilling 3 mm disks from the wafer. Their thickness was reduced from the rear with a jet etching apparatus, using chlorinated methanol as etchant. Cross-sectional specimens of (110)-orientation were made by first cutting two bars out of the wafer along the [110]-direction and gluing them together with the epilayers facing each other. Slices were then cut off the bar and mechanically polished and thinned by Ar-ion milling at 5 kV on a rotating sample stage. The sputter etch rates of GaP and Si differ by an order of magnitude. Nevertheless, thin interfaces could be obtained by first milling at an angle of incidence of 25°, followed by a final milling at grazing incidence and reduced voltage (1-2 kV). TEM micrographs we obtained with a Philips EM420ST microscope, equipped with a double tilt goniometer stage and LaB6 filament, operated at 120 kV.

RESULTS AND DISCUSSION

The crystallographic quality of the interface region between Si and GaP was studied by High Resolution Electron Microscopy (HREM) of a cross-sectional specimen. An example is shown in Fig. 1.

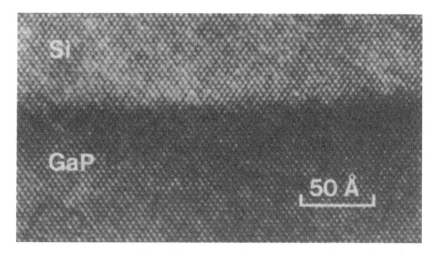

Figure 1. High resolution lattice image along the [110] direction of the GaP-Si interfacial region.

Although the interface seems woolly, with an asperity of about 10 Å, the lattices are completely matched. It is hard to determine whether the interface region, which is about 4 monolayers thick, is best described as an intermixing of both components, or as an abrupt transition at an interface, which in-

corporates atomic steps and terraces. In any case the width of
the terraces, if they are present at all, should be significantly smaller than the thickness of the specimen, which is
about 500 Å. The interface roughness is possibly due to the
method of cleaning the GaP surface prior to MBE growth. This
consisted of a sputter treatment followed by thermal annealing
at 570°C.

In plane-view specimens of a layer of Si 1000 Å thick on
(001) GaP we observed:
(i) A dense dislocation network, located near the interface.
Individual dislocations run exactly parallel to the [110] and
[1$\bar{1}$0] directions, forming a square pattern. The contrast of
the dislocations did not disappear when the diffraction vectors g=220 and g=2$\bar{2}$0 were used for imaging. This means that
the associated Burgers vectors b of the type 1/2a[110] do not
lie in the interface.
(ii) Stacking faults on the four {111} planes and extending in
the two [110] directions, thus also forming a square pattern.
When g=220 was used, only one of the two sets of stacking
faults showed contrast, and with g=2$\bar{2}$0 the other set showed
contrast.
Examples of the dislocations and stacking faults are shown in
Fig. 2.

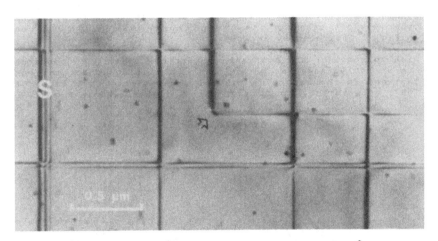

Figure 2. Transmission electron micrographs of 1000 Å of MBE:
Si on GaP, showing dislocations along the interface and a
stacking fault (S) intersecting the Si layer.

A complete characterization, using the extinction criterion g.b=0, revealed that all the dislocations are of 60-de-
gree type. This is a common configuration for dislocations,
which nucleate at the surface of the strained layer and glide
towards the interface along {111} slip planes (5). These slip
planes intersect the interface along the [110] and [1$\bar{1}$0] directions consistent with the direction of the dislocations.

The lattice displacement of 1/2a √2 associated with an
individual dislocation is thus directed at an angle of 45° to
the (001) interface, as illustrated in Fig. 3. The
out-of-plane components average out, however, so that the
average displacement is again located in the interfacial
plane. Its magnitude is 1/4a √2, and is directed perpendicular
to the dislocations.

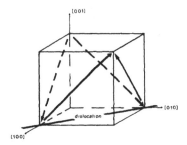

Figure 3. Schematic illustration
of the 4 possible Burgers vectors of a 60-degree dislocation, running in one of the
two <110> directions along the
(001) interface.

Each of the dislocations may also glide along two (111)
planes, which intersect the interface at right angles. Thus a
single dislocation may change from one [110] direction to the
other without changing its character. An example can be seen
in Fig. 2 indicated by an arrow.
    In order to explain the presence of stacking faults, we
wish to consider an important geometrical effect. It concerns
the atomic arrangement on the |111| slip planes. For an fcc-
lattice, in the case of a biaxial strain field in a layer of
(001) orientation, the arrangement is illustrated in Fig. 4.

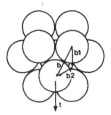

Figure 4. Slip in fcc crystals
(6). b1 = 1/6a[11$\bar{2}$],
b2 = 1/6a[2$\bar{1}\bar{1}$], b=½a[10$\bar{1}$],
t is along [11$\bar{2}$].

The resolved shear stress t on a (111) plane is in the [11$\bar{2}$]
direction. It can be seen from Fig. 4 that the unit displacement b, which results in a 60-degree dislocation, is achieved
by two movements: first b1 yielding a 90-degree Shockley partial dislocation, and then b2 yielding a 30-degree Shockley
partial dislocation. The force, acting on the dislocations, is

given by F=b.t . Consequently, the 90-degree partial (with b parallel to t) experiences a force twice that experienced by the 30-degree partial (with an angle of 60 degrees between b and t). The principle just outlined is the same for all four |111| planes. It is applicable to Si (7). In other words, the particular geometry of Si on (001) GaP causes the 90-degree partial dislocation, which experiences the larger force, to nucleate first. When it moves towards the interface it produces a stacking fault. The stacking fault energy then provides a force which tends to reduce the effect of the shear stress on the 90-degree partial, but to add to the shear stress on the 30-degree partial dislocation, which may then nucleate as well. We therefore suggest that the defect structure in a growing layer of Si on (001) GaP first consists of stacking faults, bounded by a 90-degree Shockley partial dislocation at the interface. When the layer grows thicker the stacking faults should collapse because of the nucleation of 30-degree Shockley partial dislocations, which also move to the interface, resulting in the rectangular network of 60-degree dislocations, which was observed indeed. The stacking fault seen in Fig. 2, may be associated with such a case, where the 30-degree partial has not yet been nucleated. A preliminary examination of a thinner layer of Si (750 Å) on (001) GaP indeed revealed that the relative number of stacking faults was larger (by a factor of 5).

Close to the interface it was found that each pair of Shockley partials was still separated by a distance of about 100 Å, as can be seen in the HREM image of Fig. 5.

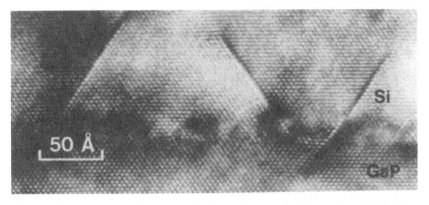

Figure 5. High resolution lattice image of dissociated 60-degree dislocations close to the GaP/Si interface.

The separation is substantially larger than the equilibrium distance at 880°C (8), in spite of the stress field which would be expected to move them closer together. This may indicate that local strains in the interface region differ from the average strain throughout the bulk of the layer.

Finally we note that the geometrical effect discussed above, depends on the crystallographic orientation of the strained film. It also depends on the sign of the stress field. If, for instance, it were compressive, then the 30-degree partial would have to nucleate first, instead of the 90-degree, because of the atomic arrangement. Since the 30-degree partial experiences the smaller force from the shear stress, there may be a greater barrier to its nucleation than in the tensile case. Thus, pseudomorphic layers of Si(Ge) on (001) Si, which are compressed, may be thicker than pseudomorphic layer of Si on (001) GaP, in which the stess field is tensile. Once the 30-degree partial has been nucleated in a compressed layer, we expect that the 90-degree partial will follow immediately, because it experiences the larger force from the shear stress.

ACKNOWLEDGEMENT
    Valuable discussions with J. Hornstra and F.W. Saris are gratefully acknowledged.

REFERENCES

1. T.de Jong, W.A.S. Douma, J.F. van der Veen, F.W. Saris and J. Haisma, Appl.Phys.Lett. 42, 1037 (1983);
   T. de Jong, F.W. Saris, Y. Tamminga and J. Haisma, Appl. Phys.Lett. 44, 445 (1984);
   C.W.T. Bullę Lieuwma, T. de Jong and J. Haisma, Proc. 8th Europ. Congress on Electron Microscopy, Budapest, Hungary, Aug. 1984
2. E. Kasper, H.J. Herzog and H. Kibbel, Appl.Phys. 8, 199 (1975);
   E. Kasper and H.J. Herzog, Thin Solid Films, 44, 357 (1977)
3. J.C. Bean, T.T. Sheng, L.C. Feldman, A.T. Fiory and R.T. Lynch, Appl.Phys.Lett. 44, 102 (1984);
   A.T. Fiory, J.C. Bean, L.C. Feldman and I.K. Robinson, J.Appl. Phys. 56 1227 (1984)
   J.C. Bean, L.C. Feldman, A.T. Fiory, S. Nakahara and I.K. Robinson, J.Vac.Sci.Technol. 2 436 (1984)
4. J.H. van der Merwe and C.A.B. Ball, in Epitaxial Growth, Part B, Academic Press, 1975, J.W. Matthews, Ed.
5. J.W. Matthews, in Epitaxial Growth, Part B, Academic Press, 1975, J.W. Matthews, Ed.
6. A.H. Cottrel, Dislocations and Plastic Flow in Crystals, Oxford University Press (1953)
7. J. Hornstra, J.Phys.Chem.Sol. 5, 129 (1958)
8. A. Gomez, D.J.H. Cockayne, P.B. Hirsch and V. Vitek, Phil. Mag. 1974 (1975)

LOW TEMPERATURE REORDERING OF IMPLANTED AMORPHOUS Si WITH
Al SURFACE LAYERS

L. S. HUNG, S. H. CHEN, AND J. W. MAYER
Department of Materials Science and Engineering
Cornell University, Ithaca, NY 14853

ABSTRACT

Ion backscattering and channeling techniques, transmission and scanning
electron microscopy, and secondary ion mass spectroscopy were used to investigate the reordering characteristics of implanted amorphous Si in the presence of an Al surface layer. It was found that reordering takes place at the
temperature of about 400°C and is associated with an interfacial migration
between Al and Si. The regrowth behavior appears to be a function of the
initial annealing temperature and annealing sequence. High density of twin
faults and substantial concentration of Al are observed in the regrown layers.
We believe that the low temperature reordering is due to processes analogous
to solid epitaxial growth with transport media.

INTRODUCTION

Ion implanted amorphous Si in the presence of Al films has been found
to reorder at the temperature of about 400°C, which is significantly below
the temperature ($\sim$550°C) normally encountered in the epitaxial regrowth of
amorphous Si layers [1,2]. In our early study, the low temperature reordering was investigated in-situ with transmission electron microscopy using a
heating stage [2]. It was found that the recrystallization takes place and
proceeds during thermal treatment with substantial concentration of residual
damage in the regrown layer. However, it was not clear what mechanism is
responsible for the low temperature reordering. In this paper we report a
set of systematic studies of the Si(100)/Si(amorphous)/Al system with an aim
to elucidate the mechanisms which enhance the regrowth rate.

EXPERIMENTAL

Commercially available Si wafers with <100> orientation were used for
the experiments. Following standard cleaning, Al films of 1500 A in thickness were evaporated onto the Si substrates with an e-gun at pressures of
$\sim$1x10^{-6} Torr. A sequence of Si implants was used to ensure a relatively
flat implant profile in the samples. The samples were then annealed in a
vacuum furnace or heat-treated in-situ using a heating stage on the goniometer during channeling measurements.
    The reordering and interfacial migration between Al and Si were measured
by ion backscattering and channeling techniques. A backscattering angle of
θ = 170° was selected. In some cases the detector was set at an angle of θ
= 120° with respect to the impinging beam in order to improve the depth resolution. The residual damage and interfacial morphology were studied by transmission electron microscopy (TEM) using cross-section techniques. The surface
topology of the regrown layer was examined by scanning electron microscopy
(SEM) and the Al profile by secondary ion mass spectrometry (SIMS).

RESULTS

For a sample of Al on Si(100) after multiple irradiations at energies
between 100 and 300 keV with a total dose of 6x10^{15} Si atoms cm^{-2}, the re-

338

growth behavior is shown in Fig. 1. The as-implanted samples had an amorphous layer extending from the Si/Al interface to a depth of around 3000 Å. The peak at about 1.28 MeV is due to the overlapping of Si and Al signals near the Si/Al interface. After anneal at 420°C for 10 min, a decrease in the aligned spectrum indicates the reordering of the amorphous layer. The high dechanneling yield below the regrown layer reveals substantial concentration of residual damage in the regrown layer. The aligned spectrum of an unimplanted sample is presented in the figure for comparison.

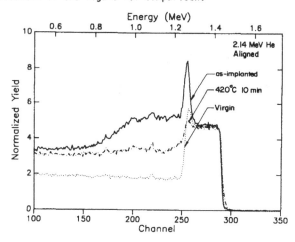

Fig. 1. Channeling-effect measurements of an implanted sample before and after annealing at 420°C for 10 min.

In-situ channeling measurements with the detector at 170° relative to the incident beam were performed on a Si(100)/Al sample after multiple irradiations at energies between 100 and 200 keV. The as-implanted sample contained an amorphous layer of about 2000 Å in thickness. The height and shape of the peak at about 0.65 MeV (in Fig. 2) provide an indication of the interfacial mixing between the two elements. As the temperature is raised to 280°C, interfacial migration of Al and Si atoms takes place, as shown by the decrease of the peak height. The change in the backscattering profile at energies below 0.65 MeV suggests a deep penetration of Al into the implanted Si layer. When the temperature is continuously increased to 380°C, the aligned spectrum shows a uniform drop in yield, which suggests the transition of amorphous Si to single-crystalline layers. These observations indicate that the reordering occurs upon annealing in good agreement with our previous in-situ TEM study [2] and that the interfacial migration begins in the initial stage of the regrowth process.

Annealing at different temperatures was performed to determine whether the regrowth depends on the initial annealing temperature. The effect is demonstrated in Fig. 3, which shows the backscattering spectrum from an implanted sample with a configuration of Si(100)/amorphous Si(2000 Å) after anneal at 330°C for 10 min. The Si migration toward the surface and Al penetration into the original amorphous region are observed. Such movement is revealed by the disappearance of the peak at 1.15 MeV and by the presence of Si at its surface energy position. The valley at 1.15 MeV comes from the separation of the surface Si and the Al-rich Si layer, which is represented by a broad peak at low energies. The combination of channeling and transmission electron diffraction techniques reveals a transition of amorphous to polycrystalline structures in the surface Si layer and a partial reordering of Si in the Al-Si mixture.

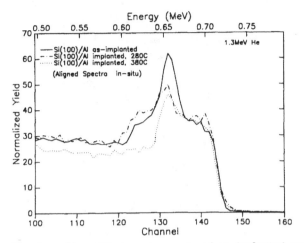

Fig. 2. In-situ channeling-effect measurements of an implanted sample.

After additional annealing at 420°C for 30 min, the backscattering spectrum indicates that the Al has been displaced toward high energies with a thin silicon layer on top. The Al and the rest of the implanted Si were separated with a relative abrupt interface. It is found from channeling measurements that the surface Si retains its polycrystalline structure, while the rest of the implanted Si is oriented with respect to the substrate. From these results it appears that the structure of the implanted Si layer and the extent of the intermixing is determined by the initial annealing temperature and the annealing sequence. It is also evident that the regrowth process is associated with Al migration.

The samples consisted of Si(100)/amorphous Si(2000 Å)/Al(1500 Å) were annealed at different temperatures. The surface topology of the annealed samples after removing Al is presented in Fig. 4. SEM measurements show a

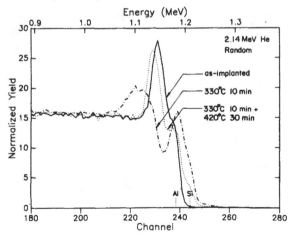

Fig. 3. Backscattering measurements of an implanted sample before and after different thermal treatments.

340

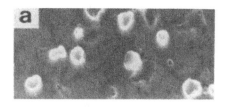

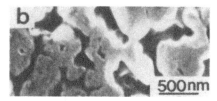

Fig. 4. SEM micrographs for an implanted sample after 10 min anneal at a) 420°C and b) 330°C. The surface Al layer was removed before SEM measurements.

relatively smooth surface with a number of precipitates in an annealed sample (420°C, 10 min). The micrograph of a lower-temperature annealed sample (330°C, 10 min) reveals a columnar structure, in agreement with the backscattering data of Fig. 3 which indicate massive transport of Al into the implanted Si.

The epitaxial nature of the regrown layer and the morphology of the Si/Al interface were examined by cross-sectional TEM. Fig. 5 is the dark field image taken of an annealed sample (420°C, 10 min) showing a relatively smooth interface between the regrown layer and the Al overlayer, whereas the interaction between Al and Si is evident from the penetration of Al into the regrown layer at several sites. A clear single-crystal structure is visible in the

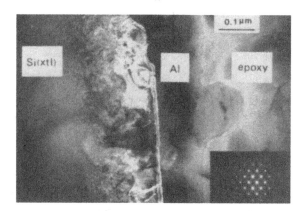

Fig. 5. Cross-section TEM micrographs and diffraction pattern for an implanted sample after 420°C anneal for 10 min.

diffraction pattern of the regrown layer. The dominant defects in that layer are twin faults as indicated by the weak spots in the pattern.

The Al profiling was made by SIMS. It was found that Al is present in the regrown layer with a concentration of 0.4 ∿ 0.8 at.%, which is a factor of 20 higher than the solid solubility of Al in Si at eutectic. It appears that part of the penetrating Al is trapped in the Si matrix during the epitaxial growth process.

DISCUSSION AND CONCLUSION

Several mechanisms can be suggested to account for the low temperature reordering process: 1) Si dissolves into Al at annealing temperatures and epitaxially grows on the single-crystal substrates upon cooling, 2) the presence of Al as impurity enhances the growth rate [3], and 3) the reordering is due to the processes analogous to solid epitaxial growth (SPE) with transport media [4,5].

The in-situ channeling measurements show that the reordering takes place during annealing in disagreement with the first explanation.

The impurity effects on the regrowth rate has been extensively studied [3,6,7]. Although the presence of high concentrations of electrically active dopants might increase the growth rate, the magnitude of the enhancement does not suffice to explain the present observations [3]. It appears that the Al doping has little effect on the low temperature reordering. Additional support comes from the following experiment. The surface layer of the regrown sample was re-amorphized after removing Al. Subsequent annealing did not result in the epitaxial growth of amorphous Si at temperatures below 550°C, even though high concentration of Al was present in that amorphous region.

The results presented so far show that the reordering is associated with the Al migration into the original amorphous Si. This is evident from the shift of the Al profile to lower energies in the initial stage of in-situ channeling measurements, the massive transport of Al and Si upon annealing at 330°C and the high concentration of Al in the regrown layers. The rapid migration of Al into amorphous Si has also been reported in the early study of SPE with Al as transport medium [8]. We believe in our case that a certain amount of Al moves into the implanted Si and partially distributes at the original Si(100)/Si(amorphous) interface in the initial annealing stage. The subsequent process includes transport of the Si atoms through the metal layer, arrival at the single-crystal surface and finally growth. The situation may be more complicated, as indicated in the case of two step annealing, where the transition of polycrystalline to epitaxial layers contributes to the reordering as well. SPE with transport media has been investigated in the Si(crystal)/Al/Si(amorphous) by Ottaviani et al., who found that the epitaxial growth is achieved by dissolution and transport of amorphous Si through Al [9]. In both cases, the epitaxial growth occurs at temperatures below 400°C and is characterized by high density of twin faults and substantial concentration of Al in the epitaxial layers. These similarities provide further evidence to our argument that the same mechanism is responsible for the epitaxial growth in the Si(100)/Al/Si(amorphous) and in the Si(100)/Si(amorphous)/Al system.

In conclusion, the reordering of implanted amorphous Si on Si(100) substrates with a surface Al layer takes place at temperatures of about 400°C. We attribute the low temperature reordering to the migration of Al toward the original Si(100)/Si(amorphous) interface and subsequent epitaxial growth of Si through the Al-Si transport medium.

342

ACKNOWLEDGMENTS

This work was supported in part by SRC. The authors are grateful to S. S. Lau (UCSD, La Jolla, CA) and C. Palmstrom (Cornell) for helpful discussion, and Charles Magee (RCA Laboratories, Princeton, NJ) for SIMS measurements. Ion implantation was carried out in the National Research and Resource Facility for Submicron Structures supported by NSF.

REFERENCES

1. D. H. Lee, R. R. Hart, and O. J. Marsh, Appl. Phys. Lett., 20, 73 (1972).
2. L. S. Hung, S. H. Chen, and J. W. Mayer, in Thin Films and Interfaces II, eds. J.E.E. Baglin, D. R. Campbell, and W. K. Chu (Elsevier, NY, 1984), p. 253.
3. L. Csepregi, E. F. Kennedy, T. J. Gallagher, J. W. Mayer, and T. W. Sigmon, J. Appl. Phys., 48, 4234 (1977).
4. S. S. Lau and W. F. Van der Weg, in Thin Films--Interdiffusion and Reactions, eds. J. M. Poate, K. N. Tu, and J. W. Mayer (John Wiley, NY, 1978), Chap. 12.
5. S. S. Lau, J. W. Mayer, and W. Tseng, in Handbook on Semiconductors, Vol. 3, Materials Properties and Preparation, ed. S. P. Keller (North Holland Amsterdam, 1980), Chap. 8.
6. E. F. Kennedy, L. Csepregi, J. W. Mayer, and T. W. Sigmon, J. Appl. Phys., 48, 4241 (1977).
7. I. Suni, G. Goltz, M. G. Grimaldi, M-A. Nicolet and S. S. Lau, Appl. Phys. Lett., 40 (1982).
8. G. Ottaviani and G. Majni, J. Appl. Phys., 50, 6865 (1979).
9. G. Majni, G. Ottaviani, and R. Stuck, Thin Solid Films, 55, 235 (1978).

Epitaxial Metals
on Semiconductors

SCHOTTKY BARRIER HEIGHTS AT SINGE CRYSTAL METAL SEMICONDUCTOR INTERFACES

R. T. TUNG
AT&T Bell Laboratories, Murray Hill, New Jersey 07974

ABSTRACT

Electrical behavior at single crystal silicide-silicon interfaces was studied. Schottky barrier heights were determined for epitaxial $NiSi_2$ and $CoSi_2$ layers grown under ultrahigh vacuum conditions on (111), (100) and (110) surfaces of Si. A dependence of Schottky barrier heights on interface structure was observed. These results favor intrinsic mechanisms for Schottky barrier formation. The advantages of having homogeneous metal-semiconductor interfaces for the study of Schottky barrier mechanisms are pointed out. In particular, the present epitaxial silicide-silicon interfaces represent ideal candidates for detailed theoretical investigations based on experimentally obtained atomic structures.

INTRODUCTION

Every semiconductor device requires some form of electrical conduction between a metal and a semiconductor [1]. Reliable ohmic contacts and Schottky barrier (SB) junctions are necessary for the successful performance of any semiconductor device. Technologically, it would be desirable to predict and tailor Schottky barrier heights (SBH). However, the exact relationship between electrical properties such as the SBH and physical properties of the metal-semiconductor interface, such as composition, atomic structure, defects and impurity content, is not known. This problem has been an interesting challenge to scientists for several decades. Considerable effort has been undertaken to correlate the experimentally observed SBH's with specific parameters of the given metal-semiconductor system. Many interesting, and occasionally puzzling, matches were found when selected pieces of the experimental data were considered. Theories and models were developed accordingly, yet the proposed microscopic SB mechanisms still remain vague and speculative, at best. Several articles exist which review the current situation [2-5].

To go beyond phenomenological theories is difficult. Experimentally, detailed structural information about the physical interface is not easily obtained. With a few exceptions, almost all metal films grown on semiconductors are multicrystalline. Variation of the interface structure is expected from diode to diode, from grain to grain, or even from different regions in the same grain. Such complicated structures make reliable theoretical calculation of the electronic structures extremely difficult. It's apparent that to gain insight of the microscopic SB mechanism one needs simple, controllable metal-semiconductor systems as a testing ground. In this paper, we show that the interfaces between epitaxial silicides and silicon are ideal candidates for such a purpose.

Disilicides of nickel and cobalt have the cubic CaF_2 structure and lattice constants within 0.5% and 1.2%, respectively, of that of Si. As a result of such favorable conditions for epitaxy, thin films of single crystal $NiSi_2$ and $CoSi_2$ can be grown with a very high degree of perfection on silicon under ultrahigh vacuum (UHV) conditions [6-7]. Lattice imaging of such silicide-silicon interfaces by high resolution transmission electron microscopy (TEM) has demonstrated the sharpness of the interfaces and revealed the most likely interfacial atomic structures [8]. Moreover, the orientation of the $NiSi_2$ films grown on Si (111) can be controlled to be either identical to that of the substrate (type A orientation), or to be rotated 180° with respect to the substrate about the surface normal (type B orientation) [9], thus providing a unique opportunity to examine SBH in systems identical except for interface structure. The comparison of SBH of the same silicide on different Si orientations can also be quite illuminating.

EXPERIMENTAL PROCEDURES

Polished Si (111), (100) and (110) wafers were used in these studies, with resistivities ranging from 0.1 to 0.7 Ωcm (n-type) and 0.4 to 3 Ωcm (p-type). Arrays of circular windows with diameters of 127-635 μm were photolithographically defined in 3000-5000Å thick SiO_2 layers grown on these wafers. Large areas of bare Si were also available on the patterned wafers for various structural characterizations. Samples were degreased and chemically cleaned by repeated oxide growth and removal. A final thin oxide layer [10] was grown before the samples were loaded in a UHV chamber with base pressure of 1×10^{-10} torr. The oxide layer was evaporated off the surface at ~ 850°C. The exposed Si regions displayed reconstructed LEED patterns indicative of clean surfaces. On Si (111) samples, thin (60-70Å thick) single crystal $NiSi_2$ templates were then grown in exposed Si regions by either deposition of 17-19Å Ni at room temperature followed by annealing to 500°C for type A orientation, or by 4 or 5 repeated steps of room temperature deposition of Ni, ~ 4Å at a time, followed by annealing to 500°C for type B orientation [9]. Single crystal type B $NiSi_2$ layers were also grown by deposition at room temperature of ~ 16Å Ni, followed by ~ 30Å Si, and then annealing to ~ 500°C. Thick (700-1000Å) $NiSi_2$ type A and type B layers were grown by deposition of nickel onto the template layers of the corresponding orientation while the sample temperature was kept at ~ 650-775°C [7]. Single crystal type B $CoSi_2$ layers were grown by deposition of ~ 200Å of cobalt in UHV and annealing to ~ 950°C for 30 minutes [6]. Much thinner $CoSi_2$ layers were also grown at a lower temperature. The orientations of these silicides were determined by LEED, RBS and channeling. Single crystal $NiSi_2$ layers were grown on Si (100) and (110) by the use of thin $NiSi_2$ templates. The perfection of these single crystal layers has already been described [8]. The unreacted Ni and Co on the SiO_2 was removed by a chemical etch. Ohmic contacts were made to the back sides of the samples without heating the bulk samples above room temperature by deposition of ~ 150Å Sb, pulsed laser melting (30 ns Ruby, 1.5 J/cm^2) of the Sb into the Si, and finally an aluminum deposition. Metallized openings were probed with a thin gold wire to avoid puncturing the thin layers.

SBH's were determined by I-V and C-V scans recorded on a $x-y$ plotter. Capacitances were measured mainly at 1 MHz with a capacitance meter; a few select samples were studied with different frequencies ranging from 20 kHz to 4 MHz using a bridge. Extrapolation of the linear portions of the semilogarithmic I-V curves was used to determine the SBH's [11]. The effective Richardson's constants were taken to be 112 and 32 Amp K^{-2} cm^{-2}, respectively, for electrons and holes [12]. On high resistivity substrates, occasionally it became necessary to correct for the series resistance. Such series resistances can be estimated by plotting the high-current portions of the curves in a fashion suggested by Norde [13].

RESULTS AND DISCUSSIONS

$NiSi_2$ on Si (111)

Forward current characteristics of a type A and a type B template layer are shown in Fig. 1 [14]. These data are for junctions which show close to ideal electrical behavior, approximately 50% of the samples studied. All junctions which show a linear log I to V dependence display a good ideality factor (< 1.06) and extrapolate to yield the same barrier height (± 0.01 eV) of the corresponding orientation. Linear C^{-2} to V behavior was found for almost all diodes studied, with the slope agreeing well with the nominal doping concentration. For a given orientation, extrapolations consistently yielded the same SBH as that obtained by I-V method. Thick (700 ~ 1000Å) and thin (~ 70Å) diodes of the same orientation yield the same SBH (± 0.01 eV); they have identical capacitance scans for the same kind of substrate. This demonstrates that misfit dislocations, which exist at the interfaces of the thick layer (~$10^5/cm^2$), but not at the thin pseudomorphic template layer interfaces, do not contribute actively to the formation of the SB. This is in contrast to semiconductor-semiconductor interfaces [15] where misfit dislocations were thought to pin the position of the Fermi level locally. As shown in Fig. 2, SIMS analysis of a thick A and a thick B layer showed identical levels of Ni impurities in the Si substrate within ~ 1000Å of the interfaces. The Ni levels in Si are much lower for thin A and B templates which were grown at lower temperatures. These results suggest metal (interstitial) impurities are not the main SB mechanism. Values of asymptotic SBH's (ϕ_{BO}^n), which correspond to the flat band situation, are summarized in Table I. C-V measurements give ϕ_{BO}^n directly, while

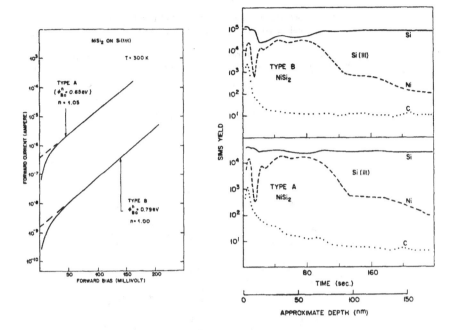

Fig. 1.   Forward current characteristics of thin (∼ 70Å) type A and type B NiSi₂ layers on n-type Si (111). Spot size is 394 μm.

Fig. 2.   Unnormalized secondary ion mass spectrometry signals of two 750Å thick NiSi₂ layers on Si (111) (a) type B orientation (b) type A orientation.

I-V measurements give ϕ_B^n. In comparing values listed in Table I with published SBH's (mostly ϕ_B^n's), one should allow for the small image-force lowering ($\Delta\phi \approx$ 10-40 meV depending on doping and on the magnitude of the SBH).

Possible effects of steps and/or microfacets on the SBH's were studied with misoriented wafers. No change was observed for the type A SBH as wafer misorientation of up to 4° was introduced. For type B NiSi₂, a small decrease (∼ 0.03 eV) in SBH was observed. At present it is not known whether a small amount of A grains existed in these B films (because the growth conditions were slightly different due to the misorientation) which caused this drop in SBH, or if this drop was from the increased density of steps and microfacets. In either case, it can be concluded that steps have a very small effect on the SBH's.

The difference between SBH's of A and B is rather significant considering the subtle difference between the atomic structures at the two interfaces. Atomic structures of the two 7-fold coordinated interfaces on Si (111) are depicted in Fig. 3(a) and (b). Computer simulations based on these unrelaxed structures agree well with high resolution TEM images [8]. The structures of these two interfaces are very similar, differing only in the positions of the third and higher nearest neighbors to the last nickel layer. The key feature is the unpaired electrons which due to Ni-Si hybridization are directional, similar to dangling bonds [16].

Results were also obtained for NiSi₂ layers grown by annealing deposited ∼ 200Å thick nickel films. The SBH's showed some scatter among different diodes, but C-V measurements consistently yielded higher values than those obtained by I-V methods. Typical SBH's were 0.66 eV and 0.72 eV from I-V and C-V methods, respectively. These are in agreement with

Table I. Schottky barrier heights (ϕ_{BO}^n) of thin ($< 80\text{Å}$)
single crystal silicide layers to n-type Si.

Silicide	Silicide Orientation	Substrate Orientation	Interface Structure	ϕ_{BO}^n (eV) I-V	C-V
$NiSi_2$	A	(111)	7 fold	0.65	0.65
$NiSi_2$	B	(111)	7 fold	0.79	0.79
$NiSi_2$		(100)	6 fold	—	0.48
$NiSi_2$		(110)	7 fold	0.65	—

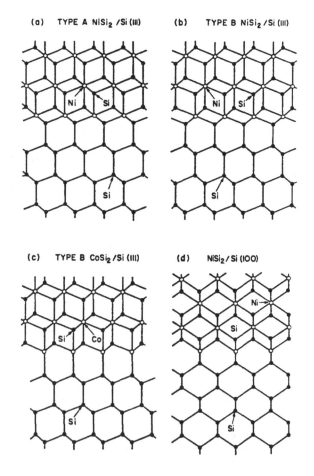

(a)   TYPE A $NiSi_2$ /Si (III)   (b)   TYPE B $NiSi_2$ /Si (III)

(c)   TYPE B $CoSi_2$ /Si (III)   (d) $NiSi_2$/ Si (IOO)

Fig. 3.   Balls and sticks models of silicide-silicon interfaces viewed in the [110] direction.

published values of ϕ_B^n for non-UHV grown NiSi$_2$ layers on Si (111), which lie in the range of 0.64-0.7 eV [17-18]. Electron microscopy and channeling have shown that such layers have large grains ($\geq 2\ \mu$m) of both type A and type B orientations [19]. An interface which consists of regions of high barrier and regions of low barrier would display capacitance characteristics intermediate of the two and current characteristics resembling those for the lower barrier. For instance, the calculated capacitance of a diode with half its area having a SBH of 0.79 eV and the other half 0.65 eV gives an apparent ϕ_{BO} of 0.73 eV. Such a diode would render a ϕ_{BO} of 0.67 eV with I-V analysis. This example can explain the origin of the difference between the C-V and the I-V measured SBH's of reacted NiSi$_2$, which consists of roughly equal portions of A and B grains. Hence the reported SBH's of non-UHV films are easily explained as the result of orientational averaging. Had both C-V and I-V measurements been done in those studies, discrepancies presumably would have been observed. Such diodes with large patches of different SBH's can be conveniently examined by EBIC. Preliminary studies showed just the kind of patterns expected of such diodes, namely dark and bright regions [20].

SBH's of type A and type B NiSi$_2$ on p-type Si (111) were also studied. Results from I-V measurement showed that $\phi_{BO}^p = 0.47 \pm 0.02$ eV for type A NiSi$_2$ (Fig. 4). Capacitances were found to be frequency dependent and hence could not be used for SBH determination. Ohmic behavior was found for all B samples on p-substrate. For type A NiSi$_2$, the sum of n and p type SBH's is close to the room temperature Si band gap, showing a single position of the Fermi level at the interface.

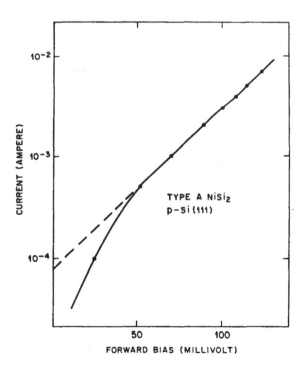

Fig. 4.   Forward current characteristics of type A single crystal NiSi$_2$/p-Si (111) interface. The curve has been corrected for a series resistance.

NiSi$_2$ on Si (100) and (110)

Electrical properties of single crystal NiSi$_2$ layers on Si (100) were also studied [21]. The SBH of thin template layers ($\leq$ 60Å thick) was consistently measured to be $\phi_{BO}^n = 0.48 \pm 0.02$ eV by C-V analyses. An example is shown in Fig. 5. The forward I-V characteristics lacked long and straight sections in log I plots. The situation was much improved with the use of $n^+ - n$ epitaxial layers; however the ideality factor was high ($n > 1.1$) and hence no reliable SBH can be deduced from I-V measurements. Nevertheless, one notices that the absolute current level, forward or reverse, is close to what thermionic emission theory would predict for a SBH of ~ 0.47 eV. The atomic structure at these (100) interfaces, as determined from high resolution TEM, is depicted in Fig. 3(d). This sixfold structure [8] is entirely different from either of the (111) interfaces. The fact that the observed SBH differs appreciably from the (111) results tends to support the close tie between SBH and the interface structure suggested by the (111) results.

Junctions of thick NiSi$_2$ layers ($\sim$ 1000Å) grown by the use of templates on n-type Si (100) generally had ohmic characteristics. The interface atomic structure is not expected to vary as film thickness increases. It is known, however, that thicker films have interfaces with a higher density of steps, facets, dislocations, and, because of the higher temperature during growth, metal impurities in Si. It is far too early to speculate on the reasons behind the decrease of SBH. Work is continuing with emphasis placed on the evolution of the junction characteristic and the interface morphology. Due to the consistency of the observed data on thin NiSi$_2$, it is not likely that their SBH (0.48 eV) is, for instance, a result of similar lowering from the 0.65 eV level associated with polycrystalline nickel silicide layers.

The interfaces of NiSi$_2$ on Si (110) consist exclusively of {111} type inclined facets [22]. Analyses of dislocations at facet boundaries indirectly suggest that at the inclined {111} interfaces the atomic structure has the 7-fold configuration. The interface morphology changes as growth conditions are varied. As a result, a significant variation of SBH is observed with sample preparation. However, for a certain set of growth conditions, the SBH's are reasonably consistent and reproducible. The I-V determined SBH for thin template layers ($\leq$ 80Å) is $\phi_{BO}^n = 0.65 \pm 0.02$ eV. The I-V characteristics of these layers show close to ideal behaviors ($n \leq 1.02$). For thicker layers grown by the use of thin NiSi$_2$ templates, the SBH's found by I-V analyses decreased by ~ 0.1 eV with an accompanying increase in ideality factor.

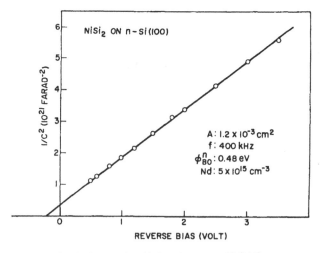

Fig. 5.   C-V plot of a single crystal NiSi$_2$ layer on n-type Si (100).

Type B CoSi$_2$ on Si (111)

Single crystal CoSi$_2$ layers can be grown on Si (111) under UHV conditions. Both reaction and molecular beam epitaxy produce type B oriented epitaxial films.*ref* (6) Shown in Fig. 6 is the C-V plot of two ~ 800Å thick single crystal CoSi$_2$ layers grown by the UHV reaction techniques. These measurements and the I-V method both yield a ϕ_{BO}^n of ~ 0.64 eV. The interface atomic structure of the type B CoSi$_2$ has been determined from high resolution TEM to be consistent with a 5-fold coordinated structure*ref* (23) (Fig. 3(c)), different from that of type B NiSi$_2$. Also present at these interfaces are misfit dislocations of higher density than at NiSi$_2$/Si (111) interfaces, and a network of coarse defects whose nature remains unknown.*ref* (6) At present, it is not clear how these line defects will affect the SBH. If they have only a secondary effect on the magnitude of the SBH, then the large difference between SBH's of type B CoSi$_2$ and type B NiSi$_2$ (.64 to .79 eV) seems to suggest a large change in an important parameter in SBH mechanism. At present, it is more reasonable to assume that this parameter is related to the interface structure rather than to bulk electronic properties.

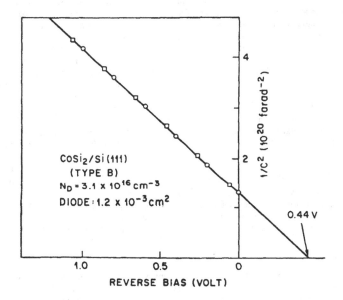

Fig. 6.   C-V plot of single crystal CoSi$_2$ layers on n-type Si (111).

POSSIBLE MECHANISMS

The dependence of SBH's on orientation is usually very weak for polycrystalline systems. SBH's of metal silicide/Si were previously reported to be nearly constant for a given metal, independent of the structure and stoichiometry of the silicides.*ref* (17,24) For instance, Ni$_2$Si, NiSi, and NiSi$_2$ all have the same reported ϕ_B^n, 0.66 ± 0.02 eV, on (111) and (100) substrates. The present study of single crystal SBH's, on the other hand, reveals a very strong dependence on orientation and structure. This apparent contradiction is quite intriguing and deserves some attention. There are many mechanisms which contribute to the interface electronic states. It is possible that different mechanisms determine the SBH under different circumstances. This is particularly possible when one considers the vast difference in metallurgical behaviors between, say, group IV and III-V semiconductors. It is possible that different mechanisms are responsible for SB of single crystal and polycrystalline interfaces, or even for SB of different single crystal interfaces. But to keep discussions more tractable, we shall assume one main SB mechanism for all metal-Si interfaces and ignore perturbations from less important mechanisms. The question remains as to why single crystal systems show different behaviors from polycrystalline contacts.

One possibility is that the constancy of SBH as observed in previous studies is not a direct consequence of the underlying SB mechanism. It is apparent that SBH's deduced from electrical measurements reflect only the average behavior of the interfaces. In the case of a single crystal interface, a translational symmetry exists; the measurements directly probe the homogeneous electrical property. A change in the local electronic structure, such as due to a change in the bonding configuration, can readily be detected. At a polycrystalline interface, the measured SBH is likely to be some weighted average of a spectrum of barrier heights associated with differently-structured portions of the interface. Moreover, for a non-epitaxial system, the interface is usually incommensurate; variation of the atomic structure is likely to occur on the scale of a few atomic spacings. The interface Fermi level position should be determined on a much longer lateral scale. It then follows that the "averaged" local electronic structure could be very similar from region to region. Since atomic structure is randomized for every polycrystalline interface, a constant SBH may result regardless of orientation and morphology. This interpretation of the invariance of apparent (polycrystalline) SBH's is only speculative. However, it obviates the need for mechanisms which have a built-in "SBH constancy feature" to explain Si SB formation. In the above example, the SB mechanism was assumed to be the interplay of the interface electronic structure and the Fermi level position. This mechanism, by itself, may or may not have a strong dependence on the atomic structure — a question that the polycrystalline results can not directly answer. This last point illustrates the merits of having simple, homogeneous metal-semiconductor interfaces for the study of SB's.

Looking at the present results alone, it is tempting to relate the differences in SBH to the differences in the interface atomic structure and the associated intrinsic electronic states. Metal induced gap states ref (25—26) which penetrate several atomic layers into the semiconductor are an attractive candidate for SB formation. An alternative explanation ref (27) is that the difference in SBH between A and B $NiSi_2$ might be due to electronic properties characteristic of hexagonal Si at the type B interface. The observation of a SBH difference between $NiSi_2$ on (100) and type A $NiSi_2$ on Si (111) does not support this explanation. Deep level defects a few interatomic distances away from the interface should have identical energy levels regardless of the detailed interface structure and hence cannot account for the present results. So the most reasonable implication of the present work is that SBH mechanism for Si is intrinsic and structure sensitive.

CONCLUSIONS

Despite enormous investigative efforts in recent years on phenomena occuring at metal-semiconductor interfaces, the microscopic formation mechanism of the SB still remains an unresolved focus of attention. It is clear that to shed light on this problem, one needs well-characterized, controllable, homogeneous metal-semiconductor systems. With the discovery of the stabilization of ultra-thin silicide-Si epitaxy under UHV conditions, sharp and homogeneous metal-semiconductor interfaces are now available. Moreover, the interface structure has already been determined by high resolution TEM. Thus, these are ideal systems to better our understanding of the SB. We have performed the first measurements of barrier heights of such single crystal metal layers on silicon. Experimental results indicate the importance of crystallographic orientation and, hence, the interface structure. Interface steps and microfacets seem to be of negligible significance. This dependence does not favor metal interstitials, Si vacancies, or the presence of an interfacial layer as major determinants of the SB, but suggests instead intrinsic mechanisms for barrier formation.

ACKNOWLEDGEMENTS

The author is grateful to J. M. Gibson, K. K. Ng, J. M. Poate, D. C. Joy, J. C. Bean and D. R. Hamann for helpful discussions and various contributions to this project.

References

[1]   E. H. Rhoderick, *Metal-Semiconductor Contacts* (Clarendon, New York, 1978).

[2]   L. J. Brillson, Surf. Sci. Rep. *2*, 123 (1982).

[3]   R. H. Williams, Contemp. Phys. *23*, 329 (1982).

[4]   M. Schlüter, Thin Solid Films *93*, 3 (1982).

[5]   W. E. Spicer, I. Lindau, P. Skeath and C. Y. Su, J. Vac. Sci. Technol. *17*, 1019 (1980).

[6]   R. T. Tung, J. C. Bean, J. M. Gibson, J. M. Poate and D. C. Jacobson, Appl. Phys. Lett. *40*, 684 (1982); and Thin Solid Films *93*, 77 (1982).

[7]   R. T. Tung, J. M. Gibson, and J. M. Poate, Appl. Phys. Lett. *42*, 888 (1983).

[8]   J. M. Gibson, R. T. Tung and J. M. Poate, Mat. Res. Soc. Symp. Proc. *14*, 395 (1983).

[9]   R. T. Tung, J. M. Gibson and J. M. Poate, Phys. Rev. Lett. *50*, 429 (1983).

[10]   A. Ishizaka, K. Nakagawa and Y. Shiraki, 2nd International Symposium on Molecular Beam Epitaxy, Tokyo 1982, p. 183.

[11]   S. M. Sze, *Physics of Semiconductor Devices* (Wiley-Interscience, New York, 1981).

[12]   J. M. Andrews and M. P. Lepselter, Solid State Electron. *13*, 1011 (1970).

[13]   H. Norde, J. Appl. Phys. *50*, 5052 (1979).

[14]   R. T. Tung, Phys. Rev. Lett. *52*, 461 (1984) and J. Vac. Sci. Technol. B *2*, 465 (1984).

[15]   J. M. Woodall, G. D. Pettit, T. N. Jackson, C. Lanza, K. L. Kavanagh and J. W. Mayer, Phys. Rev. Lett. *51*, 1783 (1983).

[16]   Y. J. Chabal, D. R. Hamann, J. E. Rowe and M. Schlüter, Phys. Rev. B *25*, 7598 (1982).

[17]   P. E. Schmid, P. S. Ho, H. Föll and T. Y. Tan, Phys. Rev. B *28*, 4593 (1983).

[18]   T. R. Harrison, A. J. Johnson, P. K. Tien and A. H. Dayem, Appl. Phys. Lett. *41*, 734 (1982).

[19]   K. C. R. Chiu, J. M. Poate, J. E. Rowe, T. T. Sheng and A. G. Cullis, Appl. Phys. Lett. *38*, 988 (1981).

[20]   C. Pimentel, J. M. Gibson and R. T. Tung, to be published.

[21]   R. T. Tung, K. K. Ng and J. M. Gibson, to be published.

[22]   R. T. Tung, S. Nakahara and T. Boone, Appl. Phys. Lett. (1985).

[23]   J. M. Gibson, J. C. Bean, J. M. Poate and R. T. Tung, Appl. Phys. Lett. *41*, 818 (1982).

[24]   M. Iwami, K. Okuno, S. Kamei, T. Ito, and A. Hiraki, Electrochem. Soc. Symp. Proc. *80-2*, 102 (1980).

[25]   V. Heine, Phys. Rev. A *138*, 1689 (1965).

[26]   S. G. Louie and M. L. Cohen, Phys. Rev. Lett. *35*, 866 (1975).

[27]   J. Tersoff, Phys. Rev. Lett. *52*, 465 (1984).

SPECULAR SCATTERING IN ELECTRICAL TRANSPORT IN THE THIN FILM SYSTEM CoSi$_2$/Si

J. C. HENSEL, R. T. TUNG, J. M. POATE, AND F. C. UNTERWALD
AT&T Bell Laboratories, Murray Hill, New Jersey 07974

ABSTRACT

We have investigated electrical transport in thin films of CoSi$_2$ at low temperatures as a function of film thickness and observe in conductivity a size effect much smaller than seen heretofore indicative of a high degree of specularity in the boundary scattering. This in large part owes to the unique characteristics of these films, i.e., they are single crystal and continuous down to ~60Å thickness with long bulk scattering lengths ($\approx$1000Å) in transport at liquid He temperatures and have nearly atomically perfect interfaces.

INTRODUCTION

In electrical transport in very thin metal films "size effects" are expected to show up when the thickness becomes comparable to the bulk carrier scattering length and boundary scattering becomes appreciable. But this is only true if the boundary scattering is largely diffuse, i.e., random; if it were specular in nature, then a size effect would be nonexistent as the carriers would reflect coherently from the boundaries. Although it was almost a century ago that the question of size effects was first raised (by J. J. Thomson [1]), the subject has been slow in developing. Examples of size effects are fairly common, but documented cases where specularity is clearly dominant are rare, if they exist at all. The problem is materials related; for in essentially every case investigated heretofore, other kinds of scattering are present, most notably that due to grain boundaries, which tend to mask the sought-after effects [2]. In the present work we have performed transport experiments in a new system, epitaxial, single-crystal films of CoSi$_2$, the results of which show very small size effects, unambiguous evidence for a high degree of specularity.

EXPERIMENTAL

Recent advances [3] have produced single-crystal thin films of CoSi$_2$ and NiSi$_2$ of extraordinary quality and nearly atomically perfect interfaces eminently suitable for size-effect experiments. Previous transport studies [4] revealed that of the two materials CoSi$_2$ is preferable inasmuch as it exhibits an exceptionally long scattering length. Thin CoSi$_2$ samples (~60 to ~500Å thick) were prepared by UHV deposition of Co on atomically clean, n-type (111)Si wafers followed by an anneal. These films are single crystals having the type B orientation [3]. Thick, polycrystalline films ($\geq$1000Å thick) were prepared under less stringent conditions. Transport measurements were conducted as described earlier [4].

GALVANOMAGNETIC PROPERTIES OF CoSi$_2$

As a first step a determination needs to be made of a key quantity which sets the length scale of the problem, the low temperature bulk scattering length ℓ_e. Typically, the band structure of a metal is of such complexity that little information can be gleaned as regards to carrier densities, scattering lengths, etc. from conductivity and Hall measurements at low fields. However, it is possible to make some headway with the help of galvanomagnetic measurements at high fields. Such results obtained at 4.2K are shown in Figs. 1 through 3. The magnetoresistance in Fig. 1 exhibits an H^2 behavior at low magnetic fields with a negative H^4 contribution appearing at the higher fields. The Hall effect in Fig. 2 is essentially linear and characterized by a *positive* Hall coefficient, $R_H = 2.4 \times 10^{-24}$ gaussian units; however, at high fields there appears a small, positive H^3 contribution (best seen by sighting along the curve). This shows up more clearly in the trace in Fig. 3 recorded with increased sensitivity and with the linear part partially bucked out.

A number of conclusions can be drawn immediately:

(1)   The existence of a magnetoresistance implies that the number of bands contributing to conduction is no less than two; accordingly we shall henceforth assume the minimal, two-band model.

(2)   The positive sign of the cubic Hall term implies that the charges of the two carrier species are of opposite signs.

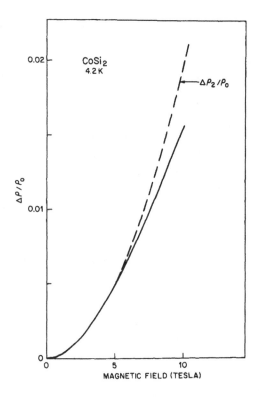

Fig. 1.   Experimental trace of the magnetoresistance $\Delta\rho/\rho_0$ of a "thick" (1100Å) $CoSi_2$ film. The dashed curve is the H^2 component.

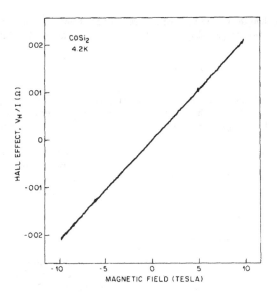

Fig. 2. Experimental trace of the Hall voltage/sampling current for a "thick" (1100Å) CoSi$_2$ film.

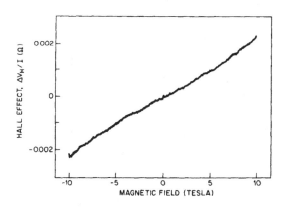

Fig. 3. Same as Fig. 2 with scale expanded and linear signal partially bucked out.

(3) The positive sign of the Hall coefficient signifies that *holes* are the majority carriers.

A detailed analysis based on a two-band model [5] of the data in Figs. 1 through 3 yields the carrier density ratio $x = n_e/n_h \approx 5\%$ with a hole density of $n_h \simeq 2\times10^{22}$ cm^{-3} and the mobility ratio $y = \mu_e/\mu_h \approx 2$.

The scattering length is most directly ascertained from the magnetoresistance, specifically the second-order part $\Delta\rho_2/\rho_0$ (cf. Fig. 1). Within the two-band model [5] we can write

$$\Delta\rho_2/\rho_0 = (\omega_{c1}\tau_1)^2 yf(x,y)$$

where $f(x,y) = x(sy-1)^2/(xy+1)^2$ is a slowly varying function of order unity magnitude and ω_{c1} and τ_1 are, respectively, the cyclotron frequency and scattering time of species 1 (holes). [In $f(x,y)$ $s = sgn(e_2/e_1)$ where e_2 and e_1 are carrier charges.] The aforementioned results give $yf(x,y) \approx 0.7$ whence it follows from the value $\Delta\rho_2/\rho_0 \approx 2.1\%$ (an average of 4 runs, cf. Fig. 1) that $\omega_{c1}\tau_1 = 0.17$ at 10T. Finally, noting the relation $\ell_e = k_F\ell_c^2 \cdot (\omega_c\tau)$ where k_F is the Fermi wavevector and ℓ_c is the magnetic length $(\ell_c^2 = \hbar c/eH)$ we obtain $\ell_e = 970$Å, a surprisingly large value! Note that this procedure does not depend on ones knowing the effective mass and only weakly on the density $[k_F = (3\pi^2 n)^{1/3}]$.

SIZE EFFECTS IN TRANSPORT

Figure 4 shows the behavior of the resistivity as a function of temperature when the thickness of the film is reduced to much less than the bulk scattering length, nominally ≈ 1000Å. A comparison of measurements on films of thickness 125Å and 197Å with that for a "thick" $(\approx 1100$Å) film yields the astonishing result that there is little effect,

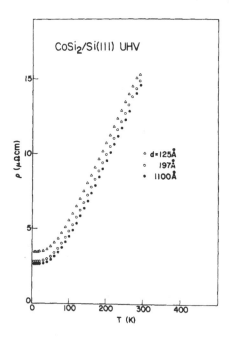

Fig. 4. Temperature dependence of resistivities of thick and thin CoSi$_2$ films.

even down to $d \approx (1/8)\ell_e$ where the maximum fractional change in ρ (at liquid He temperatures) is a mere 30%. The results for all samples examined are summarized in Fig. 5 wherein we plot the residual resistivity ρ_0 (4.2K data) relative to the value in an "infinite" specimen, $\rho_{0\infty} = 2.6 \, \mu\Omega\text{cm}$, vs the relative thickness d/ℓ_e. This plot serves to re-emphasize the fact that there is little change from bulk behavior down to $d/\ell_e \sim 0.1$; below this value there is an accelerated upturn possibly indicating some deterioration in the quality of the thinnest films. It should be emphasized that these are the *raw* data representing an upper limit in $\rho_0/\rho_{0\infty}$; corrections for any defects in the films would tend to depress the points towards the specular limit $\rho_0/\rho_{0\infty} = 1$. Even so to the best of our knowledge the data represent the smallest size effects for a given relative thickness seen to date and unequivocally imply an exceedingly high degree of specularity.

Included in Fig. 5 are curves calculated from the theory of Fuchs [6] for a range of values of the specularity parameter p (p = 0 represents purely diffuse scattering, p = 1 purely specular). Although there is not an exact fit, it is reasonably clear that p is of the order of 80 to 90%, probably closer to the latter. This includes the effect of the free surface which is decidedly less perfect than the silicide/Si interface. For the interface the specularity is plausibly even nearer 100%.

CONCLUSIONS

We have observed in transport in $CoSi_2$ thin films an unprecedently small size effect in the resistivity which we believe results from the bulk and interface perfection of epitaxial $CoSi_2$ grown on single crystal Si. The results would imply an overall specularity of close to 90% (and perhaps even higher for the silicide/Si interface) demonstrating that in longitudinal transport in the present system the carriers reflect from the boundaries with little loss of phase coherence.

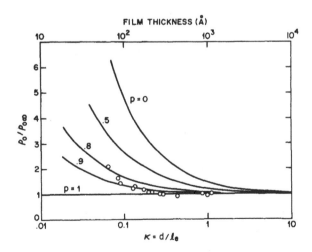

Fig. 5.   Size effect in thin films of $CoSi_2$: residual resistivities relative to the bulk value $\rho_{0\infty} = 2.6 \, \mu\Omega\text{cm}$ for 14 specimens vs relative thickness κ.

ACKNOWLEDGEMENT

We are grateful to J. M. Gibson for providing invaluable TEM analysis of the silicides and their interfaces.

REFERENCES

1. J. J. Thomson, Proc. Camb. Phil. Soc. *11*, 120 (1901).

2. See critical review by J. R. Sambles, Thin Solid Films *106*, 321 (1983).

3. R. T. Tung, J. M. Gibson, and J. M. Poate, Phys. Rev. Letters *50*, 429 (1983) and references therein.

4. J. C. Hensel, R. T. Tung, J. M. Poate, and F. C. Unterwald, Appl. Phys. Lett. *44*, 913 (1984).

5. The fact that the bands near the Fermi energy are free-electron-like (s and p character) affords some hope that such a simple analysis might work. See, e.g., J. Tersoff and D. R. Hamann, Phys. Rev. B*28*, 1168 (1983).

6. K. Fuchs, Proc. Camb. Phil. Soc. *34*, 100 (1938); also see E. H. Sondheimer, Adv. Phys. *1*, 1 (1952).

ION CHANNELING AND BLOCKING STUDY OF ULTRA-THIN NiSi₂ FILMS GROWN ON ATOMICALLY CLEAN Si(111) SURFACES

E.J. VAN LOENEN AND J.F. VAN DER VEEN
FOM-Institute for Atomic and Molecular Physics, Kruislaan 407, 1098 SJ Amsterdam, The Netherlands

ABSTRACT

Epitaxial NiSi₂ films of less than 40 Å thickness have been grown on atomically clean Si(111) surfaces. Ion channeling and blocking have been used to determine the morphology and orientation of these films, as well as the strain resulting from the lattice mismatch with the Si substrate. The films are found to be continuous for Ni coverages above 5×10^{15} atoms/cm². For all coverages the films are (111) oriented but rotated 180° around the surface normal with respect to the substrate. The NiSi₂-Si(111) interface has been probed directly by the ion beam and is found to be well ordered and atomically abrupt.

Silicides are very promising materials for future metallization schemes in device fabrication because of the good adhesion and low resistivity that can be obtained [1]. The use of epitaxial silicides may offer special advantages. Firstly, their resistivities will be lower than those of the corresponding polycrystalline silicides, and secondly one can study actual interface structures, which is essential for understanding Schottky Barrier formation on Si. Finally, provided films of high crystalline quality can be obtained, epitaxial Si or insulator films can be grown on top, thus allowing for buried interconnects and possibly three-dimensional device structures [2].

Of the known epitaxial silicides, NiSi₂ has the smallest lattice mismatch with Si and single-crystal films of high quality can be grown on Si(111) by using a 'template' method as described by Tung et al. [3]. The orientation of the NiSi₂ films can be controlled to be identical to that of the substrate (type-A orientation) or to be rotated 180° about the surface normal with respect to the substrate (type-B orientation). Here we demonstrate that ion channeling and blocking can be used very effectively to determine not only the crystalline order and orientation of such thin epitaxial films but also their morphology and lattice strain. Furthermore, the arrangement of atoms at the NiSi₂ - Si interface can be established.

The experiments were performed in a UHV chamber, coupled to a 200 keV ion accelerator. Backscattered ions were analysed with a toroidal electrostatic analyser [4], capable of measuring energy and direction of backscattered ions within a 20° range of backscattering angles simultaneously. Si(111) substrates were cleaned in UHV by mild sputtering and annealing. Ultra-thin NiSi₂ films were then grown on top by deposition of Ni at room temperature, followed by annealing at relatively low temperatures of 720 to 820 K. In this manner various films were prepared, with the deposited thickness of Ni ranging from 1.7 to 12.8 monolayers (one monolayer here is defined as the number of atoms in one Si(111) layer, which is equal to 7.84×10^{14} atoms/cm²). It was checked that the films had the correct stoichiometry NiSi₂.

The orientation of the films was determined as follows. A 175 keV He⁺ beam was aligned with the [001] direction of the substrate (fig.1a) and energy spectra were taken, which exhibited well-resolved backscattering peaks from the exposed Ni and Si atoms in the surface region. By use of a standard calibration method [5] the integrated peak areas were converted into the number of Ni and Si monolayers visible to beam and detector. The Ni peak area, expressed in monolayer units, is plotted in fig. 1b versus the exit angle β with respect to the surface plane. The curves (a) to (d) refer to films containing an increasing amount of Ni. Distinct minima (e.g. the ones labeled 1 to 4 in fig. 1) are observed, which for all films are found at the exit angles expected for

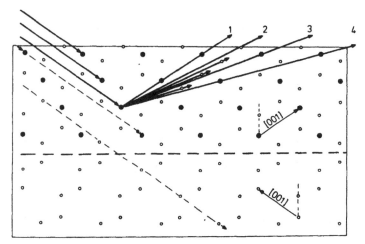

Si (111)+B-type NiSi₂
(1̄10)-plane

Fig.1a. Backscattering geometry for channeling along the [00$\bar{1}$] axis in the Si substrate, corresponding to the [$\bar{2}\bar{2}$1] direction in the NiSi$_2$ overlayer. Open circles denote Si atoms, filled circles denote Ni atoms.

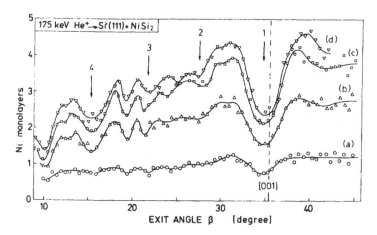

Fig.1b. Blocking profiles from NiSi$_2$ films prepared from (a) 1.7, (b) 3.3, (c) 6.3 - 10.4, and (d) 12.8 monolayer of deposited Ni. The scattering geometry is that of fig. 1a. The [001] blocking direction for bulk NiSi$_2$ is indicated.

purely B-type NiSi$_2$. The minima are due to blocking of backscattering ions by (near)-surface atoms in the silicide. Note that the presence of minimum 2 implies termination of the NiSi$_2$ crystal by an⁻Si plane.

The morphology of the films was investigated by comparing the measured number of visible Ni monolayers with the number expected for a uniform and continuous film. The chosen direction of the incident beam corresponds with the [$\bar{2}\bar{2}1$] crystal axis in the silicide. For this direction, and in the absence of blocking effects, only the first 3 atomic layers of Ni are fully visible to beam and detector (fig.1a), while atoms in deeper layers are shadowed. Monte Carlo computer simulations of shadowing along the thermally vibrating [$\bar{2}\bar{2}1$] atomic rows indeed predict 3.7 monolayers to be visible. A uniform and continuous film therefore should exhibit as many visible Ni layers as its Ni content up to ≈4 monolayer coverage, above which the visible number should level off at ≈3.7 monolayer. If, however, the film is clustered in islands, which contain at least 4 Ni layers, then on average it will expose a smaller number of Ni atoms to the beam than the actual number it contains. Turning now to the experimental data in fig.1b we see that the films containing 1.7 and 3.3 monolayer equivalents of Ni (curves a and b in fig.1b) exhibit only 1.2 and 2.7 visible layers at exit angles where blocking effects are absent (e.g. for 30° < β < 32°). Thus parts of the film are at least 4 Ni layers thick, which amounts to 4 Si - Ni - Si triple-layers of silicide. Analysis of backscattering peak widths shows that the films indeed consist of islands [6]. Thicker films, containing more than 6.4 monolayer equivalents of Ni (curve (c) in fig.1b), are continuous since they show a yield which levels off at the predicted 3.7 monolayer. This conclusion is confirmed by TEM and additional peak width analyses [6] (the slightly higher yield in curve (d) indicates some disorder).

The lattice constant of NiSi$_2$ is 0.46% smaller than that of Si. Tung et al. [3] have shown that ultra-thin epitaxial films contain no dislocations, which means that the overlayer must be stretched to match the Si substrate. As a result, the crystal axes should be tilted toward smaller angles with the surface plane. The corresponding blocking directions should therefore be shifted toward smaller angles β, the shift being largest for exit angles close to 45°. Such a shift is indeed observed, as is most clearly visible for the strong [001] minimum near β = 35°. The angle expected for the bulk [001] direction is indicated by a broken line (β = 35.26°). All [001] minima are shifted by 0.4°. A more accurate determination has been made by measuring the angle between two directions in the silicide. From such measurements and from the shifts observed in figure 1b it is concluded that, knowing the lateral distances to be expanded by 0.46%, the normal distances are contracted by 0.85 ± 0.2%. We note that these values for the expansions and contractions in the overlayer correspond to a unit cell volume which is almost equal to that of a non-strained film.

Knowing the film to be pseudomorphic and the orientation of the silicide to be B-type, we now investigate the crystallographic structure at the very interface. The ion beam was aligned with the [$\bar{1}\bar{1}1$] direction of the silicide and backscattering energy spectra were taken in a 20° range of exit angles β centered around the [00$\bar{1}$] axis of the silicide (fig.2). Sharp Ni and Si surface peaks are observed, which are due to backscattering from the unshadowed atoms in the first triple-layer of the silicide. The shift in energy with increasing exit angle is simply due to the elastic energy loss in the collision being larger for larger scattering angle, and the monotonic decrease of the peak intensities directly reflects the angular dependence of the Rutherford cross section. At lower energies the Si yield increases strongly due to dechanneling in the substrate. Between this substrate yield and the Si surface peak a small peak is observed in a limited angular range, centered around the [001] axis of the silicide. This contribution corresponds to backscattering from the second and fourth monolayer of the substrate: these two layers are well visible to the beam and are visible to the detector through the [001] channels only (atoms indicated by diamonds in inset). The [001] blocking direction itself is discernable in fig.2 as a weak depression in the backscattering yield behind the Ni peak. The following conclusions can be drawn from fig.2.

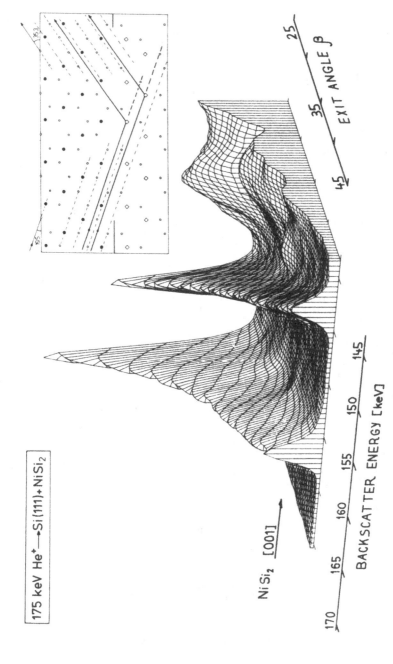

175 keV He⁺ ──► Si(111)+NiSi₂

NiSi₂ [001]

BACKSCATTER ENERGY [keV]

EXIT ANGLE β

Fig.2. Backscattering energy spectra taken in the scattering geometry of inset and shown for a 20° range of exit angles centered around the silicide [001] axis.

Firstly, at the energy (depth) corresponding to the interface no Ni peak is observed for any of the exit angles. Thus, contrary to what has been proposed [7,8], there is no intermediate layer between silicide and silicon with a structure different from NiSi$_2$. One monolayer of Ni atoms at disordered positions near the interface would result in a second Ni peak with almost the same area as the Ni surface peak. By integrating the area behind the surface peaks, we find that fewer than 7.5×10^{13} disordered Ni atoms/cm^2 are present near the interface. These results set a 20 times lower limit on the number of disordered interface atoms than an earlier ion channeling study by Chiu et al. [9].

Secondly, surface and interface signals are well separated even though this film is only 24 Å thick. This is a manifestation of the very high depth resolution that is obtained by use of an electrostatic analyser. The peak separation is even somewhat smaller than expected on the basis of random stopping powers, because of the anomalously small inelastic energy loss of the projectiles on their paths in and out through the centers of the silicide channels.

The intensity and angular profile of the Si interface signal is sensitively dependent on the amount of lattice relaxation at the interface. The data were therefore compared with Monte Carlo computer simulations of backscattering from differently relaxed interfaces. The best fit is obtained for a slight relaxation of the entire overlayer toward the substrate by 0.06 ± 0.08 Å. Similar analyses for other scattering geometries show furthermore, that the Ni atoms at the interface are 7-fold coordinated, as had in fact been assumed throughout this work. The alternative possibility of 5-fold coordinated Ni atoms at the interface can be ruled out. A detailed account of these results will be given elsewhere.

This work is part of the research program of the Stichting voor Fundamenteel Onderzoek der Materie (Foundation for Fundamental Research on Matter) and was made possible by financial support from the Nederlandse Organisatie voor Zuiver-Wetenschappelijk Onderzoek (Netherlands Organisation for the Advancement of Pure Research).

REFERENCES
[1] S.P. Murarka, J.Vac.Sci.Technol. 17, 775 (1980).
[2] S. Saito, H. Ishiwara and S. Furukawa, Jap.J.Appl.Phys. 20, 49 (1981).
[3] R.T. Tung, J.M. Gibson and J.M. Poate, Phys.Rev.Lett. 50, 429 (1983).
[4] R.G. Smeenk, R.M. Tromp, H.H. Kersten, A.J.H. Boerboom and F.W. Saris, Nucl.Instrum.and Methods 195, 581 (1982).
[5] E.J. van Loenen, M. Iwami, R.M. Tromp and J.F. van der Veen, Surface Sci. 137, 1 (1984), and references therein.
[6] E.J. van Loenen, A.E.M.J. Fischer, J.F. van der Veen and F. Legoues, to be published.
[7] M. Iwami, K. Okuno, S. Kamei, T. Ito and A. Hiraki, J.Electr.Soc. 127, 1542 (1980).
[8] J. Freeouf, Solid State Commun. 33, 1059 (1980).
[9] K.C.R. Chiu, J.M. Poate, L.C. Feldman and C.J. Doherty, Appl.Phys.Lett. 36, 544 (1980).

THE ROLE OF LATTICE MISMATCH IN GROWTH OF EPITAXIAL CUBIC SILICIDES ON SILICON

R. A. HAMM, J. M. VANDENBERG, J. M. GIBSON, AND R. T. TUNG
AT&T Bell Laboratories, Murray Hill, NJ 07974

ABSTRACT

The lattice mismatch of single-crystal epitaxial nickel disilicide of orientation A and B on Si(111) has been measured in and out of the <111> growth direction using a four-circle X-ray diffractometer. In both directions the mismatch of the as-grown B-type film was found to be larger than for the A-type; after annealing at 800°C the mismatch of the A-type approached the value of the B-type, while the latter remained the same. This result suggests that the B-type NiSi$_2$ film has relaxed more during growth than the A-type which can only be relieved by subsequent annealing. This difference in A- and B-type mismatch is correlated with TEM studies. The combined results provide evidence that the growth of either A or B orientation is strongly affected by the nature and density of the misfit dislocations.

INTRODUCTION

There is considerable interest in the growth of the epitaxial cubic silicides CoSi$_2$ and NiSi$_2$ on silicon, due to their high quality metallic properties [1,2]. These materials can now be grown under UHV conditions as continuous single-crystalline films. High-resolution transmission electron microscopy (TEM) as well as Rutherford backscattering and channeling revealed that single-crystalline NiSi$_2$ films grow on Si(111) with either type A or type B orientation using template layers [3], while CoSi$_2$ films grow with the B orientation. The A orientation films show all major axes aligned with the substrate whereas the B orientation films are rotated 180° about the (111) surface normal.

It has been suggested that the ease of growth of the B orientation in the higher mismatch silicide CoSi$_2$, paralleling observations of epitaxial alkaline-earth fluoride films [4], is due to the role of misfit dislocations in stress relief during growth [5,6]. In order to better understand the role of misfit dislocations and defects, X-ray diffraction was used to study the lattice mismatch of NiSi$_2$ films of both orientations A and B on Si(111) as well as mixed orientation films.

EXPERIMENTAL

Single crystal NiSi$_2$ films of ~1000Å thickness were grown under UHV conditions as described in Ref. 3; after deposition the samples were cooled at a rate of ~1°C/sec.

The lattice mismatch was measured along and out of the <111> growth direction. For this purpose a Huber 424/511 four-circle X-ray diffractometer was used which enables us to scan along various directions in reciprocal space. The diffraction system is operated in a non-dispersive mode and utilizes a pair of symmetrically cut Ge(111) crystals (SCC) to collimate and detect X-rays with the sample inserted between the two SCC's. The lattice mismatch was measured as a function of angle α with the growth direction ($\alpha = 0°$) up to $\alpha = 80°$, which approaches the lateral direction ($\alpha = 90°$) of the film. Only a small number of strong (hkℓ) reflections from NiSi₂ were suitable since for many reflections the X-ray intensities were too weak; also in some cases, in particular around $\alpha \sim 65°$, the separation from the Si reflections became too small to be resolved.

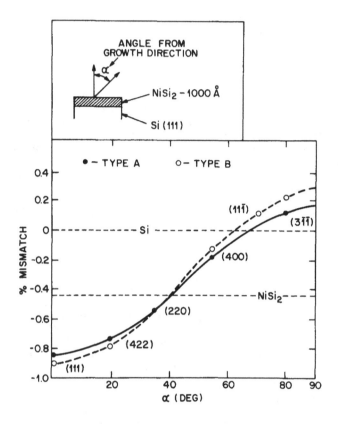

Figure 1. The lattice mismatch of as-grown single crystal NiSi₂ films (#1) on Si(111) as a function of angle α.

In Figure 1 the measured lattice mismatch is plotted as a function of α for single crystal films grown at a temperature $\leq 775°C$ (set #1) with orientation A and B. Another X-ray diffraction experiment carried out for a mixed A/B orientation film resulted in a similar plot, which indicates that the mismatch is not significantly affected by limited grain size (~ 1–$10,000$Å). Therefore our experiments and interpretation were pursued for films of predominantly A or B orientation. The A and B films, as measured in Figure 1, were subsequently subjected to a high-vacuum annealing treatment of 2 hrs at 800°C and then rapidly cooled (6°C/sec). The lattice mismatch was measured as a function of α for the same number of (hkℓ) reflections; only the mismatch for the growth direction and $\alpha = 80°$ is plotted together with those of the as-grown A and B films (Fig. 2).

In order to verify that the plots of lattice mismatch vs. angle α, as shown in Figure 1, are truly characteristic for the growth of single crystal A- and B-type films, another X-ray diffraction experiment was carried out for films grown at a temperature $\geq 650°C$ (set #2), approximately 100°C lower than the previous set #1. This second set was then subjected to a 15 min anneal at 800°C and measured again by X-ray diffraction.

RESULTS AND DISCUSSION

The plots of the as-grown films (#1) in Figure 1 show that both types of films expand in the lateral and contract in the growth direction. However, in the lateral direction the lattice mismatch of the B-type is distinctively larger than for the A-type, while in the growth direction this difference in mismatch is somewhat less pronounced.

For the calculation of the mismatch the cubic cell parameter $a_o = 5.431Å$ for Si is used [7]. By extrapolation in Figure 1 we find that the mismatch in the lateral direction ($\alpha = 90°$) is +0.17% for type A and +0.3% for type B with a standard deviation of ± 0.015, while in the growth direction we measured −0.85% for A and −0.91% for B with a standard deviation of ± 0.02. This result suggests that the B-type film has relaxed more at the growth temperature than the A-type, resulting in a larger lateral lattice mismatch and a higher more saturated dislocation density in the interface than for the A-type film. On cooling to room temperature additional contraction for both A- and B-type occurs in the growth direction due to differential thermal stresses; this effect may explain why the difference between the A and B mismatch is relatively small in the growth direction. For A- and B-type films (#2) grown at a temperature $\sim 100°C$ lower, essentially the same dependence of lattice mismatch as a function of α was observed (Fig. 2). The values of the mismatch for $\alpha = 0°$ and 80°, corresponding to the reflections (111) and (3$\bar{1}\bar{1}$) respectively, are listed in Table I together with the extrapolated values for $\alpha = 90°$. Considering the standard deviations, the values of the

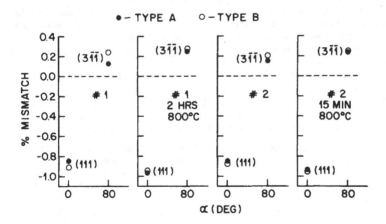

Figure 2. The lattice mismatch measured in ($\alpha = 0°$) and out ($\alpha = 80°$) of the growth direction for single crystal $NiSi_2$ films (#1 and #2) of orientation A and B.

mismatch are very close for the two sets of films. However it is interesting to note that for the second set (#2) the difference in lattice mismatch between A- and B-type is smaller than for the first set #1. These results suggest that the mismatch is 'locked in' not at the growth temperature but at a slightly lower one, presumably a temperature at which dislocations become immobile.

Table I. Values of lattice mismatch (%) of single crystal $NiSi_2$ films.

angle α		0°	80°	90°
type A	#1	−0.85	+0.13	+0.17
type B		−0.91	+0.24	+0.30
type A	2 hrs at	−0.97	+0.24	+0.32
type B	800°C (#1)	−0.94	+0.27	+0.34
type A	#2	−0.86	+0.15	+0.20
type B		−0.87	+0.2	+0.25
type A	15 min at	−0.94	+0.23	+0.30
type B	800°C (#2)	−0.95	+0.24	+0.30

The mismatch, as calculated from thermal expansion coefficients [8], is close to zero at the growth temperature (~400°C) of the templates and the $NiSi_2$ lattice expands in the lateral direction during further growth at higher temperature; it accommodates this strain by contraction in the growth direction. This results in a hexagonal distortion along <111>. For comparison the mismatch -0.44% of bulk $NiSi_2$ (a_0 = 5.407Å [9]) is included in Figure 1. It is interesting to note that the hexagonal volume of the as-grown single crystal $NiSi_2$ films (#1 and #2) is approximately the same for both orientations. Based on a pseudo-cubic cell volume, derived from a hexagonal unit cell, we calculate a_0 = 5.422Å for A and a_0 = 5.425Å for B, which approaches the cell parameter a_0 = 5.431Å for Si. This implies that the single crystal $NiSi_2$ films accommodate to the Si lattice by elastic strain. This furthermore proves that the only difference in stress between as-grown A- and B-type films is associated with the misfit dislocations.

The degree of lattice mismatch in the lateral direction is affected by the interfacial structure and the nature of the misfit dislocations, and is therefore of particular interest. The type and separation of misfit dislocations was measured in the second set (#2) of films (grown ~100°C lower) from plan-view diffraction contrast TEM. The type A film exhibits a hexagonal array of Burgers vector $\underline{b}$ = 1/2 <110> type edge dislocations of mean spacing 1660±250Å (Fig. 3a). The type B film exhibits a hexagonal array of $\underline{b}$ = 1/6 <112> type edge dislocations of mean spacing 880±150Å (Fig. 3b). From the X-ray diffraction data of set #2 (Table I) we calculate a dislocation spacing of 1900±150Å for A and 870±100Å for B. This is in good agreement with the spacings obtained from TEM, which has less accuracy in that it does not average over a large area.

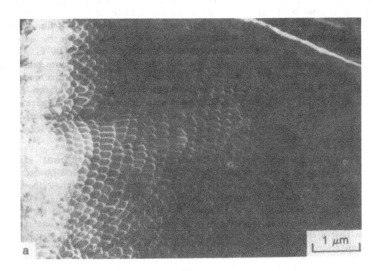

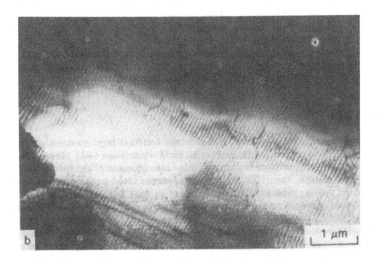

Figure 3. Dark field TEM images with $(2\bar{2}0)$ reflections of the misfit dislocations at the interfaces of as-grown (111) NiSi$_2$ films (#2) on Si (a) type A and (b) type B.

After a 2 hr annealing treatment at 800°C of the first set (#1) of as-grown films we find in comparing (Fig. 2) that while the mismatch of the B-type does show only a small increase, the mismatch of the A-type increases much more and becomes close to the values of B (Table 1). A similar result was observed for the A and B film of set #2, which was annealed at 800°C for a short time of 15 min (Fig. 2). On the assumption that the annealed films exhibit the equilibrium mismatch at or near 800°C, it is clear that for both sets of as-grown films (#1 and #2), the B-type film is more relaxed than the A-type; only by subsequent annealing it is possible to relieve the lattice mismatch of the A type. The fact that, after annealing, the mismatch of type A approaches that of B in the lateral direction, while the B mismatch remains almost the same, suggests that the value of the mismatch of B which is ~+0.32 on the average, represents the equilibrium pseudo-cubic lattice parameter $a_o = 5.448Å$ for single crystal $NiSi_2$ films at (or near) the growth temperature.

From published values of the lattice parameters [9] and thermal expansion coefficients [8] of $NiSi_2$ and Si, we calculate a lattice mismatch range of 0 to +0.57% for a temperature range of 400-800°C, where the zero mismatch corresponds to the template temperature [3]. The growth temperature of the two sets (#1 and #2) is in the range 650-775°C which corresponds to a relieved or equilibrium mismatch in the +0.41 to +0.54% range. We therefore conclude that the value +0.32% of the B lateral mismatch is close to that of the relieved mismatch and that the B-type is nearly relaxed at the growth temperature.

Our data provide evidence that the relaxation time of B-oriented $NiSi_2$ films at elevated temperatures is significantly shorter than for A-oriented films, as the latter need post-annealing to relieve mismatch. This can be explained by the nature of misfit dislocations at the interfaces [6]. Type A interfaces exhibit $\underline{b} = 1/2 <110>$ type dislocations whereas B interfaces contain only $\underline{b} = 1/6 <112>$ type. The latter dislocations have considerably lower energy ($E \propto b^2$) yet are prohibited at type A interfaces without a stacking fault. The lower energy of type B dislocations could explain the greater rapidity of mismatch relief. Indeed from more detailed studies it should be possible to measure activation energies for dislocation generation.

CONCLUDING REMARKS

The importance of this observation lies in the fact that large mismatch silicides and alkaline-earth fluorides grow exclusively with the B orientation [4-6]. Presumably the B orientation is able to grow more rapidly in these cases because of the ease of misfit relief. For larger mismatches than $NiSi_2$ this effect appears to lead to the complete dominance of B over A during growth. The assumption is that both A and B occur during nucleation and the dominance of B occurs during growth only for high mismatch. A recent result of Charles Tu et al. (private communication), that epitaxial mixed alkaline-earth fluorides with zero mismatch on GaAs grow with A orientation, is consistent with this theory and suggests that for this case at least, A has the lower interface energy. The template phenomena appear to occur at a temperature where the mismatch is close to zero and either A or B can be stabilized by as yet unfamiliar mechanisms.

Our data provide a natural explanation for the exceptionally high single crystal quality of moderate mismatch silicides such as $CoSi_2$ on Si, which have exciting physical properties. It also shows a new direction for improving crystal quality in doubly-positioned (111) thin film systems by increasing lattice mismatch, which may be considered an unusual, counter-intuitive result that is not untypical of nature.

REFERENCES

1. R. T. Tung, and J. C. Hensel, R. T. Tung, J. M. Poate and F. C. Unterwald, in this volume.

2. J. M. Gibson, R. T. Tung and J. M. Poate, Proc. Mat. Res. Soc. *14*, 395 (1984).

3. R. T. Tung, J. M. Gibson and J. M. Poate, Appl. Phys. Lett. *42*, 888 (1983).

4. J. M. Phillips and J. M. Gibson, Proc. Mat. Res. Soc. *25*, 381 (1984).

5. J. M. Gibson, J. C. Bean, J. M. Poate and R. T. Tung, Thin Solid Films *93*, 99 (1982).

6. J. M. Gibson, R. T. Tung, J. M. Phillips and J. M. Poate, Proc. Mat. Res. Soc. *25*, 405 (1984).

7. O. Yasumasa and Y. Tokumaru, J. Appl. Phys. *56*, 314 (1984).

8. S. S. Lau and N. W. Cheung, Thin Solid Films *71*, 117 (1980).

9. Crystal Data, vol: 4: Inorganic Compounds 1967-1969, JCPDS, Third Edition.

EPITAXIAL GROWTH OF REFRACTORY SILICIDES ON SILICON

L.J. CHEN, H.C. CHENG, W.T. LIN, L.J. CHOU AND M.S. FUNG
Department of Materials Science and Engineering and Materials Science
Center, National Tsing Hua University, Hsinchu, Taiwan, R.O.C.

ABSTRACT

Epitaxial refractory silicides were grown on silicon by solid phase epitaxy method. Transmission electron microscopy has been performed to study the microstructures of epitaxial layers and their orientation relationships with respect to substrate Si.

Metal thin films, electron-gun deposited, or sputtered metal-silicon films were annealed in N_2 ambient or in vacuum at 200°C-1100°C. Substrate heating, two step annealing and ion beam mixing were applied to induce the growth and improve the quality of epitaxial films. In this paper, formation and structures of epitaxial $CrSi_2$, VSi_2, $ZrSi_2$, $MoSi_2$ and WSi_2 are presented. Preliminary results of the epitaxial growth of $TiSi_2$, $TaSi_2$ and $NbSi_2$ are reported.

I. INTRODUCTION

Epitaxial silicides are known to possess low resistivity, low stress and high thermal stability. Regular atomic arrangements of the silicides/ silicon interfaces permit basic understanding of the electronic properties associated with silicides/Si interfaces. In addition, the growth of epi-Si/ epi-silicide/Si double heteroepitaxy structure facilitates the fabrication of novel classes of devices. Following the first report of epitaxial growth of PtSi on silicon in 1967, epitaxial Pd_2Si, $NiSi_2$ and $CoSi_2$ were found before the end of 1975. $FeSi_2$ was reported to grow epitaxially on Si in 1983 [1]. In the past one year or so, $CrSi_2$, VSi_2, $ZrSi_2$, $MoSi_2$, WSi_2, $TiSi_2$, $TaSi_2$ and $NbSi_2$ were grown epitaxially on silicon by solid phase epitaxy method [2-5]. All the recently discovered epitaxial silicides are refractory silicides. Crystal structures and atomic arrangements of selected planes of epitaxial refractory silicides are listed and shown in Table I and Fig. 1, respectively. Refractory silicides have been used in various aspects of VLSI devices because of their low resistivity and the ability to withstand high temperatures required for IC processing [6]. In this paper, epitaxial growth of refractory silicides are reported.

II. EXPERIMENTAL PROCEDURES

3-5 Ω-cm, n or p type (111), (001) or (011) Si wafers were first cleaned chemically with the usual procedures. The samples were etched in buffered HF solution ($HF:H_2O = 1:50$) for 2 minutes then dried with N_2 gun immediately before loading into an electron gun deposition system.

Metal thin films, 100-300Å in thickness, were electron gun deposited onto Si substrates maintained at room temperature, 300°C or 400°C. The depositions were performed in vacua better than 2×10^{-6} torr. An amorphous silicon layer, 200-300Å in thickness, was then deposited onto metal layer to protect the layer from oxidation during heat treatments. The deposition rate was about 1Å/sec. For W/Si system, tungsten-silicon films were also deposited using a DC magnetron sputtered system. A hot-pressed target of homogeneous composition with Si to W atom ratio 1.9 was used. The thickness of the sputtered layer was 700Å.

Heat treatments at temperatures 200-1100°C were performed in N_2 ambient with a three zone diffusion furnace or in an oil free vacuum furnace. For

Table I. Crystal structures of epitaxial refractory silicides.

	IV A		V A	VI A	
4	**Ti**[22] TiSi$_2$(C54) a=8.253Å {b=4.783Å c=8.540Å	TiSi$_2$(C49) a=3.62Å {b=13.76Å c=3.60Å	**V**[23] VSi$_2$(C40) {a=4.571Å c=6.372Å	**Cr**[24] CrSi$_2$(C40) {a=4.428Å c=6.363Å	
5	**Zr**[40] ZrSi$_2$(C49) a=3.721Å {b=14.68Å c=3.683Å		**Nb**[41] NbSi$_2$(C40) {a=4.7971Å c=6.592Å	**Mo**[42] MoSi$_2$(C40) {a=4.613Å c=6.424Å	MoSi$_2$(C11$_b$) a=3.203Å {c=7.855Å
6	**Hf**[72]		**Ta**[73] TaSi$_2$(C40) {a=4.7821Å c=6.5695Å	**W**[74] WSi$_2$(C40) {a=4.614Å c=6.414Å	WSi$_2$(C11$_b$) a=3.211Å {c=7.868Å

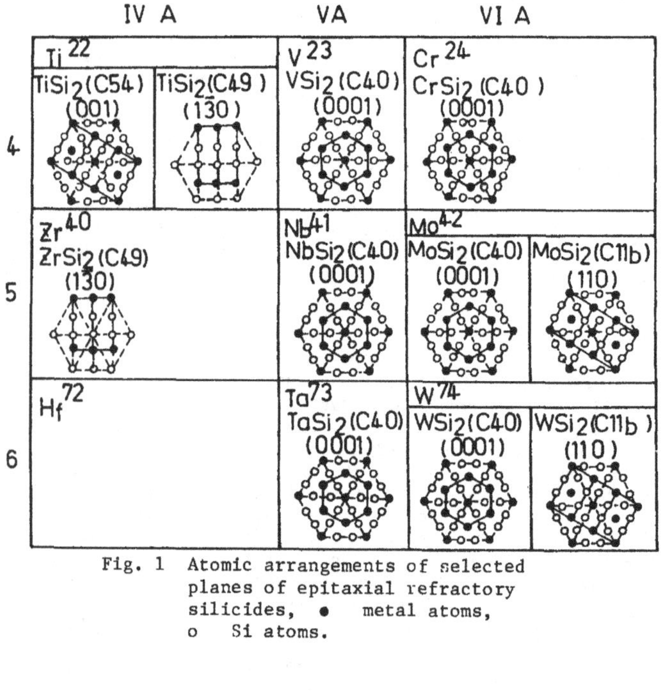

	IV A		V A	VI A	
4	**Ti**[22] TiSi$_2$(C54) (001)	TiSi$_2$(C49) (130)	**V**[23] VSi$_2$(C40) (0001)	**Cr**[24] CrSi$_2$(C40) (0001)	
5	**Zr**[40] ZrSi$_2$(C49) (130)		**Nb**[41] NbSi$_2$(C40) (0001)	**Mo**[42] MoSi$_2$(C40) (0001)	MoSi$_2$(C11$_b$) (110)
6	**Hf**[72]		**Ta**[73] TaSi$_2$(C40) (0001)	**W**[74] WSi$_2$(C40) (0001)	WSi$_2$(C11$_b$) (110)

Fig. 1 Atomic arrangements of selected planes of epitaxial refractory silicides, ● metal atoms, o Si atoms.

diffusion furnace annealing, high purity N_2 gas was first passed through a titanium getter tube maintained at 800°C to reduce the O_2 content. For vacuum annealing, the vacuum was maintained to be better than 3×10^{-7} torr during heat treatments. Two step annealings with first step temperature 300°C or 400°C followed by second step at 1000°C or 1100°C were also conducted. The annealing time at each temperature was 1 hr.

For ion beam mixing, As^+ ions, 70-170 KeV in energy, were implanted into the samples to a dose of $1 \times 10^{16}/cm^2$ at room temperature. The dose rate was kept to be less than 1 $\mu A/cm^2$ to minimize the heating effect during ion bombardment. Post-implantation annealings at 1000°C or 1100°C were also carried out in vacuum.

III. FORMATION AND STRUCTURE

1. CrSi$_2$

CrSi$_2$ was found to grow epitaxially on (111), (001) and (011)Si [4]. Substrate heating at 300°C or 400°C during Cr deposition was found to be more effective than two step annealing in promoting the epitaxial growth and improving the quality of epitaxial CrSi$_2$. The best epitaxy was obtained when sample substrates were heated at 300°C or 400°C followed by 1000°C-1100°C annealing for 1 hr. The orientation relationships were found to be $[0001]CrSi_2//[111]Si$, $[22\bar{4}0]CrSi_2//(22\bar{4})Si$, $(20\bar{2}0)CrSi_2//(20\bar{2})Si$, and $[1\bar{2}13]CrSi_2//[101]Si$. Dislocations present in the regular interfacial dislocation network were found to be of edge or 60° type with 1/6<112> Burgers vectors. The average dislocation spacings were measured to be 270-320 Å [2].

Epitaxial regions were observed in α-Si(200Å)/Cr(300Å)/Si samples implanted by 100-120 KeV As^+ to doses $1 \times 10^{15}-1 \times 10^{16}/cm^2$ followed by 1000°C annealing in vacuum. The effects of ion beam mixing in inducing the epitaxial growth and improving the quality of epitaxial layer were not particularly pronounced [5].

2. VSi$_2$

Epitaxial VSi$_2$ was grown on (111)Si. Best epitaxy was found in samples substrate heated at 400°C, followed by 400°C-1000°C two step annealing in vacuum. The orientation relationships were the same as those in CrSi$_2$/Si system. Interfacial dislocations were identified to be of edge type with 1/6<112> Burgers vectors. The average spacing was measured to be about 250Å [3].

Samples were As^+ implanted with different energies (130-160 KeV) and doses ($5 \times 10^{14}-1 \times 10^{16}/cm^2$) at room temperature. Post-implantation annealings were performed at 600-1000°C for 1 hr in vacuum furnace. No epitaxy was detected in all cases [5].

3. ZrSi$_2$

Ion beam mixing has been demonstrated to play a critical role in growing refractory silicide epitaxially on silicon. Epitaxial ZrSi$_2$ has been successfully grown on (111) and (001)Si.

Best epitaxy was obtained in samples irradiated by As^+ with ion range close to the original Zr/Si interface to a dose of $1 \times 10^{16}/cm^2$ followed by 1100°C annealing. The orientation relationships between epitaxial ZrSi$_2$ and Si were found to be $[010]ZrSi_2//[001]Si$, $(00\bar{2})ZrSi_2//(2\bar{2}0)Si$ (with about 1° misorientation), $[3\bar{1}0]ZrSi_2//[112]Si$ and $(130)ZrSi_2//(11\bar{1})Si$. Hexagonal and square networks of interfacial dislocations were observed. They were

identified to be of edge type with 1/6<112> or 1/2<110> Burgers vectors.
The average spacings were measured to be 140Å and 800Å, correspondingly [4].
Examples are shown in Figs. 2 and 3.

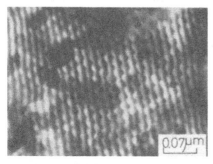

Fig. 2 Interfacial dislocations
of ZrSi$_2$/(001)Si, 110 KeV
As$^+$, 1x10^{16}/cm^2, 1100°C.

Fig. 3 Diffraction pattern
(DP) of overlapping
[010]ZrSi$_2$ and [001]Si.

4. MoSi$_2$

Both hexagonal and tetragonal MoSi$_2$ (h–MoSi$_2$ and t–MoSi$_2$) were found to
grow epitaxially on (111)Si after 1000-1100°C annealings. The orientation
relationships between epitaxial h–MoSi$_2$ and Si were identified to be
[0001]MoSi$_2$//[111]Si, (22$\bar{4}$0)MoSi$_2$//(22$\bar{4}$)Si, (20$\bar{2}$0)MoSi$_2$//(20$\bar{2}$)Si and
(1$\bar{2}$13)MoSi$_2$//(101)Si. For epitaxial t–MoSi$_2$ and Si, the orientation
relationships were found to be [110]MoSi$_2$//[111]Si, (004)MoSi$_2$//($\bar{2}$02)Si and
($\bar{2}$20)MoSi$_2$//($\bar{2}$4$\bar{2}$)Si. Regular interfacial dislocations were identified to be
of edge type with 1/6<112> Burgers vectors by diffraction contrast analysis.
The average spacing was measured to be about 100Å [5]. Figs. 4(a) and (b)
show the diffraction patterns corresponding to overlapping [110]t–MoSi$_2$ and
[0001]h–MoSi$_2$ with [111]Si, respectively. Fig. 5 is an example of
interfacial dislocation structure.

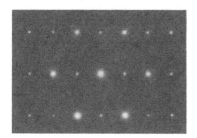

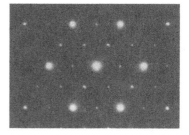

Fig. 4(a) DP of overlapping
[110]t–MoSi$_2$ and [111]Si.

Fig. 4(b) DP of overlapping
[0001]h–MoSi$_2$ and [111]Si.

5. WSi$_2$

Both hexagonal and tetragonal phases of WSi$_2$ were grown epitaxially in
(111)Si samples annealed at 1000-1100°C. The orientation relationships with
Si were found to be the same as those between MoSi$_2$ and Si. The interfacial

dislocation network appears to be extraordinarily regular. The average
spacing is about 40Å [5]. An example is shown in Fig. 6.

Fig. 5 Interfacial
dislocations of MoSi$_2$/
(111)Si, 1050°C.

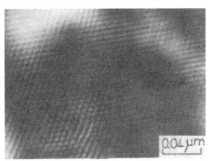

Fig. 6 Interfacial
dislocations of WSi$_2$/(111)Si,
1050°C, sputtered film.

6. TiSi$_2$, TaSi$_2$ and NbSi$_2$

Regular interfacial dislocation networks were observed for the thermally
grown TiSi$_2$, TaSi$_2$ or NbSi$_2$ on (111)Si. The orientation relationships are
yet to be determined [5]. Examples are shown in Figs. 7 and 8.

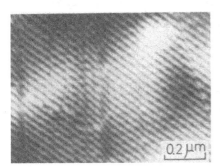

Fig. 7 Interfacial
dislocations in TiSi$_2$/
(111)Si, 1100°C.

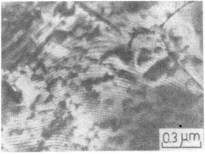

Fig. 8 Interfacial
dislocations in TaSi$_2$/
(111)Si, 300°C–1100°C.

Two step annealings were found to be effective in inducing epitaxial
growth of CoSi$_2$, FeSi$_2$, CrSi$_2$ and VSi$_2$ but not for ZrSi$_2$. The
ineffectiveness of two step annealing in Zr/Si system may be attributed to
the relatively small solubility of Si in Zr (∿1%) compared to more than 10%
of solubility of silicon atoms in metals like Co, Fe, V and about 5% in Cr.
The effect of intermixing of Zr with Si is minimal during pre-annealing.

Ion beam mixing has been demonstrated to play a critical role in
growing ZrSi$_2$ epitaxially on silicon. The breaking of metal–oxygen bonds,
dissolution of native silicon dioxide as well as dispersion of impurities
present at the interface by ion beam are thought to be beneficial in
promoting the epitaxial growth of ZrSi$_2$ on Si. The cause for the relative
ineffectiveness and even retardation by ion beam mixing in inducing

epitaxial growth in Cr/Si and V/Si systems, respectively, is not clear at
this time. The presence of appreciable amount of As atoms near the
interfaces may play an important role in the growth process.

IV. SUMMARY AND CONCLUSIONS

Epitaxial growth of refractory silicides on silicon are reported. The
formation and structures of epitaxial $CrSi_2$, VSi_2, $ZrSi_2$, $MoSi_2$ and WSi_2
are described. Preliminary results of the epitaxial growth of $TiSi_2$, $TaSi_2$
and $NbSi_2$ are presented. With the growth of eight epitaxial refractory
silicides, the roster of epitaxial silicides has now consisted of thirteen
members. Potentially, a number of other silicides may also grow
epitaxially on silicon based on the structural analysis. The properties of
silicides varied considerably from one another, e.g. sheet resistivities
ranged from 15 $\mu\Omega$-cm for $TiSi_2$ to 600 $\mu\Omega$-cm for $CrSi_2$. The swelling of the
rank of epitaxial silicides presents the exciting possibilities that novel
devices, particularly those essential for three dimensional integrated
circuits, with desirable characteristics may be realized.

REFERENCES

1. H.C. Cheng, T.R. Your and L.J. Chen, J. Appl. Phys. (In Press).
2. F.Y. Shiau, H.C. Cheng and L.J. Chen, Appl. Phys. Lett. 45, 524 (1984).
3. C.J. Chien, H.C. Cheng, C.W. Nieh and L.J. Chen, J. Appl. Phys.
   (In Press).
4. H.C. Cheng and L.J. Chen, Appl. Phys. Lett. (In Press).
5. H.C. Cheng, W.T. Lin, C.W. Nieh, F.Y. Shiau, C.J. Chien, T.R. Yew,
   L.J. Chou, M.S. Fung and L.J. Chen, unpublished work.
6. T.P. Chow and A.J. Steckl, IEEE Trans. ED-30, 1480 (1983).

COMPOUND SEMICONDUCTOR SURFACES AND INTERFACES: WHENCE FERMI LEVEL PINNING?

JERRY M. WOODALL
IBM Thomas J. Watson Research Center, P.O. Box 218 Yorktown Heights, NY, 10598

ABSTRACT

Except for a few special cases, the electrical and optical properties of most III-V semiconductor surfaces and interfaces can be explained in terms of surface Fermi level pinning. However, despite years of research there is no universal agreement on the origin of this pinning. This problem is of more than just academic interest since pinning affects optoelectronic device performance via surface recombination losses, and high speed device performance through uncontrollable ohmic and Schottky contact properties for MESFETs and through a high interface trap density for MISFETs.

This talk reviews some of the approaches to the pinning problem and presents some recent results on the role of misfit dislocations in pinning. In particular it will be shown that there are are several models which can explain the usually observed pinning positions. However, we have developed a modified work function model capable of explaining both the usual pinning positions and the experimentally observed exceptions in Schottky barrier height for some metal/semiconductor interfaces. The electrical properties of lattice mismatched GaInAs/GaAs heterojunctions suggest that Fermi level pinning occurs at misfit dislocations.

INTRODUCTION

There has been much recent progress on both the theoretical understanding of the electronic structure of idealized III-V semiconductor surfaces and interfaces and the UHV characterization of real surfaces and interfaces. However, except for a few cases, this knowledge is insufficient to adequately explain the optical and electrical behavior of surfaces and interfaces used in practical devices. For example information obtained from monolayer and submonolayer depositions in UHV conditions on cleaved (110) surfaces, although useful, does not lead to a fundamental understanding of such device properties as Schottky barrier height, ohmic contact resistance and surface recombination velocities. One reason for this is that the interpretation of UHV results is complicated by the difficulty of differentiating between metallurgical, e.g. interface chemistry, and structural, e.g. surface defects, effects which occur simultaneously during deposition. Another reason is that practical devices are made on (100) surfaces whereas most of the UHV work has been on (110) surfaces, and it is well known that the "clean" versions of these surfaces are both structurally and electronically different. Finally there are problems in relating the experimental results of measurements such as I-V, C-V, XPS, and photoemission to concepts such as barrier height, electron affinity, work function, surface states, and surface defect energy levels.

PREVIOUS WORK

Let us compare some interface models with experimental data, especially with respect to the variation of surface and interface properties with surface treatment. We know for example that GaAs (110) surfaces formed by cleaving in UHV are free of surface states which lie in the direct gap[1]. Thus the Fermi level at the surface is the same as that for the bulk. Theory and experiment are reconciled on this issue. To date it appears that all other gas or vacuum/GaAs surfaces have a surface Fermi level which is pinned roughly at mid-gap. In fact, the results of UHV experiments on (110) surfaces with sub-monolayer coverage of a variety of ad-atoms suggest that there may be two levels, a deep acceptor and a deep donor, which are associated with surface defects which are responsible for the pinning[2]. It has been known for some time that Schottky barriers to GaAs cannot be directly explained by the Schottky work function model, i.e. the barrier height is not proportional to the work function of the applied metallurgy. This coupled with the UHV pinning results has led to the hypothesis that the fundamental mechanism of Schottky barrier formation for metal/GaAs interfaces used in solid state devices is Fermi level pinning due to surface defects[3]. At first glance this model appears compelling since the pinning positions are within 0.1 eV of the barrier heights commonly quoted in textbooks. However, there is a problem with reconciling the predictions of a two energy level, i.e. two pinning position, surface defect model with experimental observations that in some cases, the sum of the p-type and n-type barrier heights equals the band gap energy. A recent attempt[4] to resolve this problem shows that in order to reconcile these observations with the defect models a surface defect density of at least 0.1 monolayers is required. At this density the barrier height is only a weak function of the metal work function. Therefore, the observed mid-gap barrier heights can be explained by juggling the deep level defect density and the metal work function.

Thus, it is beginning to appear that barrier heights at metal/GaAs interfaces can be explained in terms of a coupled interaction between the electronic properties of a dominant surface effect, e.g. vacancies, antisite defects, clusters, MIGS, surface reconstruction, etc. and the electronic properties of the applied metallurgy, e.g. the work function. The task is to identify the dominant surface effect(s) and to relate their occurance to surface preparation and metallization method. The nature of the dominant surface effect is by no means resolved. There are many observations which tend to rule out deep level surface vacancies or antisite type defects as the dominant surface effect[5]. For example the Pd/GaAs interfaces for which the Fermi level is not at pinned mid-gap[5]. Futhermore, for the two level defect models, two distinct regions of almost constant barrier height (each corresponding to one pinning defect) should be observed as a function of the metal work function[4]. This may have been seen for InP[6], but not for GaAs as yet. There are, in fact, observed barrier heights, e.g. for Al/AlAs formed by MBE[7], which are adequately explained by the simple work function model, i.e. the Al work function and AlAs electron affinity, and with no defect levels required. Also, there is the following paradox: 1) it takes 0.1 monolayers of deep level defects to pin the Fermi level at mid-gap and produce the observed barrier heights; 2) these defects are produced by metallization from thermal beams; 3) the Schottky barriers, and hence the defect density, are stable to an 800 C anneal; 4) this surface defect density corresponds to a volume density of 10^{21} cm^{-3}; 5) layers grown by the MBE method using thermal beams have volume defect densities of only 10^{13} cm^{-3} or less. How is this defect density reduced by eight orders of magnitude for MBE grown layers and not for annealed metal/GaAs interfaces? Thus, a theoretical framework which requires the surface effects to be deep, multileveled and of opposite conductivity type seems too restrictive in view of the wide variety of experimental results. Also, metal induced gap state models are not sufficiently developed to explain the wide range of observed barrier height for a given material.

EFFECTIVE WORK FUNCTION MODEL

An alternative framework is the effective work function model,[5,8] in which the Fermi energy position at the surface (or interface) is not due to or fixed by surface defects or surface states but rather it is related to the work functions of microclusters of the one or more interface phases resulting from contamination, e.g. exposure to air, or to reactions which occur during metallization. The theory requires that "pinned" surfaces already contain microclusters of interface phases. According to the model when a metal is deposited, for example to an air exposed surface, there is a region at the interface which contains a matrix of native oxide embedded with microclusters of different phases, each having its own work function. Thus, the normal work function model is rewritten as

$$\phi_{bn} = \phi_{eff} - X_{s.c.} ,$$

where ϕ_{eff} is an appropriately weighted average of the work functions of the different interface phases. Thus, the measured ϕ_{bn} can depend somewhat on the measurement technique, i.e. C-V or I-V. In other words, the interface phases comprise the Schottky barrier contact. The rest of the bulk metallurgy has little or no effect on the barrier height, except when the interface phases are predominantly the same as the applied metallurgy, e.g. Al/AlAs interfaces formed by MBE.

For most of the III-V compounds including GaAs, conventional metallization, i.e. non-UHV conditions, results in a condition in which ϕ_{eff} is due mainly to ϕ_v, the work function of the group V component, and occurs as a result of either one or both of the following reactions:

$$VO + IIIV \rightarrow V + IIIO$$

$$M + IIIV \rightarrow (V, MV_x) + (M,III)$$

where VO + IIIO are generic group V and III oxides and M is a metal. Using this framework we are able to explain the usual mid-gap pinning related Schottky barrier heights, the anomalous barriers, e.g. the

Al/AlAs interface and the Pd/GaAs interface, where the sum of the n and p type barrier heights is much less than the band gap, and the properties of MIS structures[5,8].

However, determining the proper framework is not enough. In order to impact metal/semiconductor device technology we must attend to details. A chip with an array of ten thousand gates requires a threshold voltage variation of less than 20 mV. This translates into a barrier height variation of less than 20 mV. We are not aware of any model/theory, unified or specific, capable of this resolution. A specific model with this resolution which relates I-V determined barrier heights to surface treatment and deposition condition for a single useful metallurgy to GaAs would be a major contribution and greatly welcomed by the device processing community. However, it is quite possible that this problem is too complicated to be addressed by our current knowledge and experimental capabilities. This is particularly true in view of the that fact most of our current knowledge of GaAs surfaces is based on UHV studies of (110) surfaces, whereas devices are predominantly made on (100) surfaces which have not been observed in an unpinned condition. The (110) surface is non-polar (containing equal numbers of As and Ga atoms in each plane), and is known to reconstruct by relaxation mechanisms which do not alter the surface symmetry or unit cell; the (100) surface is highly polar, possibly containing only As or Ga atoms in a single plane, and is known to exhibit many reconstructions involving large surface unit cells. Epitaxial growth is known to be easier on (100) surfaces, especially by MBE; attempts at growth on (110) surfaces typically leads to morphological and (for alloys) clustering phenomena. Such distinctions in surface structure and growth phenomena could well correlate with metallurgical interactions, which in turn would impact several of the Fermi level pinning models currently under discussion.

HETEROJUNCTIONS

With regard to semiconductor heterojunctions it has been thought for some time that from a practical point of view the common anion and electron affinity rules are adequate to explain trends in the optical and electronic behavior of properly formed lattice matched or pseudomorphic isoelectronic heterojunctions. However, recent work suggests that this approach may be inadequate for determining band alignment in the GaAs/AlAs system[9]. This has lead Tersoff[10] to suggest that the proper and unique band alignment should produce a zero interface dipole. This condition is associated with a characteristic effective midgap energy which in turn is related to the pinning position at metal/semiconductor interfaces predicted by metal induced gap state theory[11]. The electrical properties of nearly all other heterojunction interfaces used in electronic devices including those containing misfit dislocations[12] can be described in terms of a characteristic band offset and Fermi level pinning in combination with doping effects.

The role of misfit dislocations on the electrical properties of GaInAs/GaAs heterojunctions has been recently studied[12]. Since this system is not latticed matched, interfaces with misfit dislocations can be formed by the epitaxial growth of a layer whose thickness

excedes a critical thickness which is dependent on the lattice mismatch. The density of misfit dislocations is proportional to the
difference in lattice constant. A layer whose thickness is less that
the critical thickness for misfit dislocation formation will be
strained (known as pseudomorphic) and will assume the lattice constant of the substrate in directions parallel to the interface.
Thus, by a suitable choice of composition and layer thickness, it
possible to compare the effects due to band offset with those due
to band offset plus misfit dislocations. Using n+GaInAs/n-GaAs
structures it was found that a given GaInAs pseudomorphic interface
which produced ohmic electrical behavior showed rectifying behavior
when misfit dislocations were present. It was also found that n-type
pseudomorphic interfaces can be made to exhibit rectifying behavior
by proper selection of composition and doping. These results can
be explained in terms of Fermi level pinning at misfit dislocations
and a band offset which is predominately in the conduction band.

However, by device fabrication standards, adequate theoretical
tools have not yet been developed for heterojunction interfaces.
Thus, there is a need for a microscopic interface theory for device
structures with parameters which can be both measured by existing
experimental procedures and related to important device parameters.

REFERENCES

1. J. Van Laar, A. Huijser and T.L. Van Rooy, J. Vac. Sci. Technol.
   114, 894 (1977).

2. W.E. Spicer, I. Lindau, P. Skeath and C.Y. Su, J. Vac. Sci.
   Technol. 17, 1019 (1980).

3. W.E. Spicer, I. Lindau, P. Skeath and C.Y. Su, Applications of
   Surface Science, 9, 83 (1981).

4. A. Zur, T.C. Mcgill, and D.I. Smith, Surface Science, 132, 456
   (1983). A. Zur, T. C. McGill, and D. L. Smith, Physical Review B
   28, 2060 (1983).

5. J.M. Woodall and J.L. Freeouf, J. Vac. Sci. Technol., 19(3), 794
   (1981).

6. R.H. Williams, J. Vac. Sci. Technol., 18, 929 (1981).

7. K. Okamoto, C.E.C. Wood, and L.F. Eastman, Appl. Phys. Lett., 38,
   636 (1981).

8. J.L. Freeouf and J.M. Woodall, Appl. Phys. Lett., 39, 727 (1981).

9. R.C. Miller, A.C. Gossard, D.A. Klainman, and O. Munteanu, Phys.
   Rev. B 29, 3740 (1984).

10. J. Tersoff, Phys. Rev. B 30, 4874 (1984).

11. J. Tersoff, Phys. Rev. Lett. 52, 465 (1984).

12. J.M. Woodall, G.D. Pettit, T.N. Jackson, C. Lanza, K.L. Kavanagh,
    and J.W. Mayer, Phys. Rev. Lett., 51, 1783 (1981).

PINNING ACTION OF A THIN Ga INTERFACIAL LAYER IN AN Sb/Ga/GaAs SCHOTTKY BARRIER STRUCTURE GROWN BY MBE

X.-J. ZHANG,[*] H. CHENG[**] and A. G. MILNES[**]
*Fudan University, Shanghai, China
**Carnegie-Mellon University, Pittsburgh, PA 15213

ABSTRACT

The pinning action on the barrier height of a Ga interfacial layer in an Sb/Ga/GaAs structure prepared by molecular beam epitaxy has been tracked as a function of Ga layer thicknesses for both n and p-GaAs (100). The barrier height ϕ_{Bn} for n-type GaAs ($N_D = 3 \times 10^{15} cm^{-3}$), determined by thermal activation measurements and capacitance measurements, changes from 0.8 to 1.0 eV as the Ga layer is increased from zero to three monolayers. The barrier height ϕ_{Bp} for p-type GaAs decreases in a converse fashion so that the sum $\phi_{Bn} + \phi_{Bp}$ is equal to the GaAs energy gap. Up to a Ga layer thickness of about one monolayer the barrier height changes are about linearly proportional to thickness.

The action seen is compatible with barrier height changes that are seen when Sb/GaAs junctions are prepared with the original surface alternatively As or Ga rich and subjected to thermal annealing at temperatures up to 250°C.

INTRODUCTION

Fermi level pinning in Schottky barriers observed on III-V compound semiconductors is recognized to be greatly influenced by interfacial layers and the generation of defects that create surface states [1,2,3]. If these surface states are acceptor-like they may dominate the barrier height by controlling the system Fermi level if the metal Fermi level tends to lie above the acceptor energy levels, so that electrons are taken from the metal. Donor-like surface states conversely may be dominant if the metal must receive electrons to allow Fermi level equilibration. Similarly if a p-type semiconductor is involved the surface states that are dominant in determining barrier height may be either donor or acceptor states depending on the need for the metal to receive or donate electrons for equilibration.

The studies of Spicer and coworkers [1] in which submonolayers of metal were deposited on (110) surfaces showed barrier heights tending to become about 0.7 eV below the conduction band edge almost irrespective of the metal on n-type material and to become 0.5 eV above the valence band edge for p-type GaAs. Hence the observed barriers ϕ_{Bn} and ϕ_{Bp} for n and p-conductivity sum to less than the bandgap of the GaAs, with the interface donor levels states lying about 0.25 eV below the acceptor levels. Spicer, et al., speculate that these acceptor and donor surface states could be two charged states of the same lattice defect due to a missing As atom [2].

Zur, et al. [4] have shown that the behavior with thick metal layers will result in $\phi_{Bn} + \phi_{Bp}$ summing very closely to the bandgap, even if there is no pinning. A defect density of the order of $10^{14} cm^{-2}$ is necessary to pin the system Fermi level and for this condition they show that $\phi_{Bn} + \phi_{Bp}$ is even more closely equal to the bandgap. In their model the defects are considered to be in a region about 5 Å deep into the semiconductor but the exact distribution is shown not to be significant. For submonolayer coverage they conclude that $\phi_{Bn} + \phi_{Bp}$ can be less than the bandgap and that a defect density of only $10^{12} cm^{-2}$ is sufficient to pin the Fermi level.

388

The difference between these two conditions depends on the origin of the main source of charge that balances the charge in the interface layer and the depletion charge. For thick metal coverage the balancing is accomplished mainly by charge in the metal setting up a dipole layer. With submonolayer metal coverage and 10^{11} to $10^{12} cm^{-2}$ defects the charge can come almost entirely from the semiconductor bulk and lock-on of the Fermi level to the acceptor or donor states can occur for n and p-type material.

Effects associated with surface orientations and submonolayer coverages have been observed by Chiaradia, et al. [5]. Other groups have examined the effects of providing an ultra-thin metal interfacial layer between a thick Schottky barrier layer metal and the semiconductor [3,6,7]. Massies, et al. [6] show the barrier heights of Al/n-GaAs junctions become greatly lower if the surface of the GaAs is saturated with H_2S. In other work Wang [8] has shown that the barrier height of Al/n-GaAs shows a steady decrease with increasing submonolayer As coverage at the interface. The possible role of As at the interface of a GaAs Schottky barrier has been discussed by Freeouf and Woodall [9]. It is apparent that more studies of the role of interface layers are needed to further develop understanding and control of Schottky junctions.

In recent studies of Schottky barrier heights on GaAs [10] we have used Sb as the thick metal coverage since this is a metal that can be readily employed without significant cross-doping problems in a molecular beam epitaxy (MBE) system dedicated to III-V studies. An epi-layer of GaAs can be grown to establish a new clean (100) surface and then the Sb deposited in ultra-high vacuum conditions. The surface reconstruction can be varied by controlling the ratio of the As_4 to Ga just prior to the Sb deposition. We have found that for a Ga-rich surface prior to deposition the barrier height ϕ_{Bn} is 0.8 eV whereas for an As-rich surface it is 0.68 eV or less. Thermal annealing of 30 minutes at 250°C causes the barrier for the originally As-rich surface to increase, Fig. 1, to the value for the Ga-rich surface.

In view of these results, we discuss in this paper further studies of the role of interfacial layers of Ga between the Sb and the GaAs. The effect of adding the Ga layer is to increase the ϕ_{Bn} barrier height from 0.8 to 1.0 eV as the layer thickness is increased from zero to one or two monolayers. The sum $\phi_{Bn} + \phi_{Bp}$ is closely equal to the GaAs bandgap over the whole range of Ga thickness from zero to many monolayers and the barriers obtained are substantially stable with temperature up to at least 150°C.

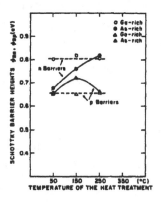

Fig. 1 Influence of thermal annealing on the 300 K Schottky barrier heights for n and p-type GaAs surfaces without (o,Δ) and with ($\bullet$,$\blacktriangle$) the As_4 treatment prior to the Sb deposition.

SPECIMEN PREPARATION AND MEASUREMENTS

The Sb/GaAs Schottky junction specimens with ultra-thin interlayers of Ga were grown in a Perkin-Elmer MBE 400 system on (100) substrates. The n-type doping level was $3x10^{15}cm^{-3}$ and direct preparation of the barrier without growth of a significant n-GaAs buffer layer was usual. When a buffer layer was provided in situ the results were unchanged. For the p-type specimens only heavily doped substrates (Zn, $10^{18}cm^{-3}$) were available and so buffer layers (3.5 μm thick, $5x10^{16}cm^{-3}$ Be doped) were grown immediately prior to the junction formation. Before loading in the system the substrates were degreased in organic solvents and etch-polished in 6:1:1 $H_2SO_4/H_2O_2/H_2O$ solution for two minutes and then In mounted on the Mo specimen block. Oxide desorption was accomplished by heating to 610°C for 10 minutes in a vacuum of about $2x10^{-10}$ Torr. A Ga-rich surface is formed by this desorption procedure. The specimen block was then allowed to cool to about 20°C and the gallium shutter opened for a time period that would provide the desired ultra-thin coverage assuming a sticking coefficient of unity. The Ga flux was estimated from quadrupole mass spectrometer data calibrated against the growth rate of GaAs. An Sb layer of about 1 μm thickness was then provided at a deposition rate of about 1 μm/hr.

Current-voltage forward characteristics were taken to obtain the ideality factor of the specimens. The Schottky barrier heights were determined by the activation energy method in which the diode current is observed as a function of temperature, from room temperature to 50°C, for a constant forward bias voltage of 0.10 to 0.15 V.

To obtain some feel for the stability of the barriers, thermal annealing tests for successive steps of temperature 100, 150, 200, and 250°C with the temperature held for 30 minutes at each step were made.

EXPERIMENTAL RESULTS AND DISCUSSION

Ga/GaAs Schottky Diodes

Relatively few results have been reported for Ga Schottky barriers on GaAs. Bachrach and Bauer [11] observed a non-Schottky interface between thick Ga metal and cleaved (110) p-type GaAs with doping concentration of 1.8 x $10^{18}cm^{-3}$. Woodall et al. [12] observed a temporary ohmic contact between liquid Ga and lightly-doped n-type GaAs and this contact became rectifying with time and exposure to air. Svensson et al. [13] in an experiment in which Ga was deposited on an As-stabilized GaAs surface in an MBE system reported a barrier height of 1.06 eV as determined by C-V measurements. In low metal coverage experiments, Skeath et al. [14] found Fermi level pinning positions CBM-0.7 eV and VBM+0.5 eV for cleaved n and p-type (110) GaAs surfaces respectively.

For our Ga/GaAs studies Ga metal of 1.5 μm in thickness was deposited on both n and p-type GaAs substrates. The I-V characteristics of Ga/n-GaAs (3 x $10^{15}cm^{-3}$) Schottky barrier diodes show good diode characteristics in both forward and reverse directions. The reverse bias voltage at a current of 100 μA exceeds 35 V with no attempt to provide a guard-ring structure to protect against edge leakage. The ideality factor calculated from the linear portion of semi-log plot is 1.09 which indicates that thermionic emission is the major current transport mechanism in the diode. The barrier height as determined by thermal activation measurements is 0.99 eV which is a high value among ϕ_B for various metals on n-type GaAs. Our result is in good agreement with the 1.06 eV barrier height reported by Svensson et al. [13] for a 2 x $10^{16}cm^{-3}$ MBE grown layer.

The I-V and ϕ_B measurements of our Ga/GaAs diodes showed no substantial change in I-V curves, ideality factors, and barrier heights with time after being kept for many weeks in a dry air ambient at room temperatures of 20-25°C which were below the melting temperature of the Ga. The solid interface between Ga and the semiconductor presumably prevents any possible causes of degradation.

Contacts between Ga and a p-type GaAs layer ($5 \times 10^{16} \text{cm}^{-3}$ MBE grown layer) also show rectification. However, the I-V curves are not as perfect as those for Schottky barriers on n-type GaAs. The increase of reverse current with increasing reverse bias voltage indicates a relatively low barrier height formed at the interface. The thermally measured barrier heights from forward characteristics were 0.44 eV and the ideality factors were 1.08, an indication of good Schottky behavior. The non-rectifying behavior observed by Bachrach and Bauer [11] is probably due to the high doping concentration in the p-type substrate they used.

In unified defect model the pinning positions of Fermi level are different for n and p-type materials. For GaAs with monolayer type metal coverage the pinning positions are 0.75 eV and 0.5 eV above the maximum of valence band for n and p-type GaAs, respectively, and therefore there is a gap of 0.25 eV between them. This is inconsistent with the results we obtain for thick Ga metal on GaAs. Despite the difference in doping concentration in the specimens, our results for barrier height (0.99 eV and 0.44 eV) sum to the bandgap of GaAs. Zur et al. [4] have shown theoretically that thin and thick metal layer coverage conditions differ because the charge balance schemes are different. Spicer et al. [14] have recently extended their theoretical and experimental studies and used different approximations for their modeling. With this approach, they can explain an observed difference of 0.1 eV for the barriers of Al on n-type GaAs with thin and thick metal coverages. No attempt has been made to apply this modeling to Ga on GaAs.

No interfacial chemical reactions are expected at Ga-GaAs interfaces because of the zero interface heat of reaction. Interdiffusion will probably be the dominant activity at the interface. For a Ga-GaAs interface the out-diffusion of As into Ga metal and the in-diffusion of Ga into the semiconductor will increase the concentration of various crystal defects, such as the Ga_{As} antisite defect, arsenic vacancies V_{As} and interstitial Ga. The increase in density of defects at various energy levels may be expected to cause a new charge balance level at the interface and change the barrier height. Svensson et al. [13] use a similar explanation for their observation of a barrier height of 1.06 eV. But exactly what defect level or levels are contributing to the surface Fermi level position for Ga/GaAs Schottky barrier diodes is unclear.

Sb/Ga/GaAs Schottky Diodes

Difference in Schottky barrier heights from 0.80 to 0.68 eV for Sb on n-GaAs surfaces with different As coverages [10] indicates that a slight change in the surface chemical condition can affect the contact behavior significantly. In these previous experiments the surface conditions were altered simply by varying desorption procedures in an MBE system. However, further changes can be achieved by intentionally depositing thin layers of other metals on the surface before deposition of thick metal. This is similar to low metal coverage experiments performed by surface scientists except a layer of thick unreactive metal is added for electrical measurements of the diode characteristic. Brucker and Brillson [7] in studies of Au-CdS barriers have observed changes of diode characteristics from strongly rectifying to ohmic as the thickness of an Al interlayer was increased from zero to 2 Å. This is the largest barrier height modification achieved by interface techniques.

All our Sb/Ga/GaAs specimens showed good I-V characteristics in both forward and reverse bias directions. The results are summarized in Table I for a selection of specimens. For n and p-type specimens with the same thicknesses of Ga interlayers, such as 0.4 ML for SGN12 and SGP12, the values of ϕ_{Bn} and ϕ_{Bp} are seen to sum to a quantity very close to the bandgap of the GaAs. Similar summation to the bandgap is observed for Ga/GaAs and

Table I   Measured Barrier Heights and Ideality Factors for the Schottky Barrier Specimens

| Specimen No. | Type | Coverage of Ga | Ideality Factor | | ϕ_{Bn} eV | ϕ_{Bp} eV | $\phi_{Bn} + \phi_{Bp}$ eV |
			n-type	p-type			
SN11	n	no	1.04		0.81		1.47
SP11	p	no		1.04		0.66	
SGN12	n	0.4 ML	1.01		0.88		1.46
SGP12	p	0.4 ML		1.02		0.58	
SGN13	n	0.8 ML	1.05		0.96		1.45
SGP13	p	0.8 ML		1.06		0.49	
SGN14	n	1.2 ML	1.01		0.98		1.45
SGP14	p	1.2 ML		1.07		0.47	
SGN15	n	3.0 ML	1.01		1.01		1.47
SGP15	p	3.0 ML		1.08		0.46	
GN16	n	∞	1.09		0.99		1.43
GP16	p	∞		1.08		0.44	

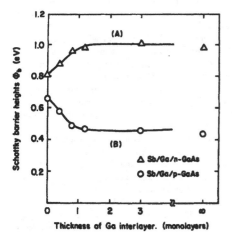

Fig. 2 Barrier heights of Sb/Ga/n-GaAs and Sb/Ga/p-GaAs Schottky diodes versus thickness of Ga interlayer.

Sb/GaAs [10] Schottky diodes on Ga-rich surfaces.

The effects of Ga interface layers of various thickness on barrier heights for typical n and p-type junctions are shown in Fig. 2. For n-type specimens (curve A), the barrier height increases almost linearly from 0.81 eV to 1.0 eV as the thickness of the Ga layer increases from zero to just over one monolayer (ML), and stabilizes at about 1.0 eV. The barrier height for p-type material (curve B) shows a change in a converse way -- decreasing from 0.66 eV for an Sb/GaAs Schottky diode to 0.46 eV and then remaining unchanged as the thickness of the Ga interfacial layer is greater than 1 ML. The results for Ga/GaAs Schottky barriers without Sb overlayer are also shown in the same figure. The barrier heights of specimens SGN14 and SGN15 are the same, within experimental error, as the height for the Ga/n-GaAs diode. This is similarly true for p-type specimens. The effect of the Ga is therefore dominant when present at a thickness of 1.2 ML or more and the Sb does not control the barrier height seen. Hence we conclude that there is no significant interaction of the Sb with the Ga or with the GaAs. We may also infer that the surface state density of the GaAs covered with a monolayer of Ga is high. If we view the first monolayer of Ga and the GaAs semiconductor as an integral part, then Ga/GaAs Schottky diode can be considered as a Ga/Ga(1ML)/GaAs structure. The difference between Ga/Ga(1ML)/GaAs and Sb/Ga(1ML)/GaAs is the different cover metals, and the difference of the work functions of the cover metals does not affect the Schottky barrier heights they form. Zur, et al. [4] have estimated that $10^{14} cm^{-2}$ states would be expected to be sufficient to pin a GaAs surface.

The rapid change of the barrier height takes place when Ga of submonolayer thickness is added to the surface. This phenomenon is similar to the Fermi level movements for low metal coverage [1,2,3,7].

In a similar experiment, Svensson et al. [13] have observed an increase of Schottky barrier height from 0.75 eV to 0.87 eV when the Ga coverage is increased from zero to about 50% in Al/Ga/n-GaAs Schottky diodes. With UPS and SEM they found that the deposited Ga forms clusters on the GaAs surface. Because of the cluster formation of Ga, they have to deposit more Ga metal in order to achieve a certain area coverage. For instance a Ga film of 50 Å nominal thickness is necessary to cover half of the surface. However, Skeath et al. [16] have reported that at submonolayer coverage Ga is most likely to form flat (one- or two-atom-thick) clusters or rafts, on the surface, and a fairly uniform Ga overlayer can be obtained for thicknesses greater than approximately 1ML. In our specimens we believe the deposited Ga forms rafts when submonolayer coverage is attempted and fairly uniform overlayers when the deposited layer is more than 1ML.

Svensson et al. [13] observed cluster charging during photoemission spectroscopy measurements. They suggest that the loss of photoelectrons cannot be balanced by the supply of the electrons from the substrate because of poor conductivity between the Ga clusters and the GaAs substrate. This poor conductivity they believe is due to a thin insulating disordered GaAs layer formed by the reaction between incoming Ga atom and excess As on the $c(2x8)$ As-stabilized surfaces they used in their experiments. These disordered GaAs conglomerates may be effective nucleation sites for new incoming Ga atoms and cause the cluster formation. On the (110) surfaces and the (100) Ga-stabilized surfaces used in our experiment there will be no excess As to react with incoming Ga atoms to form a disordered GaAs layer, and thus the formation of Ga clusters is not expected.

Because of the cluster formation in their specimens, Svensson and Andersson [17] suggest that the true structure of their specimen can be analyzed as a model of two "intermixed" diodes connected in parallel with the Schottky barrier height increasing as the Ga coverage area increases from pure Al/GaAs diode to pure Ga/GaAs diode. They argue that dominance of the low-barrier-diode on the I-V characteristic over most of the coverage range in macroscopic case will not be observed because of the depletion region edge effects around the high-barrier Ga clusters.

Another approach is that of Cowley and Sze [18] in which study of the actual small dependence of ϕ_{Bn} on the work function of the metal used leads to an expression from which a term ϕ_o is defined [19] as the Fermi energy level filling at the surface, measured with respect to E_v, the valence band edge, at the surface for a density of surface states Q_{SS}. This energy level coincides with the surface Fermi level before the metal-semiconductor contact is formed and is specified as the level below which all surface states must have been filled with electrons for charge balance conditions at the surface. Since the metal probably creates defects in the semiconductor to a depth of a few Å and these contribute to the density of states Q_{SS}, the model to be useful depends on assuming that Q_{SS} is independent of the metal used. From such studies Cowley and Sze conclude that ϕ_o is 0.53 eV for GaAs and that the density of states is $1.2 \times 10^{14} cm^{-2} eV^{-1}$. Tyagi [20] in more recent work, that assumes a 20 Å oxide interface insulating layer for Schottky barriers on chemically etched GaAs, prefers a value 0.45 eV for ϕ_o and about $3 \times 10^{14} cm^{-2} eV^{-1}$ for the density of states. This value of 0.45 eV is arrived at similarly in studies by Seiranyan and Tkhorik [21].

More recent barrier determinations (Waldrop [22]) for 14 metals on $6 \times 10^{16} cm^{-3}$ n-type GaAs show no correlation between ϕ_{Bn} and the metal work function and no very obvious correlation between ϕ_{Bn} and the known metal reactivity with GaAs. His results, however, are not incompatible with a significant density of states induced at the interface, even if this density is not a value independent of the metal used as proposed by earlier workers and as implicit in the ϕ_o model of Cowley and Sze [18]. Even though the ϕ_o model must be regarded as not applicable to barriers on GaAs, the Sb barriers for ϕ_{Bp} (0.44 to 0.47 eV in Table 1) for Ga layers in excess of 1.2 ML certainly show pinning at around an energy level that is somewhat close to the ϕ_o values mentioned above.

Bachrach and Bauer [11] find from core level photoemission that arsenic diffuses into their Ga overlayer as the layer builds up. They suggest that the loss of arsenic generates vacancies in the semiconductor surface region and these give acceptor action that contribute to the resistance being low for Ga contacts on their heavily doped p-GaAs. They also comment that for n-type GaAs the same As-loss action might produce a thin p-type converted layer. The barrier then would be a metal-p$^+$ Schottky barrier, or contact, in series with a p-n junction. If a p-n junction is indeed formed this might contribute to the barrier rise from 0.8 to 1.0 eV that we observe with a Ga interfacial layer. However, we have not explored this possibility since the pinning explanation seems more probable in view of the systematic trend seen in Fig. 2.

Table II shows the effects of room temperature, air-ambient, storage and thermal annealing of specimen SGN14 that has an interfacial layer estimated as 1.2 ML of Ga. No change is observed in the diode characteristic after a two-month dry ambient storage as the Ga/GaAs Schottky diode. The thermal annealing of the diode up to 150°C for 30 minutes in a UHV environment does not affect its properties. A slight lowering of the barrier height is observed on annealing at 200 and 250°C.

Table II. Thermal Annealing Effects on Sb/Ga(1.2 ML)/n-GaAs Schottky Diode, SGN14

	As grown	After two months in room ambient	Annealing at 100°C	Annealing at 150°C	Annealing at 200°C	Annealing at 250°C
Diode Ideality Factor	1.01	1.02	1.02	1.04	1.04	1.04
ϕ_{Bn} (eV)	1.01	1.02	1.01	1.01	0.97	0.95

ACKNOWLEDGMENTS

This work has been supported by NSF Grant ECS 82-14859 and has made use of the facilities of the Carnegie-Mellon University Center for the Joining of Materials. The cooperation of Fudan University, Shanghai, in providing a leave of absence for X.-J. Zhang is appreciated.

REFERENCES

1. W. E. Spicer, P. W. Chye, P. R. Skeath, C. Y. Su, and I. Lindau, J. Vac. Sci. Technol., 16, 1422 (1979).
2. W. E. Spicer, S. J. Eglash, P. S. Keath, P. Mahowald, I. Lindau, S. Pan, D. Mo, and D. M. Collins, 2nd Int. Symp. on Molecular Beam Epitaxy CST-2 1982, Tokyo, p. 269 Japan. Soc. of Appl. Phys.
3. L. J. Brillson and C. F. Brucker, J. Vac. Sci. Technol., 19, 661 (1981), Also L. J. Brillson, Applics. Surface Sci., 11/12, 249 (1982) and L. J. Brillson, Thin Solid Films, 89, 461 (1982).
4. A. Zur, T. C. McGill and D. L. Smith, Phys. Rev., B28, 2060 (1983).
5. P. Chiaradia, A. D. Katnani, H. W. Sang,Jr., and R. S. Bauer, Phys. Rev. Lett., 52, 1246 (1984).
6. J. Massies, F. Dezaly, and T. Linh, J. Vac. Sci. Technol., 17, 1134 (1980).
7. C. F. Brucher and L. J. Brillson, Appl. Phys. Lett., 39, 67 (1981).
8. W. I. Wang, J. Vac. Sci. Technol., 81, 574 (1983).
9. J. L. Freeouf and J. M. Woodall, Appl. Phys. Lett., 39, 727 (1981).
10. H. Cheng, X-J. Zhang and A. G. Milnes, Solid State Electronics, 27 (1984) to be published.
11. R. Z. Bachrach and R. S. Bauer, J. Vac. Sci. Technol., 16, 1149 (1979).
12. J. M. Woodall, C. Lanza, and J. Freeouf, J. Vac. Sci. Technol., 15, 1436 (1978).
13. S. P. Svensson, J. Kanski, and T. G. Andersson, Phys. Rev. B30(10), 6033 (1984).
14. P. R. Skeath, I. Lindau, C. Y. Su, P. W. Chye, and W. E. Spicer, J. Vac. Sci. Technol., 17, 511 (1980).
15. W. E. Spicer, S. Pan, D. Mo, N. Newman, D. Makowald, T. Kendelewicz, and S. Eglash, J. Vac. Sci. Technol. B2(3), 476 (1984).
16. P. Skeath, I. Lindau, C. Y. Su, and W. E. Spicer, Phys. Rev. B28(12), 7051 (1983).
17. S. P. Svensson and T. G. Andersson, Proc. 3rd Int. Conf. on Molecular Beam Epitaxy, San Francisco August (1984).
18. A. M. Cowley and S. M. Sze, J. Appl. Phys., 36, 3212 (1965).
19. S. M. Sze, Physics of Semiconductor Devices, Wiley-Interscience 2nd Ed., p. 270 et seq.
20. M. S. Tyagi, Surface Science, 64, 323 (1977).
21. G. B. Seiranyan and Y. A. Tkhorik, Phys. Status Solidi, A13, K115 (1972).
22. J. R. Waldrop, J. Vac. Sci. Technol. B2(3), 445 (1984).

Atomic Structure of the Epitaxial Al/Si Interface

F.K.LeGoues, W.Krakow and P.S.Ho

IBM Thomas J. Watson Research Center, Yorktown Heights, New York 10598, USA

Abstract

Al was deposited on Si(111) and observed by cross-sectional electron microscopy, both in the annealed and as deposited states. It is shown that Al is strongly textured when deposited on Si(111), with Al(111)//Si(111) and Al<110>//Si<110> or Al(100)//Si(111) and Al<110>//Si<110>. Annealed samples are completely epitaxial with Al(111)//Si(111) and Al<110>//Si<110>. Lattices imaging of the interfaces shows an amorphous layer (native Si oxide) between Al and Si, in the as-deposited case. The two lattices are in contact only at pinholes in the native oxide and fringes on both sides of the interface are seen to be continuous through the interface only at those points. Annealed samples do not show any amorphous or disordered layer at the interface: The two lattices are completely in contact, with lattice fringes extending from one side of the interface to the other. An atomic model of the annealed interface, based on energy considerations and consistent with TEM observations, is proposed.

INTRODUCTION

Despite the wide use of evaporated Al films on Si in the electronics industry, very little work has been done on the microscopic structure of the Al/Si interface. Such studies can now be carried out down to atomic level because of the development of very high resolution electron microscopes. The atomic structure of the interface is critical for calculating and understanding such properties as Shottky barrier heights. In the Al/Si case, it was generally assumed that the interface is completely disordered (not epitaxial). For example Louie and Cohen [1] calculated Shottky barrier heights using a jelium potential representing the Al-ion potential in contact with the Si surface. This assumption was based on the fact that there is about 30% misfit between the Al and Si lattices [2] and it thus seemed safe to assume that epitaxy was impossible. Nonetheless, there has been some evidence that this is not the case. D'Heurle et al [3] reported a strong (111) orientation of Al deposited on Si(111). Dahmen and Westmacott [4] studied precipitation of Si particles from an electron bombarded Al matrix and found that these particles where strongly faceted along the (111) planes, indicating a low energy interface in this direction as well as epitaxy. The authors proposed that this happened because four Al(111) lattice planes almost exactly match three Si(111) lattice planes, leaving only 0.4% misfit to be accommodated by stress or structural dislocations. LEED studies [5,6] showed that, in ultra high vacuum conditions, Al tends to be epitaxial on Si(111). Finally, it was proposed by Zur and McGill [7] that epitaxy should exist between two lattices if, in the plane of the interface, n lattice translations of one lattice nearly equal m lattice translations of the other one, with n and m being "small" integers. As pointed out before, this is clearly the case for the Al(111)/Si(111) interface.

High resolution, cross-sectional transmission electron microscopy has been used extensively in the recent past to investigate the microstructure of epitaxial interfaces. The cases of Ni_2Si/Si and Pd_2Si interfaces are particularly well known [8,9]. In coordination with computer simulations, such technique can describe exact atomic positions at the interface In this paper, the Al/Si interface will be studied using cross-section TEM. It will be shown that Al is completely epitaxial on Si(111). A model consistent with the TEM pictures will be presented.

DESCRIPTION OF THE EXPERIMENT

4000 Å of Al was evaporated on chemically clean Si<111> wafers at room temperature. Those samples, as well as heat treated samples (2 hours at 400°C) were then prepared for cross-sectional transmission electron microscopy (TEM). Furthermore, 250 Å of Al was evaporated on Si(111) at room temperature and annealed two hours at 400°C in order to do flat-on microscopy. Samples were studied in a Philips 400ST transmission electron microscope, at 120KV.

RESULTS

Fig. 1a and 1b show the diffraction pattern and corresponding imaging obtained from the sample on which 250 Å of Al was deposited on Si(111), as deposited and annealed 2 hours at 400°C respectively. Clearly, as deposited Al is strongly textured on Si(111): The sample shows two Al orientations: Al(100)//Si(111) and Al<110>//Si<110> or Al(111)//Si(111) and Al<110>//Si<110>. The epitaxy is not perfect , showing evidence of rotational misorientation, and the Al layer is multi-grained. The annealed sample presents only the (111) type of epitaxy. Some rotational misalignment is still present but it seems that the sample is now completely epitaxial and thus single-grained.

Fig 2 a and 2b represent the 4000 Å Al sample, as deposited on Si(111). Fig 2a is a low magnification picture, Fig 2b shows the same sample under lattice imaging conditions. As for the case shown on Fig 1a , Fig 2a and 2b show that the as-deposited sample is epitaxial or very strongly textured. On Fig 2, only the (111) orientation is visible, but that could be due to the fact that cross-section only shows a very small area of the sample (much smaller than the area covered by a flat-on sample). Fig 2b shows again that the lattice fringes in aluminum are in the same direction as the ones of the silicon lattice. Nonetheless, the interface is not very clearly defined: some native oxide is present between Al and Si (which is expected under the preparation conditions) and shows up in the image as a brighter, amorphous layer separating the two lattices. Consequently, the fringes do not extend from one side of the interface to the other. There seem to be pinholes in the native oxide, where the two lattices are indeed in contact (fringes are seen to be continuous from one side of the interface to the other). Those pinholes are probably the starting point for epitaxy.

Fig. 3a and 3b correspond to Fig 2a and 2b, for the sample annealed two hours at 400 °. The low magnification picture shows that the epitaxy is over a long range. Fig. 3b, showing the lattice image of the interface, differs from Fig. 2b because the native oxide has now disappeared and fringes extend from one side of the interface to the other. There is clearly no interface compound and/or "disordered" or amorphous layer between Al and Si. The lattice image is consistent with Si being completely coherent with Al except for an extra Al plane every fourth plane.

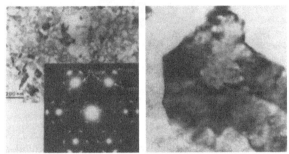

Figure 1:
a) 250 Å of Al on Si(111), as deposited. The low magnification picture shows that the sample is multi-grained, but the diffraction pattern indicates very strong texture of the Al grains along the (111) and (100) planes of Si. The higher magnification shows the Moire pattern arising from the misfit between Al and Si.

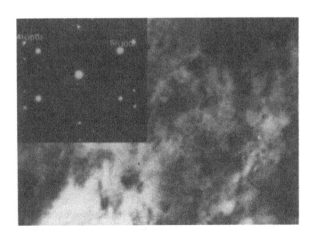

b) 250 Å of Al on Si(111), annealed 2 hours at 400°C. The sample is now completely epitaxial, with Al(111)//Si(111) and Al<110>//Si<110>.

Figure 2:
a) Low magnification, bright and dark field images of 4000 Å of Al on Si(111), as deposited. The dark field picture shows that the epitaxy is real on a relatively long range (at least a few microns).
b) High resolution picture of the same sample. Note the amorphous layer at the interface between Al and Si, which is probably native Si oxide, and the pinholes where the two lattices are in contact.

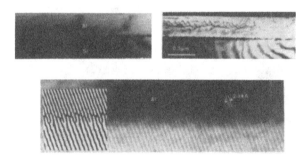

Figure 3:
a) Low magnification, bright and dark field images of 4000 Å of Al on Si(111), annealed 2 hours at 400 °C. The epitaxy is seen to be perfect over a long range.
b) High resolution picture of the same sample. One lattice fringe of Al out of each four stops at the interface. The inset picture shows an example of computer simulations done using the model presented on Fig. 4b.

DISCUSSION

Despite the very high misfit between Al and Si, it was shown that Al is epitaxial on Si(111). Even in as-deposited samples, a very strong texture was observed, with Al(111)//Si(111) and Al<110>//Si<110> or Al(100)//Si(111) and Al<110>//Si<110>. The case of Al(111) on Si(111) seems to be particularly interesting, as compared to other systems with very high misfits, because four Al(111) spacings nearly exactly equal three Si(111) spacing (with a remaining mismatch of 0.4 %). This is indeed one is the best example to illustrate the Zur and McGill proposition. The Al(100) orientation on Si(111) shown on Fig. 1a is somewhat more surprising and difficult to explain since there is about 6.5% misfit between Al<100> and Si<110> and the relative orientations of Al and Si are more difficult to visualize than for the Al(111)/Si(111) case. Indeed this (100) orientation must be energetically less favorable than the (111) type since it disappears upon annealing, leaving only the (111) type of epitaxy: Fig 1b.

Fig. 4a and 4b represent a lattice image of the annealed Al/Si(111) sample and a possible atomic model, respectively. The model is consistent with all the microscopic observations, but is clearly not the only possibility and electron microscope image computer simulations are necessary to determine its validity. Nonetheless, the model is based on one major assumption which has been successful in explaining the epitaxy of Nickel silicides on Si, namely that the number of broken bonds, on the silicon side, is minimum and the number of silicon –metal bonds is maximized: the model predicts one dislocation every 10Å, which is an extremely high density of dislocations. In order to explain such a high density (and the large elastic energy associated), a great deal of chemical energy has to be gained by forming the interface. The present model has been chosen so that the Si atoms keep their four bonds per atom structure everywhere. Furthermore, The Al bonds are only broken at the dislocations, which is unavoidable. An interesting point to note is that there is also a 30% misfit between Al and Si in the direction perpendicular to the boundary. Thus a single atomic step should not be acceptable since it would create a 30 % mismatch between the two crystals. Thus, the only possibility is to have a multi-plane step consisting of four Al or three Si planes. Indeed most steps observed at the interface (although not all of them) are similar to the one shown on Fig. 4a, which could be described as shown in Fig. 4b.

As mentioned before computer simulations are necessary in order to confirm the atomic model proposed on Fig 4b. Indeed it would have been possible to chose another model by simply "cutting" Si one quarter of a monolayer lower. In this case only one Si bond would have been cut for each Si atom at the interface which would result in two dangling Al bonds per atomic site at the interface, or, on the average, 1.5 Al-Si bond less per Al at the interface than for the model proposed on Fig. 4b. Again, because of energy considerations, this model does not seem likely. Nonetheless computer simulations using both models have been performed and compared with actual TEM pictures. These results will be published independently [10]. An example of the type of simulation obtained using the model presented on Fig. 4b has been added to Fig. 3b in order to show that it is indeed possible to

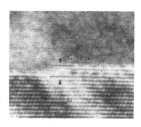 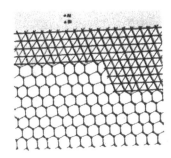

Figure 4:
a) High resolution picture of 4000 Å of Al on Si<111>, 2 hours at 400°C. Note the multi-layer step at the interface.
b) Model of the Al(111)/Si(111) interface.

simulate the microscope conditions at which we have been working, namely 120 KV and tilted beam illumination. For the actual comparison of the two models [10] , better TEM pictures (showing more than one set of lattice fringes) were obtained. Images were Fourier filtered in order to obtain as clear an image of the interface as possible and compared to computed images obtained by reproducing the microscope conditions as nearly as possible.

SUMMARY

This study demonstrated that Al evaporated on Si(111) and annealed at 400°C is completely epitaxial with Al(111)//Si(111),Al<110>//Si<110>. As deposited samples were strongly textured. An atomic model, based on energy considerations and consistent with actual TEM results was proposed. Computer simulations, necessary to uniquely prove the validity of the model, have been performed and will be published elsewhere [10].

REFERENCES
1. S. G. Louie and M. L. Cohen, Phys. Rev. B, **13**, 2461, (1976)
2. W. B. Pearson, *Handbook of Lattice Spacings and Structures of Metals and Alloys*, Pergamon Press,(1958)
3. F. d'Heurle, L. Berenbaumand R. Rosenberg, Trans. AIME, **242**, 502, (1968)
4. K. H. Westmacott and U. Dahmen, *Proceedings of the 40th Annual EMSA Meeting*, 620, (1982)
5. J. J. Lander and J. Morrison, Surf. Sci., **2**, 553, 1964
6. M. Strongtin, O. F. Kammerer, H, H, Farrell and D. L. Miller, 129, (1972)
7. A. Zur and T. C. McGill, J. Appl. Phys., **55**, 378, 1984
8. D. Cherns, D. A. Smith, W. Krakow and P. E. Batson, Phil. Mag., **45**, (1982), 107
9. W. Krakow, Thin Sol. Films, **93**, (1982), 109-125
10. F. K. LeGoues, W. Krakow and P. S. Ho, to be published

EPITAXY OF ALUMINIUM FILMS ON SEMICONDUCTORS
BY IONIZED CLUSTER BEAM

I. YAMADA*, C.J. PALMSTRØM*, E. KENNEDY*, J.W. MAYER*, H.INOKAWA** AND T. TAKAGI**
*Department of Materials Science and Engineering, Cornell University, Ithaca, NY 14853
**Ion Beam Engineering Experimental Laboratory, Kyoto University, Sakyo, Kyoto 606, Japan

ABSTRACT

Epitaxial Al films have been deposited onto the clean surface of single-crystal Si by ionized cluster beam (ICB) at room temperature. Thermal stability of the film has been examined by SEM, AES depth profiling, ion backscattering/channeling, and electrical characterization of the Al-Si interface. It was found that the ICB Al film on Si substrate was remarkably stable up to 550°C although pure Al was used. Alloy penetration at the interface, shift of barrier height, degradation of crystalline quality and development of annealing hillocks on the surface were not observed after the heat treatment. Extremely long electromigration life time was also confirmed. Epitaxial growth on GaAs(100) substrate was attempted and preliminary results are given.

INTRODUCTION

Aluminium metallization is widely used for contact electrodes and interconnects in silicon semiconductor devices [1-3]. It is important to establish the quantitative limits of the Al/Si system imposed by such problems as electromigration, hillock growth, alloy penetration and interface stability [2-5]. These problems depend critically upon film structure (primarily grain size and orientation) and interface, properties not precisely controlled under any conventional Al deposition technique. A study of epitaxial Al is thus quite important because it permits separating those limits intrinsic to Al itself from those related to a specific deposition technique.

Ionized cluster beam (ICB) deposition [6,7] permits control over the energy and ion content of the beam, and thus control over sputter cleaning, formation of nucleation sites and adatom migration. This control can be used to achieve epitaxial film growth at low substrate temperature [8]. In this paper we describe Al films grown epitaxially on (111) single crystal Si by ICB. In order to examine the thermal stability of the Al/Si system morphological, crystallographic and electrical properties are examined before and after anneals at temperatures to 550°C. A preliminary result of similar Al deposition on GaAs (100) will also be presented.

EXPERIMENTAL

The apparatus for ICB deposition and the general characteristics of the films have been already described [6,7]. Deposition conditions were chosen from previous experiments combining in situ reflection electron diffraction (RED) on the growing film and Rutherford channeling to assess crystal quality of the finished film [9]. Acceleration voltage (Va) of ionized cluster, electron current for ionization (Ie), and ionization voltage (Ve) were at 5kV, 100mA, and 500V, respectively. Experiments were performed under a vacuum of 2×10^{-7}Pa maintained by a sputter ion pump and a titanium sublimation pump. Single-crystal Si (111) and Si (100) substrates were chemically oxidized by a H_2SO_4-H_2O_2 mixture and the oxide removed by heating to 1000°C for 5 minutes in vacuum to obtain clean surfaces [10], on which pure Al was deposited epitaxially by ICB at room temperature. Deposition rate was 12nm/min, and small changes in the deposition rate did not have an effect on the quality of the film. After the deposition of about 400nm thick Al film, the sample was removed from the deposition chamber to another vacuum system and was annealed at various temperatures up to 550°C. Thermal stability of the Al-Si interface and the Al film itself was evaluated by scanning electron microscopy (SEM), Auger electron spectroscopy (AES) depth profiling, Rutherford backscattering spectroscopy (RBS), and electrical measurements of Al-Si junction. Lastly, a preliminary experiment using GaAs (100) substrates was done and the orientation of the epitaxial film was determined.

Crystalline Orientations

Crystalline orientation of the Al film on Si(111) substrate is determined by reflection electron diffraction (RED) patterns as
Al (111)//Si(111), Al[$\overline{1}$10]//Si[$\overline{1}$10].
Two orthogonal orientations,
Al($\overline{1}$10)//Si(100), Al[00$\overline{1}$]//Si[01$\overline{1}$], denoted as Al(110), and
Al (110)//Si(100), Al[$\overline{1}$10]//Si[01$\overline{1}$], denoted as Al(110)R,
are mixed together in the film deposited on Si(100) substrate [9]. It is remarkable that a lattice misfit as large as 26%, at least in one direction, permits epitaxial growth of Al on both Si(111) and Si(100) substrate.

In-situ AES and RED observation showed that application of higher acceleration voltage during the deposition resulted in nearly layer-by-layer growth and a flat film surface. RBS measurement showed that the χ min decreased with increasing acceleration voltage during deposition. In this experiment, Si(111) substrates were used.

Morphological Stability

Changes in the morphology of the surface and the interface after annealing at 450°C, 500°C, and 550°C for 30 min were examined by SEM and AES with sputter etching. Figure 1 shows the SEM images of the surface and the interface revealed by phosphoric acid etching after annealing at 550°C in comparision with those by conventional vacuum deposition. As is often observed, there are annealing hillocks (small protuberances) and valleys probably caused by the extended alloy penetration on the surface of the Al film deposited by conventional vacuum deposition. No annealing hillocks or valleys can be seen on the surface of the ICB film. At the interface of the sample prepared by conventional vacuum deposition, there are strong undulation caused by so-called alloy penetration, while no irregularity can be seen at the interface of the ICB deposited sample.

Figure 2 shows AES depth profiles of the Al-Si interface. Figure 2 (a) is of the sample deposited by ICB and annealed at 550°C for 30 min and (b) is of the sample by conventional vacuum deposition and annealed

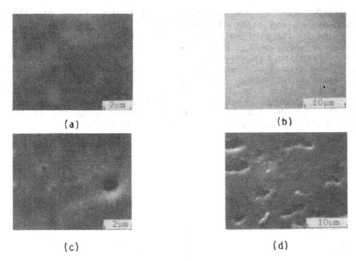

(a)   (b)

(c)   (d)

Fig.1. SEM images of the surface and the interface of Al films on Si(111), post-annealed at 550°C for 30min in vacuum. (a) surface of an Al film deposited by ICB (Va=5.0kV, Ie=100mA, and Ts=room temperature, (b) interface of (a), (c) surface of an Al film deposited by conventional vacuum deposition, and (d) interface of (c).

identically. The FWHM depth resolution of the AES system was measured to be about 20nm by using the same sample which had not been annealed. It can be seen that the interface of the sample deposited by ICB remains abrupt after annealing.

Degradation of Al-Si interface after annealing is explained as the result of the nonuniform dissolution of Si into Al [3,4]. High stability of the interface of the ICB deposited sample is probably due to the uniformity of the interface and the limited diffusion of Si in the epitaxially grown Al film.

Crystallographical Stability

Since the solid solubility of Si in Al reaches about 1.2 at.% at 550°C in thermal equilibrium [11], dissolved Si may affect the crystalline quality. 1.5 MeV He$^+$ and 165keV H$^+$ backscattering/channeling were used to evaluate the crystalline quality of the Al film before and after annealing.

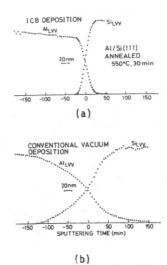

Fig.2. AES depth profiles of Al-Si interfaces after annealing at 550°C for 30 min. (a) a sample deposited by ICB (Va=5kV, Ie=100mA, Ts=room temperature, and (b) a sample by conventional vacuum deposition.

Figure 3 shows the spectra obtained by 1.5MeV He⁺ of Al <111> incidence before (a) and after (b) annealing at 500°C for 30 min. Spectra
corresponding to the Al film and the Si substrate overlap at about 180 ch.
and produce a sharp peak. Spectra before and after anneal show the same
features, indicating that no chemical reaction has occurred. The slope (a
measure of the degree of dechanneling) is reduced after anneal, indicating
crystal quality improved. Figure 4 is the dependence of the minimum
yield, X_{min}, obtained by 165keV H⁺ beam of Al <111> incidence on the
annealing temperature. From these RBS data it can be said that the
crystallinity of the ICB Al film improves after annealing at less than
500°C and degrades little even after annealing at 550°C.

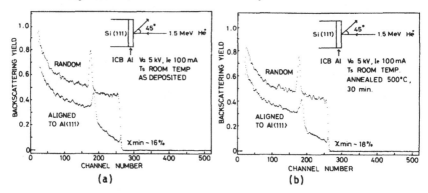

Fig.3. 1.5MeV He⁺ backscattering/channeling spectra of the ICB Al film on
Si(111) before (a) and after (b) annealing at 500°C for 30 min. Incident
ion beam is aligned to Al<111>.

Electrical Stability

Recrystallization of Si which
has been dissolved in Al causes
such problems as an increase in
contact resistance on n⁺-Si [12]
and an increase of the barrier
height of Al-n-Si junctions, [13-14].
Morphological change in the Al-Si
interface degrades or completely
destroys the device beneath the
Al film. In this experiment Al-n-
Si (0.7 ohm-cm) junctions of 1 mm
diameter were fabricated and the
current density-voltage (J-V) characteristics were measured before
and after annealing at 450°C,
500°C and 550°C. Figure 5 shows
the J-V characteristics of the
junctions. The extrapolated value
of current density at zero
voltage, which is related to barrier height remains stable after
annealing. The barrier height
and the n value are around 0.75eV

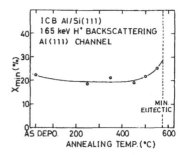

Fig.4. Dependence of the minimum
yield X_{min} on annealing temperature, measured by 165keV H⁺ beam
of Al <111> incidence. The ICB Al
films were deposited on Va=5kV,
Ie=100mA, and Ts=room temperature
and post annealed at various temperatures for 30 min.

and 1.17, respectively. Changes in the barrier height and the n value are 0.03eV and 0.02, respectively, while in the case of Al-Si junction fabricated by conventional deposition techniques, changes of more than 0.1eV and 0.1, respectively are reported [13,14]. It is estimated that the dissolution and/or the recrystallization of Si are suppressed in the epitaxial Al films.

To see whether ICB bombardment created sufficient disorder in the silicon surface to influence the Schottky barrier height we performed 0.925 MeV He$^+$ channeling experiments. The epitaxial Al films were chemically removed. As a standard sample for comparison, Ar$^+$ bombarded Si substrates were prepared. The bombardment was made at 0.5kV to dose of 2x10^{15} ions/cm^2. Figure 6 shows the surface peaks obtained from Si <111> incidence. Si surface peaks produced by Al-ICB bombardment at the energy of 5keV and at 0.2keV are much smaller than that caused by the 0.5keV Ar ion beam. Since Al etched substrates were exposed to air, a thin oxide layer was formed which might also increase the surface peak. The samples were chemically etched by diluted HF and surface peaks compared again. No difference in the peak height can be observed. Since HF might conceivably etch amorphous silicon, this does not prove that ICB bombardment produces no disorder. However, it can be concluded that the disorder induced by ICB is much smaller than that due to the naturally oxidized layer and significantly smaller than the disorder produced by atomic ion bombardment.

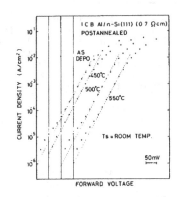

Fig.5. Current density-voltage (J-V) characteristics of ICB Al-Si junctions after annealing at a various temperatures for 30 min.

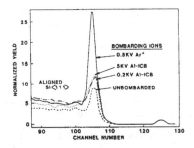

Fig.6. Comparison of surface peaks of Ar bombarded (0.5kV, 2x10^{15} ions/cm^2) and Al-ICB bombarded Si substrates.

Electromigration resistance was measured at the current density of 10^{6}A/cm^2 at 250°C. No change in the resistance of the epitaxial Al film was observed during 400 hours of operation, although similar Al film deposited by sputtering usually failed after 20 hours. Further experiments are under way to determine the limitation and dependence on the film properties

Deposition on GaAs(100)

An Eaton ICB-100 was used to deposit epitaxial GaAs on (100) LEC GaAs substrates. The epitaxial films were cleaned in $H_2SO_4:H_2O_2:H_2O$ and HCl [15] , then heated to 580°C in vacuum for 10 minutes just before deposition. The surface showed no traces of carbon and oxygen by AES, and RED indicated clear GaAs(100) 1x1 pattern. Al was deposited by ICB under the same conditions as for the Si substrate. The orientation of the film is determined as

Al(100)//GaAs(100), Al[011] //GaAs[010] .

Further investigations of the electrical, crystallographic and morphological properties are now under way.

CONCLUSIONS

Epitaxial Al films were grown on (111) and (100) Si by ICB. These films, and particularly the Al/Si interface, showed remarkable stability even when annealed to 550°C. Preliminary data show these films very resistant to electromigration. Many difficulties encountered with Al metallization for VLSI are thus apparently not intrinsic to the Al/Si system itself, but are consequences of particular deposition techniques.

ACKNOWLEDGEMENT

The authors wish to thank Drs. P. Younger and J. Blake of Eaton Ion Materials Systems for supporting parts of the experiments and providing the epitaxial GaAs films made by ICB deposition. The authors also wish to thank Dr. G. Chen of Eaton Operations and Technical Center for evaluating electromigration lifetime.

REFERENCES

1. J.L. Vossen, J. Vac. Sci. Technol. 19 761 (1981).
2. A.J. Learn, J. Electrochem. Soc. 123 894 (1976).
3. J.M. Poate, K.N. Tu, and J.W. Mayer (eds.), Thin Films-Interdiffusion and Reactions, (Wiley, NY, 1978).
4. P A. Totta and R.P. Sopher, IBM J. Res. & Dev. 13 226 (1969).
5. P. Chaudhari,J. Appl. Phy. 45 4339 (1974).
6. T. Takagi, I. Yamada, A. Sasaki J. Vac. Sci. Technol. A2 382 (1984)
7. I. Yamada, H. Takaoka, H. Inokawa, H. Usui, S.C. Cheng, and T. Takagi, Thin Solid Films 92 137 (1982).
8. I. Yamada, F.W. Saris, T. Takagi, K. Matsubara, H. Takaoka, and S. Ishiyama, Japn. J. appl. Phys. 19 L181 (1980).
9. I. Yamada, H. Inokawa, and T. Takagi, J. Appl. Phys. 56 2746 (1984).
10. M.Tabe, K.Arai, and H. Nakamura, Jpn. J. Appl. Phys. 20 703 (1981).
11. M. Hansen, Constitution of Binary Alloys, (McGraw-Hill,NY 1958).
12. M. Mori, IEEE Trans. Electron Devices ED-30 81(1983).
13. H.C. Card, in Metal Semiconductor Contacts, Conf. Ser. No. 22 129 (Institute of Physics, Manchester, England, 1974).
14. T.M. Reith and J.D. Schick, Appl. Phys. Letters 25 524 (1974).
15. A.Y. Cho, Thin Solid Films 100 291 (1983).

IN SITU INVESTIGATION OF THE PALLADIUM SILICON REACTION

D.A. Smith, P.A. Psaras, I.J. Fisher[†] and K.N. Tu
IBM Thomas J. Watson Research Center, Yorktown Heights, N.Y. 10598

ABSTRACT:

Palladium has been deposited on {100} and {111} oriented silicon wafers and also on polysilicon. Cross-sectional specimens for transmission electron microscopy were prepared and heated in-situ. The interfaces between silicide and silicon were rough and the volume changes accompanying heating and compound formation caused elastic strains in the substrates and in one case hillock formation in the products.

INTRODUCTION

Interfaces between silicides and silicon are of such great practical and scientific importance that they have been the focus of numerous recent studies [1-4]. Key issues include the atomic configuration of the interface, the kinetics of silicide formation and the electrical properties of the interface. Behavior of the palladium-silicon system has been investigated by x-ray techniques, Rutherford backscattering spectrometry, transmission electron microscopy and by Auger spectroscopy [1,2,5-8]. As a result it is known that reaction produces a series of palladium-silicon compounds as the reaction temperature is increased, palladium is thought to be the dominant diffusing species during growth of Pd_2Si, epitaxial growth of Pd_2Si can occur on {111} silicon substrates and the transition region at silicide-silicon interface, in common with other crystal-crystal boundaries is narrow. In this paper we report preliminary results of an in situ transmission electron microscopy investigation of the reaction between palladium and silicon.

EXPERIMENTAL TECHNIQUES

Palladium was deposited onto silicon substrates which had been chemically cleaned and from which the oxide film had been removed with buffered HF. The substrates were {100} and {111} silicon wafers and polycrystalline silicon grown on oxidized silicon wafers. Two thicknesses of palladium were deposited; ~500Å by electron beam evaporation onto substrates maintained at room temperature and ~2000Å by radio frequency sputtering during which the substrate temperature rose to ~150°C. In both cases the base pressure in the system before deposition was 1-2 10⁻⁷ torr. The resistivity of the electron beam deposited palladium was ~20 μm-cm and that of the sputtered palladium ~13μm-cm. Cross sectional specimens were prepared by now familiar techniques. The specimens were characterized by Rutherford backscattering spectrometry and transmission electron microscopy and then heated incrementally to temperatures up to 700°C whilst under observation in a Philips EM301 transmission electron microscope.

RESULTS

Representative images of partially reacted bilayers are shown in Fig. 1. Significant features are the roughness of the silicide interface and the appearance of contrast effects related to elastic strain fields visible in the silicon substrates. The amplitude of the silicide thickness fluctuations at the silicide-silicon interface is about 100Å on wafers but can be more on polysilicon. Such rough interfaces occurred in all the specimens examined in this investigation, but are normally not found in fully reacted epitaxial interfaces, e.g. Pd_2Si-Si {111} and $NiSi_2$-Si{100} or {111} [1,3,4]. The strain fields visible in

[†]I.J. Fisher, Fairleigh Dickenson University, Rutherford, N.J. 07080.

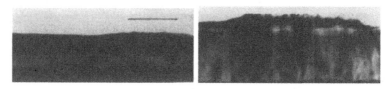

Figure 1 shows an early stage in the reaction of e-beam evaporated Pd with (a) {100} Si and (b) polysilicon. the marker represents 1000Å, note the strain effects in 1a and the interface roughness in both cases.

silicon substrates at the reaction temperature and room temperature are present even when nonepitaxial Pd_2Si has been formed and have too large a spacing, (~1000Å) to correlate with misfit dislocation. Similar features can be found in previously reported studies of cross sectional specimens of Pd_2Si-Si{100} and PtSi-Si [9] imaged at room temperature.

The progress of the metal-silicon reaction can be followed. Fig. 2 is a sequence of bright field photomicrographs showing the thickening of a silicide layer during heating. This example also serves to show two of the difficulties of the present technique (a) contamination of the specimen due to glue and environment and (b) the relative lack of electron transmission through the palladium or the silicide. Consequently it is not always possible to characterize the phases present by electron diffraction. Figure 3 shows some features of the reaction between palladium and polycrystalline silicon. The facetted interface at the left of images is a striking feature.

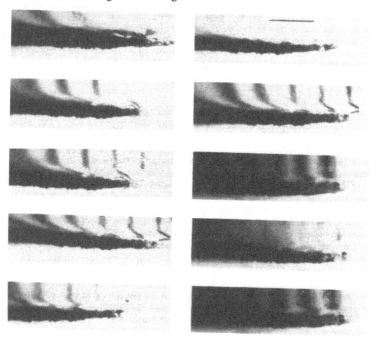

Figure 2 shows successive stages in the reaction of e-beam evaporated Pd with a {100} Si wafer. The specimen was heated incrementally to ~500°C. The marker represents 1000Å.

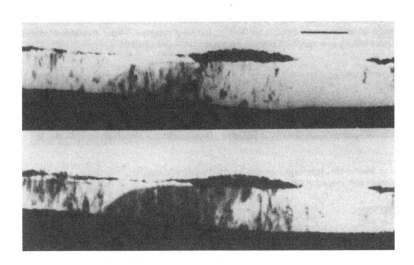

Figure 3 shows two stages in the reaction of e-beam evaporated Pd with polysilicon. The changes took place in 2 hours at 300°C. The marker represents 1000Å.

A remarkable result was obtained when palladium was heated on polycrystalline silicon from 400°-700°C, it appeared that certain grains of the polycrystalline silicon reacted preferentially (Fig. 4).

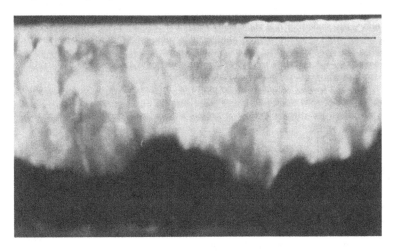

Figure 4 shows the irregular nature of the silicide-polysilicon interface resulting from a few minutes of reaction at 400°C. The marker represents 1000Å.

A further intriguing result was obtained when a {111} Si wafer with 2000Å of RF sputtered palladium was heated to 300°C for five minutes. This treatment produced numerous hillocks of epitaxial silicide which are illustrated in Fig. 5.

Figure 5 shows the formation of hillocks by Pd-Si reaction for five minutes at 300°C. The marker represents 1000Å.

DISCUSSION

The observation of the generation of stress in this and other related structures is consistent with a diffusion controlled process in which silicon transport is not the dominant process. The dominant flux of the palladium reacting at the Si-Pd$_2$Si interface (or in the silicide) gives rise to a constrained volume change which in turn results in an intrinsic stress in the film. It is emphasized that the strain effects are visible in the silicon substrate at the reaction temperatures as well as after cooling to room temperature. The specific volumes of silicon, palladium and Pd$_2$Si, are 20.02Å^3, 14.72Å^3 and 37.74Å^3 respectively i.e. Pd$_2$Si is fomed with an accompanying volume decrease of 23%. The observations of hillocks in figure 5 imply that the reaction products are in compression and are not consistent with an overall volume contraction accompanying compound formation. The rough silicide-silicon interface certainly implies local differences in reaction rate but this may be an extrinsic effect. It is clear that palladium cannot reduce the air formed SiO$_2$ film which must remain on chemically cleaned substrates. It has been previously suggested that the silicide formation reaction begins through weak spots in the oxide film; possibly defects formed in response to thermal mismatch stresses. However there does seem to be some crystallographic specificity to the silicide formation as evidenced by the facetting in Fig. 3. and the widely varying thicknesses of product in Fig. 4 both of which are on polysilicon substrates.

A puzzling feature in many of our photo micrographs is a line of contrast beneath the silicide. Dark field microscopy establishes that the contrast feature is located in the silicon but whilst its origin is unclear at present it may be better understood after STEM investigation.

CONCLUSION

In situ reaction of palladium and silicon to give Pd$_2$Si is feasible. Preliminary results concerning the topography of the silicide-silicon interface indicate that in the partially reacted condition the interface is rough on a scale varying from ~100Å for Si wafers to several hundred Å for polysilicon. The observation of hillock formation and apparent interfaces are not yet explained satisfactorily.

REFERENCES

1. H. Föll, P.S. Ho and K.N. Tu, Phil. Mag. A *45*, 31 (1982).

2. D.Cherns, D.A. Smith, W. Krakow and P.E. Batson, Phil. Mag A *45*, 107.

3. R.T. Tung, J.M. Gibson and J.M. Poate, Appl. Phys. Lett. *42*, 888 (1983).

4. R.T. Tung, Appl. Phys. Lett. *41*, 818 (1982).

5. K.N. Tu, W.K. Chu and J.W. Mayer Thin Solid Films *25*, 403 (1975).

6. R.W. Bower, D. Sigurd and R.E. Scott, Solid State Electronics *16*, 1461 (1973).

7. D.J. Fertig and G.Y. Robinson, Solid State Electronics *19*, 407 (1976).

8. M. Wittmer and K.N. Tu, Phys. Rev. B *27* 1173 (1983).

9. H. Föll, P.S. Ho and K.N. Tu, J. Appl. Phys., *52*, 250 (1981).

Interfaces

Electrical Transport through Metal Contacts to In$_{0.53}$Ga$_{0.47}$As Thin Films

Pong-Fei Lu*, D. C. Tsui* and H.M. Cox**
Department of Electrical Engineering & Computer Science, Princeton
University, Princeton, NJ 08544
**Bell Communications Research, Murry Hill, NJ 07974

Abstract

We report two phenomena observed in the electrical transport
through metal contacts to In$_{0.53}$Ga$_{0.47}$As thin films. First, a surface accumulation layer of electrons is found at the oxide-In$_{0.53}$Ga$_{0.47}$As interface of
Pb-oxide-In$_{0.53}$Ga$_{0.47}$As tunnel junctions, suggesting that the surface
Fermi level is not pinned in In$_{0.53}$Ga$_{0.47}$As, and ideal ohmic contacts to n-
In$_{0.53}$Ga$_{0.47}$As can be made by using low work function metals. Second, we
observed a strong oscillatory conductance on the I-V characteristic of
electrical transport through In- In$_{0.53}$Ga$_{0.47}$As contacts, with a period
approximating the LO-phonon energy of In$_{0.53}$Ga$_{0.47}$As. We explain the
data by successive phonon emission in the high field transport of ballistic
electrons and point out that the experiment is a solid state analogue of
the Franck-Hertz experiment.

1. Introduction

In$_{0.53}$Ga$_{0.47}$As is a promising material in device engineering. It has a
small effective mass (~0.04m$_o$) and an energy band gap of ~0.75eV (~1.65
μm), which is close to the attenuation minimum of silica fiber at 1.55 μm.
There have been numerous reports on transport in bulk In$_{0.53}$Ga$_{0.47}$As and
on its device applications. In this paper, we want to report two
phenomena observed in the electrical transport through junctions fabricated on high mobility n-type In$_{0.53}$Ga$_{0.47}$As thin films, grown by tri-
chloride vapor levitation epitaxy[1] on the (100) surface of an n$^+$-InP substrate.

First, we investigated the surface band structure of In$_{0.53}$Ga$_{0.47}$As by
using tunneling through Pb-oxide-In$_{0.53}$Ga$_{0.47}$As tunnel junctions. We
observed a surface accumulation layer of electrons at the oxide-
In$_{0.53}$Ga$_{0.47}$As interface. Our results support the notion that metal-
semiconductor contacts with low work function metals form ideal ohmic
contacts to n-In$_{0.53}$Ga$_{0.47}$As. Second, we observed an oscillatory conductance in the electrical transport through In-In$_{0.53}$Ga$_{0.47}$As contacts in both
bias polarities. It is found electron tunneling plays no role in the transport process and the data are explained by phonon emission of ballistic
electrons through micro-channels of In$_{0.53}$Ga$_{0.47}$As. A unified picture of
these oscillations due to phonon emission is proposed.

2. Tunneling spectroscopy of $In_{0.53}Ga_{0.47}As$

The surface Fermi level of $In_{0.53}Ga_{0.47}As$ is commonly believed to be pinned[2]. Only recently the report by Hsieh et al[3] showed that Al epilayer forms ideal ohmic contacts on $In_{0.53}Ga_{0.47}As$ grown by molecular beam epitaxy. They point out that the ideal ohmic contact may be due to the fact that the work function of Al (~4.2eV) is smaller than the $In_{0.53}Ga_{0.47}As$ electron affinity (~4.6eV). This observation, together with the report by Kajiyama et al[4] on Au-InGaAs contacts, suggest that the surface band bending in junctions, formed by metal-oxide-semiconductor (MOS) and metal-semiconductor (MS) contacts to $In_{0.53}Ga_{0.47}As$, is determined by the work function difference between the metal and the $In_{0.53}Ga_{0.47}As$. Henceforth, an electron accumulation layer is expected at the oxide-$In_{0.53}Ga_{0.47}As$ interface in MOS junctions with low work function metals and it can be probed by utilizing electron tunneling[5].

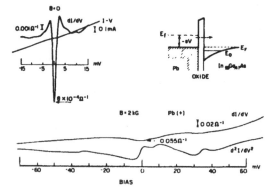

Fig.1:   dI/dV vs V and d^2I/dV^2 vs V at 4.2K, with B=2 kG to quench Pb superconductivity. The inset on the left shows the I-V and dI/dV when B=0, and that on the right is the proposed energy diagram.

Here we will briefly review the results from tunneling through Pb-oxide-$In_{0.53}Ga_{0.47}As$ junctions[6]. The quality of the tunnel junctions is demonstrated by the superconducting energy gap and the phonon structure of Pb at 4.2K, which are shown in the inset of Fig. 1. The normal-state dI/dV vs V and d^2I/dV^2 vs V as shown in Fig. 1 are obtained by applying a magnetic field B (~2kG). The data show a zero bias anomaly[7] and three structures at approximately 32 mV in the Pb negative bias and approximately 5 mV and 35 mV in the Pb positive bias. The structure at ~35 mV is not identified. The structures at V=+5 mV and -33 mV are identified, respectively, as the band edge of the two-dimensional subband in the surface electron layer and as the threshold energy for phonon emission[8] by tunneling electrons. These identifications are based on the well established model for tunneling through MOS junctions[9] and the magneto-tunneling data shown in Fig.2 and 3.

When a perpendicular B is applied, it leads to the complete quantization of the surface electrons and their density-of-states is a series of broadened Landau levels. The d^2I/dV^2 vs V at fixed B will show oscillations reflecting the Landau level structure[5]. Fig. 2(a) shows the data with B= 35

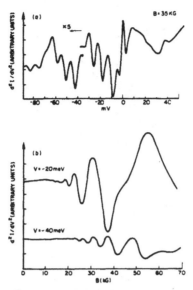

Fig.2: (a) d^2I/dV^2 vs V with B=35 kG applied perpendicular to the junction. (b) d^2I/dV^2 vs B with V=-20 mV and -40 mV.

kG. We notice that there is a sudden drop in the oscillation amplitude at the bias close to the LO-phonon energy of $In_{0.53}Ga_{0.47}As$ (i. e. ~-32 mV[10]). This drop is due to the reduction in the lifetime of injected electrons via LO-phonon emission. At fixed V, the d^2I/dV^2 vs B data show Shubnikov-de Haas type oscillations, periodic in 1/B (Fig. 2(b)). The period is related to the number of states per unit area, N_s, in the subband with $E \leq eV+E_F$ through $N_s=(e/\pi\hbar c)/\Delta(1/B)$. The N_s vs V data can be used to determine the effective mass m^*. We obtain $m^*=(0.042\pm0.001)m_0$, in good agreement with that from recent magneto-phonon experiments[11].

Fig. 3 summarizes our results. The data for V > 0 is not identified. For V < 0, the two sets of data are in agreement, confirming our model that the oscillations observed in these measurements are a manifestation of the same Landau level structure of the surface electrons at the interface. An extrapolation to B=0 gives a subband edge energy, $E_0 \simeq 5$ meV, confirming our identification in Fig. 1.

3. Ballistic transport through micro-channels of $In_{0.53}Ga_{0.47}As$

We have observed a strong oscillatory conductance in the electrical transport through $In-In_{0.53}Ga_{0.47}As$ contacts. The contacts, of size 250 μm $\times$ 250 μm, are made by thermal evaporation of a thick In film on the as-grown InGaAs crystal, which was precleaned only in organic solvents. Fig. 4 is an example of the I-V characteristic and its first and second derivatives of an $In-In_{0.53}Ga_{0.47}As$ contact at 4.2K. The oscillations, already discernible in the dI/dV data, are apparent in the d^2I/dV^2 vs V data. They are stronger in the In negative bias and up to 17 oscillations have been observed. Similar but weaker oscillations, up to 6, are observed in the In positive bias in Fig. 4. They correspond to dips in the I-V in either bias

418

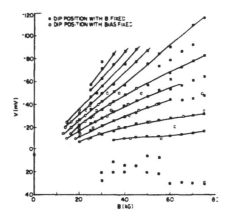

Fig.3: Bias position, at which the dips in the d^2I/dV^2 curves are observed, plotted as a function of B (O) and as a function of V (•)

polarity at biases equal to integral multiples of ~33 meV, close to the LO-phonon energy of $In_{0.53}Ga_{0.47}As$[10]. The oscillations, corresponding to a series of broadened dips in the I-V characteristic, are due to emission of LO-phonons.

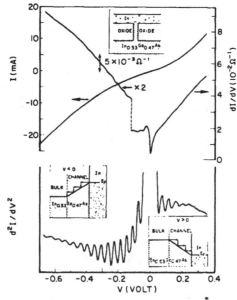

Fig.4: I-V, dI/dV vs V and d^2I/dV^2 vs V of an In-$In_{0.53}Ga_{0.47}As$ contact at 4.2K. The insets show the proposed structure of the contact and the energy band diagram under biases.

We did not observe the superconducting tunneling characteristics when the temperature is lowered below the superconducting transition temperature of In (3.4K). This experimental fact shows that electron tunneling plays no role in the transport process and there is no tunneling barrier between the In electrode and the InGaAs crystal. To explain the fact that the expected ideal ohmic contact is not observed and that the contact resistance is ~100 ohm, we propose the following model. The contact between the In and $In_{0.53}Ga_{0.47}As$ is via a large number of small channels of $In_{0.53}Ga_{0.47}As$ crystal protruding through the insulating oxide. Ideal ohmic contacts are made by In to these small channels, which must have lengths approximately the oxide thickness of several hundred angstroms and diameters probably several tens of angstroms, rendering them one-dimensional wires. This model of contact and the energy band diagram under biases are illustrated in the inset of Fig. 4.

Recently Kulik and Shekhter[12] treated theoretically the electrical transport through a narrow semiconductor channel, where the diameter of the channel d is smaller than its length L and the electron's inelastic relaxation length λ_E (not associated with the phonon emission process considered in the problem). They predicted an I-V characteristic showing singularities at $eV=n\hbar\omega_0$, where ω_0 is the optical phonon frequency of the semiconductor and n=integers. The periodic singularities result from the phonon emission events, which satisfy $L=nZ_n$, where Z_n is the distance an electron must travel along the channel to gain enough energy to emit a phonon. This commensuration condition is satisfied by those particular events of phonon emission which cause a sudden drop in the terminal velocity of the accelerating electron, as it reaches the collecting electrode, and consequently give rise to a sudden decrease in the current flow. This process is reminiscent of the successive inelastic collisions of electrons with gas atoms in the classical Franck-Hertz experiment[13]. The resulting oscillations in the conductance are expected in both bias polarities, consistent with our observation.

Similar oscillatory conductance were observed in tunnel junctions made on high purity InSb by Katayama et al[14] in 1967 and by Cavenett[15] in 1969. More recently Hickmott et al[16] made similar observations in GaAs-AlGaAs tunnel structures in the presence of a high magnetic field. These experiments differ from ours in that the oscillations are observed only when the tunneling electrons are injected into the high purity semiconductor electrode. We wish to point out that this model based on the calculation of Kulik and Shekhter, when extended to the ballistic regime, can account for our results as well as previous observations on tunnel junctions. As pointed out by Hickmott et al[16], the oscillations result from phonon emission in the electrode and the role of the tunnel junction is to supply electrons. It is also well known that tunneling is a highly directional process and the tunnel junction in fact makes the subsequent ballistic transport of the injected electrons one-dimensional. In the bias polarity corresponding to the acceleration of electrons out of the semiconductor toward the tunnel junction, the junction acts merely as a blocking contact and no oscillations are expected.

4. Conclusion

In summary, we have investigated the surface band structure of $In_{0.53}Ga_{0.47}As$ thin films using electron tunneling through Pb-oxide-$In_{0.53}Ga_{0.47}As$ tunnel junctions. The results suggest that MS contacts using low work function metals (such as In, Pb, or Mg) make ideal ohmic contacts to electrons in $In_{0.53}Ga_{0.47}As$. We also observed an oscillatory conductance in electron transport through In-In$_{0.53}$Ga$_{0.47}$As contacts in both bias polarities. A model based on the theoretical work of Kulik and Shekhter explains the phenomenon and provides an understanding of previous experiments on tunnel junctions as well. This model makes it apparent that the experiment is a solid state version of the classical Franck-Hertz experiment.

This work is supported in part by the Office of Naval Research through contract No. N00014-82-K-0450.

Reference:

(1) H. M. Cox, J. Crystal Growth 69, (1984).

(2) Donald P. Mullin & H. H. Wieder, J. Vac. Sci. Technol. B1 (3), 782 (1983).

(3) K. H. Hsieh, M. Hollis, G. Wicks, C. E. C. Wood and L. F. Eastman in "GaAs and Related Compounds 1982 " (Institute of Physics, Alberquerque, 1983), p. 165.

(4) K. Kajiyama, Y. Mizushima, and S. Sataka, Appl. Phys. Lett. 23, 458 (1973).

(5) D. C. Tsui, Phys. Rev. B 4, 4438 (1971); 8, 2657 (1973).

(6) Pong-Fei Lu, D. C. Tsui and H. M. Cox, Appl. Phys. Lett. 45, 772 (1984).

(7) R. N. Hall, J. A. Racette and H. Ehrenreich, Phys. Rev. Lett. 4, 456 (1960).

(8) J. A. Appelbaum and W. F. Brinkman, Phys. Rev. 186, 464 (1969).

(9) D. J. BenDaniel and C. B. Duke, Phys. Rev. 160, 679 (1967).

(10) A. Pinczuk, J. M. Worlock, R. E. Nahory, and M. A. Pollack, Appl. Phys. Lett. 33, 461 (1978).

(11) M. A. Brummell, R. J. Nicholas, J. C. Portal, M. Razeghi and A. Possion, Physica B 117, 118, 753 (1983).

(12) I. O. Kulik and R. I. Shekhter, Phys. Lett. 98A, 132 (1983).

(13) See, for example, D. C. Peaslee, "Elements of Atomic Physics" (Prentice Hall, 1955), p.158.

(14) Y. Katayama and K. Komatsubara, Phys. Rev. Lett. 19, 1421 (1967).

(15) C. B. Cavenett, Phys. Rev. B 5, 3049 (1972).

(16) T. W. Hickmott, P. W. Solomon, F. F. Fang, Frank Stern, R. Fischer and H. Morkoc, Phys. Rev Lett. 52, 2053 (1984).

$Hg_{1-x}Cd_xTe$-Cr INTERFACE REACTION

P.PHILIP[*], A.WALL[*], A.FRANCIOSI[*], AND D.J.PETERMAN[**]
* Department of Chemical Engineering and Material Science
  University of Minnesota, Minneapolis, MN. 55455
** Mc Donnell Douglas Research Laboratories, St.Louis, MO. 63166

ABSTRACT

We summarize photoemission studies using Synchrotron Radiation of the formation of the HgCdTe-Cr interface at room temperature on in situ-cleaved single crystal substrates. Evidence is found of a Cr-Hg exchange reaction in the subsurface region. The surface and near surface layers appear completely depleted of mercury.

Mercury cadmium telluride, a ternary semiconductor used primarily in infrared detection devices, has been much studied in recent years [1-4]. While HgTe and CdTe are both relatively stable compounds, the types of bonding in the ternary HgCdTe tend to substantially weaken the Hg-Te bond relative to that of Cd-Te [3-4]. As a result the ternary material may undergo large variations in composition, particularly at the surface, due to mechanical stresses or chemical processes following, for example, metal deposition [5-7]. Recent studies concerning HgCdTe-noble metal and HgCdTe-simple metal contacts have related the variations in the Schottky barrier height to non-stoichiometry at the semiconductor surface [7-8]. In this paper we summarize a photoemission study of a HgCdTe-refractory metal interface to gain a better understanding of the chemical processes driving the interface formation [9].

Synchrotron radiation photoemission was used to examine the $Hg_{0.78}Cd_{0.22}Te$ (110)-Cr interface. The HgCdTe single crystals were grown at Mc Donnell Douglas Research Laboratories using a modified Bridgmann method. Infrared transmission and Hall effect measurements of the material determined that the band gap of the bulk crystal was 0.175 ± 0.01 eV with a room temperature carrier concentration of $2x10^{16}$ cm^{-3}, giving p-type conductivity. Clean (110) surfaces were prepared by cleaving in an ultra-high vacuum chamber (p< $5x10^{-11}$torr). The interfaces were obtained in situ by sublimation of Cr from a tungsten boat (p< $5x10^{-10}$torr) and the Cr coverage was measured using a quartz thickness monitor.

To examine the chemical interaction between the semiconductor and metal at the interface, the relative intensities and binding energies of the Hg 5d, Cd 4d, Te 4d, and Cr 3p core levels were measured as a function of Cr coverage. In the bottom section of Fig.1, the change in concentration of the various species at the surface is shown as a function of Cr coverage through the attenuation of the corresponding core level emission. The integrated emission intensities of the Te 4d, Cd 4d, and Hg 5d levels were normalized to the initial surface emission and is shown in a semilogarithmic plot. Emissions from the Cd 4d and the Hg 5d levels appear partially superimposed in energy, such that a deconvolution was necessary [9]. Representative results of the deconvolution for the clean surface are shown in the top section of Fig.1. While the intensity of the Te peak remains fairly constant as the Cr deposition progresses, the Cd 4d and especially, the Hg 5d emissions attenuate rapidly with increased Cr coverage. These results are inconsistent [9] with what might be expected for an escape-depth-driven mechanism. The surface region appears depleted of Hg while the binding energy of the Hg 5d doublet remains the same, indicating that Hg exists in only one type of chemical environment. It is possible to show [9] that these results are quantitatively consistent with a Cr-Hg exchange

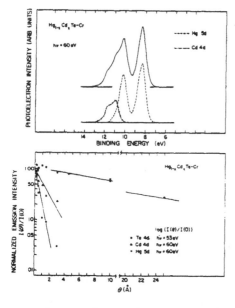

Fig. 1 Top: deconvolution of the experimental lineshape (solid line) into Hg 5d (dashed line) and Cd 4d (dot-dashed line) contributions for the HgCdTe surface [9].

Bottom: semi-logarithmic plot showing the integrated intensity of the Hg 5d, Cd 4d, and Te 4d core emission, normalized to the emission from the clean surface, as a function of metal coverage.

reaction occuring at the surface at low Cr coverage ($\Theta<2$). The attenuation of the Cd 4d is much slower, indicating that Cd is not driven from the surface in the same manner as Hg. Analysis of the Te 4d core emission proved to be essential in revealing the nature of the chemical interactions at the interface. Lineshape analysis uncovered the existence of two types of Te environments near the semiconductor surface. Two identical Te 4d lines, characteristic of the clean HgCdTe surface, were shifted ~ 0.6eV with respect to each other and scaled in intensity so that the addition of the two lines would approximate the Te 4d core emission for various Cr coverages [9]. In Fig. 2 we show the integrated emission intensity of the Te 4d I and 4d II lines after deconvolution as a function of Cr coverage. The intensities have been normalized to the initial Te 4d surface emission and are shown in a semilogarithmic plot. The Te 4d I component predominates at low Cr coverages, and its intensity attenuates exponentially, corresponding to the covering of the surface with Cr. This Te 4d I component initially has the clean surface binding energy, but shifts ~ 0.25eV to lower binding energy for $\Theta \sim 2$. This supports our picture of a Hg-Cr exchange reaction occuring at low metal coverage since the corresponding chemical shifts of the Cr 3p and Te 4d core levels [9] indicate that Cr and Te atoms interact directly in the subsurface region. Fig. 2 indicates that the Te 4d II forms for $0<\Theta<2$Å and that its intensity remains constant upon further chromium deposition. The 0.4 eV shift to higher binding energy relative to the initial 4d line, is consistent with that expected for the formation of elemental tellurium [10], so this evidence suggests that a layer of elemental tellurium has formed on the Cr surface. Escape depth-dependent studies [9] indicate that the Te 4d II emission derives from a relatively thin (~ 1 monolayer) Te layer segregated at the film-vacuum interface. Such dissociated species appear released at the surface as a by-product of the Cr-Hg exchange reaction that takes place in the near-surface layer.

While this description of the HgCdTe-Cr interface is by no means complete, it suggests that the interface formation procedes in two stages. The first stage, for $\Theta<2$Å, is characterized by a Cr-Hg exchange reaction, resulting in Hg depletion of the subsurface layer and the release of

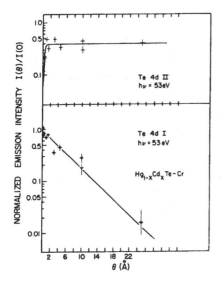

Fig.2 Semilogarithmic plot of the
integrated emission intensity
of the HgCdTe-Cr interface,
normalized to the Te 4d
emission from the clean surface.

Top: the Te 4d II component
corresponds to dissociated Te.

Bottom: the Te 4d I component
corresponds mostly to Te atoms
in the subsurface layer, where
the Cr-Hg exchange reaction
takes place.

elemental Te at the surface. The second stage, for $\Theta > 2\text{Å}$, is characterized
by the growth of a metallic Cr film, the surface of which is still covered
by a layer of elemental tellurium. These conclusions are much different
than those proposed for the HgCdTe-simple metal and HgCdTe-noble metal
interfaces [7-8]. For HgCdTe-Au, although Te outdiffusion was observed,
elemental tellurium was not seen at the interface [8]. For HgCdTe-Al [7],
an apparent change of 0.2eV in the Fermi level position at the interface
was related to the change in the Hg/Cd ratio due to Hg depletion. In our
case the constant binding energy of the residual Hg 5d core emission
indicates that no modification of the semiconductor band bending was
observed, while the apparent change in the Hg/Cd and Hg/Te surface concentration ratios were related to different chemical processes occurring in
the outermost surface layers. The composition of the HgCdTe-Cr interface
is therefore somewhat more complicated than that resulting from semiconductor interactions with simple metals or noble metals. To analyze the
interface formation process, it is necessary to distinguish the chemical
evolution of the surface layer from that of the subsurface region and the
underlying bulk. This appears to be essential to clarify the interface
morphology and to determine the relationship between metal-semiconductor
chemistry and Shottky barrier formation.

ACKNOWLEDGEMENTS

This work was supported, in part, by the Graduate School of the
University of Minnesota under Grant #0100490832, by the Office of Naval
Research under contract N00014-84-K-0545, and by Mc Donnell Douglas
Independent Research andDevelopment program. We wish to acknowledge
J.H. Hollister, B.J. Morris and C.S. Wright for their assistance in
sample preparation and characterization. We thank G.D. Davis and G.
Margaritondo for useful discussions and communication of results prior to
publication. Finally, we are grateful for the support of the staff of the
University of Wisconsin, Synchrotron Radiation Center (supported by
NSF Grant #DMR 76-15089).

REFERENCES

[1] See, for example, the Proceedings of the 1984 U.S. Mercury-Cadmium-
Telluride Workshop, J.Vac.Sci.Technol.A, (in press)

[2] W.E.Spicer, J.A.Silberman, J.Morgan, I.Lindau, J.A.Wilson, Au-Ban Chen,
and A.Sher, Phys.Rev.Lett. 49,948(1982); J.A.Silberman, D.Laser,
I.Lindau, and W.E.Spicer, J.Vac.Sci.Technol. A1,1706(1983) and
references therein.

[3] W.A.Harrison, J.Vac.Sci.Technol. A1,1672(1983)

[4] A.-B.Chen, A.Sher, W.E.Spicer, J.Vac.Sci.Technol. A1,1675(1983)

[5] H.M.Nitz, O.Ganschow, V.Kaiser, L.Wiedmann, and A.Benninghoren,
Surf.Sci. 104,365(1981)

[6] A.Lastras-Martinez, V.Lee, J.Zehnder, and P.M.Raccah, J.Vac.Sci.Technol.,
21,157(1982)

[7] G.D.Davis, N.E.Byer, R.R.Daniels, and G.Margaritondo,J.Vac.Sci.Technol.,
A1,1726(1983); R.R.Daniels, G.Margaritondo, G.D.Davis, and N.E.Byer,
Appl.Phys.Lett. 42,50(1983)

[8] G.D.Davis, W.A.Beck, N.E.Byer, R.R.Daniels, and G.Margaritondo,
(in press)

[9] A more extensive report will be presented in the near future; see
A.Franciosi, P.Philip, and D.J.Peterman, to be published.

[10] V.Solzbach andH.J.Richter, Surf.Sci. 97,191(1980); G.D.Davis, T.S.Sun,
S.P.Buchner, and N.E.Byer, J.Vac.Sci.Technol. 19,472(1981)

EFFECTS OF INCOMMENSURACY ON THE STRUCTURAL AND ELECTRICAL PROPERTIES IN PbTe-Bi SUPERLATTICE FILMS

SUNG-CHUL SHIN*
Materials Research Center, Northwestern University, Evanston, IL 60201
*Present address: Research Laboratories, Eastman Kodak Company, Rochester, NY 14650

ABSTRACT

We studied the structural and electrical transport properties of incommensurate PbTe-Bi superlattice films. The properties of those samples were noticeably different from those of commensurate samples. For the incommensurate samples, the satellite peaks, in the θ-2θ x-ray scans along the [111] growth orientation, became broader and the ratio of satellite intensities to the Bragg intensity became smaller. The resistances of incommensurate samples were about three times larger than those of corresponding commensurate ones. These features are interpreted by an enhancement of the lateral nonuniformity of interfaces in incommensurate samples. The exponential dependence of the resistance on temperature in incommensurate samples was also in contrast to the logarithmic behavior observed in commensurate ones.

INTRODUCTION

Recently we reported the structural [1] and transport [2] properties of commensurate PbTe-Bi superlattice films, which have an integer number of atomic planes of each constituent in a modulation wavelength Λ, where Λ is the repeat distance of two constituents along the [111] growth orientation. These samples had smooth interfacial structures and showed two-dimensional logarithmic transport behavior. In this paper, we present the x-ray diffraction and electrical transport results of incommensurate samples, which have a noninteger number of atomic planes of each constituent in a Λ. We will designate the samples by m/n, where m and n are the numbers of atomic planes of PbTe and Bi per Λ, respectively. (The number of molecular planes of PbTe is m/2.) The d-spacings of bulk PbTe and Bi were used in calculating m and n.

EXPERIMENTAL

Details of sample preparation were described elsewhere [1]. Briefly, samples were prepared by depositing PbTe (purity of 6N, intrinsic) and Bi (5N) alternately on mica substrates heated to 100°C in a vacuum system maintained at ~1 x 10^{-6} Torr during the deposition. A layer of 1000 Å of PbTe was predeposited to ensure a strong texture. Incommensuracy (or commensuracy) of superlattice films can be controlled only after quartz crystal sensors in a deposition unit are well calibrated. Our sensors were calibrated with an accuracy of 1% by x-ray diffraction method in conjunction with multiple-beam interferometry [3]. The deposition rate also should be well controlled to achieve the incommensuracy. A soaking time of ~30 min with ~60% of the deposition power was necessary to stabilize the deposition rate. An average fluctuation was about 1% for both constituents at the deposition rate of 5 Å/s.

Samples were structurally characterized by using a GE XRD-5 x-ray diffractometer. Cu Kα radiation, monochromatized by a pyrolytic graphite

426

crystal, was used. For the resistance measurement, the shape of the sample was defined by a mask, which consisted of a 3 x 0.7 mm rectangular strip with four tabs for current and voltage leads. By use of 4-terminal dc techniques, a resistance of $\Delta R/R \sim 10^{-5}$ could be detected with a sample current of 1 mA. Care was taken to keep the current low enough to ensure Ohmic response and negligible heating. All electrical measurements were made with a specially designed probe [4] in an S.H.E. VTS facility (800 series), which can operate in the temperature range 1.8 to 400 K.

RESULTS AND DISCUSSION

Several incommensurate samples were prepared, and we observed that satellite peaks became broader and that the ratio of satellite intensities to the Bragg intensity became smaller for the less commensurate samples. Figure 1 shows the θ-2θ x-ray scans of the 6/4 commensurate sample and the 6.5/3.5 incommensurate sample along the [111] growth orientation. Although there is nearly no difference in the Bragg intensities of two samples, the differences in the satellite peaks are noticeable; in the incommensurate sample, the integrated intensity of the first-order negative satellite is decreased by ~22% and, simultaneously, its full width at half maximum is increased by a factor of 8, compared to the commensurate sample with the same modulation wavelength. A similar result has been reported for the Cu-Ni superlattice films [5]. It is believed that the destructive interferences due to the lateral irregularity of interfaces in an incommensurate sample are responsible for the decrease of satellite intensities. The broadening of satellite peaks in an incommensurate sample is understood by the fluctuation of Λ along the growth orientation. The additional broadening $\Delta\theta_f$ due to the fluctuation is estimated by

$$\Delta\theta_f = \left| \tan\theta \frac{\Delta\Lambda}{\Lambda} \right|. \tag{1}$$

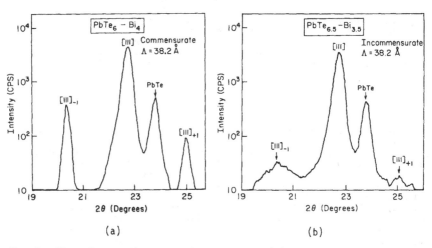

(a)   (b)

Fig. 1. The θ-2θ scanning x-ray diffraction of (a) the 6/4 commensurate sample and (b) the 6.5/3.5 incommensurate sample. The PbTe peaks for both samples arise from the predeposition layer.

Assuming this effect is simply additive, the fluctuation $\Delta\Lambda$ is estimated to be ~8Å (corresponding to ~2 atomic layers) in our incommensurate samples. Therefore, the lateral nonuniformity of composition is expected in an incommensurate sample.

Lateral nonuniformity provides an additional scattering of the carriers in the plane, and thus the in-plane resistance will be increased as the sample departs from perfect commensuracy. It is evident that the effect should be more prominent in the insulator/metal superlattice films than in metal/metal ones. Table I lists the characteristic parameters of two incommensurate samples, together with those of commensurate samples having nearly the same Λ. Here, the sheet resistance $R_\square$ is defined as the sample resistance of a square area multiplied by the total number of Λ. The $R_\square$'s of incommensurate samples are approximately three times larger than those of corresponding commensurate ones. This reflects the importance of a scattering due to the lateral irregularity of interfaces in incommensurate samples. A most interesting feature of incommensurate samples is the exponential behavior of the resistance, in contrast with the logarithmic behavior of commensurate ones reported earlier [2]. Figure 2 shows the temperature dependence of the resistance of 6.5/3.5 and 12.5/3.5 incommensurate samples between 1.8 and 25 K. The data for these samples have a better fit to a exp (1/T) than to a ln T relationship. We therefore conclude that the conduction is thermally activated. As expected, the sample with the larger resistance has a larger activation energy. This exponential behavior was observed for the incommensurate samples having resistance larger than ~8 kΩ/$\square$. We interpret an exp(1/T) behavior of incommensurate samples as a crossover from logarithmic (in a weakly localized regime) to exponential behavior (in a strongly localized regime) in two-dimensional (2D) systems with increasing the resistance, as samples depart from the perfect commensuracy. Mott [6]

Table I

Structural and Electrical Characteristics of Commensurate and Incommensurate Samples

m/n[a]	Λ(Å)	Δ[b](%)	FWHM[c](deg.)	I_-/I_B[d](%)	$R_\square$(kΩ/$\square$)	$R_\square$ vs. T
6/4	38.2	0	0.3	8.3	2.408	ln T
6.5/3.5	38.2	10	2.2	1.8	8.532	exp(0.18/T)
12/4	60.2	0	0.4	25.7	3.300	ln T
12.5/3.5	60.6	6.25	1.5	6.2	10.240	exp(0.34/T)

[a] m atomic layers of PbTe and n atomic layers of Bi per modulation wavelength Λ.

[b] Degree of departure from a perfectly commensurate structure defined in Ref 5.

[c] Full width at half maximum of the first-order negative satellite peak obtained from a θ-2θ scan.

[d] Ratio of the first-order negative satellite intensity I_- to the Bragg intensity I_B.

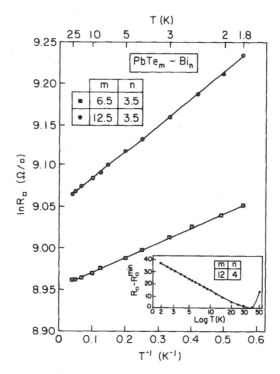

Fig. 2.   Temperature dependence of the sheet resistance $R_\square$ for the 6.5/3.5 and 12.5/3.5 incommensurate samples. The inset shows the logarithmic temperature dependence of the resistance for the 12/4 commensurate sample and is typical of the other commensurate samples; $R_\square^{min}$ is the minimum resistance of the sample.

suggested that a metal-insulator transition in 2D systems exists at the resistance $R_\square \sim 10$ k$\Omega/\square$. On the other hand, recent scaling arguments [7] based on localization concepts of Thouless [8] predicted a gradual change from logarithmic to exponential behavior at $R_\square \sim 10$ k$\Omega/\square$. Such a transition has been observed in several systems such as thin metal films [9] (by decreasing the film thickness or the grain size) and Si MOSFETs [10] (by decreasing the carrier concentration), although there is some ambiguity in the transition value of the resistance. To the best of our knowledge this is the first time the transition in the multilayered film controlling the topology of the interfacial configuration has been observed.

In conclusion, we could control the incommensuracy or commensuracy of PbTe-Bi superlattice films in a well-calibrated deposition unit. The structural and electrical transport properties were markedly changed as samples depart from perfect commensuracy.

ACKNOWLEDGEMENTS

The author would like to thank Professors J. E. Hilliard and J. B. Ketterson for helpful discussions. This work was supported by the Office of Naval Research under Grant No. N00014-82-K-5298 and the Northwestern Materials Research Center under National Science Foundation Grant No. DMR79-23573.

REFERENCES

1. S. C. Shin, J. E. Hilliard and J. B. Ketterson, Thin Solid Films $\underline{111}$, 323 (1984).
2. S. C. Shin, J. B. Ketterson and J. E. Hilliard, Phys. Rev. B $\underline{30}$, 4099 (1984); ibid., J. Vac. Sci. Technol. A $\underline{2}$, 296 (1984).
3. S. C. Shin, Ph.D. thesis, Northwestern University, 1984.
4. H. K. Wong and B. Y. Jin, S.H.E. Corporation VTS Newsletter No. 3, 1 (1984).
5. N. K. Flevaris, D. Baral, J. E. Hilliard and J. B. Ketterson, Appl. Phys. Lett. $\underline{38}$, 992 (1981).
6. N. F. Mott, Metal Insulator Transition (Taylor and Francis, London, 1974), p. 31.
7. E. Abrahams, P. W. Anderson, D. C. Licciardello and T. V. Ramakrishnan, Phys. Rev. Lett. $\underline{42}$, 693 (1979).
8. D. J. Thouless, Phys. Rept. $\underline{13C}$, 93 (1974).
9. G. J. Dolan and D. D. Osheroff, Phys. Rev. Lett. $\underline{43}$, 721 (1979); R. C. Dynes, J. P. Garno and J. H. Rowell, Phys. Rev. Lett. $\underline{40}$, 479 (1978).
10. S. J. Allen, Jr., D. C. Tsui and H. DeRosa, Phys. Rev. Lett. $\underline{35}$, 1359 (1975); D. J. Bishop, D. C. Tsui and R. C. Dynes, Phys. Rev. Lett. $\underline{44}$, 1153 (1980).

TRANSITION FROM SINGLE-LAYER TO DOUBLE-LAYER STEPS ON GaAs(110) PREPARED BY MOLECULAR BEAM EPITAXY

J. FUCHS, J. M. VAN HOVE, P. R. PUKITE, G. J. WHALEY, AND P. I. COHEN
Department of Electrical Engineering, University of Minnesota
Minneapolis, MN 55455, USA

ABSTRACT

Even though GaAs (110) is the only semiconductor whose surface structure is known with confidence, little is known about its microscopic growth mechanisms. We have used RHEED to study the role of steps in the MBE growth of GaAs on vicinal GaAs(110) surfaces which were misoriented by less than 2 mrad. After thermally desorbing the initial oxide, 20 layers of GaAs deposited at 700K produced a surface with single atomic-layer steps having an average terrace length of a few hundred Angstroms. Upon annealing to 800K, a slow mass migration occurred producing a surface with one thousand Angstrom average terrace lengths and predominantly double layer step heights. The RHEED pattern was nearly instrument limited at in-phase angles of incidence, with little background intensity and bright Kikuchi lines. Subsequent deposition showed only weak oscillations in the RHEED intensity, in contrast to growth on the (001) surface. The period of the observed oscillations indicates that the layer-by-layer growth involves single-layer steps. Growth of as little as 5 atomic layers on a surface with double steps could not be annealed to give a RHEED intensity as great as the first annealed surface. These measurements reconcile previous LEED results with the oxygen adsorption measurements of Ranke. The results clearly show the dominance of steps in the formation of RHEED streaks.

INTRODUCTION

Attempts to prepare the (110) surface of GaAs by molecular beam epitaxy (MBE) have met with varying degrees of success [1-5]. Most reports indicate that the surface morphology is much rougher than that of a (001) surface prepared under similar conditions. Low growth rates and low substrate temperatures are needed for optimum growth. This suggests that by going to low temperature to avoid Ga agglomeration, the surface diffusion of adsorbed species is also low and responsible for the defects produced. Petroff and co-workers [2] commented that smooth surfaces of the ternary GaAlAs could be prepared at more usual (001) growth conditions if the (110) surface were misoriented towards an As terminated (111)B surface, perhaps indicating that Ga clustering can be prevented by providing a sufficient density of As step sites. Despite these growth difficulties, the singular (110) is important since it connects the vast body of surface science performed on cleavage surfaces with the increasing literature of epitaxial growth on the (001). It is the only semiconductor whose surface structure is known with certainty [6], making it the best surface to test microscopic models of growth as well as, for example, dissolution processes [7]. The purpose of this work was to characterize the surface morphology of GaAs(110) prepared by MBE using reflection high-energy electron diffraction (RHEED). Especially because of the high mobility of Ga on the (110), we expect that surface steps are important in the growth and will use RHEED to study their formation.

There have been three previous low-energy diffraction (LEED) studies of steps on GaAs(110) surfaces prepared both by vacuum cleavage and by ion bombardment [8-10]. All found that surfaces with single-layer steps were formed and that the average terrace lengths would increase upon annealing. Clearfield and Lagally [10] (CL) examined the annealing rate and determined that two separate processes were involved. In addition Ranke and co-workers

[5] modeled the adsorption of oxygen at defect sites on vicinal GaAs(110) surfaces prepared both by ion bombardment followed by annealing and by MBE. On the ion-bombarded surfaces, like the LEED studies, they found single layer steps. On the MBE prepared surfaces, which were inclined by less than a few degrees toward the (111)A (surfaces that Petroff et al found were rough), double layer steps were observed. For the latter surfaces prepared by MBE, the step distribution close to the (110) and toward the (111)B were less certain. To account for the difference between the ion-bombarded and MBE surfaces they, like Clearfield and Lagally, suggested that more than one ordering mechanism operated on the surface. In the next sections we will show that on nearly singular GaAs(110) surfaces either single- or double-layer steps could be prepared by MBE, depending only on the substrate temperature. Hence the differences in the processes might not be due to the damage introduced into the lattice by ion bombardment. Further we show that GaAs(110) grows by single-layer step propagation even though the formation of double-layer steps is preferred.

EXPERIMENTAL

Sn doped GaAs(110) wafers (Morgan, nominally 10^{18} cm^{-3} Si) were prepared by the methods normally used for the (001) [11] with the exception that small samples were cleaved from the wafers after etching. The samples were about 1 cm on edge. Upon introduction into the growth chamber of the MBE apparatus, the samples were heated, under an As$_4$ flux, to 900K for less than 5 min to drive off the surface oxide left by the chemical etch. The sample was then cooled to 700K where the initial depositions were performed. The sample temperature was measured with a thermocouple pressed against the back of the sample holder. Absolute temperatures were known to within $\pm$ 20°C. The apparatus and procedures for the 10 keV RHEED measurements have been described elsewhere [11,12]. Samples were found to be misoriented by less than 2 mrad by the method of x-ray goniometry and by the RHEED method of Pukite et al. [18]. The As$_4$ flux was maintained at 3 x 10^{14} molec$\cdot$s$^{-1}\cdot$cm^{-2} throughout the experiments in order to prevent depletion of As from the surface.

DISCUSSION AND RESULTS

To characterize the step distribution on these GaAs(110) surfaces we will apply an analysis similar to that used by Henzler [8] and by Lagally and co-workers [9,10] but our measurements will be made with RHEED. The fundamental idea is that electrons scattered from the top and bottom terraces of surface steps will interfere constructively or destructively, depending upon the path length difference. In the LEED studies this path length difference is varied by changing the electron energy; in our RHEED measurements it is more convenient to vary the angle of incidence. At several points in the growth, we measure the intensity along the length of the specular RHEED streak at several angles of incidence. This is similar to LEED measurements of the intensity across a diffracted beam; the main differences are that in these RHEED measurements the component of momentum transfer perpendicular to the surface is not as constant and that RHEED is sensitive to order over much larger distances [13]. At angles of incidence where scattering is constructive (in-phase or Bragg angles) the diffraction is insensitive to steps and the diffracted beam is sharp. At angles where diffracted electrons from different surface levels are π out of phase the interference is destructive, and because of the range of terrace lengths the beam is broadened. Detailed discussions are given in refs. 14 and 15.

After the initial desorption of the surface oxide, the diffraction pattern was weak with a diffuse background evident. With the sample at 700K about 20 layers of GaAs were deposited. In agreement with Kroemer [1] we had found this temperature to give the strongest diffraction pattern during slow ($\lesssim$1/3 μm/h) steady-state growth on this nearly singular surface. At this temperature CL [10] found a noticeable change in the width of the diffracted beams on ion-damaged samples after annealing for about 10 min, indicating that surface species are mobile. After this initial short deposition the specular intensity doubled and the diffuse background became relatively weaker. Fig. 1 shows angular profiles of the specular streak from the resulting surface. These curves were measured with the incident electron beam directed along the [$\bar{1}$10] axis. For single-layer (110) steps, constructive interference should occur at glancing angles of incidence of 31 and 62 mrad (no refraction correction is needed) and destructive interference at 46 mrad. In Fig. 1 the intensity along the specular streak is plotted vs. the angular deviation from the peak. Note that close to the out-of-phase angle the beam is broad (i.e. the streak is long) and at the Bragg angles the beam is sharp. To rule out double-layer steps one in principle could fit the data to a model and then check for the different angle of incidence dependencies of random double and single layer steps [14]. Though we have not yet done this, it is clear that single steps are present. At the Bragg angle, the beam is about 0.6 mrad wide, corresponding to the instrument limit. To determine the mean terrace size from the out-of-phase profile, one needs to make some assumptions about the distribution of steps. If the steps are non-interacting and the distribution of steps is the same on each level, then the mean terrace size is of the order of the reciprocal of the half-width [16] of the diffracted beam. Taking into account the low angle at which the Ewald sphere cuts the reciprocal lattice rod, this corresponds to about 200Å. The asymmetry of the profiles is similar to what is observed for two-level systems where one can fit the data by calculating the intersection of the Ewald sphere with a step-broadened reciprocal lattice rod [15]. Thus in contrast to the MBE experiments of Ranke et al., single-layer steps can be obtained on MBE prepared material when growth takes place at low temperature. Apparently ion damage is not needed to limit the surface mobility.

The annealing behavior of the step distribution is shown in Fig. 2. Keeping the As$_4$ flux constant, the sample temperature was raised to about 790K and the intensity along the streak scanned at increasing times. For these scans the glancing angle of incidence was fixed at 46 mrad which is the out-of-phase angle for a surface with steps that are a single atomic layer high. The time required to record a scan was 10 s. As shown, the specular beam sharpens to 1.7 mrad after 35 min at this temperature. A few points are worth noting. First, we could not reach steady state at 790K quickly enough to look for the break in the time dependence of the half-width observed by CL [10]. Second, though these data were measured with the incident electron beam along a symmetry axis, similar results are also obtained a few degrees away from symmetry. Hence we do not think that dynamic effects are important in analyzing the shape of the diffracted beam (they are clearly important in analyzing the intensity). Finally, the observed sharpening of the diffracted beam during annealing at the single-layer out-of-phase angle could mean either (1) the average terrace length of the existing steps becomes large or (2) the step height changes to two (110) layers. The latter possibility arises because the single-layer out-of-phase is also a double-layer Bragg angle.

To distinguish between these two different distributions, the angular profile of the specular streak was measured for a few angles of incidence as shown in Fig. 3. For double-layer steps the Bragg angles should be near 15, 31, and 46 mrad, with out-of-phase angles halfway in between. If the surface were single-layer stepped, then 46 mrad would be an out-of-phase angle and curves on either side of it would be sharper; instead, the data clearly indicates that double-layer steps predominate. Some single layer steps remain even after the annealing procedure since the angular profile at the double-layer Bragg (single-layer out-of-phase) is broader than the instrument response of the diffractometer. It is interesting that neither

Henzler [8] nor Lagally and co-workers [9,10] saw this transition. A major difference is that the instrument response in these RHEED experiments was greater than that of their LEED instruments: the average terrace length of these double layer steps is of the order of $4/(k\cdot0\cdot\delta0)\sim1000\text{A}$ [14] which would have been unobservable within the several hundred Angstrom instrument limit of their experiments. Two other differences were that these samples were not damaged by ion-bombardment and that an external As_4 flux was present during the heat treatment so that the surface did not become As deficient. Our results are consistent with the more indirect measurements of Ranke et al. [5] who, in order to account for the orientation dependence of oxygen uptake, showed that double-layer steps were present on MBE prepared surfaces. From their data, though, the evidence on the singular surface was less conclusive than on the vicinal surfaces oriented towards the (111)A. On ion-damaged surfaces Ranke et al. had found only single-layer steps near the singular orientation; but because there was no external As flux they could not anneal to quite as high a temperature. In light of these measurements it is surprising that Lagally and co-workers did not see the formation of double-layer steps. They did observe a lengthening of the single-layer terraces.

Using RHEED one can also follow the steps during growth. Fig. 4 shows the intensity of the specular RHEED beam at the single-layer out-of-phase angle as a function of time after growth is initiated on the annealed (110). For comparison, data from (001) growth are also shown, though under slightly different growth conditions. These intensity oscillations result from the competition between nucleation and step-propagation in the layer-by-layer growth of GaAs on these surfaces [12, 14, 17]. The period of the oscillations corresponds to the time required to deposit a monolayer of GaAs. The striking features are that (1) the intensity oscillations are typically much weaker on the (110), dying out after about 5 periods, and (2) the period of the oscillations on the (110) corresponds to the deposition of single layers of GaAs, in this case 40s. These oscillations have been observed during the initial growth on annealed substrates held between 700K and 900K. Thus even at substrate temperatures where the surface prefers double-layer steps in steady state, the system tries to grow via the nucleation and propagation of single-layer steps. Even at the highest temperatures there is insufficient time for the double-layer steps to form.

If growth on this nearly singular surface is interrupted after only 5 layers are deposited and then allowed to anneal, the resulting pattern corresponds to a surface that is never as well ordered as the starting annealed surface. For example, after deposition and annealing at 850K, the 46 mrad angular profile is as sharp as the initial surface but the intensity is reduced by a factor of two, indicating that there is some random disorder over the surface other than atomic steps.

It is important to realize that there was a large difference in the quantitative scale of these measurements and those reported by Lagally and co-workers on ion-damaged GaAs(110). First, the average terrace length of the stepped surfaces prepared by deposition at low temperature began at several hundreds of Angstroms -- already above the instrument limit in the LEED experiments. The average terrace length of the annealed surface was an order of magnitude larger. Second, because of the greater sensitivity of RHEED to large distances, the annealing process could be followed for much longer times. Unfortunately, the sample heating and temperature measurements of our MBE apparatus are not yet suitable for making comparisons with either the early annealing behavior or the activation enthalpy measurements of CL. Third, the time variation of the width of the angular profiles in Fig. 2 can be seen to be far more rapid than the width variation reported by CL.

Last, we should mention two preliminary measurements of the kinetics of the surface migration processes. After the initial deposition of 20 layers at 700K the intensity increased with a $t^{1/2}$ time dependence. In addition, the average terrace length from the data of Fig. 2 increases according to

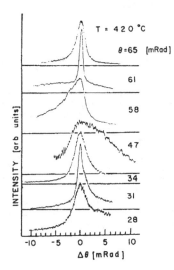

Figure 1. Angular intensity profiles of (00) beam at several incident angles. Bragg angles are multiples of 31 mrad for single layer steps.

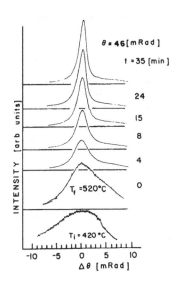

Figure 2. Intensity profiles of the (00) beam showing the transition from single- to double-layer steps at 790K.

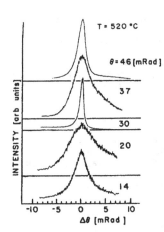

Figure 3. Intensity profiles of (00) beam after annealing at 790K. Bragg angles for double-layer steps are multiples of 15 mrad.

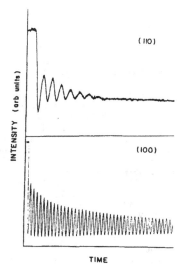

Figure 4. Typical RHEED Intensity Oscillations vs. time for the (110) and (100) surfaces. The period corresponds to the single-layer deposition time (different time scales).

$t^{0.4\pm0.1}$. Both indicate the role of surface diffusion.

CONCLUSION

The step distribution of very thin layers of GaAs deposited on nearly singular GaAs(110) surfaces by molecular beam epitaxy is shown to depend sensitively on the substrate temperature. At low temperatures random steps with several hundred Angstrom terrace lengths and single layer step heights are formed. The annealing behavior of these stepped surfaces is different than that observed by LEED measurements of ion-damaged, As deficient surfaces. At about 800K long average terrace lengths and double-layer steps are formed, directly corroborating the measurements of Ranke et al. on vicinal surfaces. Further growth on the very long terraces of these surfaces resulted in cyclic variations in the diffracted intensity, corresponding to the growth of single layer steps. After subsequent depositions disorder was present that could not be annealed, in contrast to the rapid anneal of the initial growth. The angular dependence of the shape of the RHEED streaks agrees with an analysis that emphasizes the role of steps. The shape and angular dependence are observed to change dramatically with temperature. The shape of the RHEED beams was not observed to depend on the azimuthal angle of incidence. Steps were found to be a major cause of RHEED streaks on these MBE surfaces.

ACKNOWLEDGEMENTS

This work was partially supported by the Corrosion Center of the University of MN, by the MN Microelectronics and Information Sciences Center, and by the NSF under grant DMR-8319821.

REFERENCES

1. H. Kroemer, K. J. Polasko, and S. C. Wright, Appl. Phys. Lett. 36 (1980) 763.
2. P. M. Petroff, A. Y. Cho, F. K. Reinhart, A. C. Gossard, and W. Weigmann, Phys. Rev. Lett. 48 (1982) 170.
3. J. M. Ballingall, and C. E. C. Wood, Appl. Phys. Let. 41 (1982) 947; J. Vac. Sci. Technol. B1 (1983) 162.
4. W. I. Wang, J. Vac. Sci. Technol. B1 (1983) 630.
5. W. Ranke, Physica Scripta. T4 (1983) 100; W. Ranke, Y. R. Xing, and G. D. Shen, J. Vac. Sci. Technol. 21 (1982) 426; W. Ranke, Y. R. Xing, and G. D. Shen, Surf. Sci. 120 1982) 67.
6. A. Kahm, Surf. Sci. Reports, 3 (1983).
7. H. Gerischer, J. Vac. Sci. Technol. 15 (1978) 1422.
8. M. Henzler, Surf. Sci. 22 (1970) 12.
9. D. G. Welkie and M. G. Lagally, J. Vac. Sci. Technol. 16 (1979) 784.
10. H. M. Clearfield and M. G. Lagally, J. Vac. Sci. Technol. A2 (1984) 844.
11. P. R. Pukite, J. M. Van Hove, and P. I. Cohen, J. Vac. Sci. Technol. B2 (1984) 243.
12. J. M. Van Hove, C. S. Lent, P. R. Pukite, and P. I. Cohen. J. Vac. Sci. Technol. B1 (1983) 741.
13. J. M. Van Hove, P. R. Pukite, P. I. Cohen, and C. S. Lent, J. Vac. Sci. Technol. A1 (1983) 609.
14. C. S. Lent and P. I. Cohen. Surf. Sci. 139 (1984) 121.
15. P. R. Pukite, C. S. Lent, and P. I. Cohen, to be submitted to Surf. Sci.
16. T.-M. Lu and M. G. Lagally, Surf. Sci. 120 (1982) 47.
17. J. M. Van Hove, P. R. Pukite, and P. I. Cohen, J. Vac. Sci. Technol. B (1985) in press.
18. P. R. Pukite, J. M. Van Hove, and P. I. Cohen, Appl. Phys. Lett. 44 (1984) 456.

THE CHARACTERIZATION OF THIN FILMS AND LAYERED STRUCTURES USING X-RAY ABSORPTION AND REFLECTION AT GRAZING INCIDENCE

S.M. HEALD, J.M. TRANQUADA, D.O. WELCH, AND H. CHEN
Metallurgy and Materials Science Division
Brookhaven National Laboratory, Upton, New York 11973

ABSTRACT

X-rays at grazing incidence have a short, controllable penetration depth and are well suited as a probe of surface and interface structures. This paper examines the possibility applying grazing-incidence reflectivity and Extended X-Ray Absorption Fine Structure (EXAFS) measurements to such systems. Results are presented for an Al-Cu couple for which both high resolution reflectivity and interface EXAFS measurements are made. The latter results are the first interface specific EXAFS data to be reported. Distinct changes in both signals are observed upon annealing, demonstrating the potential of the techniques.

INTRODUCTION

X-rays at glancing angles have properties which make them very useful for studying thin surface layers and interfaces. Their potential was first pointed out by Parratt in 1954 [1] and in recent years glancing angle techniques have been fruitfully applied to x-ray diffraction studies of surface systems [2]. For polycrystalline or disordered surface systems, however, diffraction techniques often have limited utility and other techniques are required. One technique which has been very successfully applied to many disordered and/or dilute systems is the Extended X-ray Absorption Fine Structure (EXAFS) [3]. In this paper we use glancing angle techniques to make the first reported measurement of EXAFS from an interface region. Such measurements can provide detailed information about the local environment of the interface atoms on an atomic scale. More macroscopic parameters such as the interface roughness can be derived from high resolution reflectivity measurements, results of which are also presented.

The system studied was an Al-Cu thin-film couple. Previous work [4, 5] has established compound formation at the interface upon annealing, with $CuAl_2$ being formed first. However, the techniques used, Rutherford backscattering and x-ray diffraction, only had a resolution of about 100A. The techniques described here have higher resolution, and we have attempted to look at the initial stages of compound formation.

X-RAYS AT GLANCING ANGLES

The index of refraction for x-rays is slightly less than one and is usually written $n=1-\delta-i\beta$. This means that at glancing angles the x-rays will undergo total external reflection. The angle below which this occurs is called the critical angle and can be found from Snell's law to be, $\theta_c = \sqrt{2\delta}$. For energies away from absorption edges, δ is proportional to the electron density which means that the total reflection region for heavy elements extends to larger angles than for light elements. This is illustrated in Fig. 1(a) where the reflectivity curves for Al and Cu are compared.

Also shown in Fig. 1(a) is the calculated reflectivity for 1000 A of Al on Cu. Striking oscillations are seen which are a result of interference between the incident and reflected beams. When the angle is above the Al critical angle the x-rays can penetrate to the Al-Cu interface

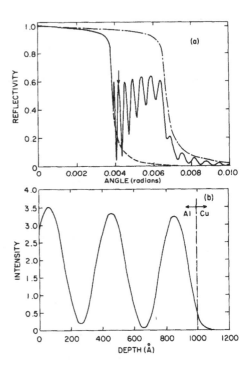

Fig. 1 a) Calculated reflectivity for 1000 A
of Al on Cu at 8.6
keV. For compar-
ison the reflectiv-
ities for pure Al
(dashed line) and
pure Cu (dot-
dashed line) are
also shown.

b) X-ray intensity as
a function of pen-
etration depth for
the system in (a)
at the angle indi-
cated by the arrow.

where they will be reflected as long as the angle remains less than the
Cu critical angle. Interference between the two waves sets up a standing
wave field in the Al layer as shown in Fig. 1(b). In this case if the
standing wave field is a maximum at the Al surface then the reflectivity
also has a maximum. Thus, the oscillations in the reflectivity are a
result of the changing standing wave fields in the Al layer as a function
of angle.

The utility of glancing angle x-rays to the study of surface and
interface systems comes from their small penetration depth which is also
illustrated in Fig.1(b). When total external reflection occurs, the
interface-penetrating evanescent wave decays exponentially with penetration depth, which for the example shown has a characteristic penetration
depth of ~25 A. The penetration can be controlled by varying the angle
and increases to several thousand Angstroms above the critical angle.
Thus, if a signal can be measured which is specific to Cu atoms then the
local environment of the interfacial Cu atoms can be probed. One such
signal is the x-ray fluorescence, and it will be shown below how this signal can be used to detect the EXAFS from an interfacial region.

SAMPLES

The samples were prepared by evaporation onto 2.5 x 5 cm float glass
substrates. A thin Cr binding layer was used between the glass and the
~2000 A Cu layer. The pressure during evaporation was about 10^{-6} Torr and
the three metals were evaporated sequentially without breaking the vacuum.
To insure uniform thickness layers a source to substrate distance of ~28
cm was used. Four samples were prepared during the same run. This paper
will present results on two: one as-prepared and one annealed at 100°C

for four hours in an N_2 atmosphere. All x-ray measurements were made at the Cornell High Energy Synchrotron Source (CHESS) using the standard EXAFS beam line C2.

REFLECTIVITY STUDIES

Figure 2(a) compares the reflectivity for the annealed sample with a calculated curve assuming 1000 Å of Al. The calculation assumes an ideally flat interface and surface with the only adjustable parameter being the layer thickness. Good agreement is found, with the experimental oscillations having a somewhat smaller amplitude than the theoretical prediction. This is not unexpected since our angular resolution was only about 0.05 mrad, and real surfaces and interfaces have roughness associated with them. Other authors [1,6] have shown the sensitivity of such measurements to roughness parameters, and we are currently attempting to extract a quantitative measure of the surface and interface roughness from the results.

The potential of such studies for revealing changes in interfacial roughness is demonstrated in Fig. 2(b) which compares the reflectivities from the as-prepared sample and the annealed sample. The first few oscillations are seen to be distinctly sharper in the annealed case indicating changes have occurred in the condition of the interface and/or surface.

Fig. 2. Top: Comparison of the reflectivity measured for the annealed sample with the theoretical calculation shown in 1(a). Bottom: Comparison of reflectivity in the oscillatory regions for the as-prepared and annealed samples.

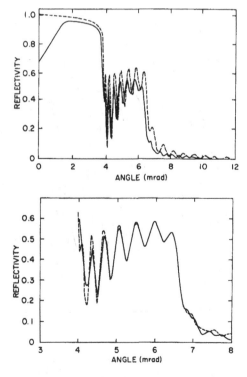

EXAFS RESULTS

EXAFS data were obtained by scanning the photon energy through the Cu K edge while maintaining a fixed angle. The changes in Cu absorption can be seen in both the reflected and fluorescence signals, but are much clearer in the fluorescence since that signal is Cu specific. Since the fluorescence probability is proportional to the Cu absorption coefficient the fluorescence can be used to obtain the Cu EXAFS. Earlier work on monolayer Au films demonstrated the superior sensitivity of this technique [7]. Figure 3 shows the EXAFS obtained after removing the slowly varying background and normalizing to the Cu edge jump. These scans each required only about 10 minutes of acquisition time and demonstrate the high signal/noise available with the technique.

Such EXAFS spectra depend on the local atomic environment of the Cu and contain information about the number, distance, type and disorder of the neighboring atoms. A detailed analysis of these spectra will be the subject of a later paper, but some features are evident even from a cursory examination. The amplitude of the EXAFS is reduced as compared with that from bulk copper for both of the interfacial systems, with that of the annealed sample being the smallest. The signal that remains, however, is very similar to that obtained for pure Cu, which indicates that there are two types of Cu atoms: those which maintain a Cu metal environment and those in a highly disordered Cu-Al environment which contribute little to the EXAFS signal. There seems to be no evidence for significant amounts

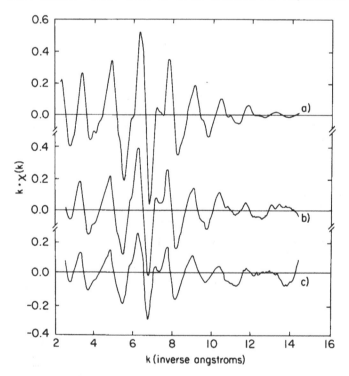

Fig. 3. Background-subtracted EXAFS signals obtained from: a) bulk copper, b) the as-prepared sample at θ=5 mrad, c) the annealed sample at θ=5 mrad.

of ordered CuAl$_2$, which implies compound formation has not yet begun. Since most of the signal comes from Cu atoms within about 50 A of the interface and there remains a significant Cu metal component, the interfacial region (in which most of the disordered Cu sites are presumably located) in both samples must be quite thin. A rough estimate obtained from comparing amplitudes gives 10-15 A for the as prepared sample, and the thickness increases to ~20 A upon annealing at 100°C. Clearly an extension of the studies to higher annealing temperatures would be interesting.

SUMMARY

In this paper studies of Al-Cu bilayers were used as an example to demonstrate the potential of glancing-angle reflectivity and EXAFS measurements in studying interfacial regions. Changes were observed in the data from both types of measurements as a function of annealing which are indicative of changes in interfacial roughness in the case of the reflectivity measurements, and of changes in the local Cu environment for the EXAFS measurements. A key feature of both techniques is that they do not require long range order, and thus can be applied equally well to polycrystalline or disordered systems. The major limitation of these techniques is that it must be possible to reflect from the interface. This means the top layer must have a smaller critical angle than that of the substrate. This still leaves a great number of systems open to study, and continued development of these techniques should make possible a large number of interesting interface experiments.

ACKNOWLEDGMENTS

We wish to thank R. DiNardo for his help in preparing the samples and the CHESS staff for their help in making the measurements. This work was supported by the Division of Materials Sciences, Office of Basic Energy Sciences of the Department of Energy under Contract Nos. DE-AS05-80-ER10742 and DE-AC02-76CH00016.

REFERENCES

1. L. G. Parratt, Phys. Rev. 95, 359 (1954).

2. W. C. Marra, P. Eisenberger, and A. Y. Cho, J. Appl. Phys. 50, 6927 (1979); S-L Weng, A. Y. Cho, W. C. Marra, and P. Eisenberger, Solid State Commun. 34, 843 (1980).

3. E. A. Stern and S. M. Heald, "Basic Principles and Applications of EXAFS" in Handbook of Synchrotron Radiation, Vol. 1, E. E. Kock, Editor (North-Holland, Amsterdam, 1983).

4. H. T. G. Henzell, R. D. Thompson, and K. N. Tu, J. Appl. Phys. 54, 6923 (1983).

5. R. A. Hamm and J. M. Vandenberg, J. Appl. Phys. 56, 293 (1984).

6. D. H. Bilderback, Reflecting Optics for Synchrotron Radiation, Proc. SPIE 315, 90 (1981); L. Nevot and P. Croce, Rev. Phys. Appl. 15, 761 (1980).

7. S. M. Heald, E. Keller, and E. A. Stern, Phys. Lett. 103A (1984).

THE XPS SEARCHLIGHT EFFECT: A NEW ANALYTICAL TOOL
FOR LAYERED STRUCTURES, EPITAXY, AND INTERFACES

W. F. Egelhoff, Jr., Surface Science Division, National Bureau of Standards,
Gaithersburg, MD 20899

ABSTRACT

Enhanced core-level peak intensities at angles corresponding to the
internuclear axes among the near surface atoms is a characteristic feature of
angle resolved XPS. This phenomenon, which is due to constructive interference in forward scattering of photoelectrons, acts, in effect, as a searchlight allowing relatively easy mapping out of the structural arrangement
atoms in the near-surface region. Examples which illustrate the usefulness
of the XPS searchlight effect are presented.

Introduction

In studies of layered structures, epitaxy, and interfaces one of the
most important types of information sought is the structural arrangement of
atoms at the surface and in the near-surface region. With this aim in mind
great efforts and progress have been made in recent years using techniques
such as LEED, SEXAFS, NEXAFS, RSB, TEM, etc. [1] However, serious shortcomings remain in present capabilities for structural analysis. Some of the
methods in current use are rather cumbersome, requiring extensive computer
analysis or long data acquisition times, some cannot be used for in situ
studies, and some provide only rather indirect structural insights. However
a recent reinterpretation of a well-known phenomenon in angular-dependent
x-ray photoelectron spectra (XPS) indicates that XPS can be used to great
advantage as an important, new structural tool for studying layered materials, epitaxy and interfaces. [2-7] Its importance derives from the fact that
it does not suffer from the shortcomings of the other techniques.

In angular-dependent XPS studies of single crystals a phenomenon has
long been known in which core-level peaks exhibit enhanced intensities along
major crystal axes. [10] The phenomenon is generally attributed to electron
channeling (the Kikuchi effect). [11] In recent work the evolution of these
core-level intensities has been studied as a function of film thickness in
the epitaxial growth of Fe, Co, Ni, and Cu on single crystal surfaces Ni(100)
and Cu(100), and in alternating epitaxial layers of these elements. [3] It
has been found that in the conventional electron channeling model the
enhancement cannot even begin to occur until a thickness of about 8 ML is
reached. [4] The actual physical basis of the phenomenon is found to be
forward scattering of the photoelectrons by overlying atoms in the
lattice.[2-7] Thus the enhancement occurs along major crystal axes only
because these coincide with the internuclear axes of the lattice. Forward
scattering is found to be the dominant photoelectron interaction with lattice
atoms whenever the photoelectron kinetic energy exceeds a few hundred eV. [3]
With this new understanding of its physical basis, the phenomenon becomes a
useful new tool. It readily provides information on structural aspects of
epitaxial growth, surface segregation, and surface alloying or interdiffusion
that would otherwise be difficult to obtain. Moreover, since changing the
angle of XPS observation permits selective enhancement of core levels from
different underlying layers near the surface, the core-level spectra of the
different layers can be separated to determine the layer-wise core-level
binding-energy shifts for thin films and interfaces. [3]

Experimental

The work reported here was performed in an extensively modified AEI-ES200 XPS instrument* under ultrahigh ($<10^{-11}$ torr) vacuum conditions. The Cu was evaporated from W filaments and the deposited thicknesses were determined using two quartz-crystal thin-film-thickness monitors (TFMs) as well as two ion gauges. One is mounted in the line of sight of the metal flux and one well out of the line of sight to measure the background pressure of gases in the chamber. Calibrating the metal flux at the ion gauge using large TFM readings (for good signal-to-noise) yields, for example, 3×10^{-7} Torr ·sec as the ion gauge flux at the sample surface producing a deposit of one monolayer of Cu [$\equiv 1.54 \times 10^{15}$ atoms/cm^2, i.e., a (100) structure]. This ion gauge calibration is particularly useful for low coverage studies where the TFMs lack precision.

The angle between the x-ray source and the entrance to the analyzer is 90°. The sample is rotated so that the surface normal can vary from pointing at the x-ray source to pointing at the analyzer entrance. The analyzer is a 180° hemisphere type and the x-rays are Mg Kα. See Ref. 2 for further details.

Results and Discussion

Figure 1 presents data of a simple but instructive example illustrating how the XPS searchlight effect is manifest. When 0.5ML of Ni is deposited on Cu(100) at 100K the Ni resides on the surface as adatoms. The Ni core-level peaks have an angular intensity distribution much like the instrument response indicating that the emission is approximately isotropic. Since, in this surface structure, there are no atoms overlying the Ni there is nothing for the photoelectron wave to scatter off on its path to detector so the distribution observed Fig. 1a would be expected to be approximately isotropic. [12] However, when a thin, epitaxial Cu film (~ 1ML) is deposited over the Ni, Fig 1b, a marked enhancement is observed in the Ni core-level peak intensities at a polar angle of 45° in the <100> surface azimuth. This angle corresponds to the Ni-Cu internuclear axis.

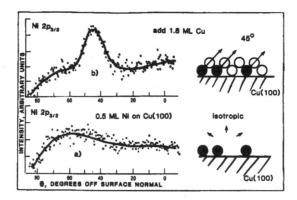

Figure 1   The dependence of the Ni $2p_{3/2}$ core-level peak intensity on polar angle (in the <100> surface azimuth) for a) 0.5ML of Ni on Cu(100) and b) after adding 1.8 ML of epitaxial Cu to this surface. The surface profiles are illustrated schematically (the mirror image arrows at 45° are not shown).

The magnitude of this enhancement at 45° depends on the coverages of Ni and Cu involved but can be as large as a factor two. [13] The physical basis for this enhacement is a constructive interference along the internuclear axis between the initial outgoing photoelectron wave and the wave forward scattered off an overlying Cu atom. [2-7] Since there is an intensity enhancement corresponding to an important direction, the internuclear axis, it seems appropriate to term this a searchlight effect.

The value of the XPS searchlight effect may be readily appreciated by reflecting on Fig. 1. First, it gives us the bond axis or angle, something rather difficult to obtain by other surface structural techniques. Second, it gives us that information very easily. No complex deconvolutions, multiple scattering calculations, or long data collection times are required. It merely involves rotating the sample while the spectrometer is fixed on a particular core-level peak. [14] Third, it gives element-specific information since the core levels are characteristic of the element. [15] This should be particularly useful in studying multielement materials. [15] This element specificity has already been found to be particularly useful for observing and getting structural information on low-concentration components in various Cu-Ni combinations. [15] Fourth, it permits observation of changes in real time (by rotating the crystal back and forth) so the surface and interface structural dynamics can be followed while varying parameters such as temperature. This is, for example, one of the few practical ways to follow the kinetics of interdiffusion or surface alloying. Fifth, it can readily answer certain important questions about nucleation morphology in the initial stages of epitaxial growth. Figure 2 illustrates such a case.

In the course of studies of the epitaxial growth of Cu on clean Ni(100) it was of interest to investigate how the Cu epitaxy would be influenced by a contaminating layer on the Ni(100) surface. The basic result is illustrated in Fig. 2.

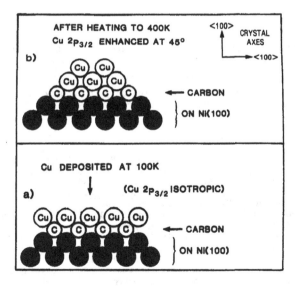

Figure 2 An illustration of how up to 1ML of Cu deposited on a carbon covered Ni(100) surface lies flat at 100K but agglomerates into a double layer upon heating to 400K.

For a Ni(100) surface with a carbon c(2x2) structure on it, up to a monolayer
of Cu could be deposited with the surface at 100K without any indication of
any Cu second layer formation (the Cu $2p_{3/2}$ intensity was isotropic). It had
been anticipated that the carbon would reduce the reactivity of the surface
and the Cu would agglomerate. However, this only occurred upon raising the
sample temperature. Upon heating the sample somewhat above room temperature
the peak at 45° began to appear and by 400K it was large enough to suggest
that all the Cu present had agglomerated into a double layer. Further
heating (to ~ 500K) produced a peak at 0° (in the Cu CVV data [16]) indicating the formation of Cu triple layers.

While this particular data on the influence of carbon on Cu/Ni(100)
epitaxy is not in itself surprising, it is significant in that it illustrates the fact that the XPS searchlight effect, as a probe of short-range
order, can be a useful, new tool for following the atomic morphology at the
initial stages of epitaxy. It can provide a kind of information about events
crucial to epitaxy that would otherwise be difficult if not impossible to
acquire. Moreover, it can do this quickly and easily.

Conclusions

The major conclusions of the work can be summarized as follows.
1) The XPS searchlight effect is a useful, new probe of short-range
   order for studying layered structures, epitaxy, and interfaces.
2) It has the advantage of providing structural information rapidly and
   easily, without complex computer calculations.
3) It may be used for real-time studies of surface structural dynamics.
4) It provides information of a type not readily accessible by other
   surface analytical techniques.

Acknowledgements

The advice, assistance, and encouragement of Dr. C. S. Fadley and
Dr. S. Y. Tong during the course of this work are gratefully acknowledged.

*This commercial instrument is identified only to specify the experimental
conditions and does not signify any endorsement by NBS.

References

1) For a good overview, see: S. Y. Tong, Physics Today, Aug. 1984, p. 50.
2) W. F. Egelhoff, Jr., J. Vac. Sci. Tech. A2, 350 (1984).
3) W. F. Egelhoff, Jr., Phys. Rev. B30, 1052 (1984).
4) W. F. Egelhoff, Jr., in Proc. Int. Conf. Str. Surf., Berkeley, 1984, M.
   A. VanHove and S. Y. Tong, Eds., in press.
5) C. H. Poon and S. Y. Tong, Phys. Rev. B 30, 6211, (1984); and E. Bullock
   and C. S. Fadley, Phys. Rev. B, in press.
6) R. A. Armstrong and W. F. Egelhoff, Jr., Surf. Sci., submitted.
7) The principles of photoelectron forward scattering have been recognized
   in previous studies of the carbon 1s polar intensity distribution in
   adsorbed molecular CO, [8] and in studies of the oxygen 1s azimuthal
   intensity distribution for oxygen adatoms. [9] However, only very
   recently has it been found to apply to angle resolved XPS of bulk single
   crystals and its great potential as a structural tool for epitaxy and
   interfaces been recognized. [3,4]
8) S. Kono, S. M. Goldberg, N. F. T. Hall and C. S. Fadley, Phys. Rev.
   Lett. 41, 1831 (1978), L. G. Petersson, S. Kono, N. F. T. Hall, C. S.
   Fadley, and J. B. Pendry, Phys. Rev. Lett. 42, 1545 (1979).

9) S. Kono, C. S. Fadley, N. F. T. Hall, and Z. Hussain, Phys. Rev. Lett. 41, 117 (1978); see especially Fig. 3 in S. Kono, C. S. Fadley, N. F. T. Hall, and Z. Hussain, Phys. Rev. B22, 6085 (1980).

10) K. Siegbahn, U. Gelius, H. Siegbahn, and E. Olson, Phys. Scr. 1, 272 (1970); C. S. Fadley and S. A. L. Bergström, Phys. Lett. 35A, 375 (1971); K. Siegbahn, U. Gelius, H. Siegbahn, and E. Olson, Phys. Lett. 32A, 221 (1970); N. E. Erickson, Phys. Scr. 16, 462 (1977); S. Evans and M. D. Scott, Surf. Interface Analysis 3, 269 (1981); S. Evans, J. M. Adams and J. M. Thomas, Phil Trans. Roy. Soc. Lond. 292, 563 (1979).

11) See for example, Practical Surface Analysis, D. Briggs and M. P. Seah, Eds., John Wiley & Sons, New York, 1983, p. 136; or Electron Spectroscopy, Vol. 2, C. R. Brundle and A. D. Baker, Eds., Academic Press, London, 1977, p. 132.

12) Note here that the core-levels, being filled shells, are sperically synmetric, and the angle between the incident photon beam and the collected photoelectrons is a constant 90° (the sample is rotated).

13) W. F. Egelhoff, Jr., J. Vac. Sci. Tech., submitted.

14) To be more precise, unless the XPS system is automated it will take two angular sweeps, one with the spectrometer set on the peak and one with it set for a few eV larger kinetic energy to record the background which should subtracted off. See Ref. 3.

15) It should be stated that this particular advantage, elemental specificity, is shared by SEXAFS and NEXAFS.[1] However, they do not provide the same kind of information as the XPS searchlight effect.

16) Information on the third layer in an epitaxial structure is not manifest in data on deep core levels (low photoelectron kinetic energy). Higher kintic energy peaks do, however, provide such information. See Refs. 3 and 4 for further details.

OBSERVATION OF THE CdTe-GaAs INTERFACE
BY HIGH RESOLUTION ELECTRON MICROSCOPY

N. Otsuka[a], L. A. Kolodziejski[b] R. L. Gunshor[b] and S. Datta[b]
Purdue University   W. Lafayette, IN 47907
R. N. Bicknell and J. F. Schetzina
Department of Physics North Carolina State University
Raleigh, North Carolina 27695-8202

ABSTRACT

CdTe films have been grown on GaAs substrates with two types of interfaces -
one with the epitaxial relation $(111)_{CdTe} || (100)_{GaAs}$ and the other with
$(100)_{CdTe} || (100)_{GaAs}$. High resolution electron microscope observation of the two
types of interfaces was carried out in order to determine the role of the substrate
surface microstructure in determining the epitaxy. The interface of the former type
shows a direct contact between the CdTe and GaAs crystals, while the interface of
the latter type has a very thin oxide layer (~ 10 Å in thickness) between the two
crystals. These observations suggest that details of the substrate preheating cycle
prior to film growth is the principle factor in determining which epitaxial relation occurs in this system. The relation between interfacial structures and the origin of the two epitaxial relations is discussed.

I. Introduction.

Recently considerable efforts have been made on the growth of CdTe by molecular beam epitaxy (MBE) in order to obtain high quality single crystalline films which
will play a key role in device applications of CdTe and CdTe-based ternary compounds [1-10]. Many growth experiments have been carried out utilizing GaAs crystals as substrates because of the availability of high quality single crystal surfaces
[4,6,7,8,9]. It has been found that single crystalline films can be grown on (100) sur-
faces of GaAs with either one of two epitaxial relations: $(111)_{CdTe} || (100)_{GaAs}$ with
$[01\bar{1}]_{CdTe} || [01\bar{1}]_{GaAs}$ [(111) epitaxial relation][4,7,9], or $(100)_{CdTe} || (100)_{GaAs}$ with
$[01\bar{1}]_{CdTe} || [01\bar{1}]_{GaAs}$ [(100) epitaxial relation] [6,8]. Despite the existence of a large
lattice mismatch, high quality single crystalline films have been obtained from both
epitaxial relations.

The two epitaxial relations reportedly occur with similar deposition rates and
substrate temperatures, which suggests that the condition of the substrate surface
may be the primary factor for the selection of these epitaxial relations. There has
been some indications that differing of preheating temperatures of substrate surfaces
affect the occurrence of these two epitaxial relations. Cheung has found in his
laser-assist deposition experiments that the (111) epitaxial relation predominantly
occurred with preheating temperatures higher than 600°C while the (100) epitaxial
relation was obtained more often with lower preheating temperatures [11]. It is
known that the preheating of GaAs substrates at higher temperatures removes oxide
layers completely while causing an As-deficiency and, hence, a change in the microscopic topography of the substrate surface [12]. To date, however, no systematic
study has been made to clarify the microscopic changes responsible for the selective

---

a) School of Materials Engineering. b) School of Electrical Engineering.

occurrence of the two epitaxial relations.

In this paper, we report the high resolution electron microscope (HREM) observation of interfaces between MBE-grown CdTe films and GaAs substrates. Two types of samples, one with a (111) epitaxial film and the other with a (100) epitaxial film have been investigated. The observations show a distinct difference in atomic structure of the interface between the two samples, which clearly indicates that the nature of the substrate surface when the film growth is initiated is a crucial factor determining which epitaxial relation occurs in this system.

2. Observation of Interfaces.

2.1. The Interface Between the (111) Epitaxial Film and Substrate.

A (111) CdTe film was grown on a (100) GaAs substrate using a Perkin-Elmer model 400 MBE system. The substrate was heated at 580°C for the oxide dissolution prior to the deposition, and RHEED showed the coexistence of (2 x 6) and (3 x 6) reconstructed surface structures which is characteristic of the transition state from the Ga stabilized surface to the As stabilized one [13]. Details of the film growth and the substrate surface preparation have been described in earlier reports [9,10]. Cross-sectional specimens parallel to $(0\bar{1}1)$ and (011) planes of GaAs were prepared by mechanical grinding and argon-ion bombardment. A JEM 200 CX electron microscope with a side-entry goniometer was used at an operating voltage of 200 kV. The instrument has a resolution limit of 3.0Å at Scherzer defocus (-- 977 Å) under axial illumination.

Figure 1 is a HREM image of the interface between the CdTe film and the GaAs substrate. The beam direction is parallel to the $[01\bar{1}]$ axis of the GaAs substrate and CdTe film. The image was recorded under axial illumination at nearly the Scherzer defocus. Since the thickness of the area is about 50 Å, the observed image can be interpreted approximately as a projection of the potential distribution in the specimen (weak-phase object approximation [14]), The image shows {111} lattice fringes and weak (200) lattice fringes in both GaAs and CdTe crystals. Microtwins parallel to {111} planes are also seen in the CdTe film. Orientations of those lattice fringes exhibit the (111) epitaxial relation: (111) lattice fringes of CdTe are parallel to the interface [(100) plane of GaAs], and $(\bar{1}11)$ lattice fringes are inclined to the (111) lattice fringes by an angle of 71° but change their direction at twin boundaries.

Under the weak-phase object approximation, each bright spot in the image shown in Fig. 1 corresponds to a pair of Cd and Te atoms in CdTe and that of Ga and As atoms in GaAs [15]. The image shows a nearly perfect one to one correspondence of lattice fringes at the interface between two crystals. It is also seen that the interface is abrupt within the resolution limit of the instrument, and no abnormal lattice spacings are found in the interfacial area. These observations, therefore, indicate that the CdTe crystal has grown on the GaAs surface without forming any transitional layer (at least for the growth condition used in our experiments) which may differ from the result of a recent study suggesting the existence of a thin Te layer in this type of interface [16].

An interesting phenomenon has been observed from this interface. The interface has occasionally shown isolated micropits of the substrate surface as seen in Fig. 2. In the area of the micropit, {111} lattice fringes as well as twin boundaries appear with different orientations from those in the area of the flat portion of the substrate surface. These lattice fringes and twin boundaries are exactly parallel to {111} lattice fringes in the GaAs crystal, implying that this small area has grown with the (100) epitaxial relation. Other micropits were also accompanied by small crystals which have grown with the (100) epitaxial relation. These observations indicate that micropits on the substrate surface provide preferential nucleation sites for a (100) epitaxial crystal while the (111) epitaxial crystal tends to form at flat portions of the substrate surface. (Similar results have been observed by Cheung [11].) In this case, the film has eventually grown with the (111) epitaxial relation because these micropits are well isolated from each other.

2.2. The Interface Between the (100) Epitaxial Film and Substrate.

The growth conditions and the substrate surface preparation for (100) CdTe||(100) GaAs epitaxy have been described in an earlier report [8]. Thermocouple measurements indicated a substrate temperature of 550°C during the brief preheating (3 minutes) at the Mo substrate mount, though the actual substrate surface temperarture could be as much as 50°C lower due to transient heating effects. The TEM observation as well as cross-sectional specimen preparation have been performed using the procedure described in 2.1.

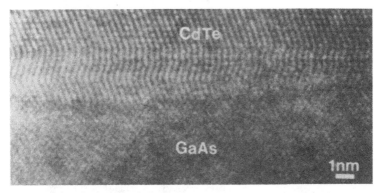

Fig. 1.   High resolution image of the interface between the (111) CdTe epitaxial film and GaAs substrate.

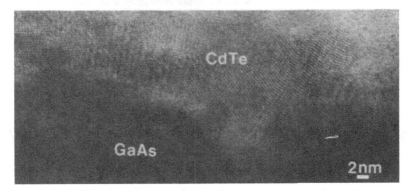

Fig. 2.   High resolution image showing a micropit on the GaAs substrate surface.

The interface between the (100) CdTe film and the (100) GaAs substrate appears as a very flat and smooth plane at medium magnifications. However, at higher magnifications with careful alignment of the orientation of the interface against the electron beam, fine structures of the interface are observed. Figure 3 is a high magnification bright field image of an interfacial area with the beam direction parallel to the interface. In the image, a very narrow white band (~10 Å in width) is seen at the interface. This band image indicates the existence of a very thin layer in the interface between the epitaxial film and substrate. The disappearance of the band in the thicker area is due to the overlapping of the CdTe and GaAs crystals along the imaging direction. It is unlikely that these bands are artifacts such as those due to the interference effect of the electron beam or those produced by the ion-thinning, since no such band images have been observed at interfaces between the (111) epitaxial film and substrate.

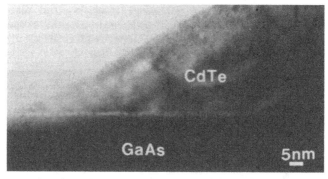

Fig. 3.   Bright field image of the interface between the (100) epitaxial film and GaAs substrate.

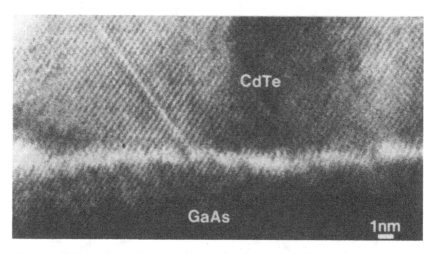

Fig. 4.   High resolution image of the interface between the (100) epitaxial film and GaAs substrate.

Figure 4 is a high resolution image of an area near the edge of the specimen. The beam direction is parallel to [1̄10] of the CdTe and GaAs crystal, and the focusing condition is close to that used for the image of Fig. 1. In the CdTe crystal, both (111) and (11̄1̄) lattice fringes are seen, while the GaAs crystal shows only (111) lattice fringes due to the residual astigmatism of the objective lens. The orientations of these lattice fringes confirm the (100) epitaxial relation. Two important features regarding the interfacial layer are noted from this image. First, the area of the interfacial layer does not have lattice fringes or shows only low-contrast fringes. Second, boundaries between the interfacial layer and the GaAs or CdTe crystals show very fine irregularities similar to those of boundaries between Si and SiO_2 [17]. Both of these features suggest an amorphous structure of the interfacial layer. The presence of lattice fringes in some areas of the interfacial layer is explained by the overlapping of the crystals and the layers along the projecting direction, since the thickness of the observed area is $50 \sim 100$ Å and boundaries between the interfacial layer and crystals are irregular as mentioned above.

3. Discussion.

The interface associated with the (111) epitaxial relation shows a nearly perfect one to one correspondence of lattice fringes between the GaAs and CdTe crystals. For this epitaxial relationship, the interatomic spacing mismatch is only 0.7% along the $[\bar{2}11]_{CdTe}$ axis but is 14.6% along the $[01\bar{1}]_{CdTe}$ axis. Despite the large mismatch along the $[01\bar{1}]_{CdTe}$ axis, no extrinsic layer is found at the interface. Weak beam dark field images of the interfce in the cross-sectional specimen parallel to $(\bar{2}11)_{CdTe}$ were observed. The images show the existence of misfit dislocations in the interface with a spacing of about 150 Å which is far greater than that expected from the 14.6% interatomic spacing mismatch (16 Å). These observations, therefore, indicate that the large lattice mismatch for this epitaxial relation may effectively be accommodated by some mechanism other than the formation of a high density misfit dislocation or an extrinsic layer.

The very thin layer in the interface between the (100) epitaxial film and substrate is the most striking feature found in the present observation. From TEM image information only, it is difficult to identify the material in the interfacial layer. However, the bright contrast of the layer suggests that the density of this material is considerably smaller than that of GaAs or CdTe. For this reason, and because of the lower substrate preheating temperature used for the growth of the (100) film, it may be concluded that the interfacial layer is an oxide layer remaining on the substrate surface after the preheating. The thickness of the oxide layer is only a few tens of angstroms, but it continuously covers the substrate surface. Based on the presence of the interfacial layer, one can readily give an explanation for the occurrence of the (100) epitaxial relation. Because of the presence of the interfacial layer, which may be amorphous, the large lattice mismatch is no longer the obstacle against the growth of an epitaxial film. The observed density of dislocation in the interfacial region is similar to that obtained for the (111) epitaxial relation. The (100) epitaxy is thought to occur as a result of the interaction of the substrate surface symmetry through the interfacial layer instead of a critical dependence on the lattice matching; however, the nature of the interaction is not known at present. The presence of the very thin and continuous interfacial layer may be an essential factor to obtain (100) single crystalline films. In the earlier study [8], the preheating at considerably lower temperatures ($\sim 400°C$), which should leave a much thicker surface oxide layer (80–100 Å), produced polycrystalline films. Based on these observations, therefore, it is suggested that details of the preheating cycle prior to film growth is crucial to obtain the desired epitaxial film orientation in this system.

In conclusion, the present observation has revealed one of the most remarkable phenomena of epitaxy: The occurrence of two different epitaxial relations is determined by the presence, or absence, of a very thin oxide layer on the substrate surface. For both epitaxial relations, high quality single crystalline films have been obtained. Some important questions such as the nature of the interaction across the oxide layer and the effect of the reconstruction of the GaAs surface on epitaxy remain for future studies.

The authors wish to acknowledge their sincere thanks to Dr. J. T. Cheung, Professors H. Morkoc, T. Sakamoto, J. K. Furdyna, H. Sato and G. L. Liedl for their valuable discussions, and C. Choi for the assistance of the TEM observation.

This work was supported by Office of Naval Research Contract No. 014-82-K-0563 (Purdue), the NSF-MRL Program under Grant No. DMR-83-16999 (Purdue). NSF Grant DMR 83-13036 (N. C. State), and by DARPA/ARO Contract DAAG 29-83-K-0102 (N.C. State).

REFERENCES

1. R.F.C. Farrow, G. R. Jones, G. M. Williams, and I. M. Young: Appl. Phys. Lett., 39, 954 (1981).

2. T.H. Myers, Yowcheng Lo, J. F. Schetzina, and S. R. Jost: J. Appl. Phys., 53, 8232 (1982).

3. J. P. Faurie, A. Million and J. Piagnet, Appl. Phys. Lett., 41, 713 (1982).

4. P. P. Chow, D. K. Greenlaw, and D. Johnson: J. Vac. Sci. Technol., A1, 562 (1983).

5. T. H. Myers, Yowcheng Lo, R. N. Bicknell, and J. F. Schetzina: Appl. Phys. Lett., 42, 247 (1983).

6. K. Nishitani, R. Okhata, and T. Murotani: J. Electron. Mater., 12, 619 (1983).

7. H. A. Mar, K. T. Chee, and N. Salansky: Appl. Phys. Lett., 44, 237 (1984).

8. R. N. Bicknell, R. W. Yanka, N. C. Giles, J. F. Schetzina, T. J. Magee, C. Leung, and H. Kawayoshi: Appl. Phys. Lett., 44, 313 (1984).

9. L. A. Kolodziejski, T. Sakamoto, R. L. Gunshor, and S. Datta: Appl. Phys. Lett., 44, 799 (1984).

10. L. A. Kolodziejski, T. C. Bonsett, R. L. Gunshor, S. Datta, R. B. Bylsma, W. M. Becker, and N. Otsuka: Appl. Phys. Lett., 45, 440 (1984).

11. J. T. Cheung, private communication.

12. K. Ploog: Ann. Rev. Mater. Sci., 11, 171 (1981).

13. A. Y. Cho: J. Appl. Phys., 47, 2841 (1976).

14. J. M. Cowley: "Diffraction Physics" (North Holland, American Elsevier, Amsterdam, 1975).

15. T. Yamashita, F. A. Ponce, P. Pirouz, and R. Sinclair: Phil. Mag., A45, 693 (1982).

16. H. A. Mar, N. Salansky, and K. T. Chee: Appl. Phys. Lett., 44, 898 (1984).

17. S. M. Goodnick, R. G. Gann, J. R. Sites, D. K. Ferry, C. W. Wilmsen, D. Fathy, and O. L. Krivanek: J. Vac. Sci. Technol., B1, 803 (1983).

A CRITICAL COMPARISON OF THE TECHNIQUES USED TO CHARACTERIZE THE
CRYSTALLOGRAPHY OF AN INTERFACE : Pd ON MBE GROWN GaAs(100)

J.P. DELRUE*, M. Wittmer, T.S. Kuan, and R. Ludeke
IBM Thomas J. Watson Research Center, P.O. Box 218 Yorktown Heights, NY,
10598

ABSTRACT

In situ Reflection High Energy Electron Diffraction and ex-situ Transmission
Electron Diffraction and Ion Channeling have been applied to a reacted
Pd-GaAs interface and the results obtained are critically compared. The
investigation has been done on the stabilized c(2x8) surface obtained by
MBE on GaAs(100) substrates. Smooth surface epitaxial growth has been observed by RHEED as soon as a few monolayers of Pd are deposited at a
substrate temperature of about 325°C. TEM diffraction studies indicate the
presence of an intermetallic hexagonal structure similar to the orthorhombic
Pd_5Ga_2 but with slightly different lattice parameters due to the possible
incorporation of As. A less abundant phase was also identified as an
hexagonal structure similar to Pd_8As_2. Ion Channeling indicates pronounced
reduction in scattering yield when the [100] axis of the substrate was
aligned with the impinging beam, thus supporting the RHEED analysis. The
three techniques listed above were found to be useful for the determination
of the epitaxial relationship between the identified phases and the
substrate.

INTRODUCTION

Various techniques can be used to characterize the structure of an
interface. The probe may consist of particles (electrons, atoms, ions,
neutrons) or radiation (photons) of preselected energy and momentum determined from the scattering geometry. The detection equipment has to be
sensitive to the interaction (elastic or inelastic scattering) between the
probing particles and the surface atoms. Eisenberger et al. [1] have reviewed this question in a paper in which they report both on the well established techniques (e.g. LEED) and new techniques based on the use of
photons (SEXAFS), high energy ions (ion scattering) and atoms (helium
scattering). Among all the techniques available, four have been selected
and applied to the same samples of the reacted Pd-GaAs interface. They are
in-situ reflection high energy election diffraction (RHEED), used during
the growth of the interface, and the ex-situ methods of diffraction using
either a transmission electron microscope (TEM) or the glancing-angle x-ray
technique (Seemann-Bohlin). The fourth technique employed was high energy
ion channeling. This paper reports in some detail the results from these
methods with the exception of x-ray diffraction which turned out to be not
sensitive enough. Step-wise evaporations of Pd done at the substrate temperature of 325°C have been selected because they resulted in unique crystalline phases, not found when the metal deposition is made at room
temperature followed by annealing at 325°C [2].

EXPERIMENTAL

For this study samples were prepared using molecular beam epitaxy in
equipment described previously [3]. A GaAs buffer layer of about 100 nm to

200 nm was grown on the (100) substrate between 580°C and 600°C in conditions for which the As stabilized C(2x8) surface prevailed. After this growth, As flux was turned off and the sample temperature was decreased to the desired one. In this particular study, stepwise evaporations were made up to a thickness of about 30 nm at a substrate temperature of about 325°C. The Pd was evaporated from a boron nitride crucible at a rate of evaporation of about 0.1 nm/min. with a background pressure of $\pm2.10^{-9}$ Torr. In situ, high energy electron diffraction (20 kev) was used at glancing incidence to monitor the buffer layer and to study the metal-GaAs interface without interrupting the growth.

Samples prepared for TEM analysis were back-thinned by etching in an $(H_2O_2)_{10}(H_2O)_5(H_2SO_4)_2$ solution. The Pd-GaAs interface was examined in a JEOL JEM-200 CX electron microscope. Ion channeling measurements with 4 He$^+$ particles of 2.3 MeV were used to determine the quality of the epitaxial Pd-compound overlayer. The surface barrier detector was set at a scattering angle of $\theta = 170°$ and channeling was obtained by aligning the major crystal axis of the GaAs sample with the impinging 4He$^+$ beam. Random alignment of the sample with respect to the analysis beam was achieved by offsetting the major crystal axis by 7° and continuously rotating the sample around that axis. To minimize beam-induced disorder we moved the beam to a fresh spot after alignment of the sample and analyzed the sample in random alignment subsequently. The total accumulated charge was kept below 10 C/mm2 [7].

RHEED RESULTS

Crystallographic information on the growth process has been obtained from the 20 KeV glancing incidence RHEED patterns in the [110], [1̄10] and [100] azimuths of the GaAs (100) substrate. The 1/4 order diffraction streaks characteristic of the [01̄1] azimuth (c(2x8) surface) is shown as a reference in figure 1a. Figure 1b corresponds to the diffraction pattern along the same azimuth after the deposition of 0.08 nm of Pd. On this figure the 1/4 order streaks have already disappeared. This means that approximately 1/4 of a Pd monolayer interacts with the substrate sufficiently to destroy the surface periodicity (reconstruction) and leave a quasi amorphous surface layer. No crystalline phase is present at this stage of the interface formation, which contrasts with Ag-GaAs case (4). Figures 1c and 1d show the evolution of the diffraction pattern for respectively 0.5 nm and 1.2 nm of Pd. From these patterns, two sets of streaks, apparently independent of each other, can be simultaneously observed. One of them, which is already present for a deposition of 0.5 nm, dominates at 1.2 nm and becomes the only visible set at higher coverage (Fig.1d). These streaks are rather sharp and stable even after a deposition of 30 nm at 325°C. The diffraction pattern along the [100] azimuth is shown in figure 1e and exhibits the same features observed for the [1̄10] one. The lattice separation corresponding to the space between those streaks has been evaluated to be 0.4 nm. The second set of streaks represent a diamond-shaped spot pattern. Present only between 0.5 nm and 1.5 nm of Pd deposition, it corresponds to an intermediate phase which has not yet been identified. The space between the streaks of this set corresponds to an interatomic distance of 0.43 nm. Although RHEED permits an in-situ crystallographic characterization of the formation of an epitaxial phase, it is not able to give us more details about the nature of the compounds that are formed for highly reactive interface. This explains the necessity of the TEM diffraction analysis presented below.

ELECTRON MICROSCOPY RESULTS

Two different phases are observed from the 10 nm-thick Pd film deposited on (100) GaAs at 325°C. A bright-field image of this sample is shown in

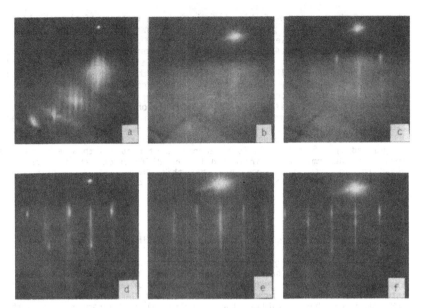

Figure 1. RHEED patterns (20 KeV) of Pd deposited on the GaAs (100) -c 2x8) surface at 325°C (a) clean surface, (b) after 0.08 nm. deposition of Pd, (c) after 0.5 nm of Pd, (d) after 1.2 nm of Pd, (e) after 10 nm of Pd all five along the (011) azimuth of GaAs; (f) shows the (011) azimuth of GaAs after 10 nm of Pd.

Fig. 2. The major phase covers about 80% of the total film area as marked by "A" in Fig. 2. This phase has a hexagonal structure with $a_0 = b_0 = 1.032$ nm and $c_0 = 0.395$ nm, which does not belong to any Pd-Ga or Pd-As binary

Figure 2. Bright field image of a 10 nm-thick Pd films deposited on (100) GaAs at 325°C.

phase. If the structure of this phase is indexed in terms of an orthorhombic cell, then the lattice parameters (a_0 = 1.032 nm, b_0 = 1.787 nm, and c_0 = 0.395 nm) are very close to those of orthorhombic Pd_5Ga_2 (a_0 = 0.5485 nm, b_0 = 1.8396 nm and c_0 = 0.4083 nm) [5]. We therefore propose that this major phase is a new ternary phase with composition similar to Pd_5Ga_2, possibly Pd_5GaAs. An electron diffraction pattern from this sample (Fig. 3A) indicates that this ternary phase maintains a texture orientation with respect to the (100) GaAs substrate. The c_0 axis of the ternary is always parallel to the [011] or [0$\bar{1}$1] axis of the GaAs substrate on the film plane, probably because of the near-perfect lattice match between the c_0 of the ternary phase (=0.395 nm) and the spacing of 022 planes in the GaAs substrate (=0.40 nm).

The less abundant phase (which might be on top of the ternary phase A), marked by "B" in Fig. 2, is hexagonal in structure with a_0 = b_0 = 0.730 nm (c_0 not determined) as indicated by the diffraction pattern from this phase in Fig. 3B. it is possible that this phase is Pd_8As_3, which has a hexagonal structure with a_0 = b_0 = 0.7399 nm and c_0 = 1.0311 nm [6].

ION CHANNELING RESULTS

Ion channeling measurements are presented in Fig. 4 and 5. Shown in Fig. 4 are spectra from random and [100] alignment of the GaAs sample with respect to the impinging beam. The [100] alignment of the sample resulted in a decrease of the backscattering yield of both the GaAs and the Pd signals. This is clear proof of the presences of epitaxial relationships between the hexagonal Pd-compound overlayer and the GaAs substrate. From TEM analysis we have established a few epitaxial relationships. Among them the most dominant one is $(110)_{hex}\parallel(100)_{GaAs}$; $[001]_{hex}\parallel[0\bar{1}1]$ GaAs or $[001]_{hex}\parallel$ [011] GaAs.

If the epitaxy of the hexagonal Pd-compound layers would be perfect then the decrease in backscattering yield would be much more pronounced than in Fig. 4. Thus, we conclude that a certain degree of imperfection is present which could be caused, for example, by different epitaxial domains. To further elucidate the epitaxial quality of the Pd-compound overlayer we have performed channeling in different crystallographic directions of the GaAs substrate. In Fig. 5, the normalized scattering yields of GaAs and Pd are plotted as a function of the tilt angle around the [100], [110] and [111] axes. The angular dips of GaAs signal decreases for [100] and [111] alignment because the thickness of the Pd-compound layers seen by the $^4H^+$ particles increase with increasing tilt angle. This, in turn, increases the amount of dechanneling caused by the epitaxial overlayer. Angular dips of the Pd signal are observed in Fig. 5 along with the [100] and [110] axes but not along the [111] axis. The dip along the [110] axis indicates that other types of domains with the $[001]_{hex}$ axis parallel to four other <110> GaAs directions (e.g. 110, 101, 1$\bar{1}$0, 10$\bar{1}$) inclined to the film plane also exist. This alignment can only be detected with TEM in a cross-sectional sample. This epitaxial relationship produces channeling in the [110] direction.

CONCLUSIONS

The aim of this paper was to demonstrate the importance of techniques one should conjugate and apply to the same samples in order to characterize the crystallography of a highly reactive interface (which can form unexpected binary or ternary compounds as in this case). The three techniques listed in the paper were complementary for the determination of the epitaxial relationship between the phases and the substrate. RHEED allowed us to observe the epitaxial growth of the interface (in situ) and was useful to define and monitor the conditions of the Pd-GaAs growth. By a careful

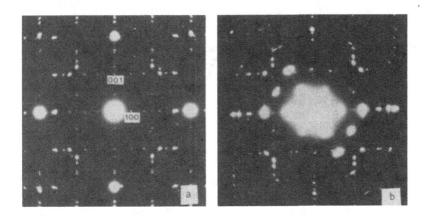

Figure 3. Figure 3a : Electron diffraction patterns from the sample
shown in Fig. 2. Figure 3a and 3b correspond respectively
to the major and the minor black phases.

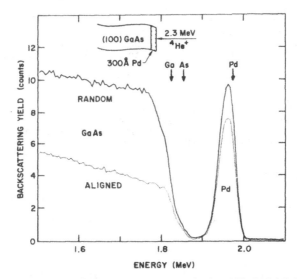

Figure 4. 2.3MeV ^{4}He$^+$ channeling spectra of the 300 A Pd/GaAs (100)
sample obtained for random and (100) alignment. The vertical
arrows indicate the surface position of the corresponding
elements.

observation of the spectra, TEM diffraction was not only important to define
the epitaxial or the textured phases but also permits us to identify two
ternary phases. Compared to the glancing angle X-ray diffraction technique,
TEM diffraction is not only much more sensitive (higher scattering factor)
but permits spatial analysis with good resolution which is necessary when
several phases coexist (Figure 2).

460

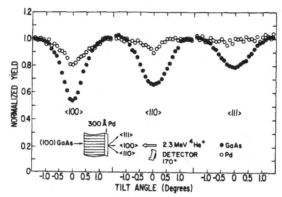

Figure 5. Angular scan curves for GaAs and Pd along the [100], [110], and
[111] axes of the GaAs substrate. The scattering yield has
been normalized with respect to random alignment.

High energy ion channeling has been applied for the first time to the
Pd-GaAs (100) interface and can be compared with the analysis previously
done by Gossmann et al. on the Pd-GaAs (110) [7], who conclude that Pd deposition disorders substantially the GaAs substrate. In our case, ion
channeling has shown that domains of the hexagonal Pd-compound are aligned
with the different [110] type axis of the GaAs substrate. This is also in
good agreement with the 0.40 nm interatomic distance detected by RHEED and
the $\bar{c}$ axis (0.40 nm) of the major hexagonal structure obtained by TEM
diffraction. The limitations of these techniques arise when one needs to
assess the chemical composition of these phases, especially when the deposited material strongly reacts with both gallium and arsenic. Auger
Electron Spectroscopy (AES), Scanning Auger Microscopy (SAM) and Energy
Dispersive X-ray Spectroscopy (EDX) can be used to determine such a composition. The auger data taken at the surface of this epitaxial growth are
in good agreement with the TEM results.

ACKNOWLEDGEMENTS

We would like to thank A. Segmuller for X-ray studies, M. Prikas for
technical assistance, and J. Freeouf for discussions. This work was partially supported by the Army Research Contract No. GAAG29-83-C-0026.

REFERENCES

1. P. Eisenberger and L.C. Feldman, Science, 214, 300 (1981).
2. P. Oelhafen, J.L. Freeouf, T.S. Kuan, T.N. Jackson and P.E. Batson,
   J. Vac. Sci. Technology B1 (3) 588 (1983).
3. R. Ludeke and S.Landgren, J. Vac. Sci. Technology 19(3) 667 (1981)
4. R. Ludeke, T-C Chiang and D.E. Eastman, J. Vac. Sci. Technology 21 (2)
   599 (1982).
5. Khalaff and Schubert, J. Less-Common Metals, 37, 129 (1974).
6. L. Cabri, et al., Can. Mineral, 13, 321 (1975).
7. H.-J. Grossmann and W.M. Gibson, Surface Science 139, 239 (1984).

*Perm. Addr.: Lab. Spectroscopie Electronique, FNDP, Namur, Belgium.

MORPHOLOGICAL STUDIES OF
POLYSILICON EMITTER CONTACTS

JOHN C. BRAVMAN, GARY L. PATTON*, ROBERT SINCLAIR AND JAMES D. PLUMMER*
Materials Science and Engineering, Stanford University, Stanford, CA 94305
*Integrated Circuits Laboratory, Stanford University, Stanford, CA 94305

ABSTRACT

Using high resolution lattice imaging techniques, the morphology of the polycrystalline silicon · single crystal silicon interface has been correlated to (1) the surface treatment used prior to polysilicon deposition, (2) the polysilicon implant dose, and (3) high temperature annealing. Specimens which were chemically oxidized prior to deposition exhibited a continuous layer of amorphous oxide ≈1.5nm thick. High temperature annealing produces small discontinuities in this oxide which allow the polysilicon to make direct contact with, and become epitaxially aligned to, the substrate. Specimens which were etched in HF prior to deposition were characterized by nearly oxide-free interfaces, which, following implantation and annealing, exhibited regions of epitaxial realignment significantly larger than those found in the chemically oxidized films. Heavily implanted films annealed at high temperature displayed almost complete epitaxial realignment.

INTRODUCTION

The polysilicon · silicon interface plays an important role in the electrical behavior of both MOS transistors fabricated with polysilicon buried contacts and bipolar junction transistors fabricated with polysilicon emitter contacts. In MOS technology, the buried contact is simply a convenient ohmic connection to the source or drain region of the device. For this application, it is thus desirable to reduce the contact resistance to as a low a value as possible. In shallow-emitter bipolar technology, the nature of the emitter contact can actually control, to a large degree, the overall electrical characteristics of the device. The use of conventional metallization schemes, for instance, significantly degrades the emitter injection efficiency, and hence the device current gain, due to the nearly infinite recombination velocity at the metal contact.

Several workers [1-4] have demonstrated that this problem is greatly reduced through the use of a layer of doped polysilicon in place of the metal as the emitter contact material. This behavior has generated a fair degree of interest in these devices, since polysilicon emitter contact technology offers a simple method for the fabrication of high gain transistors. An additional benefit stems from the elimination of any emitter metallization problems, such as aluminum "spiking," which can plague conventional contacting schemes. Often, the gain enhancement is sacrificed in a process design which includes a higher base doping level (which reduces the gain), in that this results in an increase in both the current carrying capability as well as the switching speed of the device. Significant from a commercial viewpoint, polysilicon emitter contact technology also offers a higher degree of process controllability than does standard metal contact technology, as the depth of the emitter region can be held at a value comparable to the base width. Problems associated with the lateral diffusion of the emitter dopant are thereby reduced, as well.

Several theories have been advanced to explain the gain enhancement obtained with polysilicon emitter-contact bipolar transistors. These include: reduced bandgap narrowing in the active emitter region, which leads to a reduction in the emitter Gummel number [1], asymmetric tunneling transport of electrons and holes across a thin interfacial oxide layer which grows between the polysilicon and single crystal silicon regions as an inevitable consequence of device processing [2], and minority carrier transport effects within the heavily doped polysilicon contact [3]. Other work has suggested that the electrical characteristics are sensitive to the type of surface treatment used prior to polysilicon deposition [4].

The work reported here summarizes part of a research effort which was undertaken in order to distinguish between these theories. In particular, the morphology of the polysilicon - silicon interface has been related to the surface treatment used prior to polysilicon deposition and to the specifics of the polysilicon implantation and annealing cycles.

MATERIALS PROCESSING

Phosphorus doped (100), 0.1-0.9 Ω-cm silicon wafers were used. Following one of two surface treatments which are described below, 400nm of LPCVD polysilicon was deposited at 620°C and subsequently capped with a 40nm layer of dry thermal oxide grown at 800°C. Arsenic doses of $4 \times 10^{15} cm^{-2}$, $7 \times 10^{15} cm^{-2}$, or $4 \times 10^{16} cm^{-2}$ were then implanted at 150 keV. Most samples received a two-step anneal, consisting of 60 minutes at 900°C followed by 45 minutes at 1000°C, while the remainder received 60 minute anneals at 1100°C.

The most important aspect of the present work concerned the effects of two different surface treatments used prior to polysilicon deposition. These treatments correspond to the wafer cleaning procedures used in many processing laboratories. The first treatment, which all wafers received, consisted of an RCA clean followed by an HF-dip etch. This three step procedure makes use of 70°C solutions of 1:1:5 $NH_4OH:H_2O_2:H_2O$, 1:1:5 $HCl:H_2O_2:H_2O$, and of room temperature 1:50 $HF:H_2O$. Wafer immersion times were 10 minutes, 10 minutes, and 30 seconds, respectively. For the second treatment, some of these wafers were then chemically oxidized via a second immersion in the the 1:1:5 $NH_4OH:H_2O_2:H_2O$ solution. Varying the immersion time (5 - 30 minutes), solution temperature (30°C - 90°C), substrate doping concentration ($5 \times 10^{14} cm^3$ - $10^{20} cm^3$), and substrate orientation (<100> and <111>) had no discernible effect on the chemical oxide thickness, which in each case was measured via ellipsometry and found to be between 1.6nm and 1.8nm. As will be seen, the accuracy of these measurements has been corroborated by high resolution imaging of cross-section TEM specimens. The processing independence of the oxide thickness suggests that this particular chemical oxidation is a self limiting process, much the same as the thermal nitridation of silicon. Thus, once a thin layer of oxide has grown, the rate of oxidant (H_2O_2) transport, and hence the oxide growth rate, drops rapidly to near zero.

CHEMICAL OXIDE INTERFACES

The basic structure of the silicon - chemical oxide - polysilicon interface is illustrated in Figure 1 for the case of an as-deposited polysilicon layer. This sample was not exposed to any temperature higher than 620°C (the polysilicon deposition temperature). The chemical oxide layer exhibits an appearance consistent with an amorphous material, and is continuous across the wafer. Note that the silicon (111) lattice fringe lines are resolved in both the single crystal substrate and in the polysilicon. Using the 0.314nm spacing of the (111) fringes as a magnification calibration, the thickness of the oxide was determined to be near 0.15nm, very close to the value obtained ellipsometrically. Except for a few random bulges, the oxide exhibits a marked uniformity with regard to this thickness. Viewed at lower magnification, the as-deposited polysilicon film exhibits the expected columnar grain structure.

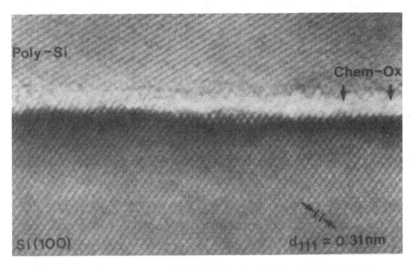

Figure 1: Basic structure of the Si - Chemical Oxide - Poly-Si interface.

Following arsenic implantations of either 4×10^{15} or 7×10^{15} and the two step anneal described above, the morphology of the interface was not significantly altered. As one would predict, the grains within the implanted and recrystallized polysilicon grew substantially, reaching diameters on the order of 200nm, but did not epitaxially realign with the single crystal substrate. Over approximately 5% of the wafer surface, however, the structural integrity of the oxide layer was lost, allowing the polysilicon to make direct contact with the substrate. In these regions, which varied between 3nm and 10nm in width, the polysilicon did realign with the substrate. One such region is illustrated in Figure 2. These "realignment structures" protrude 2nm to 3nm above the interface, but do not seed any long range regrowth. The reasons behind this "pinhole" formation are not clear, but this behavior could be a precursor to an agglomeration or "balling-up" of the oxide, as discussed in the next section.

HF-DIP ETCHED INTERFACES

Polysilicon films deposited on HF-dip etched wafers showed no evidence of long range epitaxial regrowth when examined prior to implantation and annealing. The polysilicon layer retained, for instance, the standard columnar grain structure, similar to the film morphologies already discussed. When imaged at lattice resolution, however, as in Figure 3, the interface does appear to be somewhat less planar than those processed with a chemical oxide interlayer, exhibiting non-periodic undulations with 1nm to 2nm amplitudes. This indicates that over some fraction of the wafer's surface the first few atomic layers were deposited epitaxially, but that there was insufficient thermal energy to produce an epitaxial layer. It should be noted that defect free epitaxial films are generally grown at temperatures greater than $1100^{\circ}C$.

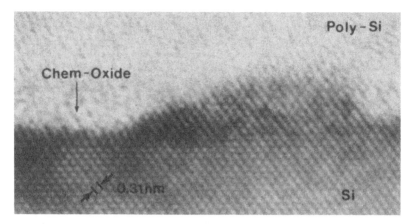

Figure 2: Small realignment structure which formed during implantation annealing.

Figure 3: HF-dip etched interface following polysilicon deposition.

Polysilicon films which were implanted with arsenic doses of either $4 \times 10^{15} \mathrm{cm}^{-2}$ or $7 \times 10^{15} \mathrm{cm}^{-2}$ and then two-step annealed displayed realignment structures qualitatively similar to those cited earlier, differing mainly in their areal coverage (now between 20% and 30%) and their size (up to 50nm along the interface and 20nm vertically into the polysilicon layer). One of these structures is illustrated in Figure 4, in which its epitaxial nature is clearly demonstrated. In several instances it was possible to obtain fringe images from within the polysilicon grains, where the misalignment between these fringes and those in the substrate indicated that full regrowth of the film had not occurred. This was confirmed by the low magnification appearance of the film, which was characterized by an equiaxed grain structure typical of implanted and annealed polycrystalline silicon.

Small spheroidal inclusions of an amorphous material - remnants of the chemical oxide, a native oxide, or both - are sometimes found within a realignment structure. Since these polysilicon layers were found to be inclusion-free prior to implantation and annealing, it is reasonable to conclude that these inclusions formed by some process of

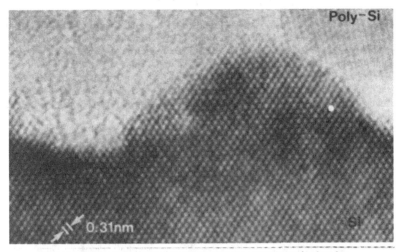

Figure 4: Large realignment structure on an HF-dip etched wafer.

oxide agglomeration, similar to that which accompanies oxygen precipitation and denuded zone formation in Czochralski silicon [5].

By implanting these films with a heavy, $4 \times 10^{16} cm^{-2}$ arsenic dose, and then annealing them at a high temperature ($1100^\circ C$), it was possible to achieve a nearly complete epitaxial realignment. This is probably due to the enhancment of the silicon atom mobility that results from both the heavy doping and the high temperature anneal. A region of a film processed in this manner is shown in Figure 5, where the position of the original polysilicon - silicon interface is marked by a linear array of oxide inclusions.

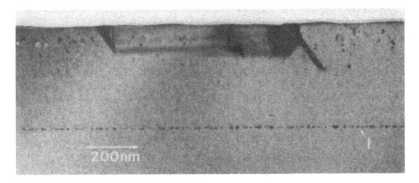

Figure 5: Heavily implanted polysilicon film showing nearly complete epitaxial realignment; the original interface (I) is marked by an array of oxide inclusions.

From these low magnification images it appears as if these inclusions constitute a nearly continuous layer of oxide, but it must be remembered that a transmission electron micrograph is a two dimensional projected image of a three dimensional object. Thus, what appears to be a linear array in projection is really a planar array viewed edge-on. By examining the thinnest regions of the TEM specimen (i.e., those nearest to the hole) the projected spacing of the inclusions decreased sharply. When examined at lattice

466

resolution, the oxide inclusions were found to be of similar proportions (2nm -10nm) to those observed in the more lightly doped films. One difference, however, was that some of these inclusions had begun to adopt a faceted, polyhedral morphology, sometimes being bounded by the (111) and (100) planes of the silicon. This is illustrated in Figure 6. This morphological transformation is again consistent with the behavior observed for oxide precipitates in Czochralski silicon, where the thermodynamically stable morphology, achieved only as a result of high temperature annealing, is also a polyhedral unit with facets on the (111) and (100) planes [6].

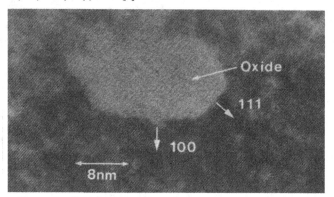

Figure 6: Oxide inclusion exhibiting a polyhedral morphology.

The oxide inclusions near the top of the film most likely formed as a consequence of the oxygen recoil events that attend heavy-ion implantation [7], given that the arsenic was implanted through a 40nm oxide cap.

SUMMARY

Our work to date has demonstrated that the morphology of the polysilicon - silicon interface is strongly related to (1) the pre-deposition surface treatment, (2) the polysilicon implant dose, and (3) high temperature annealing. Wafers chemically oxidized prior to deposition exhibited a thin layer of oxide whose continuity was slighty reduced following implantation and annealing. Epitaxial realignment of the polysilicon was limited to very small regions which covered only a small percentage of the wafer surface. Polysilicon films which were deposited onto HF-dipped wafers displayed a significant epitaxial realignment, the extent of which increased with implantation dose.

REFERENCES

[1] J. Graul, A. Glasl, and H. Murrman, IEEE J. Solid State Circuits, SC-11, 491 (1976).
[2] H.C. deGraff and J.G. de Groot, IEEE Trans. Electon Devices, ED-26, 1771 (1979).
[3] T.H. Ning and R.D. Issac, IEEE Trans. Electron Devices, ED-27, 2051 (1980).
[4] P. Ashburn and B. Soerowirdjo, IEEE Trans. Electron Devices, ED-31, 853 (1984).
[5] W.A. Tiller, S. Hahn, and F. Ponce, J. Electrochem. Soc. (submitted).
[6] F. Ponce, T. Yamashita, and S. Hahn, Appl. Phys. Lett., 43, 1051 (1983).
[7] D.K. Sadana, N.R. Wu, J. Washburn, M. Current, A. Morgan, D. Reed, and M. Maenpaa, Nuc. Inst. Meth., 209/210, 743 (1983).

ELECTRICAL CONTACT AND ADHESION MODIFICATION PRODUCED BY HIGH ENERGY
HEAVY ION BOMBARDMENT OF Au FILMS ON GaAs[a]

R. P. LIVI[b], S. PAINE[c], C. R. WIE, M. H. MENDENHALL[d], J. Y. TANG[e],
T. VREELAND, JR., AND T. A. TOMBRELLO
Divisions of Physics, Mathematics, and Astronomy and Engineering and Applied
Science, California Institute of Technology, Pasadena, California 91125

ABSTRACT

    Thin gold films over GaAs wafers with different dopants (Cr, Si, Te, and
Zn) were used to study the role of the substrate electronic properties in the
electrical contact and adhesion modification induced by MeV/nucleon heavy ion
bombardment. The enhanced adhesion was studied using a scratch test; the results show very different modifications of adhesion depending on the bulk
electronic properties of the substrate. The sample with a Cr compensation
doped substrate showed enhancement in adhesion for beam doses as low as 10^{12}
ions/cm^2, but Si and Te doped (n-type) substrates showed a sudden enhancement
in adhesion for doses around 10^{14} ions/cm^2. Samples with Si and Te doped substrates were used to study the bombarding ion dE/dx dependence of the induced
adhesion for ^{19}F and ^{35}Cl ions with electronic stopping power ranging from
161 eV/Å to 506 eV/Å. In this range the dose threshold for the onset of induced adhesion has a power law dependence, $D = D_0(dE/dx)^{-(1.90 \pm 1.0)}$.

INTRODUCTION

    The enhancement in adhesion produced by MeV/amu heavy ions bombardment
on a variety of thin metallic film-substrate combinations has been studied at
Caltech in the last few years [1-3]. This bonding occurs without the production of a significant atomic mixing layer and depends on the electronic
stopping power of the bombarding ions [2,3]. The electronic nature of the
bonding process was confirmed by Mitchell et al. by using 5-30 keV electrons
to produce enhanced adhesion [4].
    Initially the studies were made by using the "Scotch Tape" test but re-
cently more quantitative methods like the "Scratch" [5] and "Peel" [6] tests
are being used. In the present work a "Scratch" test was used to study the
role of the substrate (GaAs) electronic properties in the MeV ion-induced-
adhesion of thin Au films. Two n-type GaAs wafers were used also to study the
bombarding ion energy dependence of the process.

EXPERIMENT AND RESULTS

    In order to study the role of electronic properties on the enhanced-
adhesion with very little change in the substrate chemical composition or
crystalographic properties, GaAs wafers with four different dopants were used
as shown in Table I. The substrates were cleaned in Alconox detergent and
warm water in a ultrasonic bath, rinsed in warm water, rinsed in methanol,
etched in a solution of four drops of bromine in 100 ml methanol, rinsed in
methanol, and loaded in a diffusion-pumped vacuum chamber where 500 Å Au films
were deposited by evaporation. The Caltech EN-tandem Van de Graaf accelerator

TABLE I

Properties of the GaAs substrates used to study the enhancement in adhesion of Au films by MeV ion bombardment. All wafers had <100> cut.

Dopant	Concentration(cm^{-3})	Mobility ($cm^2/V \cdot s$)	Resistivity (OHM·cm)
Si(n)	3.5×10^{18}	1350	1.0×10^{-3}
	$*7.0 \times 10^{17}$	1100	8.0×10^{-3}
Te(n)	5.0×10^{17}	2100	3.0×10^{-3}
	$*8.0 \times 10^{17}$	2540	1.4×10^{-3}
Cr(p)	Compensation Doped		
Zn(p)	7.0×10^{17}	1000	9.6×10^{-3}

*Used for the energy dependence measurements.

was used for the bombardments with ^{19}F or ^{35}Cl ions for energies between 2 and 18 MeV and doses ranging from 8×10^{11} to 2×10^{15} ions/cm^2. After the bombardment the film surfaces were cleaned by plasma etching to avoid the influence of hydro-carbon contaminants on the "Scratch" tests. Those tests were performed using a Leitz micro-hardness tester equipped with either a 1 mm in diameter chrome plated steel ball or a 0.8 mm in diameter tungsten carbide ball loaded from 5g to 500g. Fig. 1 shows a SEM picture of scratches going across several 2.4 mm in diameter bombarded zones. The letters a, b, c, and d indicate the position of bombardment with 25, 10, 5, and 2.5×10^{13} ions/cm^2 doses of 18 MeV ^{35}Cl, respectively. As can be seen the film is removed even with the lowest applied load (5g) at the 2.5×10^{13} ions/cm^2 dose beam spot (d) and the largest available load (500g) does not completely remove the film from the 5.0×10^{13} ions/cm^2 dose beam spot (c). The scratches shown in Fig. 1 were made with the 0.8 mm diameter tungsten carbide ball. Fig. 2 shows a SEM picture of scratches in the top right hand side of the 10×10^{13} ions/cm^2

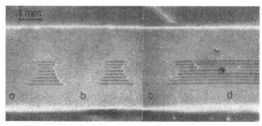

Fig. 1. Scratches made with a 0.8 mm in diameter WC ball. The letters a, b, c, and d indicate, respectively, 25, 10, 5, and 2.5×10^{13} ions/cm^2 doses of 18 MeV $^{35}Cl^+$.

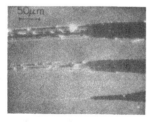

Fig. 2. Scratches going from the 10^{14} ions/cm^2 dose spot (Fig. 1b) to the non-bombarded area. The letters a, b, and c indicate, respectively, 500, 300, and 5g loads.

dose beam spot shown in Fig. 1 (b). The scratch (a) was made with a 500g load
and only partially removes the Au film from the GaAs substrate in the bombarded area. With the 5g load applied (scratch c) we can see just a fine
plastic deformation in the Au surface and the film is not removed at all in
the bombarded zone. Fig. 3 shows the "Scratch" test results for four different GaAs substrates when bombarded with 18 MeV $^{35}Cl^{4+}$. The scratches were
made in this case with the 1 mm diameter chrome plated steel ball. The lower
end of the bars indicates the load where the film starts to be partially removed and the top end the load where the film is totally removed (critical
load). In this limit the scratch width is the same in the bombarded and non-
bombarded areas. In some cases the film could not be removed completely even
with 500g, our maximum available load. In all samples the film could be removed in the non-bombarded areas with the 5g load. The Cr doped GaAs substrate sample shows very high adhesion improvement even at very low bombardment doses like 10^{12} ions/cm^2. The Zn doped p-type substrate sample showed
small improvement in adhesion for doses higher than 10^{13} ions/cm^2. The n-type
Si and Te doped substrate samples have the same behavior, representing very
high adhesion enhancement for doses higher than 10^{14} ions/cm^2. We could not
find a good correlation between the enhanced-adhesion behavior and the substrate properties shown in Table I, but we can see that the bulk electronic
properties should have a very important role in the effect, as indicated by
the very different improvements in adhesion shown by the various samples. We
believe also that the major enhancement in adhesion in this case is not produced by residues from the cleaning or by a thin native oxide layer on the
film-substrate interface, because the substrate preparation and film deposition followed the same procedures for all the samples, and the adhesion modification was very different for different substrates. Using n-type GaAs substrates we have started to study the stopping power dependence of the enhanced
adhesion by bombarding the samples with different ions over a range of
energies. The "Scratch" tests results can be seen in Fig. 4. Substrate and
bombarding ion characteristics, electronic dE/dx at the interface [7] and dose

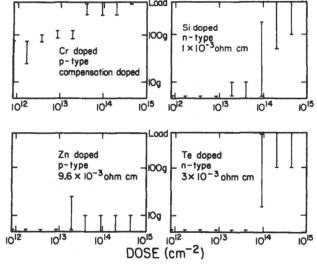

Fig. 3. Plot of scratch test load that causes film
stripping from GaAs substrate as a function of 18 MeV
$^{35}Cl^{4+}$ bombardment dose. The bars indicate the load
range, from initiating of partial stripping (bottom of
bar) to total stripping (top of bar).

thresholds for the onset of MeV ion induced-adhesion are summarized in Table II. As can be seen in Fig. 4 the effect of ^{35}Cl ion bombardment is a sudden increase in adhesion above a certain dose, depending on the electronic stopping power. For ^{19}F ion bombardments the adhesion changes more gradually but still shows the dE/dx dependence. The points in Fig. 4 show doses where 5g is already the critical load. Fig. 5 shows a plot of the doses where we can see the large increase in adhesion for ^{35}Cl ion bombardment or the dose ranges where the adhesion increases for ^{19}F ion bombardment, as a function of the ion electronic stopping power. A fit to the three points representing the sharp change in adhesion under ^{35}Cl ion bombardment shows a dE/dx dependence on the threshold dose given by $D = D_0 (dE/dx)^{-(1.9 \pm 1.0)}$. The ranges where the adhesion increases more slowly for ^{19}F ion bombardments are also intercepted by the fitting line. In Fig. 6 we can see the reverse breakdown char-

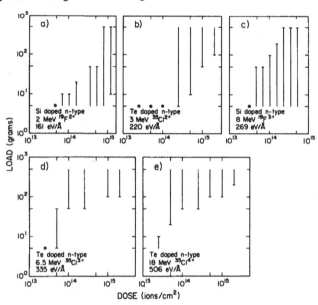

Fig. 4. Plot of scratch test load that causes film stripping from GaAs substrate as a function of ion bombardment dose.

TABLE II

GaAs and bombarding ion characteristics, electronic stopping power of bombarding ion at film-substrate interface, and the threshold for the onset of MeV ion induced adhesion.

GaAs Substrate	Bombarding Ion Characteristics	Electronic dE/dx (eV/A)	Induced Adhesion Threshold(ions/cm^2)
Si doped	2.0 MeV, ^{19}F^{2+}	161	$(3.5 - 12) \times 10^{14}$
Te doped	3.0 MeV, ^{35}Cl^{2+}	220	2.5×10^{14}
Si doped	8.0 MeV, ^{19}F^{3+}	269	$(4.7 - 25) \times 10^{13}$
Te doped	6.5 MeV, ^{35}Cl^{3+}	335	1.0×10^{14}
Te doped	18.0 Mev, ^{35}Cl^{4+}	506	5.0×10^{13}

acteristic for the electrical contact between the Au film and three different
GaAs substrates before (left) and after (right) 18 MeV $^{35}Cl^{4+}$ ion bombardment
at 10^{14} ions/cm^2 dose. The n-type substrates show diode-like characteristics
with some degradation after bombardment. The p-type substrate exhibited ohmic
contact with a one order of magnitude increase in the contact resistance after
bombardment.

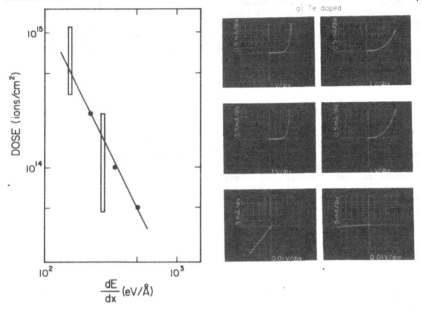

Fig. 5. Plot of dose threshold
indicating sudden increase in
adhesion (points) or dose ranges
where adhesion increases gradually
(bars) as a function of bombarding
ion electronic dE/dx.

Fig. 6. Reverse break-down characteristics for the electrical contact
between the Au film and the GaAs
substrate before bombardment (left)
and after bombardment (right) with
10^{14} ions/cm^2, 18 MeV, $^{35}Cl^{4+}$.

CONCLUSIONS

The "Scratch" test was shown to be a useful tool for the study of ion-
beam-enhanced adhesion when comparing results on substrates with the same
hardness. The results for the different GaAs substrates showed that more than
one mechanism may be responsible for the very different modifications found.
For n-type substrates we found dE/dx power law dependence for the dose threshold where the adhesion suddenly becomes very strong. Within the error bars
the result is the same as found previously for the Au-Ta system [2,3]. More
work is needed over a larger dE/dx range and with closer dose intervals to
decrease the error in the exponent. The experimental results showed that the
ionization and electronic excitation by the bombarding ions triggered very
different adhesion enhancement at the interface depending on the bulk electronic properties of the GaAs substrates. The Cr doped substrate is virtually
an insulator; thus, in this case the electronic relaxation time could be long

472

enough to allow transfer of energy to the ions via Coulomb repulsion, inducing some mixing at the interface. In the other cases the resistivity is low and the mechanism should be different; enhanced adhesion occurs only at much higher bombardment doses. Perhaps an important contribution to the enhanced adhesion comes from electrostatic forces orginating from charge distributions in the depletion region in the semi-conductor close to the film-substrate interface. The adhesion enhancement was very small for the p-type substrate where the electrical contact was ohmic and much stronger for the n-type substrates where the electrical contact between the film and substrate was diode-like. The strain/damage produced in p- and n-type GaAs by MeV ion bombardment was studied by using the Bragg x-ray rocking curve technique and showed similar results for both [8], indicating that at the Au-GaAs interface the kinematics could be very different than for the bulk semiconductor.

[a]Work supported in part by the National Science Foundation (DMR83-18274) and the IBM Corporation.
[b]Permanent address: Instituto de Fisica, Universidade Federal do Rio Grande do Sul, 90000-Porto Alegre-Brasil, acknowledges a Fellowship from CNPq-Brasil.
[c]Permanent address: Department of Electrical Engineering, Massachusetts Institute of Technology, Cambridge, Massachusetts 02139.
[d]Permanent address: Physics Department, Vanderbilt University, Nashville, Tennessee 37235.
[e]Permanent address: Nuclear Physics Division, Fudan University, Shanghai, People's Republic of China.

REFERENCES

1. J. E. Griffith, Y. Qiu, and T. A. Tombrello, Nucl. Instr. Meth. 198, 607 (1982).
2. M. H. Mendenhall, Ph.D. Thesis, California Institute of Technology (1983).
3. T. A. Tombrello, Mat. Res. Soc. Symp. Proc. 25, 173 (1984).
4. I. V. Mitchell, J. S. Williams, P. Smith, and R. G. Elliman, Appl. Phys. Lett. 44, 193 (1984).
5. P. Benjamin and C. Weaver, Proc. Roy. Soc. London A254, 163 (1960).
6. J. E. E. Baglin, G. J. Clark, and J. Bøttiger, Mat. Res. Soc. Symp. Proc. 25, 179 (1984).
7. L. C. Northcliffe and R. F. Schilling, Range and Stopping Power Tables for Heavy Ions, Nuclear Data Tables 7, No. 3-4 (Academic Press, New York, 1970).
8. C. R. Wie, T. Vreeland, Jr., and T. A. Tombrello, MRS Proc., submitted (1984).

IMPROVED UNIFORMITY OF REACTED GaAs CONTACTS BY INTERFACE MIXING

C.J. PALMSTRØM[*], K.L. KAVANAGH[*], M.J. HOLLIS[**+], S.D.
MUKHERJEE[**] and J.W. MAYER[*].
*Materials Science and Engineering, Bard Hall, Cornell
University, Ithaca, N.Y. 14853
**School of Electrical Engineering, Phillips Hall, Cornell
University, Ithaca, N.Y. 14853
+present address MIT Lincoln Labs.

ABSTRACT

Ion beam mixing is used to produce more uniform Au, Au-Ge
and Au-Ge-Ni reacted contacts to GaAs.

INTRODUCTION

Conventional alloyed ohmic contacts to GaAs suffer from two
problems: uniformity and reliability. The contacts tend to be
nonuniform with balling up and spiking of the metallization into
the GaAs[1-3]. The normal thermal annealing process for the
contacts is relatively rapid (~30sec), which results in an
incomplete reaction between the metallization and GaAs.
However, longer annealing times results in higher specific
contact resistivity. Hence, it is desirable to find a
contacting procedure which will produce more uniform and stable
contacts.

If no surface cleaning is performed in the deposition
chamber (such as heating or sputtering), there will always be
some interfacial layer between the metallization and GaAs. This
interfacial oxide layer, as it will be referred to in this
paper, may consist of oxides and hydrocarbons. The uniformity
and composition of this layer is not well controlled. During
subsequent metallization-GaAs reactions preferential reactions
will occur at weak areas in the oxide, resulting in a very
nonuniform reacted contact. This is shown schematically in
figure 1. Au/GaAs is known to react very nonuniformly with
spiking of the Au into the GaAs[2], similar behavior to
Al/Si[4]. If the interface is mixed using an ion beam prior to
annealing, the interface oxide can be dispersed and hence a more
uniform reaction may occur (see fig. 1). This has been
demonstrated to work for Al/Si contacts[5]. Recently, interface
mixing was reported to improve the uniformity of alloyed Au-Ge
contacts to GaAs[6]. However, no electrical evaluations were
made in this study. Studies on interface mixing and annealing
of Pt/GaAs[7] and Ni/GaAs[8] have also been made. The specific
contact resistances were reported to be ~$5 \times 10^{-4} \Omega\text{-cm}^2$ for
Si implanted Pt/GaAs[7]. In this paper the effect of ion beam
interface mixing followed by annealing is discussed for Au,
Au-Ge, and Au-Ge-Ni layers on GaAs.

EXPERIMENTAL

The substrates used were (100) semi-insulating(SI) GaAs. For electrical measurements samples with an n-type epitaxial layer grown by MBE on top of a SI-GaAs substrate were used. The GaAs was degreased, etched in a mixture of $NH_4OH.H_2O_2.H_2O$ (1:1:40 by volume) and rinsed in a mixture of $NH_4OH.H_2O$ (1:10 by volume) prior to loading into the evaporator. The Au($\sim$500A), Au-Ge(eutectic, 12wt% Ge)($\sim$1000A) and Au-Ge-Ni(Au-Ge eutectic with 10wt% Ni) ($\sim$400A) layers were deposited in an oil-free evaporator with a base pressure of $\sim$5x10^{-8} torr. The implantation was carried out with a 300keV ion implanter with a liquid nitrogen trapped diffusion pumped end chamber. The pressure during implantation was typically $\sim$2x10^{-7} torr for both liquid nitrogen and room temperature implants. Annealing of the Au, Au-Ge and Au-Ge-Ni contacts was done in a flowing gas furnace (H_2-N_2). JEOL JSM35 and JEOL 200CX microscopes were used for SEM and TEM analysis respectively.

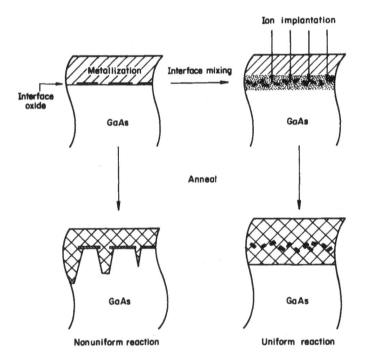

Fig. 1   Schematic of interface mixing used for improved uniformity of reacted contacts.

RESULTS AND DISCUSSION

Au/GaAs and Au-Ge/GaAs

Figure 2 shows SEM micrographs of both unimplanted and Ar ion implanted Au-Ge/GaAs structures after annealing at ~450°C for 30sec. The unimplanted contact shows the nonuniformity typical of alloyed contacts[1-3]. whereas the implanted sample surface remains much more uniform. This result is similar to that reported by Inada et al.[6]. Interface mixed Au/GaAs contacts also showed more uniform reactions than unimplanted contacts after annealing at 500°C.

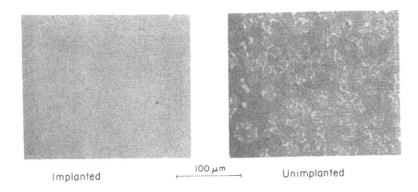

Implanted |————— 100 μm —————| Unimplanted

Fig. 2 SEM micrographs of unimplanted and implanted with 340keV 2×10^{15} Ar ions/cm^2 Au-Ge/GaAs structures after annealing at ~450°C 30sec.

Au-Ge-Ni/GaAs

After annealing at ~420°C for 30sec both unimplanted and As ion implanted contacts showed a similar structure (Fig. 3a and b). The unimplanted contact showed ohmic behavior while the implanted contacts were rectifying. The rectifying behavior probably resulted from residual ion beam induced damage in the GaAs beneath the metallization. Annealing at 460°C 15min did not result in the implanted contacts becoming ohmic. Ohmic behavior was observed for the liquid nitrogen temperature implanted contact after a 510°C 15min anneal. Both the room temperature and liquid nitrogen temperature implanted contacts were ohmic after a further anneal at 560°C for 15min. An additional 2000A Au layer was deposited on the contacts of a transmission line pattern in order to reduce the sheet resistance of the contacts and measure the specific contact resistance. The specific contact resistances for the unimplanted, room temperature and liquid nitrogen implanted contacts were 1.5×10^{-5}, 1.5×10^{-3} and 1×10^{-4} Ω-cm^2 respectively. The micrographs in figures 3c and d show the surface of the metallization after the last annealing cycle.

Unimplanted   Implanted

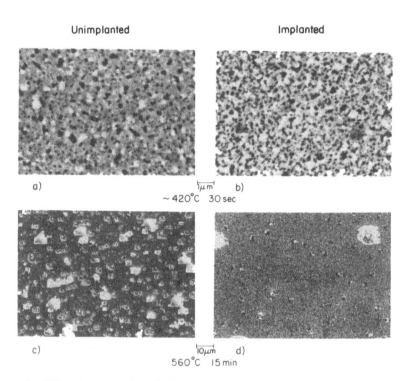

a) |μm| b)
              ~ 420°C 30 sec

c) |10μm| d)
              560°C 15 min

Fig. 3   SEM micrographs of Au-Ge-Ni/GaAs structures after
         annealing comparing unimplanted (a,c), and implanted
         150keV 1x10^{15} As ions/cm^2 at liquid nitrogen
         temperature (b,d).
         a) and b) after annealing at ~420°C 30sec.
         c) and d) after sequencial annealing at ~420°C 30sec,
         460°C 15min, 510°C 15min and 560°C 15min.

The unimplanted sample shows severe balling and pitting of the
GaAs. The implanted contact is much more uniform although some
balling of the metallization is evident. Figure 4 shows
cross-sectional TEM micrographs of the contacts shown in figures
3c and d. The unimplanted contact has balled up resulting in a
thin discontinuous layer of metallization covering the majority
of the GaAs (Fig. 4a). Some of the balls of metallization do
not penetrate deeply into the GaAs; one of these is visible at
the edge of micrograph 4a. However, in other regions deep
pitting is evident as can be seen in micrograph 4b (note the
lower magnification in this micrograph). From this micrograph
it can be deduced that pits ~2000Å can result from alloying a
400Å thick Au-Ge-Ni film. The implanted contacts show much
more uniform reaction and a relatively thick continuous layer of
metallization covers the GaAs after annealing (Figs. 4c and d).
However, some balls of contact metallization are evident even

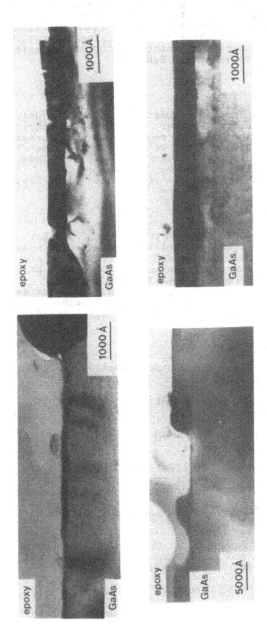

Fig. 4   Cross-sectional TEM micrographs of Au-Ge-Ni/GaAs contacts. a) and b) from sample shown in fig. 3c, and c) and d) from sample shown in fig. 3d.

for this contact (Fig. 4d). Furthermore, it can be seen that although the GaAs beneath the contact is good single crystal there remain residual defects in this layer (Figs. 4c and d). This damage may be responsible for the high specific contact resistance measured in these contacts. The defects' presence could also explain why the contacts had to be annealed at such a high temperature to get ohmic behavior.

CONCLUSIONS

Interface mixing can be used to improve the uniformity of reacted GaAs contacts. However, the ion beam induced damage in the GaAs beneath the contacts requires higher processing temperatures to form contacts with electrical properties approaching the unimplanted layers. Either the damage must be annealed out or else it must be consumed in the metallization/GaAs reaction to obtain good electrical behavior.

ACKNOWLEDGEMENTS

The authors would like to acknowledge J.C. Barbour, S.H. Chen and R.A. Hamilton for assistance in the TEM measurements. Discussions with G.J. Galvin and D.A. Lilienfeld are also appreciated. This project was supported in part by the Defense Advanced Research Projects Agency under contract #N66001-83-C-0304 monitored by the Naval Ocean System Center. Ion implantation was done at the National Research and Resource Facility for Submicron Structures (NSF grant ECS-8200312).

REFERENCES

1.   N. Braslau, J.B. Gunn and J.L. Staples, Solid-State Electron., 10, 381 (1967).

2.   J. Gyulai, J.W. Mayer, V. Rodriguez, A.Y.C. Yu and H.J. Gopen, J. Appl. Phys., 42, 3578 (1971).

3.   C.J. Palmstrom, D.V. Morgan and M.J. Howes, Nucl. Inst. and Meth., 150, 305 (1978).

4.   R. Rosenberg, M.J. Sullivan and J.K. Howard, in Thin Films: Interdiffusion and Reaction, J.M. Poate, K.N. Tu and J.W. Mayer Eds., (Wiley-Interscience, N.Y., 1978), p 13.

5.   L.S. Hung, J.W. Mayer, M. Zhang and E.D. Wolf, Appl. Phys. Lett. 43, 1123 (1983).

6.   T. Inada, H. Kakinuma, A. Shirota, J. Matsumoto, M. Ishikiriyama and Y. Funaki, Proc. Ion Beam Modification of Materials 1984 (to be published).

7.   K. Tsutsui and S. Furukawa, J. Appl. Phys., 56, 560 (1984).

8.   S. Smith and J. Solomon, private communication.

REACTION OF AMORPHOUS NiNb FILMS WITH CRYSTALLINE METAL OVERLAYERS

E. A. DOBISZ*, B. L. DOYLE**, J. H. PEREPEZKO*, J. D. WILEY* and
P. S. PEERCY**
*University of Wisconsin-Madison, 1500 Johnson Drive, Madison, WI   53706,
**Sandia National Laboratories, Albuquerque, NM 87185

ABSTRACT

In many cases the stability of amorphous films is influenced by interaction with metallic crystalline overlayers. Such interactions between
Au, Ni, Nb and Ta overlayers and a-(Ni-Nb) films are reported. During
interdiffusion Au overlayers reacted with a-(Ni-Nb) to form two different
adjacent crystalline layers. In order to study the influence of relaxation
of the amorphous film on overlayer reaction several a-(Ni-Nb) samples were
pre-annealed prior to Au deposition. High depth resolution Rutherford
Backscattering Spectrometry (RBS) demonstrates that preannealing lowers
the diffusion coefficient of Au in a-(Ni-Nb) at 450°C from 7.5×10^{-22} m^2/s
to 8.7×10^{-23} m^2/s. During interdiffusion Ta was discovered to be substantially more inert than Au. For example, negligible interdiffusion between
Ta and a-(Ni-Nb) at 505°C after 25 hours implies a diffusivity of less
than 5×10^{-24} m^2/s. These observations allow assessment of some of the
requirements for increasing the stability of crystalline-amorphous metal
film layered structures.

INTRODUCTION

In a complete assessment of the potential of amorphous films to function as metallizations or diffusion barriers in high temperature solid
state devices, it is important not only to establish the intrinsic stability of the amorphous layer, but also to identify any modification of the
stability by overlayer deposits and/or substrates. For example, in previous
work, the influence of several substrate types on the thermal stability of
amorphous NiNb, a-(Ni-Nb), alloys was established to be relatively modest
in some cases [1-3]. With substrates such as sapphire and GaP crystallization occurred at temperatures near the 575°C T_c value found for the case of
an Si substrate; however, with a GaAs substrate, a significant intermixing
of As and Ni occurred above 450°C, thereby degrading the stability [3].
With continued annealing, crystallization of a-(Ni-Nb) also resulted in
reaction with sapphire and GaP substrates. Similar observations on thermal
stability have also been reported for amorphous Ni-Mo and Fe-W films on
silicon [4-5]. The most robust amorphous film so far reported is based on
a Ta-Ir alloy [5-6] which retains its integrity following a 24 hour anneal
up to 600°C on Si or GaAs.
Moreover, it appears that metallic overlayer films can have a specific, and in some cases, dramatic influence on the thermal stability of
amorphous films. With a-(Ni-Nb) films polycrystalline Au and Ni overlayers
have been found to lower the onset temperature for crystallization in a 1
hour anneal by 100 and 200°C respectively [7]. Since the reaction of a
metallic overlayer with a-(Ni-Nb) proceeds by the progress of an interfacial
reaction front, it is of interest to determine if the reaction can be controlled to retain the intrinsic thermal stability. The amorphous film
represents a metastable single phase and the devitrified film can exhibit
a metastable two-phase glass-crystal equilibrium. In the current work the
effects of changes in the film composition, annealing state and the overlayer metal on thermal stability have been examined. These results suggest
some guidelines for optimizing the thermal stability of amorphous films.

Experimental Methods

Amorphous layers of Ni-40 a/o Nb and Ni-60 a/o Nb were deposited on single crystal Si or sapphire substrates by DC magnetron sputtering as described elsewhere [8]. The 1 μm thick Ni-Nb films were sufficiently thick that possible interactions at the substrate interface did not influence the reactions at the metal overlayer-amorphous film interface. Overlayer films of Au, Nb, Ni or Ta were deposited by evaporation (Au) or DC magnetron sputtering (Ni, Nb, Ta) to a thickness of about 100 nm for X-ray diffractometry (XRD) and 20-30 nm for RBS. As deposited samples were first checked for crystallinity using an x-ray diffractometer. Individual samples for XRD were encapsulated in quartz tubes, which were multiply flushed with Ar and sealed under vacuum with a small Ti getter chip, which was then heated. Samples for RBS studies were annealed in a UHV annealing furnace at 2×10^{-8} torr.

Details of the interdiffusion between the Ni-Nb on sapphire substrates and the Au and Ta overlayers were investigated by RBS, using 3 MeV He provided by the Sandia EN tandem. High depth resolution was obtained by a grazing angle geometry. The sample was tilted so that the beam was incident at a 55° angle with the sample normal and backscattered particles that were detected at a 164° scattering angle exited the sample at a 71° angle to the normal. The angles and reproducibility were verified with a laser and by monitoring the standard sample spectrum of 20 nm of Au on quartz that was always in the sample holder. With a detector resolution of 15 keV, interfacial motion of 1.5 nm (1/2 channel) could be detected.

RESULTS AND DISCUSSION

Overlayer-Induced Crystallization

In the absence of a crystalline overlayer, the initially amorphous NiNb films on Si crystallize on a 1 hour time scale with crystallization temperature (T_c) of about 575-600°C [1,2]. Both Au and Ni overlayers result in lower 1 hour crystallization temperatures than that of the uncoated film, as shown in Table 1. For example, for Ni, the T_c is lowered by 200°C. However, overlayers of Ta or Nb were not observed to produce any significant effect on the T_c value.

An illustration of overlayer reaction of Ni and Nb with a Ni-60 a/o Nb amorphous film is presented in Figure 1. The Ni overlayer is identified by the (111) reflection superimposed upon the broad scattering from the amorphous film in Figure 1a. With progressive annealing, the Ni reflection remains and additional reflections develop with the onset of partial crystallization. It is interesting to note that the initial crystallization

Table I
Overlayer-Induced Crystallization of a-(Ni-Nb)

| Sample | Temperature for 1-hour Anneals | | | | |
	400°C	450°C	500°C	550°C	600°C
Ni-Nb/Si	A*	A	A	A	C*
Ni-Nb/Al₂O₃	A	A	A	A	C
Au/Ni-Nb/Si	A	A	C	–	–
Ni/Ni-Nb/Si	C	–	–	–	–
Nb/Ni-Nb/Si	A	A	A	A	C
Ta/Ni-Nb/Al₂O₃	A	A	A	A	C

*Entries A or C denote samples in which the initially amorphous Ni-Nb film was found to be amorphous or crystalline, respectively, after annealing.

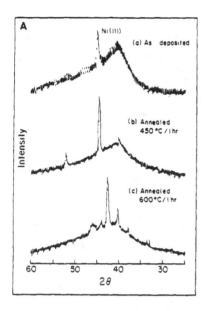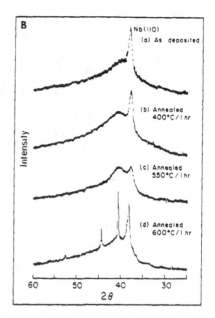

Figure 1.   XRD scans for a Ni-60 a/o Nb amorphous film following annealing
with (a) Ni overlayer and   (b) Nb overlayer.   The substrate reflections
have been substracted from the scans.

products are metastable Ni-rich phases which are replaced by other phases
with prolonged annealing at higher temperature [7]. In contrast the scan
in Figure 1b reveals a continuous decay of the Nb (110) reflection with
progressive annealing up to a crystallization onset with a phase mixture
which is not the same as in Figure 1a.

In examining the specific influence of different overlayer materials
there are a number of kinetic paths that may be considered. In one path,
material in the overlayer can diffuse partially into the amorphous film to
form an alloy with a reduced T_c which would then crystallize. The results
presented in Figure 1 do not support a simple diffusion-induced crystallization onset. While the Nb reflection decreased in intensity implying a
solution with the amorphous film, the Nb overlayer did not lower the T_c.
Alternatively, the crystalline overlayer can provide nucleation sites for
the initiation of crystallization in the amorphous film. Perhaps the
strong influence of Ni and the effect of Au are related to this process
since the polycrystalline FCC overlayer could act as an effective catalyst
in promoting the formation of an Ni rich phase from the amorphous film.
Similarly, the apparent ineffectiveness of Ta and Nb overlayers in promoting crystallization may be related to the absence of nucleation sites.
Further, since the initial amorphous layer is a metastable phase, the
beginning stages of crystallization reactions are likely to involve a
metastable mixture of amorphous and crystalline phases. In following such
a reaction, it is useful to consider the evolution of the constitution in
terms of a metastable phase diagram involving both metastable extensions
of stable phases as well as metastable intermediate phases [10].

Rutherford Backscattering Measurements

The high depth resolution of RBS to small compositional adjustments was used to study the interaction of a-(Ni-60 a/o Nb) with Ta and Au over-layers and the influence of low temperature structural relaxation annealing treatments on the interaction. Gold and Ta were chosen because the high atomic numbers and masses gave good depth resolution, sensitivity to the diffusing species signal, and a backscattered peak well-separated from the Nb edge. The Au interacted with the a-(Ni-Nb) at temperatures as low as 450°C, whereas no chemical interaction was observed with Ta overlayers at 505°C and 525°C. The effects of the Au at 450°C and Ta at 505°C are illustrated in Figures 2 and 3.

With the Au overlayer, the first reaction step is the penetration of the overlayer by a small amount of Nb. This was accompanied by movement of Au into the Nb depleted region in the Ni-Nb layer. Any depletion of Ni is not observed until very late in the diffusion process. Based upon these observations, it appears that the diffusion involves the formation of two compounds: one within the Au overlayer and one at the Au-(Ni-Nb) interface with a relative composition of about $Au_{.4} Ni_{.4} Nb_{.2}$. It is interesting to note that, although in binary thin film layers only one compound phase is observed at any given stage of the interdiffusion [11], in a ternary thin film diffusion couple this constraint appears to be removed to allow the possibility of two compound phases forming during the diffusion interaction [12, 13]. This development complicates the kinetic analysis and requires careful identification of the phase formation sequences.

In contrast, very little penetration occurred with Ta overlayers. A small amount of motion (~ 1.6 nm) was observed during the first seven hours, after which no further motion was detected up to 25 hours. The amount of penetration implied an average diffusivity $<5 \times 10^{-24} m^2/s$.

Structural relaxation of a-(Ni-Nb) was found to affect the penetration of the Au in the overlayer reaction. Following relaxation treatments that were based upon the findings of in-situ resistance measurements [14], 20 nm of Au was deposited on the samples and also on films that had not been given a relaxation treatment. Based on the RBS scans of the Au peaks before and after annealing at 450°C, the penetration distance of Au into a-(Ni-Nb) was determined [14] and is shown as a function of annealing time in Figure 4. The unrelaxed sample exhibits far more diffusion than either of the relaxed samples. Little difference was observed between the R-510/3 (i.e., relaxation anneal at 510°C for three hours) and the R-540/3 samples. Even up to 23 hours, the penetration rate between the unrelaxed and relaxed samples is different. This makes uncertain the previous assumption that Au diffusion in unrelaxed a-(Ni-Nb) would approach that of the relaxed isoconfigurational state if sufficient time for relaxation were allowed after deposition of Au [2]. Furthermore, these results indicate that the initial processes occurring during the early stages of the diffusion anneal influence strongly the behavior of the sample throughout its time of observation. For the purposes of comparison, if the penetration at 450°C is considered as diffusion limited [2], the unrelaxed state yields an average diffusion coefficient of $7.5 \times 10^{-22} m^2/s$ and the relaxed state is an order of magnitude lower at $8.7 \times 10^{-23} m^2/s$. An analysis of the penetration to include the effect of interfacial reactions is in progress.

Overlayer Reactions

The application of amorphous films in high temperature solid state devices is controlled by the diffusive interactions between the film and contact overlayers and substrate crystals. Within the wide range of interactions some patterns appear to be emerging with regard to overlayers and amorphous films. For a-(Ni-Nb) a suitable overlayer material is one in

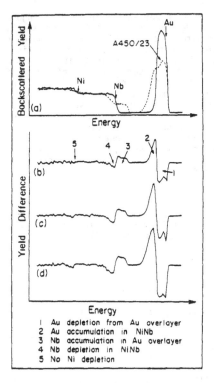

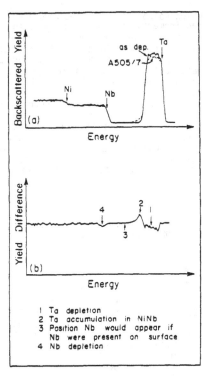

1 Au depletion from Au overlayer
2 Au accumulation in NiNb
3 Nb accumulation in Au overlayer
4 Nb depletion in NiNb
5 No Ni depletion

1 Ta depletion
2 Ta accumulation in NiNb
3 Position Nb would appear if
   Nb were present on surface
4 Nb depletion

Fig. 2. (a) RBS spectrum of Au/a-(Ni-Nb) as deposited. Figures (b)-(d) are plots of the difference between RBS spectra before and after anneal at 450°C for 6 hrs., 11 hrs. and 23 hrs. respectively.

Fig. 3. RBS spectra before and after anneal at 505°C for Ta/a-(Ni-Nb). (a) RBS Spectra, (b) Difference Spectra.

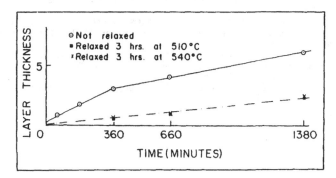

Fig. 4. Effect of structural relaxation on Au penetration into amorphous Ni-60 a/o Nb. The amount of penetration is determined by the thickness of Au (in number of channels) in the ternary Au-Ni-Nb region of the RBS scans [14].

which solubility with the amorphous film does not appear to be a stringent requirement. Niobium is soluble in a-(Ni-Nb) and Ta is not readily soluble, but both overlayers do not induce premature crystallization. Gold is readily soluble in a-(Ni-Nb) and does induce premature crystallization. More specific consideration of these effects requires information on the metastable phase equilibria pertinent to the system and the influence of solubility on the nucleation kinetics. Aside from the destabliization induced by chemical interactions, it appears possible that surface coatings may act to stabilize and perhaps enhance the intrinsic stability. In general, it appears that a relaxation treatment in the amorphous film prior to overlayer deposition is effective in reducing the level and rate of interaction with the overlayer metal. Overall, it is apparent that although the diffusive interactions can be relatively complex, the opportunities that the interactions present for controlling the thermal integrity of amorphous metallizations are also quite attractive.

Acknowledgement

The support of the DOE under (DE-FG02-84ER45096) at UW-Madison and (DE-AC04-76D00789) at Sandia National Laboratories is gratefully acknowledged.

References

1. J. D. Wiley, J. H. Perepezko, J. E. Nordman, R. E. Thomas and D. E. Madison, Bull. Am. Phys. Soc. 26, 454 (1981).
2. B. L. Doyle, P. S. Peercy, J. D. Wiley, J. H. Perepezko and J. E. Nordman, J. Appl. Phys. 53, 6186 (1982).
3. B. L. Doyle, P. S. Peercy, R. E. Thomas, J. H. Perepezko and J. D. Wiley, Thin Solid Films 104, 69 (1983).
4. M. Finetti, E. T-S Pan, I. Suni and M-A Nicolet, Appl. Phys. Lett. 42, 987 (1983).
5. K. T-Y Kung, I. Suni and M-A Nicolet, J. Appl. Phys. 55, 3882 (1984).
6. A. G. Todd, P. G. Harris, I. H. Scobey and M. J. Kelly, Solid-State Electron, 27, 507 (1984).
7. D. K. Wickanden, M. J. Sisson, A. G. Todd and M. J. Kelly, Solid-State Electron. 27, 515 (1984).
8. E. A. Dobisz, D. B. Aaron, K. J. Guo, J. H. Perepezko, R. E. Thomas and J. D. Wiley, J. Non-Crystalline Solids 62, 901 (1984).
9. J. S. Williams and W. Möller, Nucl. Instrum. Methods 157, 213 (1978).
10. J. H. Perepezko and W. J. Boettinger, Mat. Res. Soc. Symp. Proc. 19, 223 (1983).
11. K. N. Tu and J. W. Mayer, in "Thin Films-Interdiffusion and Reactions" ed. J. M. Poate, K. N. Tu and J. W. Mayer (Wiley-Interscience, NY 1968) Chap. 10.
12. U. Köster, P. S. Ho and J. E. Lewis, J. Appl. Phys. 53 7436 (1982).
13. G. Ottaviani, K. N. Tu, R. D. Thompson, J. W. Mayer and S. S. Lau, J. Appl. Phys. 54 4614 (1983).
14. E. A. Dobisz, J. H. Perepezko, B. L. Doyle, P. S. Peercy and J. D. Wiley, to be published.

ELECTRON - BEAM MEASUREMENTS OF EPITAXIAL GaAs LAYERS

O. PAZ* AND J.M. BORREGO**

* IBM East Fishkill, General Technology Division, Hopewell Junction, NY
  12533
** Electrical and Systems Engineering Department, Rensselaer Polytechnic
   Institute, Troy, New York, 12181

ABSTRACT

In studying the growth of epitaxial layers in III-V compounds, it is
advantageous to obtain information on the early stage of the growth that
takes place at the substrate-epitaxial layer interface. In this paper,
we describe a method of measuring the recombination velocity at this
interface, as well as the minority carrier diffusion length in the bulk
of the epitaxial layer. A bi-directional diffusion of excess carriers
was modeled, with layer thickness, diffusion length and recombination
velocity as variables. Layer thickness was determined using spreading
resistance, the other two variables can be separated by measuring layers
of several thicknesses. Some carriers diffuse from the point of generation
toward the surface and are collected at a Schottky barrier depletion layer.
The remaining carriers diffuse from the point of generation toward the
epi-substrate interface where they recombine.

INTRODUCTION

Several techniques are utilized at the present time to measure diffusion length, lifetime and surface recombination velocity in semiconductors
using electron - beam excitation (for example, Refs. 1-3). Previous studies
were aimed mostly at bulk properties of single crystal material. The
electron beam was incident either parallel to the collecting junction[1] or
normal to it[3] and the diffusion length was obtained by plotting the
beam-induced-current decay vs. distance when slowly moving the beam away
from the junction, with the two geometries being described by different

relationships. Wu and Wittry[4] used a stationary electron beam to penetrate
a Schottky contact that was normal to it and made their results insensitive
to surface recombination velocity by assuming infinite collection velocity
at the depletion layer edge. Their method was suitable for short diffusion

length but also aimed at bulk properties, rather than local sites, and an unfocused beam was used. In a previous paper[5] the present authors reported a method of measuring diffusion length by bombarding a Schottky contact normal to the beam with a focused electron-beam. The generation rate was varied and the diffusion length was obtained from the relationship between the collection current I_c and the specimen current I_s. The technique was applied to single crystal bulk material as well as to polycrystalline GaAs grains whose dimensions were larger than the size of the generation function. In the present paper the technique is extended for the case that the generation volume is intercepted by another interface, in addition to the depletion layer edge of the collecting junction, and is applied to an

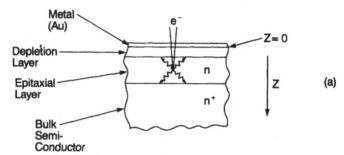

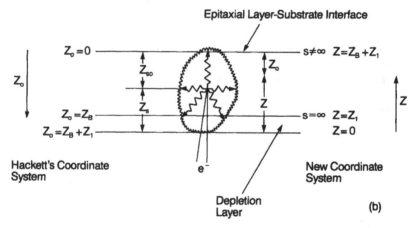

Fig. 1   (a) Present Geometry

       (b) Present geometry redrawn using two coordinate systems. z_o agrees with Hackett's coordinate system. z represents the coordinate of the system after reversing the direction of excitation. z_{so}, z_s represent the locations of a point source.

epitaxial GaAs layer (8 μm thick, n_D = 2-3x10^{16}cm^{-3}) grown, using the AsCl$_3$ Process, on an n+ substrate (n_D = 10^{18}CM^{-3}) as illustrated in Fig. 1a.

DESCRIPTION OF THE TECHNIQUES

The excess carrier distribution for electron - beam excitation when the generation volume extends through the depletion layer of a Schottky contact, through the bulk region and through the epitaxial layer - substrate interface was derived by solving the continuity equation and two boundary conditions: infinite collection velocity at the depletion layer edge and finite recombination velocity at the epi-substrate interface. A result obtained by Hackett[2], for a p-n junction, was used in obtaining the solution, after reversing the direction of excitation (Fig. 1b). As explained in ref. 5, the function that is being used to extract the diffusion length L_p from the measurements of the collection and specimen currents is N, the amount of collected charge per incident electron, with contributions from the depletion layer (N_1) and from the bulk (N_2), that give for the present geometry:

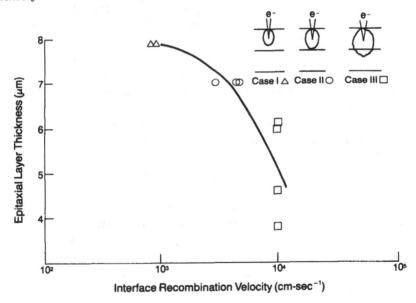

Fig. 2   A plot of interface recombination velocity as a function of epitaxial layer thickness.

$$N_1 = \int_0^{z_1} g(z) \, dz \tag{1}$$

$$N_2 = [1 + (1-S)/(1+S) \exp (-2z_B/L_p)]^{-1} \int_{z_1}^{z_1 + z_B} g(z)\{(\exp (z_1/L_p) \exp (-z/L_p) \tag{2}$$

$$+ (1-S)/(1+S) \exp (-z_1/L_p) \exp [(-2z_B + z)/L_p] \} \, dz$$

$$N = N_1 + N_2 \tag{3}$$

where z_1 and $(z_B + z_1)$ are the coordinates of the depletion layer edge and the epi-substrate interface, respectively, $g(z)$ is the electron-beam generation function[5], S is the reduced surface recombination velocity $S = s/ \sqrt{D/\tau_p}$, s is the surface recombination velocity (cm sec^{-1}), D is the diffusion length and τ_p is the minority carriers lifetime.

RESULTS AND DISCUSSION

Equation (2) contains three variables: S, L_p and z_B out of which only the thickness of the epitaxial layer $z_B + z_1$ can be independently determined. The other two variables were separated by measuring several samples that were etched prior to device fabrication, with thicknesses that vary from the original value (8.0 μm) to 5.8 μm (as measured using spreading resistance). In fitting the data to the model, the same bulk diffusion length was assumed for all the samples. The value of the L_p that was initially selected was quite close to the L_p value for an unetched sample, assuming zero interface recombination velocity. Then different values of s were assigned to all nine samples and these values were checked by minimizing (N-N $_{measured}$) where N $_{measured} = I_c/I_s$. This procedure was repeated until a self-consistent set of values was obtained for all the samples. The results of this iterative procedure were plotted (Fig. 2). The bulk diffusion length obtained (17 μm) is exceptionally high for GaAs, it translates into a lifetime of 270 nsec.. As a confirmation, lifetimes of 200 nsec. were measured, using the transient C-t method, for MOS devices fabricated on similar material. As Fig. 2 shows, the interface recombination velocity increases with decrease in the epi layer thickness, therefore an explanation is required since we would anticipate that the interface recombination velocity is dependent on the state of the interface alone. Referring to the groups identified in Fig. 2, some carriers for Case II and more carriers for Case III are being generated in the n$^+$ substrate and recombine there. Therefore, the true interface recombination velocity for our samples is in the range of 1000-5000 cm/sec. The excess carriers concentration profile in the epitaxial layer, for these three cases is plotted in Fig. 3. The plot shows that the diffusion of the excess carriers is bi-directional. Most carriers diffuse from the point of generation to the depletion layer edge. Some carriers diffuse in the opposite direction to the epi-substrate interface.

To check our results we will calculate the approximate value of the effective interface recombination velocity. Assuming thermal equilibrium condition on each side of the interface gives:

$$n_e \ \Delta p_e = n_s \ \Delta p_s \tag{4}$$

where n and Δp represent the majority and minority carrier concentrations, respectively, and the subscripts e and s apply to the epitaxial layer and the substrate. Using the definition of s (effective) in a region long with respect to L_p, gives[6]:

$$q \ s_{eff} \ \Delta p_e = qD \ \partial(\Delta p_s) / \partial z \Big|_{z=z_B} = q(\sqrt{D/\tau_s}) \ \Delta p_s \tag{5}$$

so that

$$s_{eff} = \sqrt{D/\tau_s} \ (n_e/n_s) = \sqrt{10.4/10^{-9}} \ (3 \times 10^{16}/10^{18}) = 3 \times 10^3 \ cm/sec$$

where τ_s is the band to band recombination lifetime. This result is reasonably close to the measured value.

CONCLUSIONS

We found that the diffusion length in an epitaxial layer, grown using the $A_s Cl_3$ Process, was approximately 17 μm, an exceedingly high value for GaAs, this translates to a lifetime of 270 nsec. The interface recombination velocity was in the range of 1000-5000 cm/sec, a measured value that agrees with an independent approximate calculation. Higher measured values of the recombination velocity are most likely due to generation and recombination of carriers in the highly doped bulk.

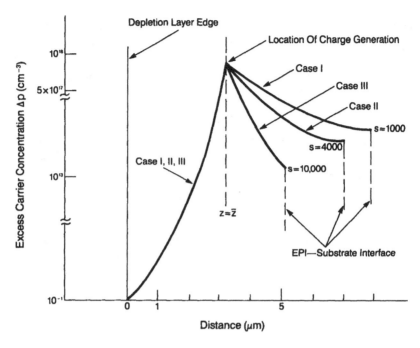

Fig. 3   The excess carrier concentration profile in the epitaxial
layer. The location of charge generation is at $z = \bar{z}$, an
average value defined as the depth at the mean value of
$g(z)$.

REFERENCES

1.   F. Berz and H. K. Kuiken, Solid-State Electronics, 19, 437,
(1976)

2.   W. H. Hackett, Jr., J. Appl. Phys., 43, 1649, (1972).

3.   D.E. Ioannou and S. M. Davidson, J. Phys. D:
Appl. Phys., 12, 1339, (1979).

4.   C. J. Wu and D. B. Wittry, J. Appl. Phys. 49, 2827, (1978).

5.   O. Paz and J. M. Borrego, Appl. Phys. Lett. 42, 958, (1983).

6.   B. G. Streetman, "Solid State Electronics Devices", 2nd ed.,
(Prentice - Hall, New Jersey, 1980), P. 154.

Metallic Superlattices

STRUCTURE AND COHERENCE OF METALLIC SUPERLATTICES

D. B. McWhan

AT&T Bell Laboratories
Murray Hill, New Jersey 07974

Most of the papers in this symposium have dealt with superlattices formed from semiconductors with the diamond or zinc blende structure or derivative structures such as fluorite in which one sublattice has the diamond structure. Such superlattices are often called compositionally

modulated alloys (CMA) because the two components have the same or related structures and because the two components have some mutual solubility. In these cases, depending on the differences in the lattice parameters, the resulting CMA will be almost a perfect single crystal, and the sharpness of the interfaces between the two alternating components will depend on the amount of interdiffusion during growth. In recent years attempts have been made to grow superlattices in which the two components have different crystal structures, and these are referred to as layered ultrathin coherent structures (LUCS) [1]. A typical structure is Nb_nCu_m where n (110) layers of body-centered cubic Nb alternate with m (111) layers of face-centered cubic Cu. In this early work, the term coherent structure (or what might be called a monolith) is meant to imply an ordered stack of thin crystals with no long range registry or coherence in the plane of the film between successive thin Nb or Cu crystals. Clearly, the next challenge to the materials scientist is to grow a LUCS in the form of a true single crystal where there is coherence between successive Nb or Cu regions. Such a crystal might be called a layered ultrathin crystal, or if the slang expression XTAL is used, then the acronym becomes LUX. The bulk of this paper traces some of the recent progress being made in growing LUX and some of the geometrical arguments why such a material is theoretically possible. Most of the experimental evidence to date is indirect and based to a large extent on geometrical modeling. The growth of LUX is an important challenge, and one which will have to be met before meaningful discussions of the effect of interfaces on various transport and magnetic properties can be made. Below a brief summary of X-ray diffraction as a probe of superlattices is followed by a discussion of Nb based superlattices. This discussion starts with the control of the epitaxy on different crystallographic planes of sapphire and progresses to Nb_nTa_m CMA. This is followed by the Nb_nZr_m system in which a transition from CMA to LUCS is observed with increasing modulation wavelengths and then the Nb_nAl_m system where the possibility of a LUX exists. Finally, the Nb_nLu_m system is discussed.

X-RAY DIFFRACTION FROM MULTILAYERS

X-ray diffraction is sensitive to periodicities in a material and to the interference between periodicities of different wavelength or of differing origin. In a CMA there are three basic periodicities. First is the average repeat distance or lattice in the structure. The second and third periodicities are the composition modulation and the associated strainwave modulation. These periodicities lead to a reciprocal lattice in which the Bragg peaks are given by $Q (h,k,l,m) = ha* + kb* + lc* + mk$ where $a*$, $b*$ and $c*$ define the average reciprocal lattice, k is the modulation wave vector, and h,k,l and m are integers. From the positions and the integrated intensities of the Bragg reflections, the crystal structure, (including the composition and strain modulation) can be determined. The results for several systems and the techniques have been discussed by a number of authors and do not need to be repeated here [2,3,4]. It is perhaps more important to emphasize some of the detailed information contained in the scattering pattern. For example, as shown in Figure 1, the modulation wave vector is normal to the growth surface so that if the substrate is cut at an angle relative to some major crystallographic axis, say $c*$, then the harmonics around the Bragg reflections of the average structure will be at an angle relative to $c*$ as illustrated in Figure 1.

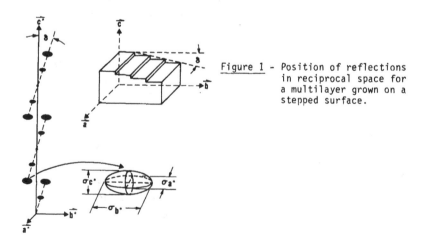

Figure 1 - Position of reflections in reciprocal space for a multilayer grown on a stepped surface.

Furthermore, the width of a reflection (in the absence of strain) is inversely proportional to the coherence length in that direction. Defining the full width at half maximum as an ellipsoid of revolution, as illustrated in Figure 1, it has been possible to show that there is a large anisotropy in the coherence length for samples grown on stepped surfaces with a longer coherence length parallel to the steps and a coherence length comparable to the average step separation perpendicular to the steps. Similar results have been obtained in independent studies of two different GaAs based CMA (when put in comparable units) [5,6]. It is usually found that the widths of the harmonics parallel to k increase linearly with m, and this gives a measure of the variation in the average modulation wavelength over the area of the sample illuminated by the X-ray beam [3]. This variation is typically a few percent and may originate in the placement of the sources relative to the substrate or in fluctuations in the source flux during deposition.

In the study of CMA and LUX the degree of coherence is given by the coherency strain, i.e., the observed average reciprocal lattice relative to that of the two components of the multilayer. In the case of a CMA where both components have the same crystal structure, a^* and b^* will be approximately the average of the values for the two components; while along the modulation direction, c^*, the amplitude of the strain wave will reflect the fact that the interplanar spacing in this direction will be larger or smaller than that of the pure components, depending on whether the component is under uniaxial tension or compression in order to achieve a single a^* and b^* in the plane. The relative strain in each component depends on the details of the appropriate elastic tensor, and calculations have only been reported for relatively simple configurations. In the case of LUX, reflections in the plane of the film can be resolved for each of the components. As discussed below for the Nb_nAl_m system, there will be strains of orthorhombic or lower symmetry as each component adjusts to form a coherent interface.

The diffuse scattering near the Bragg reflections contains information about average grain sizes, fluctuations in grain size or composition modulation, and short range order in the interfacial regions. This information is more difficult to analyze quantitatively, but the effects of a distribution of grains of finite size are evident in the diffraction patterns of most metallic superlattices. If a perfect, thin crystal is grown as an overlayer on a substrate, then the scattering from a finite crystal is given by the standard relation $SIN^2\pi N\ell/SIN^2\pi\ell$ where N is the number of layers of atoms in the thin crystal and the scattering vector is defined as $\ell\ c^*$. This relation has a maximum at integral values of ℓ and subsidiary maxima at values of $\ell = P \pm \frac{m}{N}$ which fall off in intensity as $1/SIN^2\pi\ell$. This relation is beautifully demonstrated by the overlayer of $NiSi_2$ grown on Si shown in Figure 2 [7]. This Figure shows the (111) of the Si substrate and the (111) of the $NiSi_2$ overlayer. The separation of the two reflections is a direct measure of the coherency strain, and the positions of the subsidiary oscillations around the $NiSi_2$ (111) give N. The clear observation of these oscillations shows that the $NiSi_2$ overlayer is a near perfect crystal of uniform thickness. In contrast to the near perfect overlayer, the average metallic multilayer shows a distribution of grain sizes and fluctuations in modulation wavelength. This is illustrated for a Nb_nLu_m multilayer in Figure 3. The two sharp, narrow reflections are

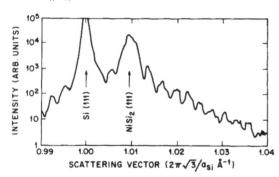

Figure 2 - Diffraction pattern of a ~1100 Å thick single crystal of $NiSi_2$ grown on a Si (111) substrate (Ref. 7).

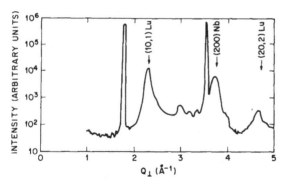

Figure 3 - Diffraction pattern for a Nb_nLu_m multilayer grown on (01.2) Al_2O_3 (Ref. 15).

the (01$\bar{1}$2) and (02$\bar{2}$4) of the Al_2O_3 substrate and the two other major reflections are the (10$\bar{1}$1) of the Lu and the (002) of the Nb. The $1/SIN^2\pi\ell$ wings which are characteristic of finite grain are clearly seen. The oscillations are averaged out because there is a distribution of grain sizes, and the average grain size is given by the ratio of the peak intensity to the $1/SIN^2\pi\ell$ tails. In this film which is 4000 Å thick the average grain size normal to the film is a few hundred Angstroms. Results were also reported for a Cu_4Ni_4 multilayer where an average grain size of ~1000 Å was observed in a 50,000 Å film [8]. The diffuse scattering is thus a useful probe of average grain size and therefore the overall quality of a multilayer.

NIOBIUM BASED MULTILAYERS

In this section the structure of a series of Nb based multilayers is traced in order to illustrate the problems associated with the synthesis of LUX. The first problem is the control of the epitaxy on a substrate or a single crystal of one of the components. In the seminal paper by Durbin, Cunningham, Mochel and Flynn, a series of Nb_nTa_m multilayers were grown at high temperatures on different planes of Al_2O_3 and NaF [9]. These authors demonstrated that the modulation direction in a multilayer could be controlled by the choice of substrate. Our work at AT&T Bell Laboratories has concentrated on using the (01$\bar{1}$2) of Al_2O_3 as the substrate, and these studies demonstrate two of the general features of epitaxial growth [10]; first the existence of a critical temperature for epitaxial growth, and second the importance of symmetry. The first is illustrated by the data in Figure 4 where longitudinal scans and rocking curves are compared for Nb_nTa_m multilayers grown on (01$\bar{1}$2) sapphire at substrate temperatures of $810^\circ C$ and $850^\circ C$. At the lower temperature the multilayer has the default texture, i.e., (110) planes which are the close-packed planes for a body centered cubic structure. The main peak in the lower right figure is at $Q=2.68Å^{-1}=2\pi/d$ (110) and there are weak harmonics resulting from the modulation. Holding the detector at the angle corresponding to $Q=2.68$ and rotating the sample yields the rocking curve in the lower left. There is rather poor texture with grains oriented over a range of 8°. Raising the substrate temperature to $850^\circ C$ results in epitaxial growth with [001] texture as illustrated in the top of Figure 4. A [002] reflection is

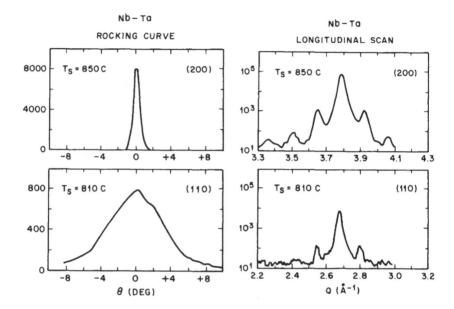

Figure 4 - Comparison of Nb_nTa_m multilayers grown on sapphire (01.2) at different substrate temperatures (Ref. 10).

observed at $Q=3.79\overset{o}{A}$ with at least three harmonies. The rocking curve is now less than 1°.

A second important feature of the films grown epitaxially at $T_s=850^oC$ is that the (001) planes of the multilayer are not parallel to the $(01\bar{1}2)$ planes of the substrate in contrast to the (110) planes of the film grown at $T_s=810^oC$ which are, on average, parallel to the $(01\bar{1}2)$. The (001) planes are consistently found to be rotated 2.6^o away from the $(01\bar{1}2)$. The rotation takes place in the $(2\bar{1}\bar{1}0)$ plane of the substrate as illustrated in Figure 5. Normally, films are grown on high symmetry planes of cubic crystals so that there is a rectangular or hexagonal two-dimensional surface symmetry, and the epitaxial film grows with planes of similar symmetry parallel to the surface. For example, (110) Nb on $(1\bar{1}20)$ of Al_2O_3

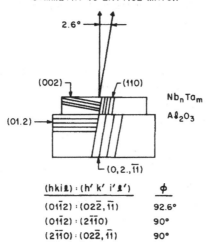

Nb ON SAPPHIRE (01.2)

SYMMETRY vs LATTICE MATCH

$(hki\ell) : (h' k' i' \ell')$	ϕ
$(01\bar{1}2) : (02\bar{2}, \bar{1}1)$	92.6°
$(01\bar{1}2) : (2\bar{1}\bar{1}0)$	90°
$(2\bar{1}\bar{1}0) : (02\bar{2}, \bar{1}1)$	90°

Figure 5 - Geometric model to demonstrate the observed 2.6°
rotation of the $Nb_n Ta_m$ [002] with respect to the (01.2) surface
of the Al_2O_3.

or (111) Nb on (0001) of Al_2O_3. For the case of $(01\bar{1}2)$ Al_2O_3 as
illustrated in Figure 5, the $(2\bar{1}\bar{1}0)$ are perpendicular to the surface and
$(0,2,\bar{2},\bar{1}1)$ planes are the next planes closest to being 90° from the
surface, and they are in fact 92.6° from the $(01\bar{1}2)$ surface. The Nb grows
so that the (110)Nb are parallel to the $(0,2,\bar{2},\bar{1}1)$ Al_2O_3 and the $(1\bar{1}0)$ Nb
|| $(21\bar{1}0)$ Al_2O_3. This suggests that the growth occurs along a zone axis of
the substrate and that it reflects the three-dimensional symmetry of the
substrate and not the symmetry of the surface layer.

A $Nb_n Ta_m$ multilayer is a CMA as both components are bcc with a good
lattice match and the interfaces are 10-12Å thick as a result of
interdiffusion. A $Nb_n Zr_m$ multilayer is a transitional case. Both
components are bcc at high temperatures but at room temperature Zr is
hexagonal close packed or has the structure of the ω-phase, depending on
the pressure. At short modulation wavelengths the multilayer is a CMA with
the bcc structure and with interfaces somewhat thicker than those observed
in $Nb_n Ta_m$ [11]. It is most likely that the interdiffusion of Nb into Zr

stabilizes the bcc phase. At longer wavelengths, after a transitional
region that has not been well characterized, the Zr reverts to the normal
hcp structure. In this regime the multilayers are best characterized as
LUCS, i.e., alternating regions of bcc Nb and hcp Zr which are not
correlated in the plane.

Turning to the layered ultrathin crystals (LUX), what are the
conceptual requirements for the synthesis of a LUX? It is a widely held
misconception that materials should be tried in which the areal density of
the (110) planes of the bcc component is close to that of the (111) planes
of the fcc component or the (0001) planes of the hcp component. It is not
at all obvious that this comparison is relevant. To form a coherent
interface a supercell must be found in which there exists subcells
corresponding to each of the components. To have a perfect match the two
subcells will usually have to be strained anisotropically with respect to
the higher symmetry of the pure components. In the case of a (111) plane
with hexagonal symmetry, the resulting orthorhombic strain will break the
hexagonal symmetry. If the structure retains this orthorhombic coherency
strain through the "fcc" region then the sixfold degeneracy will be lifted
at the next "fcc"-"bcc" interface so that the next "bcc" region will grow
coherently with respect to the previous "bcc" region. The orthorhombic
coherency strain provides a mechanism for the synthesis of LUX. The
existence of an orthorhombic strain has been demonstrated for the Nb in
Nb_nAl_m multilayers [12]. The Nb (110) planes parallel to the film are 1.7%
larger than in bulk Nb, but in the plane of the film the orthogonal

directions are smaller by -0.1% ($1\bar{1}0$) and -1.5% (002). A possible
orientation of the (110) Nb and (111) Al planes is shown in Figure 6a. The
observed strains are consistent with a partial approach to a coherent
interface. The diffraction data for the Al are not good enough to
demonstrate the breaking of the hexagonal symmetry of the (111) planes, but
an expansion of the lattice perpendicular to the (111) planes by several
percent is inferred from measurements of the Al (200) reflection in a
series of multilayers. Unfortunately, in the Nb-Al system compound
formation occurs at higher temperatures so that epitaxial growth was not
achieved in the samples available to date. Another possible candidate for
LUX is the Mo_nNi_m system [13]. Geometric arguments for coherency strains
have been presented to account for the shift in the average lattice
constant with wavelength [14], and a possible interface configuration is
shown in Fig. 6a.

Finally, the unusual texture in Nb_nLu_m multilayers is considered.[15] The expected default texture is with (110) Nb and (0001) Lu planes parallel with the substrate. However, the data in Figure 3 show reflections which correspond to (002) Nb and (10$\bar{1}$1) Lu. That this texture was correct was established by observing the (110) Nb and the (10$\bar{1}$0), (0002) and (10$\bar{1}$$\bar{1}$)

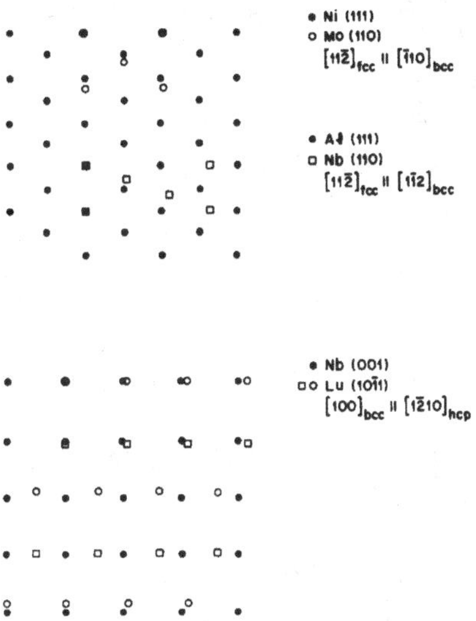

• Ni (111)
o Mo (110)
$[11\bar{2}]_{fcc}$ ‖ $[\bar{1}10]_{bcc}$

• Al (111)
□ Nb (110)
$[11\bar{2}]_{fcc}$ ‖ $[1\bar{1}2]_{bcc}$

Figure 6 - Geometric models for the interfaces in Nb_nAl_m, Mo_nNi_m, and Nb_nLu_m multilayers.

• Nb (001)
□○ Lu (10$\bar{1}$1)
$[100]_{bcc}$ ‖ $[1\bar{2}10]_{hcp}$

Lu reflections at the appropriate angles with respect to the film normal. A possible geometry for the (002) Nb - (10$\bar{1}$1) Lu interface is shown in Figure 6b, and the overall match is not very good. Examination of the lattice parameters of other hcp metals suggests that a Nb_nSc_m multilayer would have a much better lattice match and might be a good candidate for a LUX with this unusual texture.

In conclusion, the high quality of semiconductor superlattices was the result of many years of research and only recently have the limits of lattice mismatch been tested in strained superlattices. The giant step to the successful synthesis of layered ultrathin crystals with components of different structure will undoubtedly require many more years of research, but from the discussion above on Nb based superlattices, it is clear that the pieces are beginning to fall into place.

References

[1] I. K. Schuller, Phys. Rev. Letters 44, 1597 (1980).

[2] D. DeFontaine in Local Atomic Arrangements Studied by X-ray Diffraction, J. B. Cohen and J. E. Hilliard, eds., p. 51, Gordon and Breach, New York 1966.

[3] A. Segmuller, A. E. Blakesly, J. Appl. Cryst. 6, 19 (1973).

[4] R. M. Fleming, D. B. McWhan, A. C. Gossard, W. Wiegmann, and R. A. Logan, J. Appl. Phys. 51, 357 (1980).

[5] D. B. McWhan in Synthetic Modulated Structures, eds. L. L. Chang and B. C. Giessen, Academic Press, in press.

[6] D. A. Neumann, H. Zabel, H. Morkos, Appl. Phys. Letters, 43, 59 (1983).

[7] Sample supplied by R. Tung.

[8] E. M. Gyorgy, D. B. McWhan, J. R. Dillon, Jr., L. R. Walker, and J. V. Waszczak, Phys. Rev. B25, 6739 (1982).

[9] S. M. Durbin, J. E. Cunningham, C. P. Flynn, J. Phys. F. 12, L75 (1982).

[10] G. Hertel, D. B. McWhan, J. M. Rowell, Superconductivity in d- and f-band Metals, (1982), W. Buckel and N. Weber eds., Kernforschungszentrum, Karlsruhe 1982, p. 299.

[11] W. P. Lowe, T. H. Geballe, Phys. Rev. B29, 4961 (1984).

[12] D. B. McWhan, M. Gurvitch, J. M. Rowell, L. R. Walker, J. Appl. Physics, 54, 3886 (1983).

[13] M. R. Khan, C. S. L. Chun, G. P. Felcher, M. Grimsditch, A. Kueny, C. M. Falco and I. K. Schuller, Phys. Rev. B27, 7186 (1983).

[14] D. B. McWhan in High Pressure in Science and Technology, MRS symposium, Vol. 22, C. Homan, R. K. MacCrone and E. Whalley, eds., North Holland, New York 1984, p. 131.

[15] L. H. Greene, W. L. Feldmann, J. M. Rowell, B. Batlogg, E. M. Gyorgy, W. P. Lowe, and D. B. McWhan, Proc. of First "Notre Dame" International Conference on Superlattices, Microstructures, and Microdevices, to be published.

SUPERCONDUCTING PROPERTIES OF AMORPHOUS MULTILAYER METAL-SEMICONDUCTOR COMPOSITES

A.M. Kadin*, R.W. Burkhardt*, J.T. Chen+, J.E. Keem*, and S.R. Ovshinsky*
* Energy Conversion Devices, 1675 West Maple Road, Troy, Michigan 48084,
+ Department of Physics, Wayne State University, Detroit, Michigan

ABSTRACT

Following the earlier multilayer work of Ovshinsky and colleagues, we have fabricated thin-film samples consisting of alternating periodic layers of a transition metal (Nb, Mo, W) and a semiconducting element (Si, Ge, C) by sequential sputtering from two targets onto room-temperature substrates. The regular repeat spacing has been varied from 10 Å to more than 100 Å, with as many as several hundred layer pairs. Crystalline epitaxy was not required or even desired; many samples were largely amorphous as determined from x-ray scattering. Electrical transport measurements of superconducting properties have been carried out parallel to the layers. Samples exhibited highly anisotropic superconducting critical magnetic fields, with some values in excess of 200kG parallel to the layers. Evidence suggesting an asymmetric interface profile will be presented.

INTRODUCTION

There has been considerable interest in the past decade in developing novel materials using precision microlayering techniques. Much of this has been oriented towards crystalline epitaxy, using pairs of materials with lattice spacings that almost match. However, for most other material pairs, layer interfaces tend to be dominated by dislocations, agglomeration, and interdiffusion, if allowed to thermally equilibrate.

Our approach [1] has been quite different. Recognizing the difficulties and limitations in obtaining perfect crystalline multilayer films, we have attempted to make multilayer films that are amorphous, not only at the interfaces, but within the layers themselves. Since we can then deposit the metastable layers onto substrates held at room temperature, we can make very smooth and repeatable layers of many diverse materials. Because of the short electronic mean free path, these multilayer structures (which are not properly superlattices because the crystalline lattice itself is not well-defined) exhibit a different set of phenomena than those in the crystalline superlattices.

Although such an amorphous multilayer structure may show no sharp x-ray structure indicative of long range atomic order within the component materials, it can exhibit very sharp x-ray diffraction peaks corresponding to the repeat spacing d of the amorphous layers. In fact, using this principle together with the alternation of light and heavy elements, ECD has developed, and is currently marketing, OVONYX™ structures that show superior performance as high-reflectivity x-ray devices, with reflectances that exceed 80% at near-normal incidence.

We have fabricated and studied similar samples for their superconducting properties. Recently, several groups [2-7] have investigated artificially layered superconducting structures, including metal-metal systems [2-4], with emphasis on effects of crystalline epitaxy, and metal-semiconductor systems [5-7]. We have chosen to study the latter type of system, consisting of a refractory transition element (e.g., Nb, Mo, W) alternating with a semiconducting element (Si, Ge, C), with repeat spacings ranging from 10Å to over 100Å. We will survey a number of such layered samples that exhibit a wide range of superconducting behavior.

SAMPLE FABRICATION AND STRUCTURE

To produce these multilayer structures, an unheated substrate, usually an oxidized Si wafer or glass slide, was transported periodically past two targets in either an rf magnetron sputtering system (in an Ar plasma) or an Ar-ion-beam sputtering system. Background pressures were in the low 10^{-7} torr range. All indications suggested that the deposition rate (of order 1Å/sec), once established, could remain constant to within a few percent over the several hours needed to achieve total thickness up to 1 μm.

Several analytical methods were at our disposal to determine the ordering on the scales of both the atoms and the layers. X-ray diffraction was used to determine the degree of crystalline order within the layers, as well as to look for peaks corresponding to coherent layering. X-ray fluorescence was also available, particularly with thicker samples, to determine the elemental composition. Finally, samples were analyzed by Auger depth profiling, in a semi-quantitative way, to estimate the composition and to confirm the presence of layering.

The primary tool to investigate the degree of layering was low-angle x-ray diffraction, usually using a Cu K-α source. Since many of our layered samples had interfaces that were sharp on the scale of the layer spacing, they diffract up to many orders, as the example in Fig. 1a indicates. This sample, consisting of approximately 70Å of Mo alternating with about 35Å of Si, exhibited peaks up to about 15th order in diffraction. This suggests that the interface width is sharper than about 7Å, and a simple model of absolute x-ray reflectance has suggested comparable interface mixing or roughness for similar samples.

An Auger depth profile is shown for the same sample in Fig. 1b. Note that the apparent compositional modulation here is incomplete, and gradually diminishes in each successive layer. We believe that this is an artifact of the Auger depth profile itself. First, the Auger electrons have an escape depth of 20 to 30Å, limiting the resolution. Second, the depth profile involves sputter etching by Ar ions (of energy 1.5 keV in Fig. 1b), which may produce uneven etching and ion mixing. Higher energy Ar ions can obscure the layering almost completely.

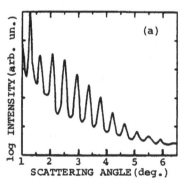 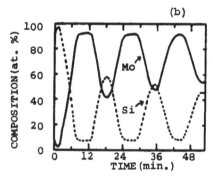

Fig. 1: Structural data for Mo-Si multilayer #302 with d_s=70Å, d_i=35Å, and 160 bilayers. a) X-ray diffraction spectrum for λ=1.54Å. b) Auger depth profile for 1.5 keV Ar ion sputter beam.

The semiconducting element, particularly Si, normally deposits in the amorphous state, even for rather thick films, when the substrate is held at room temperature. This is not always true, however, for the metallic

layers; for layer thicknesses greater than about 40Å, even refractory
metallic films come down as disordered, but rather oriented microcrystalline
films. For thinner metallic layers, however, strains at interfaces may tend
to make the amorphous metallic structure relatively more stable. Some
evidence for this was taken on a set of Nb-Si and W-Si multilayers, using
x-ray diffraction in the conventional reflection mode, which probes
structure perpendicular to the layers. The evidence clearly indicates that
there is no crystalline coherence from one metal layer to the next, and that
for the thinner layers, the broadening of the crystalline peaks seems to be
somewhat greater than that directly attributable to the size of a single
layer. The amount of disorder, and perhaps the development of amorphous
structure, does appear to increase substantially as the thickness of the
metallic layer is decreased below about 30Å. Furthermore, for the thinnest
layers, without epitaxy, it would certainly be surprising if crystalline
order continued to exist along the planes. However, a more definitive probe
of crystallinity within a layer could be made by directly probing structure
within the plane, perhaps by transmission, and we are pursuing this
approach. Finally, there seems to be very little evidence of metal-silicide
formation, except perhaps for the very smallest layer spacing, where the
atoms are largely mixed.

CRITICAL TEMPERATURE AND NORMAL-STATE PROPERTIES

Samples were generally measured in a four-probe configuration using Cu
contacts pressed into the films, with sample geometry (defined by
photolithographic lift-off) consisting of wide pads connected by a long
narrow section 3.8mm x 0.3mm. The resistivity of these multilayers,
measured parallel to the film and attributed to the metallic layers, is of
order 100 μohm-cm or less, and either falls or is essentially constant
(rising at most a few percent) down to low temperatures. Even single layers
down to 10Å appear to be continuous, although we cannot rule out the
presence of defects (e.g., holes) in such thin layers. Generally,
resistivity increases as the thickness of the metal layer is decreased,
suggesting an electronic mean free path that is limited in part by film
thickness.

As we have noted above, the structure of the very thin metallic layers
is highly disordered, perhaps even amorphous. This raises the critical
temperature of the Mo and W composites substantially (from less than 1K to
greater than 4K), while lowering that in the Nb multilayers. Both of these
trends are consistent with other studies of T_c in bulk amorphous or
disordered samples [8]. It is not totally clear, however, whether these
effects in the present case are a consequence more directly of structural
modification or of compositional mixing that certainly exists to some degree
at layer interfaces.

Taking a set of Nb-Si multilayer samples as a particular example,
several generalizations can be made concerning the dependence of T_c on the
three independent parameters d_s, d_i (the thickness of the superconducting
and insulating layers, neglecting to first order the interface smearing),
and the number of bilayers N. First, holding d_s and N constant and varying
d_i, larger d_i tends to be associated with lower values of T_c and with
broader transitions. This can be understood by noting that large d_i yields
essentially decoupled thin superconducting films, and either intrinsic 2-D
fluctuations or film defects might be expected to broaden the transition.
If d_i and N are held constant, then increasing d_s tends to raise T_c for Nb,
a feature we believe is a consequence of the effect of order on T_c.
Finally, for d_s and d_i constant, increasing N also tends to increase T_c
slightly, together with a slightly decreasing normal state resistivity.

506

This result may be due to a tendency toward improved layer smoothness and order as the number of layers increases.

CRITICAL FIELDS AND CURRENTS

Superconducting critical fields were measured in-house in a superconducting magnet dewar with fields up to 90 kG and temperatures down to 1.4K, and some samples were also taken to the Francis Bitter National Magnet Laboratory to allow fields up to 190 kG. In both cases, samples could be mounted with the film (and the layers) oriented either parallel or perpendicular to the direction of the field. We define the critical field as that field for which the resistance has reached one-half its normal-state value.

In general, the parallel upper critical magnetic field of these metal-semiconductor multilayer films is substantially larger than that for the field perpendicular to the layers. Essentially, this occurs because screening supercurrents act as pair-breaking perturbations, and the presence of the insulating layers tends to restrict the flow of supercurrent perpendicular to the layers in response to a parallel field. The perpendicular critical field is then the same as the usual upper critical field of a superconductor

$$H_{c\perp} = H_{c2} = \phi_0/2\pi\xi^2(T) \tag{1}$$

where $\phi_0 = hc/2e \dot= 2.07 \times 10^{-7}$ gauss-cm^2 is the flux quantum, and $\xi(T) \sim (T_c - T)^{-1/2}$ is the Ginzburg-Landau coherence length, typically of order 100Å or greater in our films. This implies that $H_{c\perp}$ should go to zero linearly at T_c.

For the parallel critical field, there are several relevant regimes. In the "2-D" case, the coupling between adjacent metallic layers is so weak that the critical field is that which corresponds to a single metallic layer. For a layer of thickness $d_s < \xi$,

$$H_{c\parallel} = \sqrt{3} \, \phi_0/\pi\xi d_s \tag{2}$$

Eq. (2) is an upper limit; for very small d_s, spin paramagnetism may produce a lower critical field [9].

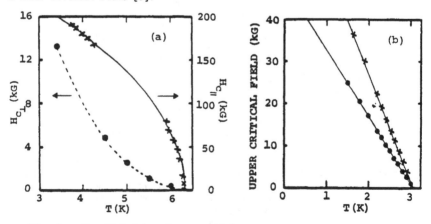

Fig. 2: Parallel (x) and perpendicular (dots) critical magnetic fields for a) Mo-Si #302 and b) Nb-Si #920-18.

The critical fields for Mo-Si sample #302 are presented in Fig. 2a. The solid curve fit to the parallel field data is of the form $(T_c-T)^{1/2}$, in agreement with Eq. (2). Note the very high critical field for this orientation (greater than 200 kG at T=0), which exceeds substantially the paramagnetic limiting field of 120 kG and indicates the presence of substantial spin-orbit scattering [9]. A parallel field of this dependence was the general rule for samples with insulator thickness d_i greater than about 30Å. However, this sample exhibits some discrepancies with the theory. The perpendicular critical field, instead of being linear near T_c, curves concave upward. If we apply Eq. (1) anyway at T=4K, one obtains $\xi \doteq$ 200Å, whereas if we take a value $d_s \doteq 70$Å, then from Eq. (2) and the parallel field, we obtain $\xi(4K)\doteq90$Å. Some similar deviations were seen in Refs. [5] and [6], and they remain somewhat of a puzzle.

The other limit, the anisotropic 3D case, has been treated in terms of a model of Josephson coupling between the metallic layers, which is then mapped onto an anisotropic Ginzburg-Landau equation [10]. The result is an effective coherence length $\xi_\perp$ that is reduced from the value $\xi \doteq \xi_\parallel$ within the plane by a factor proportional to the square root of the interfilm coupling, or roughly to the square root of the anisotropy in the normal state resistivity. This yields a parallel critical field

$$H_{c\parallel} = \phi_0/2\pi\xi_\parallel\xi_\perp \tag{3}$$

which is larger than the perpendicular critical field, but exhibits the same linear temperature dependence.

Fig. 2b shows the critical field behavior for a Nb-Si sample with $d_i=d_s=20$Å. The critical fields are far less anisotropic than those in Fig. 2a, and follow straight lines. Using Eqs. (1) and (3), we infer values of $\xi_\parallel$ and $\xi_\perp$ of 115 and 80Å respectively, both substantially larger than the repeat spacing d=40Å and thus in the anisotropic 3D regime. A dependence like this was generally seen for insulating layers thinner than about 20Å, as long as d_s was not too great.

Qualitatively, one might expect Eq. (3) to remain valid as long as $\xi_\perp$ was much greater than the metal film thickness. According to Ref. [10], the crossover between the two regimes of Eqs. (2) and (3) should occur when $\xi_\perp=d/\sqrt{2}$. We have seen some evidence of such crossover behavior for insulating layers of intermediate thickness (or for thicker metal layers).

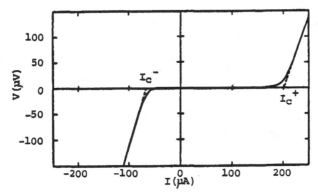

Fig. 3: Asymmetric I-V curve for Nb-Si #920-18 for a parallel field H=20kG at T=1.5K.

For subcritical magnetic fields applied parallel to the layers and perpendicular to the flow of current, some of these films exhibit behavior that appears very much like classic flux-flow characteristics. This corresponds to vortices moving perpendicular to the layers. An interesting feature of these characteristics that we have recently reported [7] is that they may be strongly asymmetric, and in particular that the critical current may be much larger (by up to a factor of 5) for one sign of the current than for the other sign (see Fig. 3b). Furthermore, this asymmetry reverses exactly if the sign of the magnetic field is reversed. This can be understood in terms of a preferred direction of vortex flow, either toward or away from the substrate. It is still unclear whether this effect is due to asymmetry of vortex entry at the film surfaces, or an asymmetric compositional profile at layer interfaces, or some alternative asymmetry. It has appeared in a number of layered samples, and only for this particular flux flow configuration of currents and fields. Investigation is continuing.

CONCLUSIONS

Finally, we believe that layering may be a useful tool to fabricate new and metastable materials, even without crystalline epitaxy, particularly in the limit of small, nearly atomic layer thicknesses. Since many superconducting compounds with high critical temperatures are metastable at best, this approach may be a fruitful one to investigate new higher - T_c superconductors.

ACKNOWLEDGEMENTS

We wish to thank J. Wood of ECD, and the staff of the National Magnet Laboratory, for assistance with some of the high-field measurements. Some of the early work was funded in part by ARCO.

REFERENCES

1. S.R. Ovshinsky, U.S. Patent # 4,342,044 (1982), and others pending.
2. I.K. Schuller and C.M. Falco, Thin Solid Films, 90, 221 (1982); C.S.L. Chun, C.G. Zheng, J. Vincent, and I.K. Schuller, Phys. Rev. B, 29, 4915 (1984).
3. D.B. McWhan, M. Gurvitch, J.M. Rowell, and L.R. Walker, J. Appl. Phys., 54, 3886 (1983).
4. W.P. Lowe and T.H. Geballe, Phys. Rev. B, 29, 4961 (1984).
5. T.W. Haywood and D.G. Ast, Phys. Rev. B, 18, 2225 (1978).
6. S.T. Ruggiero, T.W. Barbee, and M.R. Beasley, Phys. Rev. Lett., 45, 1299 (1980); Phys. Rev. B, 26, 4894 (1982).
7. A.M. Kadin, R.W. Burkhardt, J.T. Chen, J.E. Keem, and S.R. Ovshinsky, Proc. 17th Int. Conf. on Low Temp. Phys., Karlsruhe, West Germany, 1984, ed. by U. Eckern et al., p. 579 (Elsevier North-Holland, Amsterdam, 1984).
8. M.M. Collver and R.H. Hammond, Phys. Rev. Lett., 30, 92 (1973).
9. N.R. Werthamer, E. Helfand, and P.C. Hohenberg, Phys. Rev., 147, 295 (1966).
10. R.A. Klemm, A. Luther, and M.R. Beasley, Phys. Rev. B, 12, 877 (1975).

STRUCTURAL PROPERTIES OF SINGLE CRYSTAL RARE-EARTH THIN FILMS Y AND Gd GROWN BY MOLECULAR BEAM EPITAXY

J. KWO*, D. B. McWHAN*, M. HONG*, E. M. GYORGY*, L. C. FELDMAN* AND J. E. CUNNINGHAM**
*AT&T Bell Laboratories, 600 Mountain Avenue, Murray Hill, New Jersey 07974
**University of Illinois at Urbana-Champaign, Urbana Illinois 61801

ABSTRACT

By means of metal MBE technique with in-situ RHEED characterization, high-quality single crystal rare earth metal films of yttrium and gadolinium were grown as a necessary requirement for studying Gd/Y superlattices. The key step of this successful growth is the employment of the single-crystal Nb film as a buffer layer to eliminate the interaction of rare earth metals with most substrates. Structural analyses by X-ray diffraction and ion channeling show that these crystals exhibit not only complete texture of [00.1], but also narrow rocking curves both perpendicular (00.2) and parallel (10.0) to the film.

INTRODUCTION

For the past half decade significant efforts have been made in the metal superlattice field [1] with the anticipation to explore new physical phenomena occurring on these artificially man-made materials. In our research program of metal superlattices grown by Molecular Beam Epitaxy (MBE) technique, we decided to undertake the studies of rare earth (RE) metal superlattices [2] for the following reasons. Most RE metals are of similar crystal structures with close lattice constant match, and *coherent* metallic superlattices have been grown [3]. The long-range nature of the RKKY interaction existing in most magnetic RE elements [4] is expected to give rise to strong modulation in the magnetic properties as a result of artificial layering structures. The first RE superlattice of our investigation is the magnetic Gd layer modulated with the nonmagnetic Y layer. Besides their strong structural similarities, the simple ferromagnetic behavior of Gd with little anisotropy makes this combination most attractive for investigating their magnetic properties in the ultrathin multilayer configuration.

In this work we report the initial success of growing, for the first time, single crystal RE thin films of Y and Gd, as the necessary requirement for fabricating superlattice structures of Gd/Y. Difficulties of substrate interaction [2,3,5] previously encountered with RE metal films and superlattice depositions have been entirely eliminated. The structural perfections of these single crystal RE films were examined by RHEED, X-ray diffraction, and Rutherford backscattering with ion channeling.

METAL MBE GROWTH TECHNIQUE

All RE metal films of Y and Gd in this work were prepared in a versatile UHV deposition system designed for metal epitaxy and superlattice work [2]. The Y and Gd films were evaporated from 99.9999% pure sources of individual electron guns with independent rate monitoring and loop control. The typical deposition rate varies from 1.0 to 3.6 Å/sec with short-term fluctuation of about 2-3%, and with virtually no long-term drift. Each source is equipped with an individual shutter to facilitate superlattice and compound depositions. With a fast-entry load lock, the background pressure of the main chamber is kept as low as 4×10^{-11} torr during the RE metal depositions. Since low-rate deposition is generally preferred in order to control the epitaxial growth near the interface, maintaining an ultra-pure vacuum during growth is crucial for RE metal depositions because of their extreme sensitivity to residual gas-impurity contamination. With in-situ Reflection High-Energy Electron Diffraction (RHEED) of 10 KV electrons, the entire film growth process was monitored and studied layer by layer.

EXPERIMENTAL RESULTS AND DISCUSSIONS

A. Thin Film Growth of Y and Gd by RHEED Characterization

Earlier studies [2] of growing the RE thin films of Y and Gd showed that direct deposition of the RE metals onto most insulating substrates invariably resulted in chemical reaction with the substrate species at substrate temperatures greater than 300°C. Such substrate-reaction processes hinder the possibility of depositions at elevated temperatures (600-800°C), which is a commonly adopted approach in case of a large lattice mismatch between the film and the substrate. Additionally, it has adverse effects on the crystallinity of the as-grown films, including the poor rocking curve of the crystallographic axis in the growth direction and little orientation order of the grains in the plane. Phase-contrast optical microscopy revealed a rather rough surface morphology on the order of 5000 Å.

One simple approach of circumventing the above difficulty is, prior to the RE deposition, to grow a buffer layer which does not react with either the substrates or the RE metals, yet permitting a highly ordered growth of [00.1] texture of the RE films. We chose to use the Nb film as the buffer layer because in the entire periodic table, only the VB and VIB elements including V, Nb, Ta, Cr, Mo and W do not react with the RE elements. Based on earlier work [6] on thin metal overlayers grown on Nb, it was found that RE metals like Y and Er wet the Nb surface completely, yet form little solid solution at the interface. Another advantage of this approach derives from the realization that high-quality single crystal films of Nb can be grown on sapphire substrates [7,8] with rocking curves less than 0.1°, and with in-plane domain coherency over 1000 Å.

The close-packed plane (110) of bcc Nb single crystal can be grown either ($11\bar{2}0$) or (000.1) sapphire (Al_2O_3) substrates [8]. A 4500 Å thick Nb film of (110) orientation was deposited separately by a MBE system at the University of Illinois [7,8]. Then, the Nb film surface was outgas-cleaned in our UHV chamber by heating up to 900° for 1 hour. The reflection high energy electron diffraction (RHEED) pattern started to show clear and well defined streaky behavior as in Fig. 1(a), with slightly diffused background, presumably due to some residues of Nb-oxides on the surface.

(a)

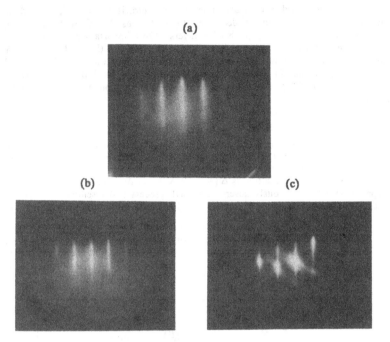

(b)   **(c)**

Fig. 1.   RHEED patterns of (a) Nb (110) film grown on sapphire
(000.1) after outgas-cleaning, (b) Y film of 100 Å thick
deposited on Nb (110), and (c) Y film of 150 Å thick and
greater deposited on Nb (110).

With the presence of a Nb (110) buffer layer, the RE films of Y and Gd can be
grown at much higher substrate temperatures like 700°C in order to obtain sufficient
surface mobility required for an optimal film growth. As monitored by in-situ RHEED
during growth, the initial deposition of Y films from 3 to 30 Å thick resulted in the
disappearance of the Nb RHEED pattern. It was presumably due to the reaction of
the Y atoms with the Nb-oxides on the surface and the formation of some amorphous-
like Y-oxide layers at the interface. Continuing the deposition of Y film up to 100 Å,
the RHEED pattern, as shown in Fig. 1(b), started to develop into streaks, indicating
an atomically smooth growth is commencing. There is some superstructure-like faint
lines between the streaks, possibly due to some ordered structures of Y-oxides grown.
The exact nature of the interfacial growth awaits a complete in-situ deposition of Nb
and Y films in our system in the near future. For the Y film thicker than 150 Å,
diffraction patterns of sharp streaks with symmetry identifiable as the (000.1)
orientation is completely developed, and is shown in Fig. 1(c). Such pattern persists
throughout the rest of the film growth for a total thickness of 5000 Å. We further
notice that this diffraction pattern repeats itself every 60° as the film is rotated 360°
around the normal axis of the sample plane. This indicates that an orientation order of
the in-plane domains of the six-fold hexagonal symmetry has been achieved.

512

Once such good growth condition of the Y film is established, we find that the substrate temperature can be decreased to as low as 300°C without causing any degradation of the further growth of Y layers. Most importantly, a single crystal Gd film of 1850 Å thick was grown epitaxially on the surface of the Y film of 5000 Å thick with a deposition rate of 1.0 Å/sec at deposition temperatures ranging from 300 to 500°C. The evidence for epitaxial growth of Gd on Y is based on the observation that RHEED pattern of Fig. 1(c) maintained the same throughout the entire growth of Gd films.

B. X-ray Structural Analysis

A detailed structural characterization on the RE films of Y and Gd grown on Nb buffer layers was carried out on a three crystal diffractometer on a rotating anode X-ray source using a Cu target. A graphite monochromator and analyzer were used, and the instrumental resolution is $\Delta q = 0.02$ Å^{-1} and $\Delta\theta = 0.13°$. All Y and Gd films prepared with the previously described growth process had essentially complete [00.1] texture, i.e., the c axis of the hcp structure coincides with the film growth direction.

The films were found to approach single crystals with quite good rocking curves. For one sample with the deposition sequence $Y_{200Å}/Gd_{1850Å}/Y_{5000Å}/Nb_{4500Å}$/sapphire (000.1) the rocking curves perpendicular and parallel to the film are shown in Fig. 2.

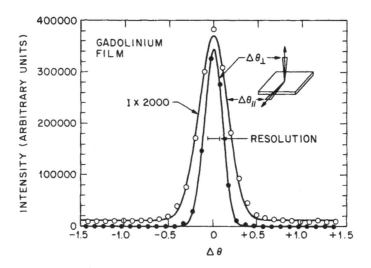

Fig. 2. Rocking curves perpendicular (00.2) and parallel (10.0) to the film for a sample of the deposition sequence of $Y_{200Å}/Gd_{1850Å}/Y_{5000Å}/Nb_{4500Å}$/sapphire.

The full widths at half maximum are 0.28° and 0.42° for the (00.2) and (10.0) reflections, respectively. We note that these rocking curves approach those of the Nb buffer layer, namely 0.21° and 0.41° for the (110) and (1$\bar{1}$0) reflections, respectively. The orientation of the RE, Nb and sapphire crystals is [10.0] of the RE aligned with the [1$\bar{1}$0] of the Nb which in turn is aligned with the [1120] of the sapphire.

Longitudinal scans through the (10.0) and (00.2) reflections gave FWHM of $\Delta q = 0.03 \text{Å}^{-1}$. These values are only slightly wider than the resolutions. Scans through higher order reflections of (00.2), namely (00.4) and (00.6) showed little increase in Δq. This result suggests that there is a negligible strain contribution. A scan was made along the [10.ℓ] direction through the (10.0), (10.1), (10.2), etc. reflections and no reflections other than those expected for the hcp structure were observed at an intensity of $10^{-4} \times 1(10.1)$. This indicates a negligible concentration of stacking faults.

Attempts to grow Y films on Nb (111) single crystal films of six-fold symmetry in its atomic structure proved to be less successful. The texture and orientation order in the plane are greatly reduced, and substantial Y-oxides (Y_2O_3) reflections were observed. Apparently, the slight inferiority of the in-plane domain coherency of the Nb (111) films has resulted in the chemical reaction of the Y atoms with oxygen of the sapphire substrates through the grain boundaries of the Nb films.

C. Rutherford Backscattering and Ion Channeling

The ion channeling technique gives information about the composition depth profile and the crystallinity of the as-grown film. The Rutherford backscattering and ion channeling spectrum measured for the same sample in Fig. 2 is shown in Fig. 3, with 1.8 MeV He$^+$ ion beams of a dose of $6 \times 10^{13}/\text{cm}^2$ at the normal incidence to the substrate plane.

The sharp rise and fall-off of the Y and Gd peaks indicates the film growth at 500°C is free of atomic interdiffusion between the Gd and Y interface at the instrumental resolution of 100 Å. The ion channeling χ_{min} determined for the Gd films is small, about 6%. For an ideally ordered single crystal, χ_{min} should approach a value as low as 3%. Notice that χ_{min} for the front Y protecting layer (~200 Å thick) is poor, presumably caused by oxidation in room air. χ_{min} for the Nb buffer layer is found to be comparable to that of the RE films by etching the RE films away and retaking the RBS measurement. This result suggests that the crystal perfection presently achieved in our RE films is largely determined by the underlying Nb substrate films. Such conclusion is in accord with the rocking curve studies of the X-ray work. Further perfecting the Nb film growth would be most important for improving the RE film qualities.

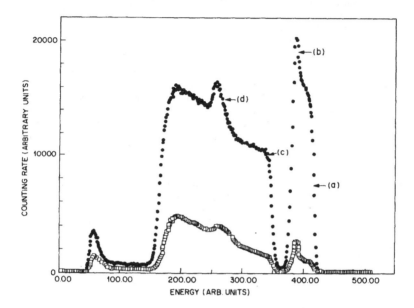

Fig. 3. Rutherford backscattering spectrum (upper curve) and ion channeling minimum yield (lower curve) for the same sample in Fig. 2. (a), (b), (c) and (d) denote the peaks for the Gd layer, upper protecting Y layer, lower seed Y layer, and Nb buffer layer.

CONCLUSIONS

The employment of the single crystal Nb (110) film as a buffer layer to eliminate the substrate interaction problem has led to the successful growth of high quality RE single crystal films of Y and Gd. The structural characterizations carried out on these films showed a high degree of crystallinity with narrow rocking curves both perpendicular and parallel to the film substrate, and complete orientation order of domains in the sample plane. The optimal film growth condition established from this work are crucial for preparing Gd/Y superlattice structures that are currently underway.

REFERENCES

[1] For a complete review of metal superlattices, see D. B. McWhan in *Synthetic Modulated Structures*, eds. L. L. Chang and B. L. Giessen, Academic Press, in press.

[2] J. Kwo, M. Gurvitch, E. M. Gyorgy, M. Hong, W. P. Lowe, D. B. McWhan and R. Superfine, *First International Conference on Superlattices, Microstructures, and Microdevices,* August 13-16, 1984, Conference Proceedings.

[3] W. P. Lowe, L. H. Greene, W. L. Feldman, M. Gurvitch, E. M. Gyorgy, D. B. McWhan, and J. M. Rowell, *Bull. Am. Phys. 29, 545 (1984).*

[4] *Magnetic Properties of Rare Earth Metals,* edited by R. J. Elliot, Plenum Press, New York, 1972.

[5] L. H. Greene, W. L. Feldman, J. M. Rowell, B. Batlogg, E. M. Gyorgy, W. P. Lowe and D. B. McWhan, *First International Conference on Superlattices, Microstructures, and Microdevices,* August 13-16, 1984, Conference Proceedings.

[6] J. Kwo, G. K. Wertheim, M. Gurvitch and D. N. E. Buchanan, *IEEE Trans. Mag., MAG-19,* No. 3, 795 (1983).

[7] S. M. Durbin, J. E. Cunningham, M. E. Mochel, and C. P. Flynn, *J. Phys. F: Metal Phys., 11,* L223 (1981).

[8] S. M. Durbin, J. E. Cunningham, and C. P. Flynn, *J. Phys. F: Metal Phys., 12,* L75 (1982).

EPITAXY OF Nb₃Sn FILMS ON SAPPHIRE

A. F. MARSHALL, F. HELLMAN AND B. OH
Center for Materials Research, Stanford University, Stanford, CA 94305

ABSTRACT

Films of Nb_3Sn vapor deposited at low rates and high temperatures on ($1\bar{1}02$) sapphire form an epitaxial <100> single crystal matrix with a domain structure of misoriented regions bounded by low-angle dislocation boundaries. Nucleation of other orientations at the interface result in a highly oriented but polycrystalline film through approximately the first thousand Angstroms of film thickness. After this point random orientations become overgrown by epitaxial <100> regions. At slightly lower tempera- tures many small <100> grains with a second epitaxial relationship also nucleate at the interface. These rotated grains persist through greater thicknesses than random orientations. The misorientation defect structure of the single crystal matrix is analyzed by transmission electron microscopy.

INTRODUCTION

Nb_3Sn and other Al5 superconducting compounds can be made in thin film form by vapor deposition processes such as electron beam evaporation and sputter deposition. Previous studies of material evaporated at tempera- tures up to 800°C have always shown polycrystalline films with a tempera- ture and rate-dependent grain size [1,2]. Low deposition rates and high deposition temperatures given optimally ordered Nb_3Sn for characterization of superconducting phenomena. We have have recently observed Nb_3Sn formed as a <100> single crystal matrix containing defects when deposited above 800°C at very low rates onto ($1\bar{1}02$) sapphire. This matrix has a domain structure of slightly misoriented regions separated by low angle disloca- tion boundaries. Small isolated grains of <100> Nb_3Sn, rotated through an angle of approximately 30° with respect to the predominant epitaxial orien- tation, also occur as defects in the single crystal matrix. It is inter- esting to note that toward the substrate/film interface the structure appears polycrystalline but with a predominance of epitaxially oriented <100> grains. Thus the formation of the single crystal matrix depends on grain growth phenomenon as the film increases in thickness as well as on nucleation phenomenon at the interface and the epitaxial structure is more complete away from the interface. This has been observed in other systems, e.g. <100> Si on ($1\bar{1}02$) sapphire where a small fraction of <110> domains nucleate at the interface but are subsequently overgrown by the predominant <100> domains [3].

The growth of Nb_3Sn as a single crystal does not directly enhance bulk superconducting properties such as the critical temperature, T_c, (the coherence length for Nb_3Sn is ~50 Å) but may be useful in the develop- ment of thin film devices such as Josephson junctions or in the measurement of such effects as anisotropic gaps. Furthermore epitaxial or polycrystal- line growth characteristics may correlate with other superconducting phenomena of these vapor deposited films such as a drop in T_c at the interface and the spread in T_c in off-stoichiometric films. These aspects will be discussed further in this paper.

MATERIALS AND METHODS

Films of Al5 Nb-Sn ranging in composition from 21-25 at% Sn were co-deposited from separate Nb and Sn sources by both electron beam evaporation and magnetron sputtering. Rates of 7 to 30 Å/sec and substrate temperatures of 800-900°C were used. Film thickness was 2-3 μm with the exception of one 5000 Å thick film as referred to in the next section.

The microstructure was characterized by X-ray Read camera and transmission electron microscopy (TEM). Specimens were thinned for TEM by ion milling from the substrate side until perforation. This allows observation of the top surface of the film. The change in microstructure with thickness and the interface structure were observed by subsequently milling away the film surface, several thousand Angstroms at a time, until the interface was reached. TEM studies were carried out in a Philips 400 TEM/STEM.

RESULTS

The domain structure of the single crystal matrix is shown in Fig. 1 for an off-stoichiometric film (22% Sn) deposited at 830°C. The corresponding selected area diffraction (SAD) pattern is a <100> pattern; the spread of the diffraction spots into arcs indicates a rotational misorientation within the selected area of approximately 5 degrees. The image

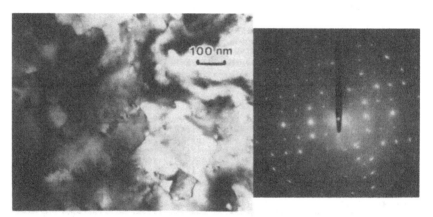

Fig. 1: (a) Small misoriented domains of <100> Nb₃Sn are separated by low-angle dislocation boundaries. Diffuse contrast features are due to single crystal diffraction effects such as bend contours and thickness fringes. (b) The corresponding SAD pattern.

shows that this misorientation is due to domains of several thousand Angstroms separated by end-on dislocation boundaries. The dislocations can be clearly seen by tilting the specimen through a large angle. The in-plane rotational misalignment indicated by the SAD pattern suggests that these boundaries are low-angle tilt boundaries comprised of end-on edge dislocations. Diffraction contrast analysis of several boundaries parallel to (010) planes is consistent with a [010] Burgers vector perpendicular to the boundary plane and the dislocations as expected. These tilt boundaries can also be identified by lattice images as shown in Fig. 2. The change in direction across the boundary and the extra half plane occurring on one side of the dislocation in traversing the boundary can be seen.

Another characteristic defect of the single crystal matrix was the high angle in-plane rotation of small <100> grains as shown in Fig. 3. The

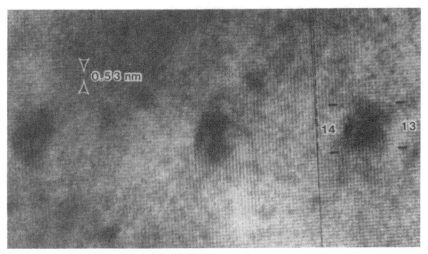

Fig. 2: Lattice image of |100| planes showing end-on dislocations forming
a tilt boundary. An extra plane occurs to the left-hand side of each
dislocation as compared to the right-hand side in traversing the boundary.

SAD pattern shows weaker |100| reflections at approximately 30° and 60°
degrees to the matrix reflections. These rotated grains were originally
observed in a 5000 Å thick on-stoichiometry film and were thought to be

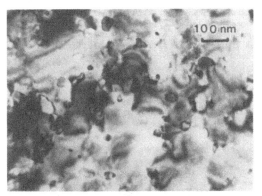
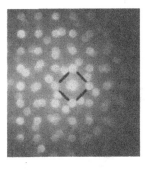

Fig. 3: Small rotated <100> grains in the single crystal matrix. Diffuse
features occur as in Fig. 1. Microdiffraction of one small grain and the
matrix shows two sets of |100| reflections (one marked by a square).

more prevalent in the on-stoichiometry composition. However, recent observations suggest that their occurrence is more dependent on film thickness
and deposition temperature than composition. A 2.5 μm thick off-stoichiometric film annealed at the same temperature as the specimen in Fig. 3
showed fewer rotated grains at the surface but these appeared to increase
as the film was ion milled toward the interface. On the other hand an off-
stoichiometric film annealed at a higher temperature (900°C) showed no
evidence of rotated grains in the SAD pattern throughout the film thickness
and at the interface.

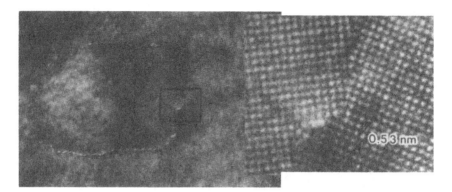

Fig. 4: Lattice image ({100} planes) of a rotated grain. The boundaries
are coherent twin boundaries with {210} twin planes.

Fig. 4 is a lattice image of the {100} planes of a rotated grain. The
fringes change orientation but maintain continuity across the interface.
The rotation angle is measured to be 29°. This is comparable to an angular
relationship between <100> grains of Nb₃Ge of 28° reported by Kitano et
al. [4]. The image, angular relationship, and atomic structure of A15
alloys in the {100} planes are consistent with a coherent twin relationship
across the rotated grain/matrix interface, with {210} planes as the twin
planes.

An off-stoichiometric 2.5 μm thick film made by magnetron sputter
deposition at 900°C showed a domain structure similar to the evaporated
films but occasionally interspersed by polycrystalline areas with a preferred <100> orientation. As the specimen was milled toward the substrate,
the polycrystalline clusters appeared to be elongated always in one direction, and to increase in lateral thickness with depth in the film. This
suggests a random nucleation effect due to directional defects in the substrate, such as scratches resulting from polishing.

When milled to the interface, the polycrystalline appearance of all
the films increased markedly although the SAD patterns continued to show a
high degree of epitaxial orientation. Fig. 5 is a comparison of the
surface and interface region of the off-stoichiometric film evaporated at
900°C. Large grains are epitaxial <100> regions. In between these are

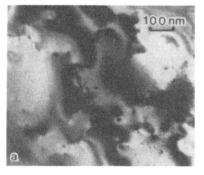

Fig. 5: Comparison of the surface (a) and interface (b) regions of a
2.5 μm thick off-stoichiometric film evaporated at 900°C.

polycrystalline clusters, also containing small epitaxial grains as well as other grain orientations. No rotated <100> grains were evident by SAD in this specimen. The size and number of epitaxial, rotated, and randomly nucleated grains at the interface depends on temperature and possibly (although to a lesser extent) on composition.

The epitaxial relationship is shown in Fig. 6. The cube axes of Nb$_3$Sn ([010],[001]) are aligned with the orthogonal axes of the sapphire orientation which has twofold symmetry. It is worth noting that the zone axis of the sapphire with which the Nb$_3$Sn appears to align, is at six degrees tilt to the (1$\bar{1}$02) sapphire orientation. The lattice mismatch in

 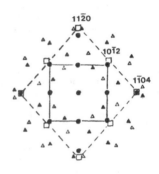

Fig. 6: SAD pattern at the interface of an on-stoichiometric film evaporated at 830°C showing the epitaxial relationship between the substrate and film. ● epitaxial Nb$_3$Sn △▲ rotated Nb$_3$Sn □ sapphire. The sapphire reflections are indexed.

one direction is large and the (002)$_{Nb_3Sn}$ and (11$\bar{2}$0)$_{sapphire}$ reflections are clearly separate. In the perpendicular direction where the lattice mismatch is about 3.5% substrate and film reflections are not clearly separate and there appears to be coherence. This is also suggested by Moiré patterns at the interface which occur in the direction of larger mismatch and have the expected spacing. However, no Moirés are seen in the perpendicular direction. Such fringes, if they did occur, would have a larger spacing due to the smaller mismatch.

DISCUSSION

The growth of epitaxial <100> Nb$_3$Sn on (1$\bar{1}$02) sapphire appears to depend on film growth phenomena during the two-dimensional deposition process as well as on nucleation phenomena at the interface. This results in a more perfect epitaxially oriented film at some distance away from the interface. The nucleation at the interface is dominated by epitaxial <100> as evidenced by SAD. However, other orientations nucleate also and persist through approximately the first 1000 Å, becoming overgrown as deposition continues. The suppression of other orientations is presumably due to the faster growth rate of the close-packed (100) surface when normal to the deposition sources. The rotated (100) grains are also in a favorable orientation for growth and persist through much greater film thicknesses. Nucleation of these grains depends on temperature and decreases markedly in going from 830°C to 900°C deposition temperature. Randomly oriented <100> grains are not observed and presumably rotate into one of the epitaxial positions during initial growth.

522

The small misorientations of the thick films resulting in low angle dislocation boundaries may be related to the different symmetries of the substrate and film orientations. The ($1\bar{1}02$) face of sapphire has only twofold symmetry. The cube axes of Nb_3Sn align along the twofold orthogonal axes of the sapphire as seen in Fig. 6. However, the <110> axes of Nb_3Sn will be angularly mismatched with the sapphire by a few degrees. The attempt to align symmetrically in these directions as well may lead to the misorientations observed. Such misorientations, as well as angular distortions of the cubic lattice at early stages of growth, have been reported for <100> Si on ($1\bar{1}02$) sapphire [3]. No angular distortions have been have been observed by electron diffraction for Nb_3Sn. The Moiré fringes and SAD patterns at the interface indicate coherency with the substrate along [010] but not [001], suggesting nonsymmetrical strains in these directions.

The growth of Nb_3Sn as a single crystal by vapor deposition processes has not been previously observed and may have applications both for thin film devices and for measurement of anisotropic superconducting effects. Two interesting phenomena of these films are a spread in T_c in off-stoichiometric films, as compared with a sharp transition in on-stoichiometric films (5), and a drop in T_c when the deposited thickness is below 1000 Å. The former has no clear relationship with the microstructure. Rather, the similarity of structures for on- and off-stoichiometry suggests that the spread in T_c is related to a tendency for compositional separation (within the A15 phase field) that is structure independent. The drop in T_c near the interface may be related to strains in the A15 structure due to both coherency of the epitaxial regions and to thermal stresses induced upon cooling due to the differences in thermal expansion coefficients of Nb_3Sn and sapphire.

This work was supported by the NSF-MRL Program through the Center for Materials Research at Stanford University.

REFERENCES

1) J. Talvacchio, Ph.D. Thesis, Stanford University (1982).
2) B. E. Jacobson, R. H. Hammond, T. H. Geballe, and J. R. Salem, J. Less Common Metals 62 (1978) 59.
3) S. Hamar-Thibault and J. Trilhe, J. Electrochem. Soc.: Solid State Science & Tech., 128 (1981) 581.
4) Y. Kitano, H-U. Nissen, W. Schauer, and D. Yin, Proc. of 17th Intl. Conf. on Low Temp. Phys. (1984) 615.
5) F. Hellman, D. A. Rudman, R. H. Hammond and T. H. Geballe, Bull. Am. Phys. Soc. 28 (1983) 262.
6) F. Hellman and T. H. Geballe, Bull. Am. Phys. Soc. 29 (1984) 385.

Influence of Modulation Wavelength Induced Order on the Physical Properties of Nb/Rare Earth Superlattices

L. H. Greene*, W. L. Feldmann*, J. M. Rowell*,
B. Batlogg**, R. Hull**, and D. B. McWhan**
*Bell Communications Research, 600 Mountain Ave., Murray Hill, NJ 07974
**AT&T Bell Labs, 600 Mountain Ave., Murray Hill, NJ 07974

ABSTRACT

We report the observation of a higher degree of preferred crystalline orientation in Nb/rare earth superlattices for modulation wavelengths in the range of 200 Å to 500 Å than that exhibited by single component films. All films and multilayers are sputter deposited onto room temperature sapphire substrates. Electronic transport measurements also show that the residual resistance ratio is higher and the room temperature resistivity is lower than for multilayers of either greater or lower periodicities. Transmission electron micrographs (TEM) showing excellent layering, grain size comparable to the layer thickness, and evidence of some degree of epitaxy are presented.

INTRODUCTION

In both semiconducting and metallic superlattices, it has been reported that the bulk properties of single films of the constituent materials are not necessarily retained in the layered structures. For example, a modification of physical properties as a function of modulation wavelength, Λ, has been observed in the strained semiconductor superlattice [1] and in the metallic case of Nb/Al a similar change of lattice structure of both Nb and Al has been demonstrated [2]. It is also known that for very small modulation wavelengths, semiconductor superlattices will approach the limit of the ordered alloy of the same average composition, whereas in many metallic systems made of components of limited solid solubility (i.e. excluding the Nb/Ta and Ni/Cu cases) the onset of structural disorder at small modulation wavelengths has been observed [3-5]. Here we report the observation of induced disorder in room temperature sputter deposited Nb/rare earth superlattices both at small $(\Lambda < 50\ \text{Å})$ and also at large $(\Lambda > 500\ \text{Å})$ periodicities. This unusual behavior at large wavelengths may be explained by noting that the orientation of the Nb and Er layers in superlattices is not that which is commonly observed in single thick films grown at the same temperature [5]. Thus we suspect that the disorder at long wavelengths occurs as the layers begin to change their growth morphology away from the orientation seen in the layers towards that of single films. This observed disorder at both short and long wavelengths implies that the physical properties of Nb/Er multilayers will be optimized for intermediate wavelengths, and we present evidence here that this is indeed the case for wavelengths from about 200 to 500 Å.

REVIEW OF PREVIOUS RESULTS

Our earlier studies on the growth and characterization of single rare earth (RE) films and RE/Nb superlattices have been reported elsewhere [5]. Our interest in these systems arises from the unusual magnetic structures exhibited by RE metals and the possibility of creating re-entrant superconductivity in a layered material. We observed that an elevated deposition temperature, T_D, required for the growth of highly textured (crystallographically oriented) films, of both Nb and RE, gave rise to an increased chemical interaction between the RE film and substrate material. By choosing sapphire as an optimum substrate and a compromise $T_D \sim 425°C$, highly ordered RE films with X-ray rocking curves of less than 1° were produced. The orientation of these films was [00.1] (or [002] in the three index notation) i.e. with the c-axis normal to the sapphire (01.2) substrate. Room temperature grown RE films are polycrystalline with no preferred orientation. Similarly, single Nb films grown on these substrates at $T_D < 810°C$ exhibit the close-packed [110] texture and a [100] orientation for deposition temperatures greater than $810°C$ [7].

For RE/RE and RE/Nb multilayers, the optimum T_D was found to be lower than that for single films. Textured Nb/RE superlattices exhibiting sharp $(< 10\ \text{Å})$ interfaces were grown at room temperature. The texture exhibited by each component was not that observed for the single component room temperature deposited films. In these multilayers, the Nb exhibits a [100] texture and the RE is [10.1]. Finally, the superconducting properties of the superlattice films we reported earlier demonstrated the proximity effect for Nb layered with chemically equivalent magnetic and non-magnetic RE's in this multilayer geometry.

ELECTRONIC TRANSPORT

The simplest measurements that can be made to determine the perfection of a metallic film are of its transport properties as a function of temperature. The residual resistance ratio, $R_{300\,K}/R_{T>T_c}$ and the room temperature resistivity ρ, are plotted as a function of Λ in Fig. 1 for

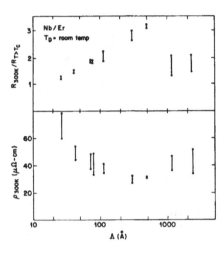

Fig. 1: Residual resistance ratio, $R_{300\,K}/R_{T>T_c}$, and room temperature resistivity, $\rho_{300\,K}$, are plotted as a function of the modulation wavelength, Λ, for Nb/Er multilayers sputtered onto room temperature sapphire substrates. Points connected by vertical lines represent multilayers of equal Λ but different Nb to Er ratios, as discussed in the text.

Nb/Er multilayers grown on room temperature sapphire substrates. The points connected by vertical lines correspond to multilayers grown at the same time, and therefore of equal periodicities. However, due to the sample preparation techniques (the substrates oscillate between sputter targets [5]), the ratio of Nb to Er layer thicknesses differ slightly. For each Λ, the Nb-rich sample exhibits the highest residual resistance ratio and the lowest room temperature resistivity. In general, for films of the same composition, a low resistance ratio and high resistivity results from a short electronic mean free path and therefore indicates disorder.

It is clear from Fig. 1 that the properties of these Nb/Er superlattices are optimized in the wavelength range from 200 to 500 Å, where the resistance ratio reaches ~ 3.2. Given the variation of resistance ratio with wavelength exhibited in Fig. 1, the variation of resistivity can be explained by considering the conductivity of the Nb alone. As the bulk resistivities of the Nb and Er are 14.5 and 70 $\mu\Omega$ —cm, respectively, it is reasonable to expect that the Nb will dominate the conductivity behavior, except perhaps at the shortest wavelengths.

X-RAY DIFFRACTION

The changes in orientation and grain size of the multilayers with increasing wavelength were measured with X-ray diffraction. A Read camera was used to obtain the photographs in Fig. 2. This technique is extremely useful for obtaining an overview of the sample structure. All the films used to produce these pictures were grown at room temperature. The many tiny dots are reflections from the sapphire (01.2) substrate and should be ignored for this discussion. Read photographs of Nb/Er superlattices with $\Lambda = 1000$ Å, 300 Å and 25 Å are shown in Figs. 2(b), (c) and (d), respectively, and that for a single Er film is reproduced for comparison in Fig. 2(a). In Fig. (2a), the arcs correspond to Er Bragg reflections. Most of these arcs are continuous indicating a lack of preferred crystallographic orientation. The thickness of each arc is inversely related to grain size. A comparison of Figs. 2(a) with (b), (c) and (d) shows that layering the Er with Nb, even for $\Lambda = 1000$ Å, results in a slightly increased texture as seen by the discontinuous arcs, some corresponding to Nb and some to Er reflections. Comparing Figs. 2(b) and 2(c), we find

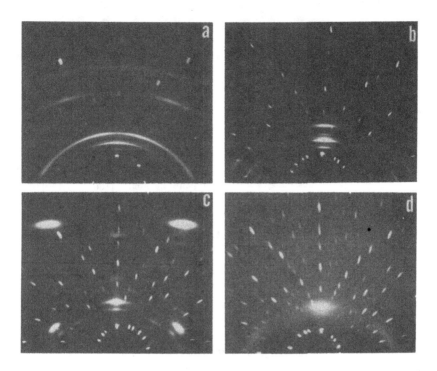

Fig. 2: Read X-ray photographs showing the effect of structural disorder in Nb/Er superlattices. The small dots are reflections from the sapphire substrate. A single component Er film was used to produce (a). Multilayers of $\Lambda = 1000$ Å, 300 Å, and 25 Å were photographed in (b), (c), and (d). Note that, in comparison to (c), (a) is disordered (smaller grain size) while retaining some texture and (d) appears more ordered (larger grain size) with significantly reduced texture. Each of these films were sputter deposited at room temperature.

that decreasing the modulation wavelengths to $\Lambda = 300$ Å results in a highly textured film. In this photograph, all the arcs are replaced by broad spots. It is interesting to note that although the $\Lambda = 300$ Å film is oriented to a significantly greater degree than the $\Lambda = 1000$ Å film, the broadness of the spots indicate a much smaller grain size in the plane of the film. Finally, the $\Lambda = 25$ Å film of Fig. 2(d) appears extremely disordered as evidenced by extremely broad spots and the disappearance of certain reflections entirely, Note however, this disordered film still exhibits a greater degree of texture than the $\Lambda = 1000$ Å multilayer. The induced disorder at small Λ can also be seen in X-ray diffraction scans around the main Bragg peaks of either the constituent materials. In Fig. 3, intensity vs. $Q = 2\pi/d$ in the vicinity of the Lu (10.1) reflection is shown for $\Lambda = 160$ Å and $\Lambda = 22$ Å Nb/Lu multilayers. Again, these structures were sputtered onto room temperature sapphire substrates. Note that for the $\Lambda = 22$ Å multilayer, the (10.1) Lu peak is significantly broader than that for $\Lambda = 160$ Å. Also a weak broad shoulder on this peak appears near $Q = 2.7$ Å^{-1} corresponding to the (011) reflection of Nb. We therefore conclude from these diffraction scans both that the grain size in the materials is decreased and that the degree of texture is slightly less in the shorter modulation wavelength material.

It is not entirely clear how the transport results of Fig. 1 can be reconciled with the X-ray results of Fig. 2 at long wavelengths. At small wavelengths, the very small grain size presumably limits the electronic mean free path, but at large wavelengths the X-ray evidence points

526

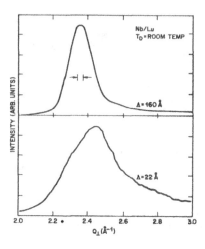

Fig. 3: Intensity vs. $Q = 2\pi/d$ in the vicinity of the (10.1) reflection of Lu for Nb/Lu multilayers. The resolution is denoted by the arrows. Sputtered onto room tempeature sapphire, (a) $\Lambda = 160$ Å and (b) $\Lambda = 22$ Å. Note that (b) exhibits a significantly broader peak and a weak shoulder near $Q = 2.7$ Å^{-1} (Nb(011)) corresponding to smaller grain size and some reduction of the texture. Textured multilayers deposited onto room temperature sapphire exhibit [100] and [10.1] orientations for the Nb and Lu, respectively.

to a polycrystalline film of larger grain size. Thus the decrease in resistance ratio for periods > 500 Å might be ascribed to the effects of strain as the orientation of the films change with increasing wavelength.

TEM STUDIES

The TEM micrograph of Fig. 4 is of a cross sectional Nb/Er superlattice with $\Lambda \sim 50$ Å. This particular multilayer was grown at room temperature on undoped (intrinsic) Si

Fig. 4: A TEM micrograph of a Nb/Er multilayer deposited on room temperature Si(100). Note the well-defined layers with $\Lambda \sim 50$ Å. The dark areas within a layer are due to strongly diffracting grains, comparable to the layer thickness.

(100) because this substrate material is easier to thin (for the preparation of TEM measurements) than sapphire. Note the obvious layering. The dark and light area within the layers correspond to the contrast between strongly and weakly diffracting grains, resulting from different orientations. It appears from this photo that the grain size is comparable to the layer thickness. In fact, in this sample the Er layers are thicker than the Nb and the grain size in the Er is correspondingly larger than in the Nb layers. Close examination of this photograph reveals fringes parallel to the layers (fine rows of dots within sections of each layer), which indicate the existence of single crystal grains and some evidence of epitaxial growth.

These TEM data, in agreement with the X-ray data, show that these Nb/RE superlattices exhibit a crystalline order in the direction normal to the film surface. In the plane of the film however, for $\Lambda \leq 500$ Å the grain size decreases with decreasing Λ and a random orientation is observed.

CONCLUSION

We have observed that the structure and physical properties of Nb/RE superlattices sputter deposited onto room temperature sapphire are optimized in the wavelength range from 200 to 500 Å. As Λ is reduced below ~ 500 Å, the grain size within the film decreases dramatically while still maintaining a high degree of texture. For $\Lambda \geq 500$ Å, the grain size remains large while the film texture is almost completely lost and the physical and structural properties of the multilayers approach those of single RE and Nb films grown at room temperature. The exciting new phenomenon observed here is that for films grown at room temperature, Nb/RE multilayers of intermediate periodicities exhibit more preferred orientation than do single component films.

REFERENCES

1. J. C. Bean, L. C. Feldman, A. T. Fiory, S. Nakahara, and I. K. Robinson, J. Vac. Sci. Technol. **A2**, 436 (1984).

2. D. B. McWhan, M. Gurvitch, J. M. Rowell, and L. R. Walker, J. Appl. Phys. **54**, 3886 (1983).

3. W. P. Lowe, T. W. Barbee, Jr., T. H. Geballe, and D. B. McWhan, Phys. Rev. B **24**, 6193 (1981).

4. S. T. Ruggiero, T. W. Barbee, Jr. and M. R. Beasley, Phys. Rev. B **26**, 4894 (1982).

5. L. H. Greene, W. L. Feldmann, J. M. Rowell, B. Batlogg, E. M. Gyorgy, W. P. Lowe, and D. B. McWhan, Proceedings of the International Conference of Superlattices, Heterostructures, and Microdevices, August 13-16, 1984, Champaign-Urbana, IL (to be published).

6. L. H. Greene, W. P. Lowe, W. L. Feldmann, B. Batlogg, D. B. McWhan, and J. M. Rowell, Bull. Am. Phys. Soc. **29**, 545 (1984).

7. G. Hertel, D. B. McWhan and J. M. Rowell, in *Superconductivity in d- and f-Band Metals 1982*, edited by W. Buckel and W. Weber, (Kernforschungszentrum Karlsruhe GmbH, Karlsruhe, Germany, 1982), p. 299ff.

MECHANICAL PROPERTIES OF COMPOSITION MODULATED
COPPER-NICKEL-IRON THIN FILMS

A. F. JANKOWSKI* AND T. TSAKALAKOS**
* ROCKWELL INTERNATIONAL, ROCKY FLATS PLANT, P.O. BOX 464, GOLDEN, CO
80401
** Dept. of Mechanics and Materials Science, Rutgers University, P.O. Box
909, Piscataway, NJ 08854

ABSTRACT

The enhanced elastic modulus effect was found in composition modulated
Cu/NiFe foils. Young's modulus was measured via tensile testing on thin
foils containing short-wavelength composition modulations of 1.4 - 10. nm.
The foils, of 53% Cu - 40% Ni - 7% Fe composition, were produced by vapor
deposition using a three-source evaporator. As compared with homogeneous
foils of the same average composition, the modulated foils exhibited an
appreciable increase (up to 300%) in modulus for two distinct ranges of
compositon wavelength: 2.1 - 2.7 nm and 3.6 - 4.1 nm. Results of x-ray
diffraction and Mossbauer spectroscopy studies suggest the concept of
interfacial coherency, accommodating the lattice misfit between the
composition layers, is the underlying structural feature responsible for
the enhanced modulus effect.

INTRODUCTION

Studies conducted on compositionally modulated binary alloy systems, such
as Cu-Ni, Cu-Pd, Ag-Pd, and Au-Ni, have shown remarkable increases in the
biaxial elastic modulus by factors of 2 to 4 [1-3]. The origin of the
"supermodulus effect" has become the subject of various theoretical
viewpoints. Explanations ranging from Fermi surface - Brillouin zone
interactions [3, 11] (accounting for critical layer thicknesses) to
coherency strain effects [4] have been suggested.

Mechanical deformation of metallic alloys can change the elastic moduli by
1 or 2% at most. Order-disorder transformations rarely change the elastic
moduli by more than 10 to 20%. Indeed, a very fundamental change on the
atomic level must be taking place to produce this dramatic increase of the
elastic modulus, previously considered as a structure insensitive
property.

In this paper, experimental results are given for the wavelength and
amplitude dependence of the elastic properties of Cu/NiFe compositionally
modulated foils.

PREPARATION OF SPECIMENS

The composition-modulated foils were prepared by thermal vapor deposition
of the three metals under high vacuum. The component metals Cu, Ni and Fe
(of 99.999% purity) were evaporated through a rotating pinwheel shutter
which alternately masked the evaporant mica substrate (held at 350°C) from
the Cu crucible, and the Ni and Fe crucibles. The modulated depostion,
consisting of alternating Cu and NiFe layers, will show a more nearly
sinusoidal than stepwise composition profile as diffusion occurs during
the deposition. The background pressure was less than 40 mPa (3x10^{-7}
Torr). By monitoring the source evaporation rates (0.04 - 0.7 nm/sec.) as
well as the deposition rates, the desired composition of the foils could
be controlled. The composition of the foils was 53 $\pm$ 2 at.% Cu, 40 $\pm$ 2

at.% Ni, and 7 + 1 at.% Fe. The wavelengths of the modulation ranged from 1.4 to 10 nm, and the foil thicknesses from 0.7 to 0.9μ m.

The amplitude of the "composition modulation" is defined as the amplitude of the sinusoidal concentration wave. The amplitude of modulation was varied by annealing. X-ray diffration, using Cu-Kα radiation, was used to determine the composition wavelength and amplitude of the modulation from the position and integrated intensity of the satellites about the (111) Bragg peak.

MECHANICAL TESTS

The Microtensile Test Method

The mechanical tests were performed on a PIEZOTRON-U Dynamic Piezoelectricity Analyzer. The foils are cut to a uniform length and width (gauge length of 12 mm and width of 4 mm) to eliminate any size or shape effect in the mechanical measurements. For safety in handling, the foils are individually mounted in a frame to prevent tearing when being clamped in the grips of the piezotron, as shown in Figure 1. The loading procedure is as follows: (1) An initial load of several grams is uniaxially applied to place the entire cross-section of the foil under tension. The gauge length is then recorded, as measured by a micrometer within the piezotron instrumentation to within 0.0005 mm, (2) The tension adjust controller is used to increase the static tension applied to the foil by gram increments. Along with each increment of applied load, the foil gauge length is recorded, (3) Once, approximately, 0.20% strain is reached, the sample is unloaded. (4) The loading procedure is then repeated. A minimum of 5 sets of loading data are obtained for each foil, in which strains beyond 0.2% are reached for the final loading data set.

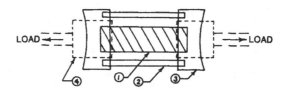

Figure 1. Tensile Sample: 1-4 mm x 12 mm x 0.8 μm foil; 2-support strap (cut after grip mounted); 3-mylar (2 single-sided adhesive sheets); 4-piezotron grips (screw tightened)

Calibration of the Microtensile Tester

Calibration tests were made, using pure copper foils and homogeneous Cu-Ni foils containing 50 at.% Cu. They were annealed, at 425°C for several hours prior to testing, to increase the degree of (111) texture. The Young's modulus of the pure Cu foil was equal to 0.16 TPa and that of the homogeneous Cu-Ni foil equaling 0.20 TPa. These values are consistent with those determined on bulk specimens.

EXPERIMENTAL RESULTS

Tests were made on 16 Cu/NiFe foils [4]. The stress-strain curves can be plotted quite easily using the loading data sets obtained. Once the sample width and thickness are measured, the necessary calculations are straightforward. The sample width is measured using an optical microsocpe to within 0.05 mm and sample thickness measurements are made using a SLOAN DEKTAK to within 5 nm. The DEKTAK provides a mechanical measurement as a 25 mg tracking force is applied to a diamond stylus transversing the foil edge at a 0.1 cm/sec. travel speed.

Stress-Strain curves for the as-deposited foils of several composition wavelengths are in Figure 2. For strains less than approximately 0.20%, a linearly elastic relationship exists. Results for all the foils are listed in Table I. The values of Young's modulus E were computed from the initial slope of the σ vs. ε curves.

The variation of Young's modulus E with composition amplitude A can be examined by performing a series of isothermal anneals on those foils exhibiting an enhanced modulus. It can be expected that as the composition amplitude decays, so will the mechanical properties [1]. A plot of E versus the square of the composition amplitude is shown in Figure 3. It can be seen that the increase of the modulus is proportional to A^2. In order to obtain the wavelength dependence of the modulus, moduli for the Cu/NiFe foils

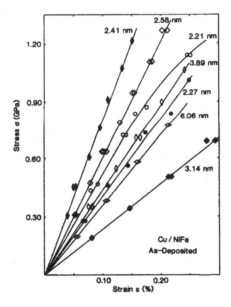

Figure 2. Stress σ (GPa) vs. Strain ε (%) curves of the as-deposited composition modulated Cu/NiFe foils wih wavelengths 2.21, 2.27, 2.41, 2.58, 3.14, 3.89, and 6.06 nm.

Table I. Young's modulus measured for as-deposited Cu/NiFe foils.

λ(nm)	%Cu	%Ni	%Fe	A(at.pct.)	E(TPa)	T anneal(°C)	t anneal(min)
1.41	50.7	39.1	10.2	.284	.216		
1.58	50.7	39.1	10.2	.309	.240		
2.16	55.3	38.0	6.7	.565	.426	400	0
				.403	.325	400	80
				.363	.308	400	160
				.292	.269	400	260
2.21	55.3	38.0	6.7	.537	.526		
2.27	52.8	41.1	6.1	.325	.402		
2.41	53.2	39.9	6.9	.643	.804		
2.58	53.2	39.9	6.9	.698	.603		
2.65	49.6	46.4	4.0	.619	.544		
3.14	53.4	39.1	7.5	.320	.231		
3.38	51.5	43.4	5.1	.502	.253		
3.68	54.5	38.1	7.4	.357	.411		
3.82	54.5	38.1	7.4	.421	.550		
3.89	54.5	38.1	7.4	.409	.435		
4.29	50.0	45.3	4.7	.574	.358	400	0
				.455	.308	400	40
				.439	.302	400	80
				.424	.293	400	160
				.369	.266	400	260
5.68	51.8	40.9	7.3	.775	.349		
6.06	49.6	42.7	7.7	.748	.361		

were extrapolated (or interpolated) to a 50% composition amplitude, assuming an A^2 dependence. Figure 4 is a plot of Young's modulus E versus the composition wavelength λ , for 50 at.% composition amplitude.

There are two distinctly different ranges of composition wavelength wherein modulus enhancement occurs. Between 2.1 and 2.7 nm, the enhancement ranges from 160% to a maximum of 260% at 2.3 nm, compared to the very short or very long wavelength modulus vlaue of 0.26 TPa (36.3 x 10^6 psi). The second wavelength range of modulus enhancement occurs between 3.6 and 4.1 nm, wherein the normalized enhancement varies from 150% to a maximum of 275% at 3.85 nm.

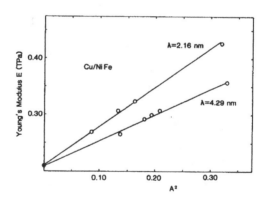

Figure 3.   Young's Modulus E (TPa) vs. the square of the composition amplitude A for Cu/NiFe foils of 2.16 and 4.29 nm wavelength.

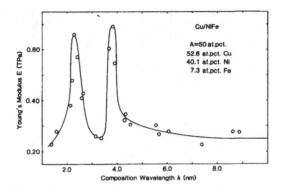

Figure 4. Young's moduls E (TPa) vs. the composition wavelength
λ(nm), normalized to a 50 at.% composition amplitude for the
Cu/NiFe foils.

The double-peak observed in Figure 4 was also discovered for the Cu-Ni
composition modulated structure by Baral [5]. Modulus enhancement for the
as-deposited Cu-Ni foils was maximum at composition wavelengths of 1.5 and
3.0 nm. The maximum increase, as shown in Figure 5, was 170% compared to
a homogeneous foil modulus of 0.19 TPa (for 50% Cu composition). The
double-peak observed for Young's modulus was shown by Baral to be
complimentary to the single peak of Biaxial Modulus Y [111] enhancement,
for the Cu/Ni system [5,9]. The presence of Fe within the Ni layer of the
Cu/NiFe layered structure shifts the wavelengths of modulus enhancement to
a higher value than for the Cu/Ni layered structure. The accommodation of
Fe substitutionally into the fcc Ni lattice distorts the layer in such a
manner that a longer wavelength is needed for a coherent interface. This

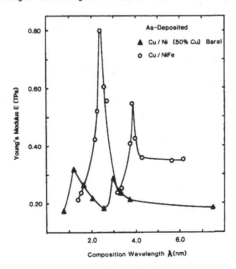

Figure 5. Young's Modulus E (TPa) vs the composition wavelength λ (nm)
for as-deposited Cu/NiFe[4] and Cu/Ni[5] composition modulated structures.

534

is consistent with the lattice spacing distortion trend, as ε_0 is smaller for Cu/NiFe [4] than for the Cu/Ni [9] system, i.e, ε_0 (Cu/NiFe) $\sim 1.3 \times 10^{-2}$ whereas ε_0 (Cu/Ni) $\sim 1.8 \times 10^{-2}$. The degree of enhancement is different for the Cu/Ni and Cu/NiFe systems. This is partially attributable to the squaring effect Fe has on the sinusoidal compositon wave profile. The larger composition amplitudes, A, observed for Cu/NiFe correspond with greater values for Young's modulus.

DISCUSSION

The enhanced Young's modulus observed for Cu/NiFe must be due to some very fundamental changes introduced by the composition modulation. The concept of interfacial coherency, accommodating the lattice misfit between the layers, is one plausible explanation. Theoretically, evidence is given in the form of a pseudopotential energy formulation of the elastic constants. A compressive biaxial elastic strain of 3% applied to a Cu lattice [7] doubles the biaxial moduls Y[100]. It is therefore suggested that the misfit is accommodated by elastic strains in each layer. Experimentally, a computer aided rocking curve analysis (CARCA) [6] provided the following information. Using a double crystal diffractometer, foils of compositon wavelength 2.41 and 3.14 nm were exposed to a highly monochromatic x-ray beam. The reflected intensity was measured as the foil was rotated (or rocked) through its range of reflection, corresponding to its textured growth. Figure 6 contains a plot of the reflected intensity, normalized to the maximum values, versus the angular position of the foils. The 3.14 nm foil rocking curve is much broader than the 2.41 nm foil rocking curve. This can be attributed to dislocatons which occupy the layer interfaces. The presence of interfacial dislocations accomodates lattice misfit, but not coherently. This result is consistent with modulus enhancement present for the 2.41 nm wavelength foil but not for the 3.14 nm wavelength foil.

A Mossbauer spectroscopy study provided further information about the microstructure within the layers of the Cu/NiFe composition modulated foils. The magnetic behavior of foils with and without enhanced elastic modulus were studied. The hyperfine pattern of magnetic α-Fe is

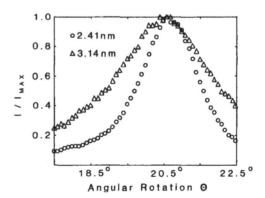

Figure 6. Reflected intensity normalized to the maximum intensity vs. the angular postion for the 2.41 and 3.14 nm composition wavelength Cu/NiFe foils.

manifested by the 6.12 nm wavelength foil in Figure 7. The 3.87 nm foil pattern (although relatively weaker due to a thinner sample examined) shows the zero velocity peak representative of the paramagnetic δ-Fe phase in addition to comparable intensity non-zero velocity peaks. The normalized Young's modulus (A=0.5) of these foils are 0.26 and 0.56 TPa, respectively. Therefore, the fcc structure of iron, significantly present within the enhanced modulus foil, is absent from the non-enhanced modulus foil. The interfacial lattice misfit appears to be accommodated incoherently as bcc iron is present whereas the presence of fcc iron lends itself to a greater extent of lattice coherency hence enhanced elastic modulus in these fcc <111> textured foils. Similar results were also obtained for a 2.64 nm wavelength foil (E = 0.43 TPa) and a 10.64 nm wavelength foil (E = 0.24 TPa), i.e., a typical α-Fe pattern manifested by the 10.64 nm foil and a zero-velocity peak present for the 2.64 nm foil (paramagnetic δ-Fe presnt). Shinjo, et. al. [12-15] have observed similar results. To study interface magnetic properties the Mossbauer isotope ^{57}Fe was thinly deposited under high vacuum on a thicker layer of pure ^{56}Fe ($\sim$10 nm) and the surface then overcoated with various non-magnetic substances. For layer thickness more than 5 nm, Fe films formed continuous layers, exhibiting bulk Mossbauer spectrum. Interface effects were appreciable only within a few atomic layers from the interface boundary. At the interface of Fe in contact with Sb [12], a decrease of the magnetic moments at the interface was apparent. Similar results were obtained in the cases of Fe-Cu or Fe-V interfaces [13-15].

The modulus enhancement in compositon modulated Au-Ni, Cu-Pd, and Ag-Pd foils was suggested to be attributable to the interaction of the Brillouin zone introduced by the modulation with the Fermi surface [1-3]. Dunaev and Zakharova have demonstrated that the singularity in the band-structure spectrum of a nonequilibrium alloy is connected with the Fermi surface [8]. The possibility of the singular part of the band-structure energy causing such a considerable change in the elastic constants has not been verified. However, the observed wavelength dependence of the modulus has

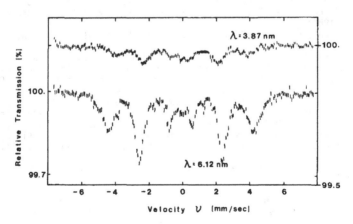

Figure 7. Relative Transmission (%) vs. Velocity (mm/sec) curves of Mossbauer Spectra for 6.12 and 3.87 nm wavelength Cu/NiFe foils.

been predicted for the Cu/Ni system [9] using positron annihilation measurements [10] for a concentration wave in contact with the Fermi surface.

SUMMARY

In summary, our experiments have shown that: (1) The Young's modulus E of composition-modulated Cu/NiFe foils were three-fold higher than those of homogeneous foils having the same compositon; (2) The wavelength dependence of the modulus had maxima at 2.3 and 3.85 nm; (3) The increase in modulus was proportional to the square of compositon modulation amplitude; (4) The deformation was Hookean for elastic strains less than 0.2%; and (5) Interfacial coherency between the layers of the composition modulated foil appear to be the underlying structural feature responsible for the extent of the enhanced modulus effect.

ACKNOWLEDGEMENT

Many thanks to A. Kostikas and S. Simopoulos of the Nuclear Research Center in Athens, Greece for performing the Mossbauer experiments. This work was supported in part by the National Science Foundation, NSF Grant DMR-78-26503.

REFERENCES

[1]   Yang, W. M. C., T. Tsakalakos, and J. E. Hilliard, J. Appl. Phys. 18, 876, (1977).
[2]   Henein, G. E. and J. E. Hilliard, J. Appl. Phys., 54, 728, (1983).
[3]   Tsakalakos, T. and J. E. Hilliard, J. Appl. Phys., 54, 734, (1983).
[4]   Jankowski, A. F., Ph.D. Thesis, Rutgers Univeristy, New Bruswick, NJ (1984).
[5]   Baral, D., Ph.D. Thesis, Northwestern Univeristy, Evanston, Illinois (1983).
[6]   Mayo, W. E., H. Y. Liu, and J. Chaudhuri, "X-Ray Topographic Methods and Application to Analysis of Electronic Materials", Proceedings of High Speed Growth and Characterization of Crystals for Solar Cells. July 1983, Port St. Lucie, Florida, U.S.A.
[7]   Tsakalakos, T., and A. F. Jankowski, "The Effect of Strain on the Elastic Constants of Copper", NATO ASI Series on Modulated Structure Materials, Maleme-Chania, Crete, Greece, Nijhoff Publishers (1984).
[8]   Dunaev, N. M. and M. I. Zakharova, JETP Lett., 20, 336, (1974).
[9]   Tsakalakos, T., Ph.D., Northwestern University, Evanston, Illinois (1977).
[10] Nanao, S., K. Kuribayashi, S. Tanigawa and M. Doyama, Phys. Lett. A, 38, 487, (1972).
[11] Purdes, A., Ph.D. Thesis, Northwestern Univeristy, Evanston, Illinois (1976).
[12] Shinjo, T., N. Hosoito, K. Kawaguchi, T. Takada, Y. Endoh, Y. Ajro, and J. M. Friedt, J. Phys. Soc. Jpm, 52, 3154, (1983).
[13] Shinjo, T., J. Lauer, and W. Keune, Physica 86-88B, 407, (1977).
[14] Shinjo, T., S. Hine, and T. Takada, J. Phys. Soc, Jpn., 47, 767, (1979).
[15] Shinjo, T., N. Hosoito, and T. Takada, J. Phys. Soc. Jpn., 50, 1903, (1981).

EFFECT OF LONG RANGE INTERACTION ON DIFFUSION IN Cu/NiFe TERNARY ALLOY COMPOSITION MODULATED THIN FILMS

JHARNA CHAUDHURI* AND THOMAS TSAKALAKOS**
*Mechanical Engineering Department., Wichita State Univ., Wichita, KS 67208
**Dept. of Mechanics and Materials Science, College of Engineering, Rutgers, The State University of New Jersey, Piscataway, NJ 08854

ABSTRACT

Composition modulated (with wavelengths between 1.6 to 4.98 nm) Cu/NiFe ternary alloy thin films containing 53 at % Cu, 40 at % Ni, and 7 at % Fe were prdduced by the vapor-phase growth technique. The inter-diffusivities $\tilde{D}_B$ in these films were measured at temperatures 320°C, 345°C, and 400°C from the growth or decay rate of satelite peak intensities. $\tilde{D}_B$ as a function of dispersion relation B^2 exhibited anomalous behavior at certain wavelengths at which an enhanced elastic modulus effect is also observed. This anomalous behavior was directly attributed to the long range interaction.

INTRODUCTION

Several years ago, a technique [1,2] was developed for producing binary alloy thin films containing short-wavelength (.8 -10 nm) composition mod-ulations. Hilliard and his co-workers [3,4] utilized such a technique successfully to check the modified continuum diffusion equation [5,6], which is the basis of the theorem of spirodal decomposition. More recently, Cook, Defontaine and Hilliard [7] derived a discrete model for the diffusion equation. In a more fundamental approach utilizing the kinetics of the Ising model, Yamauchi [8] and Khataturyan [9] presented various theoretical models of order-disorder kinetics in a unified manner. Tsakalakos [10,11] observed a minimum in the diffusivity of the Cu-Ni composition modulated foils at a wavelength of about 1.5 nm which could not be explained by either Cahn's continuum model or Hilliard's discrete model. Hence Tsakalakos developed a non-linear theory of diffusion which explains successfully the stability and Kinetics of the concentration waves. Recently, there are increasing interests in studying similar diffusional effects in ternary alloys. The purpose of the present investigation is to extend the diffusional study to ternary alloy modulated structure thin films (Cu/NiFe).

EXPERIMENTAL PROCEDURE

Composition-modulated Cu/NiFe ternary alloy thin films of [111] texture and of composition 53 at % Cu, 40 at % Ni and 7 at % Fe were prepared by co-evaporating the three components through a rotating pinwheel shutter onto a mica substrate for [111] texture [12]. For the diffusion anneals, the specimen was sandwiched between two copper blocks and placed in a vac-uum furnace. The growth and decay rate of the composition modulations were determined from the satellite intensity about the Bragg peaks on a GE XRD-5 diffractometer on a special mounting [12, 13].

EXPERIMENTAL RESULTS

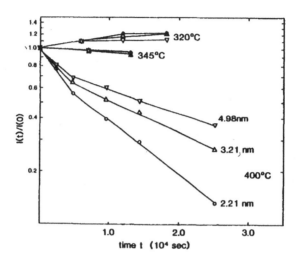

Figure 1 $\ln(I(t)/I(o))$ vs. Time for Modulations of Various Wavelengths at Different Temperatures

Figure 1 shows experimental plots of ln I(t)/I(o) vs. time at annealing temperatures 320°C, 345°C and 400°C for some of the samples. It can be seen from the figure that annealing at 345°C produced a slow decay in the satellite intensity. Further annealing at 320°C increased the satellite intensity. Lastly the satellite intensities decayed during the annealing at 400°C. This annealing behavior suggests that Cu/NiFe is a spinodal alloy and the critical temperature of the spinodal decomposition is esti-mated to be $T_c = 330°C(R_k \approx 0)$.

From the slope of the plot in the linear range the amplification fac-tor $(R(k) = \ln(I_1(t)/I(o))/2t$ was determined. The interdiffusion coeffi-cients were then calculated using the relation $R(k) = -B^2(h)\tilde{D}_B$ where the dispersion relationship $B^2(h)$ is simplified to $B^2(h) = 2/d^2(1-Cos(2\pi h))$ for the composition modulated films along [111] directions. Figures 2 and 3 show the plots of $\tilde{D}_B$ vs. B^2 at temperatures 400°C 345°C and 320°C. These curves show nonlinear behavior in which $\tilde{D}_B$ increases with the increase in wave-length, and the curve at 400°C indicates a sharp increase in $\tilde{D}_B$ at wave-lengths $\lambda = 2.4$ nm and $\lambda = 3.9$ nm.

Analysis of Data

Following the analysis of Tsakalakos, the relation between the inter-diffusivity $\tilde{D}_B$ and the diffusion coefficient $\tilde{D}$ can be approximated as the following expression [11, 14].

$$\tilde{D}_B = D[1 + K_1' B^2 + K_2' B^2 + K_2/B^4 + K_3/B^6] \text{ ---(1)}$$

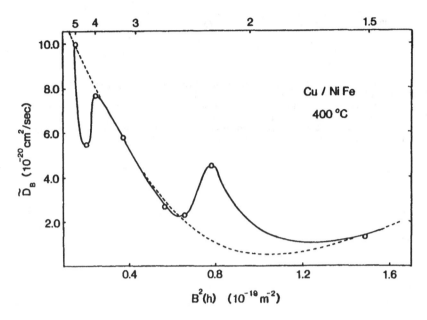

Figure 2. Observed Diffusivities $\tilde{D}_B$ vs. B^2 for 53 at %
Ni-7 at % Fe at 400°C.

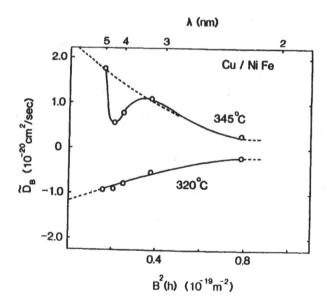

Figure 3. Observed Diffusivities $\tilde{D}_B$ ws. B^2 for 53 at %
Cu/40 at % Ni-7 at % Fe at 320°C and 345°C

where $\quad \tilde{D} = (\frac{M}{N_\nu})\ (F^{//} + 2\eta^2 y)$

and $\quad K_i^{/} = 2K_i\ (F^{//} + 2\eta^2 y)$

$F^{//}$ is the second derivative of the Helmohltz free energy per unit volume with respect to composition, M is the mobility N_ν is the number of atoms per unit volume, $2\eta^2 y[111]$ is the coherency strain energy of a coherent concentration wave in the [111] direction. The values of $\tilde{D}$ and K_i as calculated by using a nonlinear regression procedure are listed in table 1 and are plotted as dashed lines in Figures 2 and 3.

T (°C)	D ($10^{-24}m^2/s$)	$K_1^{'}$ ($10^{-19}m^2$)	$K_2^{'}$ ($10^{-38}m^4$)	$K_3^{'}$ ($10^{-57}m^6$)
400	13.8	-1.91	.97	-.065
345	2.23	-1.63	.65	-
320	-1.11	-1.07	-	-

Table I. Nonlinear Regression D_B vs. B^2 Curve Fitting for Cu/NiFe at Different Temperatures

The gradient energy coefficients K_1, K_2 and K_3 were determined according to the following procedure and are listed in table 2. For comparison, table 2 also contains the values for 50% Cu/Ni system as determined by Tsakalakos [14].

T (°C)	$f''=2\eta^2 y$ (111) ($10^8 J/m^2$)	K_1 ($10^{-20}K/NM$)	K_2 ($10^{-21}K-NM$)	K_3 ($10^{-23}K-NM$)
Cu/NiFe 400	3.54	-3.38	1.72	-1.15
345	0.76	-0.619	0.247	
320	-0.51	0.273		
Cu 50 400 at % Ni	10.43	-4.3	2.4	-4.8

Table II. Calculated Values of $f''+2\eta^2Y$ and Gradient Energy Coefficients K_1s at Different Temperatures

Utilizing a mean field theory approximation, f'' can be expressed as [14]

$$f'' = N_\nu K_B \, T/\bar{c}(1-\bar{c}) \, Tz_i V_i \quad \text{---}(2)$$

where $\bar{c}$ is the average composition of Ni, Z_i is the coordination number in the ith shell, V_i s are the interatomic potentials and $N_\nu = 8.8 \times 10^{28} M^{-3}$ for 53 Cu/40Ni7Fe. At the critical temperature T_c for a coherent spinodal alloy [15]

$$f'' + 2n^2 Y = 0 \qquad (3)$$

Hence, it follows from equations (2) and (3) that

$$f'' + 2n^2 Y = N_\nu K_B (T-T_c)/\bar{c}(1-\bar{c})$$

where $t_C(=330°c)$. The K_i values were found to increase in magnitude with temperature.

Dependence of $\tilde{D}$ ON temperature—The slope of the curve of $\ln \tilde{D}$ vs. $1/T$ is expressed as $-Q/R$, where Q is an activation energy and R is the gas constant. Hilliard [16] showed that such an assumption is approximately valid and provided an expression for Q

$$Q = Q^* + QT$$

where Q^* and Q_T are the activation energies associated with the tracer diffusivities and the thermodynamic factor expressed as

$$(1+d \ln \gamma_{cu}/d \ln C) = [C(1-C)/RT]F''.$$

T	Cu/Ni	Cu/NiFe
(°C)	$D_0(10^{-20} cm^2/sec)$	$\tilde{D}(10^{-20} cm^2/sec)$
345	-	2.23
375	0.82	-
400	4.0	13.8
450	191	-

Table III. Diffusion Coefficients D for Cu/Ni [14] and Cu/NiFe at Different Temperatures

Table 3 contains a list of values for the diffusion coefficient $\tilde{D}$ for the Cu/Ni system [14] and the Cu/NiFe system at several temperatures.

542

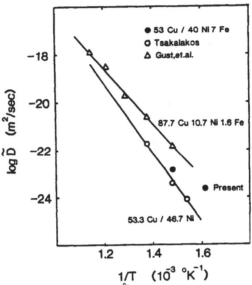

Figure IV. Arrhenius Plot of Log $\tilde{D}$ FOR Cu/Ni and Cu-Vi-Fe Alloy

Figure 4 shows an Arrhenius plot of this data. In addition, results for Cu-10.7%, Ni -1.6% Fe alloy obtained by Gust, et al. [17] are included. Whereas, Tsakalakos [10] reports an activation energy of $Q^* = 64.5$ Kcal/mol for the 50 Cu/50Ni system, an approximate value of 50 KCal/mol may be inferred from the 53-Cu/40 Ni-7Fe.

CONCLUSION

Cu/NiFe ternary alloy epitaxial thin films of composition 53 a/o Cu, 40 a/o Ni and 7 a/o Fe were produced by co-evaporating pure components in high vacuum on to mica substrates [111]. The effective inter-diffusion coefficients $\tilde{D}_B$ (111) were measured at temperatures 320°c, 345°c and 400°c. By inspecting diffusion data, spinodal temperature was concluded to be 330°c. The plot of $\tilde{D}_B$ versus dispersion relation B^2 exhibited a sharp increase at certain wave lengths at which an enhanced elastic modulus effect has been observed. The hypothesis that the Fermi surface is in critical contact with the new Brillouin zone created by the modulation is assumed to play a significant role in the theory of order-disorder transitions and long-period super structure.
    The gradient energy coefficients were calculated using a non-linear diffusion theory and the activation energy for the system was also determined.

ACKNOWLEDGEMENT

This work was supported by the National Science Foundation, NSF Grant Number DMR-78-26503

REFERENCES

1. J. Dumond and J.P. Youtz, J. Appl. Phys., 11, 357 (1940).
2. J. Dinklage and R. Frerichs, J. Appl. Phys., 34, 2633(1963)
3. H.E. Cook and J.E. Hillard, J. Appl. Phys., 40, 1797 (1964)
4. E.M. Philofsky and J.E. Hilliard, J. Appl. Phys., 55.
5. M. Hillert, Acta. Metall., 9, 525 (1961).
6. J.W. Cahn, Acta. Metall., 9, 795 (1961).
7. H.E. Cook, D. DeFontaine and J.E. Hilliard, Acta. Metall., 17, 915 (1969).
8. H. Yamauchi, Ph.D. Thesis, Northeastern Univ. (1973).
9. A.G. Khachaturyan, Prog. Mat. Sci., 22 (1978).
10. T. Tsakalakos and J.E. Hilliard, J. Appl. Physics, 55 (8), 2885 (1984).
11. T. Tsakalakos, Thin Solid Films, 86, 79-90 (1981).
12. A. Jankowski, Ph.D. Thesis, Rutgers University (1984).
13. S. Ahn, Ph.D. Thesis, Rutgers University (1984).
14. T. Tsakalakos, Ph.D. Thesis, Northwestern University (1977).
15. K.B. Rundman and J.E. Hilliard, Acta. Met. 15, 1025 (1967).
16. J.E. Reynolds, B.L. Averbach and M. Cohen with an appendix by J.E. Hilliard, Acta. Metall., 5, 59 (1957).
17. G.W.E. Wachtel, B. Frahauf and B. Predel. Phase Transformation in Solids, MRS Symposium Proceedings. North-Holland, 21, 461 (1984).

X-RAY INVESTIGATION OF STRAINED METAL-HYDROGEN SUPERLATTICES

P. F. MICELI, H. ZABEL, and J. E. CUNNINGHAM
Department of Physics and Materials Research Laboratory,
University of Illinois, Urbana, Illinois 61801

ABSTRACT

The properties of strained superlattices can be studied in a tunable fashion by dissolving hydrogen interstitially in strain free Nb/Ta metal superlattices. We report on a first investigation of a Nb/Ta 20 Å superlattice loaded with hydrogen in-situ in a high temperature x-ray furnace. After hydrogen uptake the ± 1 satellite reflections show pronounced asymmetries, indicating a hydrogen density distribution which is modulated by the superlattice periodicity.

INTRODUCTION

Many transition and rare earth metals are able to dissolve large quantities of atomic hydrogen. Depending on the H-metal and H-H interaction, the interstitial H atoms are known [1] to exhibit a large variety of gas-liquid and ordering transitions within the host frame. When the host metal composition consists of a modulation between two metals, we expect a modulation in the thermodynamic potential for an absorbed H atom. This modulation may drastically change the topology of the pristine phase diagram. Metal superlattices grown by molecular beam epitaxy offer a unique opportunity to study the effect of a composition modulation on the H solubility and phase diagram. In the following, we report the first observation of a hydrogen density modulation induced by a Nb/Ta superlattice. Part of this work has been published elsewhere [2].

LATTICE EXPANSION DUE TO HYDROGEN

Bulk quantities of Nb and Ta are known to be good absorbers of hydrogen. The dissolved hydrogen atoms occupy interstitial tetrahedral sites of the host bcc lattice and may be considered as point defects. In the context of elasticity theory, the force density due to a point defect located at the origin ($\vec{r} = 0$) is [3]:

$$\vec{f}(\vec{r}) = - \underset{\sim}{P} \cdot \vec{\nabla}\delta(\vec{r}) \quad . \tag{1}$$

Here P is the double force tensor which describes the strength and symmetry of the defect. A random distribution of interstitial hydrogen in an unmodulated cubic crystal causes a homogeneous volume expansion [4]:

$$\frac{\Delta V}{V} = \frac{1}{3} \frac{C_H}{\Omega} K_T \text{TRACE} (\underset{\sim}{P}) \quad , \tag{2}$$

where $K_T = 3/(C_{11} + 2C_{12})$ is the isothermal compressibility of the pristine metal lattice, C_H is the H to metal ratio, and Ω is the atomic volume of a metal atom. The volume expansion and therefore the H concentration can be measured via the shift in the x-ray Bragg peak.

Since the lattice expands, the interaction between two H atoms is attractive. In the context of mean field theory, this two body potential, J, combined with an entropy of mixing yields a chemical potential which requires a phase separation of the H density into high and low concentration components below a critical temperature ($T_C = J/4$) [5,6,7]

$$\mu = U + J - K_B T \ell n \ (\frac{1-c}{c}) \ . \tag{3}$$

Here, U is the H-metal binding energy which shifts the chemical potential. This lattice gas - lattice liquid transition is exactly analogous to Ising ferromagnetism.

X-RAY DIFFRACTION

The diffraction pattern for a superlattice exhibits "satellite" peaks in addition to the bcc reciprocal lattice (fundamental) peak. There is one pair (±s) of satellites symmetrically spaced about the fundamental reflection for every integer (s) multiple of the superlattice modulation wavevector ($2\pi/\Lambda$, Λ = superlattice periodicity). When the lattice parameter is constant throughout the superlattice, the +s and -s reflections due to the chemical modulation have the same intensity. This will no longer be true if the lattice parameter is modulated along the superlattice. For a chemical (metal) modulation and a strain modulation of the same wavelength, it can be shown [8] that the ±s satellite intensities are given by:

$$I_{\pm s} \propto [n_s \mp (\frac{\Lambda}{sd} \pm 1) \ \epsilon_s]^2 \ . \tag{4}$$

Here, n_s is the amplitude of the atomic scattering factor modulation arising from the artificial chemical modulation with periodicity Λ and ϵ_s is the amplitude of the strain modulation which alters the [110] interplanar spacing, d.

Nb and Ta exhibit the same lattice parameter in their bulk form as well as in the superlattice. Therefore, in the absence of H, the ± satellite intensities of the superlattice will be equal. When H is dissolved in a Nb/Ta superlattice, the presence of a H concentration modulation will result in a strain modulation with an amplitude, ϵ_s, proportional to the amplitude of the H concentration wave. By measuring the ±s satellite intensities we can obtain the induced strain modulation. Using a linear strain-concentration relationship, we can then determine the H density modulation.

HYDROGEN INDUCED STRAIN MODULATION

The Nb/Ta superlattices are grown along the [110] direction of the bcc lattice on a sapphire [1120] substrate using molecular beam epitaxy [9]. X-ray scans of the samples used here typically reveal 3 to 4 Fourier components. We have also measured reflections of the bcc reciprocal lattice (see Fig. 1) which are not along the [110] growth direction, and have found that the x-ray peak widths are comparable to those of the (110) and (220) reflections, thereby assuring that the samples are truely three dimensional crystals.

Hydrogen loading of the superlattice was accomplished in-situ using a high vcuum x-ray furnace. The superlattice was first heated to 400°C (a temperature at which no interdiffusion could ocur) in a vacuum of better than 10^{-6} Torr before introducing highly purified hydrogen from a Pd-cell.

We have measured the s=±1 satellite intensities at four temperatures for a 20Å Nb/Ta superlattice containing a constant average H concentration of 0.32 (H/Metal). Figure 2 shows that these intensities are temperature dependent. The intensity of the +1 satellite is greater than that of the -1 satellite, indicating that the Nb layers contains more H than do the Ta layers. This is expected from the known solubility isotherms [10] for bulk

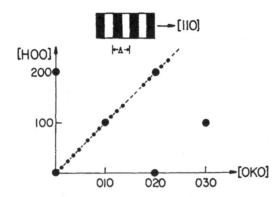

Fig. 1: Large dots indicate the measured bcc reciprocal lattice and the small dots denote the measured reciprocal lattice of the modulation. The dashed line shows the [110] direction indicating that the structure is a three dimensional crystal.

Nb and Ta. Figure 3 shows that the strain, calculated from the data in Fig. 2, is inversely proportional to the temperature. By assuming a linear strain-concentration relationship, this result implies the existence of a hydrogen concentration wave exhibiting a Curie-Weiss behavior with a Curie tempeature of 180 K. In the temperature range considered here (13°C to 300°C), we have found no evidence for a phase separation due to a hydrogen lattice gas transition or hydride formation. These phase separations would have been observed for bulk Nb and Ta at these temperatures and concentration [1].

The observed results may be explained by a spacially varying (mean field) chemical potential for a H atom in the superlattice. Since the chemical potential must be constant in thermodynamic equilibrium, the H concentration must change along the superlattice modulation. This concentration modulation must also be temperature dependent. To be more specific, we may consider the following spacially varying chemical potential:

$$\mu(\vec{x}) = U(\vec{x}) + \frac{1}{V} \int J(\vec{x}-\vec{x}') \, d^3\vec{x}' - K_B T \ell n \left(\frac{1-C(\vec{x})}{C(\vec{x})} \right) \quad , \tag{5}$$

where the first term is due to the H-metal binding, the second is the effective H-H interaction mediated by the elastic distortion field of the host lattice [7], and the third term is due to the entropy of mixing. This equation can be solved by introducing the Fourier transform for all quantities. Assuming small H concentration modulations and requiring the chemical potential to be constant we obtain:

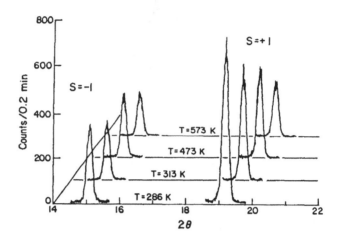

Fig. 2: X-ray scans of the ±1 satellite reflections of a 20 A Nb/Ta superlattice with an average 0.32 H/metal ratio. The relative intensities change with temperature due to changes in the hydrogen induced strain modulation.

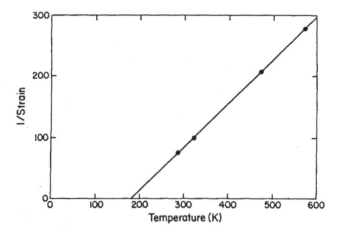

Fig. 3: The amplitude of the strain modulation is calculated from the data in Fig. 2 using equation (4). Since the strain is linearly related to the hydrogen concentration, this plot demonstrates the Curie-Weiss behavior of the hydrogen concentration modulation.

$$\vec{k} = 0, \quad \mu_0 = U(0) + J(0)C(0) - K_BT\ell n \left(\frac{1-C(0)}{C(0)}\right), \tag{6}$$

$$\vec{k} \neq 0, \quad C(\vec{k}) = C(0)(1-C(0))\left[\frac{-U(\vec{k})}{K_BT + C(0)(1-C(0))J(\vec{k})}\right] \tag{7}$$

Here, U,J and C are the Fourier transforms (with wavevector $\vec{k}$) of H-metal binding energy, H-H interaction energy, and the H/metal concentration respectively. The first equation, arising from the fact that the chemical potential must be constant, describes the average properties of the system and is important when considering the solubility of hydrogen and the lattice gas transition. The second equation results from the spacially varying properties of the host and describes the observed Curie-Weiss behaviour in Fig. 3. This is the response to an "applied" field, U(k), analogous to the way that the magnetization responds to an externally applied magnetic field in ferromagnetic systems. The only difference is that here, the one body potential is spacially varying instead of the uniform field usually assumed in magnetic systems.

THERMAL EXPANSION OF THE SUPERLATTICE

The above results assume only that the relationship between strain and H concentration is linear. To determine an absolute H concentration it is necessary to know the constant of proportionality which is, at present, unavailable. However, an x-ray measurement of the thermal expansion (Fig. 4) of the superlattice yields 7.2×10^{-6} K^{-1} which is essentially the same

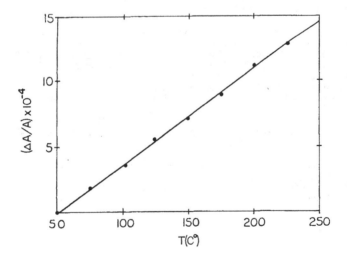

Fig. 4: Coefficient of thermal expansion for a 20 Å Nb/Ta superlattice as determined by x-ray diffraction: $\alpha_T = 7.2 \times 10^{-6}$ K^{-1}.

550

as for bulk Nb and Ta [11]. This is an indication that the elastic properties of the superlattice are similar to the bulk metals. Since the lattice expansion due to H is nearly the same for both Nb and Ta (.00058 & .00052 per %H/metal respectively), we assume that the superlattice expands by .00055 per %H/M which is the average of the bulk Nb and Ta values. Therefore, from Fig. 3 we calculate that the amplitude of the H concentration modulation changes from 20 %H/metal at 13°C to 7 %H/metal at 300°C. From the shift of the Bragg reflection to lower angle upon H uptake we determine that this sample contains an average H concentration of 32 %H/metal.

In summary, we have found a Curie-Weiss behavior of the hydrogen concentration modulation induced by a superlattice modulation and have explained the observed results in terms of a simple lattice gas model. In the light of recent interests towards strained layer superlattices, metal superlattices which absorb hydrogen provide new possibilities to investigate the structural properties of artificially modulated materials since the strain may be continuously changed by varying the hydrogen concentration and temperature.

ACKNOWLEDGEMENT

We wish to thank S. M. Durbin for the preparation of some of the superlattices. This work was supported by the US. Department of Energy, Division of Materials Sciences, under Contract DE-AC02-76ER01198.

REFERENCES

1. T. Schober, H. Wenzl, "The Systems NbH(D), TaH(D), VH(D); Structures, Phase Diagrams, Morphologies, Methods of Preparation, Volume 2", in TOPICS IN APPLIED PHYSICS, Vol. 28, ed: G. Alefeld and J. Volkl, Springer Verlag, Berlin, Heidelberg, New York: 1978.
2. P. F. Miceli, H. Zabel, J. E. Cunningham, to be published.
3. A. H. Love, "Mathematical Theory of Elasticity", Cambridge University Press, Cambridge, England: 1924.
4. H. Peisl, "Lattice Strains Due to Hydrogen in Metals", in TOPICS IN APPLIED PHYSICS, Vol. 28, edited: G. Alefeld and J. Volkl, Springer Verlag, Berlin, Heidelberg, New York: 1978.
5. G. Alefeld, "Critical Phenomena in Alloys, Magnets and Superconductors", edited: R. E. Mills, McGraw-Hill, New York: 1974.
6. G. Alefeld, Phys. Stat. Sol. 32, 67 (1969).
7. H. Wagner, H. Horner, Adv. Phys. 23, 587 (1974).
8. A. Guinier, "X-ray Diffraction in Crystals, Imperfect Crystals, and Amorphous Bodies"; Translated by P. Lorrain and D. Sainte-Marie, Lorrain Freeman and Co., San Francisco, 1963.
9. S. M. Durbin, J. E. Cunningham, M. E. Mochel, C. P. Flynn, J. Phys. F: Metal Phys. 11, L223 (1981).
10. E. Veleckis, R. K. Edwards, J. Phys. Chem. 73, 683 (1969).
11. A. Magerl, B. Berre, G. Alefeld, Phys. Stat. Sol. (a) 36, 161 (1976).

SOLID STATE REACTION OF SPUTTER DEPOSITED CRYSTALLINE MULTILAYERS OF Ni AND Zr

K. M. UNRUH, W. J. MENG, and W. L. JOHNSON*
* W. M. KECK Laboratory for Engineering Materials, California Institute of Technology, Pasadena, CA 91125
A. P. THAKOOR and S. K. KHANNA**
**Jet Propulsion Laboratory, California Institute of Technology, Pasadena, CA 91109

ABSTRACT

We have prepared sputtered metallic films comprised of alternating layers of pure Ni and Zr to a total thickness of several microns. Due to the negative heat of mixing of crystalline Ni and Zr and a significant intermixing rate we have been able to form an amorphous Ni-Zr phase by a solid state reaction. In this work we report the conditions under which a substantial fraction of amorphous material may be formed. The extent of the reaction itself as a function of time has also been monitored.

INTRODUCTION

Since their discovery nearly 25 years ago most amorphous metallic alloys have been prepared either by rapid quenching from the melt or by vapor deposition [1]. Recently, however, it has been shown that an amorphous phase can also be produced by a solid state reaction of elemental crystalline constituents [2-6]. For such a reaction to occur thermodynamics requires the amorphous phase to be lower in free energy than a simple mixture of the crystalline components. In addition, kinetic constraints must also be present which favor the growth of the amorphous phase over lower energy crystalline phases at the reaction temperature. In practice one looks for combinations of elements in which amorphous alloys are known to form with relatively high crystallization temperatures and in which one component is a relatively fast diffuser in the other. This latter requirement suggests that mixing occurs rapidly with respect to the time scale characteristic of the nucleation and growth of stable crystalline phases. Many combinations of early and late transition metals, for example, satisfy the above conditions.

In this work we have chosen to study the solid state reaction that produces an amorphous phase by annealing alternating layers of sputter deposited Ni and Zr at low

552

temperatures. As can be seen
from Fig. 1 which shows the Ni-
Zr free energy diagram,
constructed following Schwartz
and Johnson [3], the amorphous
phase is indeed lower in free
energy than a mixture of Ni and
Zr over most of the composition
range. It is also known that Ni
is an "anomalous" fast diffuser
in Zr, thereby meeting the
second criterion mentioned above
[7]. In what follows we report
on the growth kinetics of an
amorphous Ni-Zr phase during the
solid state reaction of
crystalline Ni and Zr.

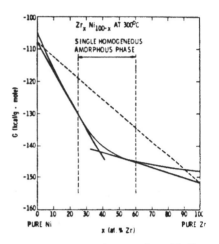

Fig.1. Gibbs free energy of a mixture of crystalline Ni
and Zr (heavy dashed line) and an amorphous Ni-Zr alloy
(solid curved line).

EXPERIMENTAL

The samples studied in this
work were prepared by a
magnetron sputtering technique
that has been described elsewhere [8]. A turbomolecular pump was
used to obtain a chamber pressure in the 10^{-7} Torr range prior
to sputtering. Alternating layers of crystalline Ni and Zr were
deposited by simultaneously sputtering pure Ni and Zr targets
while alternating the sample substrate over each individual
sputtering gun. Typical sputtering rates were several hundred
A/minute. Each sample contained a total of about 25 layers with
individual layer thicknesses between 500 and 600 A. Samples were
prepared on both glass and Be substrates at Ar pressures between
5 and 15 mTorr.

Sample heat treatments were carried out in a flowing He gas
furnace. A liquid nitrogen trap and a Ti gettering furnace
assured minimal oxygen contamination. In addition the samples
were also packed in Zr powder during annealing. When annealed in
this way no film discoloration was observed despite the fact
that in all cases a thin Zr layer formed the outermost surface.
However, similiar results were also obtained when the annealing
was carried out in a less oxygen free environment.

The x-ray measurements were performed on a conventional
Philips theta-two theta vertical diffractometer using Cu Ka
radiation.

RESULTS AND DISCUSSION

X-ray diffraction patterns of three typical as-deposited Ni-Zr multilayer samples are shown at the top of Fig. 2. Figures 2A and 2B correspond to samples containing roughly 65 and 75 atomic % Ni respectively, estimated from the individual layer thicknesses. Both of these samples were prepared under an Ar sputtering pressure of 15 mTorr. Sample 2C was prepared under a sputtering pressure of 5 mTorr and, as in the case of sample A, contains about 65 atomic % Ni. It can be seen from Fig. 2 that the Ni and Zr layers are both highly textured with the film growth normal to close packed planes, that is, <002> in Zr and <111> in Ni. This effect is commonly observed in vapor deposited thin films.

In a hexagonal close packed structure such as α-Zr one expects to find (100) and (010) Bragg planes, for example, to be equivalent, both contributing to a single diffraction line. This is not seen to be the case for samples of Ni and Zr prepared at the lower sputtering pressure of 5 mTorr. The slight inequivalence of the hexagonal basal plane lattice parameters is probably the result of the greater compressive film stresses that are characteristic of lower sputtering pressures [9].

The bottom half of Fig. 2 shows the result of a four hour anneal at 315°C of the samples whose as-deposited x-ray patterns are displayed in the upper portion of the figure. It can be seen that the sharp peaks of the as-deposited films have been greatly reduced in intensity and a broad band has appeared indicating the presence of an amorphous phase. It should be noted that the initially very strong α-Zr (002) peak has essentially disappeared while the initially weaker α-Zr (100) peak has become the dominant Zr peak. While the as-deposited films are smooth and flat the reacted films buckle from the substrate. This effect will be discussed later.

In order to better understand the kinetic processes involved in the solid state reaction of Ni and Zr we have focussed attention on one set of identically prepared films deposited at a sputtering pressure of 15 mTorr. A systematic study of the growth rate of the amorphous phase was then carried out at several different annealing temperatures. Because the amorphous Ni-Zr phase is the only new phase observed throughout the annealing process its growth has been monitored by the decrease in the Ni and Zr integrated x-ray peak intensities [10]. The results of these measurements are shown in Fig. 3 for two identically prepared films, one annealed at 250°C and the other at 315°C. In both cases the Ni and Zr total line intensities have been normalized to their initial as-deposited values. The slight increase in the Ni line area following the first 30 minute anneal at 250°C is most likely due to

554

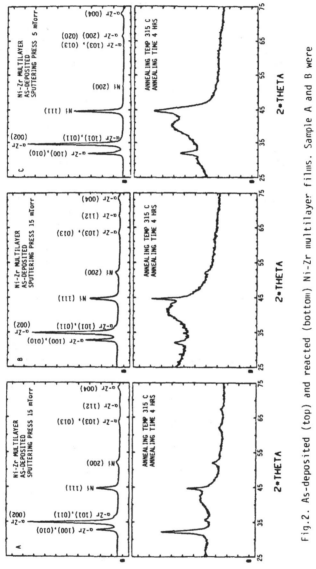

Fig.2. As-deposited (top) and reacted (bottom) Ni-Zr multilayer films. Sample A and B were prepared under an Ar sputtering pressure of 15 mTorr and are of nominal composition $Ni_{65}Zr_{35}$ and $Ni_{75}Zr_{25}$ respectively. Sample C was prepared under an Ar sputtering pressure of 5 mTorr and is of nominal composition $Ni_{65}Zr_{35}$. All samples were annealed for four hours at $315^{\circ}C$.

recrystallization of some Ni crystallites. There is no evidence for further recrystallization during subsequent anneals.

At both temperatures an initial decrease in the Ni and Zr line areas is followed by a region of less rapid decrease. In addition, the 250°C annealing data exhibits a break in the line areas curve between these two regions. This break is the result of a smaller total scattering area due to the buckling of the film from the substrate. A similar effect is not observed in the case of the 315°C anneal because the film buckling has already occured during the first 30 minute annealing period.

The fact that the consumption rate of the crystalline Ni and Zr follows a $t^{1/2}$ time

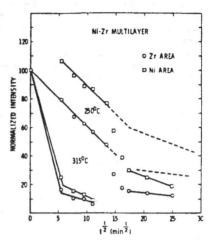

Fig.3. Normalized Ni and Zr integrated crystalline x-ray line intensities versus $t^{1/2}$ at 250°C and 315°C (see text for further explanation).

dependence implies a diffusion rather than interface limited growth process. Our data suggests two different diffusion limited growth regimes characterized by two different effective diffusion constants, D1 and D2. Based on the slopes of the straight lines shown in Fig. 3 we estimate $D1(250°C) = 2 \times 10^{-19}$ m^2/sec and $D1(315°C) = 4 \times 10^{-18}$ m^2/sec. An activation energy of 1.2 eV has been estimated from this data if an Arrhenius type behavior is assumed for the diffusion constant. We have also estimated the second stage diffusion constants to be about one order of magnitude less than that of the respective first stage constants. It is important to note that our estimated effective interdiffusion constant is at least two orders of magnitude less than that of Ni tracer diffusion in Zr extrapolated to the reaction temperature [11].

The existance of two diffusion regimes could result from one of several different physical mechanisms. The smaller diffusion constant found in the second stage of the reaction could, for example, be due to effects which appear when the supply of crystalline material is nearly exhausted [6]. On the other hand, it is observed that the change in the diffusion rate corresponds to the buckling of the film from the substrate. The resulting in-plane stress relief may then be responsible for the decreased diffusion rate.

Corresponding to the decrease in the Ni and Zr integrated

556

line intensities shown in Fig.
3, there is a shift in the
corresponding line positions
from their as-deposited values.
This effect can be seen in Fig.
4. Measured in reflection both
the Zr (100) and (002) lines, in
both cases corresponding to
lattice planes parallel to the
sample surface, exhibit a rapid
initial increase in d-spacing
followed by a decrease. The Ni
(111) line position, on the
other hand, remains essentially
unchanged throughout the entire
annealing period.

The dilation of Zr
structural cells in a direction
perpendicular to the film
surface can arise due to an in-
plane compressive stress at the
Zr-amorphous layer interface.
Such an interpretation would
require a contraction of in-

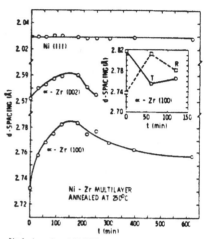

Fig.4. d-spacing of Ni (111), Zr (002), and Zr (100) Bragg
peaks as a function of time in reflection geometry. The
insert shows the α-Zr (100) d-spacing as a function of time
in reflection (R) and transmission (T).

plane Zr lattice spacings and can be verified by transmission x-
ray measurements. We have chosen to use Be substrates for this
purpose due to their low x-ray absorptance. The results of these
measurements are shown in the insert of Fig. 4 for the Zr (100)
line positions. It can be seen that the Zr cell dilation
perpendicular to the sample surface is accompanied by a
contraction parallel to the film surface. The effect of an
amorphous layer is therefore seen to result in a compressive
stress on the unreacted crystalline material. It should also be
noted that even though the amorphous layer also stresses the Ni
layer, little relative strain results due to the substantially
greater bulk modulus of Ni [12]. It is this building up of in-
plane compressive stress that eventually results in the buckling
of the film from the substrate.

SUMMARY AND CONCLUSIONS

The major results of this work can be summarized as
follows. Crystalline Ni and Zr multilayer films have been shown
to form an amorphous phase by a solid state reaction at
relatively low annealing temperatures from about 250-350°C. This
process is consistant with the semi-quantitative free energy
diagram of Fig. 1. As the reaction proceeds two different growth

regimes are observed. Both regimes, however, follow a $t^{1/2}$ time dependence indicating diffusion limited growth. The growth of the amorphous layer results in a compressive stress at the Ni and Zr-amorphous phase interface which builds up as the reaction proceeds until the film is torn from the substrate. This effect suggests that stress may play an important role in the physical processes involved in the solid state reaction.

REFERENCES

✦ This work is supported under a President's Fund Grant from the California Institute of Technology.

1. See e.g. B. H. Lieberman, in Amorphous Metallic Alloys, edited by F. E. Luborsky (Butterworths, London, 1983), pp. 26-41.
2. X. L. Yeh, K. Samwer, and W. L. Johnson, Appl. Phys. lett. 42, 242 (1983).
3. R. B. Schwarz and W. L. Johnson, Phys. Rev. Lett. 51, 415 (1983).
4. R. B. Schwarz, K. L. Wong, W. L. Johnson and B. M. Clemens, J. Non-Cryst. Sol. 61&62, 129 (1984).
5. B. M. Clemens, W. L. Johnson, and R. B. Schwarz, J. Non-Cryst. Sol. 61&62, 817 (1984).
6. M. Van Rossum, M.-A. Nicolet, and W. L. Johnson, Phys. Rev. B 29, 5498 (1984).
7. See e.g. A. D. Le Claire, J. Nucl. Mat. 69, 70 (1978).
8. A. P. Thakoor, S. K. Khanna, R. M. Williams, and R. F. Landel, J. Vac. Sci. Technol. A1, 520 (1983).
9. J. A. Thornton and D. W. Hoffman, J. Vac. Sci. Technol. 14, 164 (1977); D. W. Hoffman and J. A. Thornton, Thin Solid Films 45, 387 (1977).
10. A. Guinier, X-Ray Diffraction (W. H. Freeman and Company, San Francisco, 1963).
11. G. M. Hood and R. J. Schultz, Phil. Mag. 26, 329 (1972).
12. C. Kittel, Solid State Physics (John Wiley & Sons, New York, 1976).

AMORPHOUS PHASE FORMATION IN SOLID STATE REACTIONS OF
LAYERED NICKEL ZIRCONIUM FILMS

BRUCE M. CLEMENS AND JEFFREY C. BUCHHOLZ
Physics Department, General Motors Research, Warren, Mi. 48094

ABSTRACT

Formation of an amorphous zirconium-nickel phase by solid state reaction of a layered crystalline structure has been studied by in-situ resistivity, x-ray diffraction, and Auger depth profiling. The reaction was
studied as a function of layer thickness and reaction temperature.
Samples with a layer thickness of less than 4 atomic planes had x-ray
diffraction spectra with one broad maximum characteristic of amorphous material. As the layer thickness increased, the maximum broadened and separated into two resolved peaks corresponding to crystalline nickel and zirconium. These structures were transformed to an amorphous nickel-zirconium
alloy by an anneal at temperatures below the crystallization temperature of
the amorphous phase. The reaction occured by a layer growth process, where
the thickness of the layer evolved linearly with the square root of time.

INTRODUCTION

The recent discovery [1,2] that amorphous alloys can result from solid
state reactions of elemental crystalline layers has generated interest in
the fundamental nature of this process, as well as excitement over the possibility of its practical application for formation of bulk amorphous materials. Nucleation and growth of crystalline phases must be suppressed
during amorphous phase formation. Traditional methods of amorphous phase
formation use a rapid quench from a disordered phase to achieve this goal.
The requirement of rapid heat removal sets an upper limit on the thickness
of material which can be formed in this manner. Solid state reactions, on
the other hand, can take advantage of the fact that nucleation and growth of
an amorphous structure can occur faster than nucleation and growth of a
crystalline structure. Thus there is no requirement for rapid heat removal,
and no fundamental limit to the size of sample which can be prepared by this
technique. Recently, bulk amorphous samples have been formed by this process using starting composites formed by mechanical alloying of powders [3],
and foils [4].
The growth of a particular phase in a solid state reaction is due to a
balance between the chemical driving force, reaction kinetics, and relative
nucleation ease for the competing phases. The observation of amorphous
phase growth during a solid state reaction demonstrates that an amorphous
structure is stable under conditions where diffusion of at least one atomic
species occurs over lengths on the order of 100 nm. The persistence of an
amorphous phase under these conditions demonstrates that nucleation and
growth of an amorphous phase can occur preferentially to that for a crystalline compound.

EXPERIMENT

Layered samples were prepared by magnetron sputtering in 3 x 10^{-3} Torr
of argon. Zirconium and nickel were sequentially deposited onto oxidized
silicon (100) wafers. The relative thickness of the layers and thus the
average composition of the films was determined by the sputter rates for the
two guns (0.48 nm/sec. and 0.60 nm/sec. for nickel and zirconium, respectively). The compositional wavelength was controlled by the rotational

velocity of the substrate holder. The relative rates of deposition were selected to result in an equal number of close packed atomic planes of nickel and zirconium resulting in an average composition of 38 atomic % zirconium. Samples which had 1, 4, 8, 12, 40, and 200 atomic planes of each element per layer were measured.

X-ray diffraction was performed on both a Bragg–Brentano, and a Seeman–Bohlin thin film diffractometer. In the symmetric reflecting geometry of the Bragg–Brentano diffractometer, the scattering vector is perpendicular to the plane of the sample so only planes parallel to the surface of the sample diffract. In contrast, Seeman–Bohlin geometry employs a fixed angle of incidence between 1 and 10 degrees. This results in increased signal intensity from the surface region of the sample and a scattering vector with a non-zero component in the plane of the sample.

Resistivity was monitored during annealing in vacuum by a four point probe with platinum contacts. Absolute values of the resistivity were measured using a conventional four-point probe. The uncertainty in these values is about 20% due to uncertainty in the film thickness.

Concentration as function of depth into the sample was monitored by Auger depth profiling. The depth scale was determined by the concentration wavelength of the samples.

RESULTS

The results of the symmetric reflecting x-ray diffraction on the as deposited samples are shown in figure 1a. The 1 and 4 atomic layer samples show one broad maximum characteristic of amorphous material. As the number of atomic planes per layer is increased to 8 the maximum begins to split and at 12 atomic planes per layer we see two clearly resolved maxima. The diffraction data on the sample with 40 atomic planes per layer show that these maxima evolve to the (111) nickel line and the (002) zirconium line.

Shown in figure 1b is the diffraction data for these samples taken with the thin film diffractometer. We see the same trend with increasing layer thickness. The 1 and 4 atomic planes per layer samples are amorphous. The 8 atomic planes per layer sample shows a broadening of the band which is greater in the 12 atomic planes per layer sample. Note however, that the separation of the two peaks is not as great as was obvserved with the symmetric reflecting diffractometer. This indicates that the ordering into a crystalline structure occurs anisotropically, that is , crystalline planes are first formed perpendicular to the direction of growth.

Small angle x-ray diffraction on the Bragg–Brentano diffractometer determined the superlattice structure. Superlattice lines were observed in the 4, 8, and 12 atomic planes per layer samples. The number of lines observed increased with increasing compositional wavelength, with 6 orders observed in the 12 atomic plane per layer sample. The spacing derived from the position of these lines was consistent with the intended construction of the samples.

After annealing below 300 °C, the crystalline peaks of all samples decreased in intensity. In the 40 and 200 atomic planes per layer samples an amorphous phase grew at the expense of the crystalline phase as evidenced by the diffraction results which showed the appearance of a broad maximum and the disappearance of the crystalline peaks. Figure 2 shows the diffraction results for the 200 atomic plane per layer sample which has been annealed at 260 °C for 20 hours. The remnants of the crystalline peaks are due to unreacted layers on the film surface.

The resistivity of the samples measured prior to the reaction shows a decrease in resistivity with increasing layer thickness. These results are summarized in table 1. Note that the resistivity of the 1 and 4 atomic planes per layer samples are about the same as amorphous material of the same composition [5]. The resistance was measured during isothermal anneals performed at temperatures ranging form 157 °C to 300 °C. The resis-

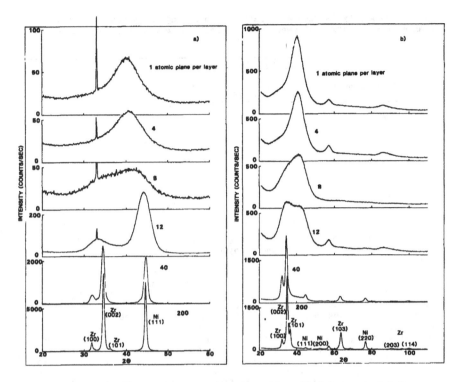

Figure 1. X-ray diffraction data for the as deposited layered nickel-zirconium samples from a) symmetric reflecting and b) thin film diffractometers.

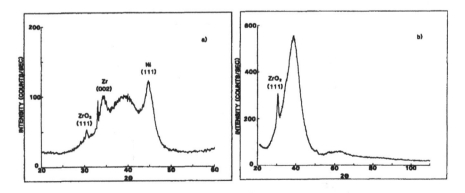

Figure 2. X-ray diffraction data for 200 atomic plane per layer sample annealed at 260 °C for 20 hours from a) symmetric reflecting and b) thin film diffractometers.

562

tivity changed by as much as a factor of 6 during the anneals. This large
change in resistivity allowed us to directly determine the reaction rate.

Table 1. Resistivity of Layered Samples.

Atomic Planes per Layer	Anneal (hours/Temp. (°C))	Resistivity ($\mu\Omega$-cm)
1	none	160
4	none	160
8	none	135
8	2/308	160
12	none	125
12	1/350	150
40	none	90
40	2/300	200
200	none	42
200	4/300	210

The diffusion controlled layer growth model [1,6,7] predicts that the
reaction will produce an amorphous layer at the interface of the two constituents and that the thickness of this layer will evolve linearly with the
square root of the time. If we assume independently conducting layers, the
change in conductivity during an anneal will be proportional to the thickness of the amorphous layer and the slope of the linear region of plots of
the change in conductivity versus the square root of time will be proportional to the square root of the diffusivity. The conductivity of the 200
atomic planes per layer samples changed in a linear fashion over a long region with the square root of time. Figure 3 is a plot of the log of the
square of the slope versus 1/T for the 200 atomic plane per layer samples
from which we can extract an activation energy. This plot is characterized
by a linear region at low temperatures with some upward curvature at higher
temperatures. The activation energy found from the linear region is 1.9 eV
which is in good agreement for diffusion in amorphous zirconium-nickel alloys [8]. The range of temperatures for which useful data could be taken was
restricted by the warm up time for the furnace. In the thinner films, the
reaction was completed or at least slowing down before the furnace warmed
up.

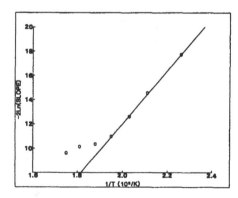

Figure 3. Arrhenius plot of the square of the slope of the change in
conductivity versus the square root of time, versus 1/T, for
annealed 200 atomic plane per layer samples.

563

Resistivity was monitored during a constant temperature ramp of 10 degrees per minute on the reacted samples. Crystallization caused a precipitous drop in resistivity which occured at 475 °C which is in good agreement with previously reported values for amorphous nickel-zirconium of this composition [5].

Auger profiling of the unannealed 40 and 200 atomic plane per layer samples showed they were compositionally modulated structures with the expected wavelength. Partially reacted films showed decreased modulation, and a compositional plateau at the position of the origional interface. Thus the growth of the amorphous phase occurs as a layer growth process as was first demonstrated by Clemens et.al. [1] and later verified by Van Rossum et.al. [9].

DISCUSSION

The results of this study demonstrate that an amorphous phase grows during anneals of layered crystalline material. The thermodynamic driving force for this reaction can be described by a simple model [1]. The large negative heat of mixing between the constituents results in a substantial lowering of free energy due to compositional mixing. This chemical energy swamps the topological energy with the result that the thermodynamic driving force is almost independent of structure of the reaction product. Thus the question of what phase will grow as a reaction product will depend more on the nucleation and reaction kinetics than the relatively small differences in thermodynamic driving force. The nucleation behavior will depend on the properties of the interfacial region and the relative ease of nucleation of the competing phases.

The fabrication of samples with varying layer thicknesses has allowed us to study the structure and properties of the interface. The interfacial region has a finite thickness. Thus, as the layer thickness decreases and the number of interfaces in the film increases, the volume fraction of the film which has the properties of the interface will increase. Our deposition conditions result in an amorphous interfacial region which extends 4 to 8 atomic planes into the layers on either side. As the layer spacing increases, the regions near the center of an individual layer are affected less and less by the interface. At a layer thickness of 12 atomic planes we see evidence that crystallinity is begining to develop. At a layer thickness of 40 atomic planes we see that the predominant structure is textured polycrystalline. The amorphous material at the interface is most likely still present but makes up only about 10% of the sample.

This amorphous interface provides a nucleus for amorphous phase growth. Growth of an intermetallic crystalline phase, such as Ni_2Zr or $Ni_{10}Zr_7$, requires formation of a nucleus layer along the interface. This requires a collective motion of atoms of both species in a spacial region extending along the interface. While we do have considerable mobility of the nickel atoms in our reaction conditions, it is evident that we do not have the mobility required for crystalline phase nucleation.

Auger depth profiling results on partially reacted films shows that the reaction occurs as a layer growth process. Resistivity measurements show that the kinetics of the reaction follow the square root of time behavior which is the expected result of the layer growth model [7] in the limit of fast interfacial reaction rates, or a thick product layer. In this limit the layer thickness x, is given by:

$$x^2 = 2G\Delta CDt. \qquad 1$$

Where G is a constant which depends on the concentrations of the three phases involved and is about 2 for the case here, ΔC is the equilibrium concentration difference across the product layer, D is the effective diffusivity, and t is the reaction time. The concentration difference can be estimated from the free energy calculation [1] to be about 30 atomic % for amorphous nickel zirconium. This is about an order of magnitude larger than that for competing crystalline phases. Thus, amorphous phases might be expected to grow faster than a crystalline phase.

The change in conductivity versus the square root of time begins to show some curvature towards the end of the reaction. This is most likely where one of the reactants, most likely zirconium is consumed. The reaction then proceeds by incorporating the remaining reactant into the amorphous phase. This may explain the curvature in the Arrhenius plot at high temperature. A second possibility is that the amorphous phase relaxes during the reaction resulting in a decrease in diffusivity. The curvature of the Arrhenius plot is the result of the crossover in the diffusivity between the two types of material. Work is in progress to determine which of these two factors is responsible in this case.

CONCLUSIONS

Amorphous nickel zirconium is formed by reaction of layered crystalline nickel and zirconium. The reaction occurs via a diffusion limited layer growth process with an activation energy of 1.9 eV. The as-deposited films with layer spacing less than 4-8 atomic planes per layer were amorphous. The mechanism leading to this amorphous phase formation in thin layers also produces an amorphous interfacial layer in the thicker film samples that are not totally amorphous as deposited. The existence of this interfacial amorphous layer eliminates the nucleation barrier for growth of amorphous phase during annealing of these thicker layer samples.

ACKNOWLEDGEMENTS

The authors gratefully acknowledge the technical support of M. Suchoski in preparing the samples for this study. We would also like to express our thanks to M. Meyer, R. Waldo, S. Gaarenstroom, and A. Dow for contributing their assistance. Helpful conversations with W. L. Johnson and R. Schulz are also acknowledged.

REFERENCES

1. B. M. Clemens, W. L. Johnson, and R. B. Schwarz, J. Non. Cryst. Solids 61&62, 817, (1984).
2. R. B. Schwarz, K. L. Wong, W. L. Johnson, and B. M. Clemens, J. Non. Cryst. Solids 61&62, 129, (1984).
3. M. Atzmon, J. D. Verhoeven, E. D. Gibson, and W. L. Johnson, DOE Report No. ER10870-151, (1984).
4. L. Schultz, Proc. MRS Europe Meeting, Strasburg, France, (1984).
5. Z. Altounian, Tu Guo-hua, and J. O. Strom-Olsen, J. Appl. Phys. 54, 3111, (1983).
6. W. L. Johnson, B. Dolgin, and M. Van Rossum, DOE Report No. ER/10870-148, (1984).
7. U. Gosele and K. N. Tu, J. Appl. Phys. 53, 3252, (1984).
8. R. W. Cahn in Physical Metallurgy, ed. by R. W. Cahn and P. Haasen (North Holland, Amsterdam, 1983) p. 487.
9. M. Van Rosum, M-A. Nicolet, and W. L. Johnson, Phys. Rev. B 29, 5498, (1984).

AMORPHOUS AND CRYSTALLINE PHASE FORMATION
BY ION-MIXING OF Ru-Zr AND Ru-Ti

Y-T.CHENG,W.L.JOHNSON,AND M-A.NICOLET
California Institute of Technology,Pasadena,California 91125

ABSTRACT

Ion mixing of multilayered Ru-Zr and Ru-Ti at different compositions and substrate temperatures produced various metastable phases. The applicability of previously proposed rules regarding amorphous phase formation are examined. Effects of point defect in amorphous phase formation by ion mixing are discussed.

INTRODUCTION

Ion mixing(IM) has been of considerable interest over the last several years [1]. It has emerged as an useful surface modification method in producing various amorphous and crystalline phases [2]. Several attempts have been made to predict the formation of amorphous phases by this technique. Liu and coworkers have formulated a rule which states that an amorphous binary alloy will be formed by IM of the multilayered sample when the two constituent metals are of different structures [3]. It has also been suggested that IM is likely to produce a crystalline phase at a composition which corresponds to a compound of simple lattice structure [4]. These rules either have limited applicability or need to be put on a more profound physical basis. Recently, the application of thermodynamic considerations to IM processes have proven fruitful [5,6]. The present authors have provided some general criteria regarding amorphous and crystalline phases formation by IM [6] of metal-metal systems based on considerations of thermodynamic free energy diagrams and the restricted growth kinetics of competing phases. In this paper we shall further develop these ideas and apply them to the metal-metal systems of Ru-Zr and Ru-Ti.

Both systems have relatively simple equilibrium phase diagrams as shown in Fig.[1a] and Fig.[1b]. Table I lists some of the relevant structural information. The Goldschmidt atomic radii are 1.34A, 1.47A, and 1.60A for Ru, Ti, and Zr respectively [7,8,9]. These two systems both have negative heats of mixing [10].

The purpose of this paper is to illustrate the relative importance of the physical parameters of the alloys in determining the result of ion mixing. Such factors as the structure of the terminal solutions, complexity of compound phases, equilibrium free energy of the various phases, effect of defects, and substrate temperature(T_{sub}) will be considered. Semi-quantitative free energy diagrams are constructed to help understand the various processes observed. Thermal stability of amorphous phase against crystallization will also be discussed.

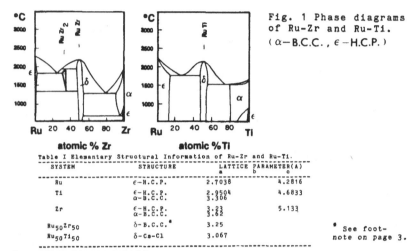

Fig. 1 Phase diagrams
of Ru-Zr and Ru-Ti.
($\alpha-$ B.C.C. , $\epsilon-$ H.C.P.)

Table I Elementary Structural Information of Ru-Zr and Ru-Ti.

SYSTEM	STRUCTURE	LATTICE PARAMETER(A) a	b	c
Ru	ϵ-H.C.P.	2.7038		4.2816
Ti	ϵ-H.C.P. α-B.C.C.	2.9504 3.306		4.6833
Zr	ϵ-H.C.P. α-B.C.C.	3.23 3.62		5.133
$Ru_{50}Zr_{50}$	δ-B.C.C.*	3.25		
$Ru_{50}Ti_{50}$	δ-Cs-Cl	3.067		

* See foot-
note on page 3.

EXPERIMENTAL PROCEDURE

Multilayered Ru-Zr(Ti) samples with the averaging compositions of $Ru_{25}Zr(Ti)_{75}$, $Ru_{50}Zr(Ti)_{50}$, $Ru_{75}Zr_{25}$, and $Ru_{70}Ti_{30}$ were prepared by e-gun evaporation onto SiO_2 substrates in an oil-free vacuum system at a pressure in the 10^{-7} torr range. The total thickness of each sample was designed to be $R_p + \Delta R_p$ for 300 keV or 600 keV Xe ions. The thickness ratio between different layers was chosen such that the required homogeneous composition would be obtained after mixing. Layer thicknesses were monitored by a quartz thickness monitor during evaporation.
The 300 keV and 600 keV Xe ion irradiations were performed at various T_{sub} from 77 K to 570 K. T_{sub} was measured in situ by a thermo-couple mounted inside the sample holder. The vacuum during the irradiations was maintained below $5 * 10^{-7}$ torr. Implantation doses ranged from $2 * 10^{15}$ to $2 * 10^{16}/cm^2$ at a flux of about 800 nA/cm^2 for 300 keV Xe^+ and 200 nA/cm^2 for 600 keV Xe^{++} ions. Backscattering spectrometry provided information on the degree of mixing and the compositional ratios (fig.[2]). Structural information was obtained by x-ray diffraction (Read camera). Isothermal annealings were performed in a vacuum with pressure below $7 * 10^{-7}$ torr for T>520 K and in a flowing He gas gettering furnace for T<520 K. Samples were annealed in sequential temperature steps. The annealing time was 30 min at each temperature step.

EXPERIMENTAL OBSERVATIONS

Ru-Zr: The general features of IM of Ru-Zr are summarized in Table IIa. At 77 K, 300 keV Xe^+ produced amorphous phases for all three compositions at doses below $5 * 10^{15}Xe^+/cm^2$. Futher increase of the dose to $2 * 10^{16}/cm^2$ did not induce other phase changes. 300 keV Xe^+ IM at a dose $8 * 10^{15}$ at different T_{sub} revealed a critical substrate temperature ($T^c_{sub} \approx$ 390 K)for

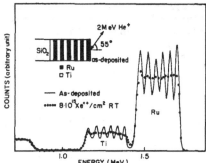

Fig. 2 Backscattering spectrum of a $Ru_{50}Ti_{50}$ multilayered sample before and after ion mixing at 300K.

$Ru_{25}Zr_{75}$ and $Ru_{50}Zr_{50}$. Below this temperature, amorphous phases were observed (Table IIa). The lattice spacing variation reflects the larger Goldschmidt radius of Zr compared with Ru (Table I and IIa). The recrystallization temperatures of amorphous $Ru_{25}Zr_{75}$ and $Ru_{50}Zr_{50}$ produced by 300 keV Xe with a dose of $8*10^{15}/cm^2$ at 77 K are 670 ± 15 K and 510 ± 15 K respectively.

Table IIa 300 keV Xe ion mixing of Ru-Zr with the dose $8*10^{15}/cm^2$ at different substrate temperatures.

T_{sub}	$Ru_{25}Zr_{75}$	$Ru_{50}Zr_{50}$	$Ru_{75}Zr_{25}$
77 K	AMOR.	AMOR.	AMOR.+H.C.P. Ru
370±10K	AMOR.	AMOR.	AMOR.+H.C.P. Ru
410±10k	H.C.P. Zr (a=3.20,c=5.15) +B.C.C.(a=3.26)	B.C.C. (a=3.23)	AMOR.+H.C.P. Ru

Ru-Ti: The general features of IM results are summarized in Table IIb. Lattice spacings were obtained from x-ray diffraction (Read camera). To study the effect of T_{sub} on amorphous phase formation, samples with an average composition $Ru_{43}Ti_{57}$ were irradiated at different T_{sub} of 120 K, 140 K, 160 K, 180 K to 210 K (error ±5 K) by 300 keV Xe^+ with the dose $8*10^{15}/cm^2$. The T^c_{sub} was again observed, and it found to be around 170 K. Below this temperature, an amorphous phase was formed. Above this temperture and up to 520 K, IM resulted in a b.c.c.phase. The recrystallization temperature of amorphous $Ru_{50}Ti_{50}$ produced by 600 keV Xe^{++} with the dose $8*10^{15}/cm^2$ at 77 K was observed to be 450 ± 20 K.

Table IIb 600 keV Xe ion mixing of Ru-Ti with the dose $8*10^{15}/cm^2$ at different substrate temperatures.

T_{sub}	$Ru_{25}Ti_{75}$	$Ru_{50}Ti_{50}$	$Ru_{70}Ti_{30}$
77 K	B.C.C. (a=3.17)	AMOR.	H.C.P. (a=2.74,c=4.40)
300 K	B.C.C. (a=3.16)	B.C.C. (a=3.10)	H.C.P. (a=2.74,c=4.39)

*The equilibrium structure of compound $Ru_{50}Zr_{50}$ is isotypic to Cs-Cl structure, though the supperlattice lines are not observed by x-ray diffraction due to the similar atomic number of Ru and Zr. We therefore were not able to determine the amount of the atomic site disorder in the $Ru_{50}Zr_{50}$ phase formed by IM. We will simply call this phase b.c.c. $Ru_{50}Zr_{50}$, since the degree of chemical ordering is unknown.

IM at Simple Compound Compositions: The b.c.c. $Ru_{50}Zr_{50}$ phase produced by IM or thermal-annealing was further irradiated by 300 keV Xe^+ to a dose of $8*10^{15}$ Xe^+/cm^2 at 77 K, and an amorphous phase was obtained. Similaly, Cs-Cl $Ru_{50}Ti_{50}$ by thermal-annealing became amorphous by following IM at 77 K.

DISCUSION

From the experimental observations (as summarized in Table II), we may conclude that the T_{sub} is of crucial importance in determining phase formation by IM. On the other hand, we see that by varying T_{sub}, we are able to identify some of the competing phases during IM. For example, at low T_{sub}(e.g. 77 K), the formation of amorphous $Ru_{25}Zr_{75}$ is more favorable than phase separation to h.c.p. Zr plus b.c.c. $Ru_{50}Zr_{50}$ (column 1, Table IIa); amorphous phase formation is more favorable than formation of the single phase crystalline $Ru_{50}Zr(Ti)_{50}$ (column 2 Table IIa,b). This situation persists up to a critical substrate temperature (T^c_{sub}) (where $T^c_{sub} \approx 170.$ K for $Ru_{50}Ti_{50}$; $T^c_{sub} \approx 390$ K for $Ru_{50}Zr_{50}$ and $Ru_{25}Zr_{75}$). Above this T^c_{sub}, formation of the competing crystalline phases become favorable. At the $Ru_{25}Ti_{75}$ and $Ru_{70}Ti_{30}$ compositions, the formation of single-phase crystalline solutions always wins over amorphous phase formation. We see that the amorphous phases compete with either one or two crystalline phases which may not depend on terminal structures. Therefore the structural difference rule [3] does not apply.

In a previous paper, we have proposed the mechanism of amorphous phase formation by compositional induced superheating [6]. The idea is that the local composition is fixed during prompt cascade mixing. Due to extremely fast effective cooling rate and in the absence of radiation enhanced long range diffusion the compositional profile does not change further following this prompt mixing. If the resulting composition is such that the free energy of an amorphous phase is lower than all other crystalline phases, then the amorphous phase will form. Similar idea has been proven succesful in predicting amorphous phase formation by solid-state reaction [11]. This mechanism requires a knowledge of free energy of each of the competing phases. Over the past fifteen years, a significant amount of work has been done in studying thermal stability of equilibrium phases in binary alloy systems [12]. Taking data from Refs [12] and [4], equilibrium free energy diagrams have been constructed in Fig.[3a] and Fig.[3b] for Ru-Zr and Ru-Ti at specified tempertures.Representing the free energy of the amorphous phase by the supercooled liquid [13], we see that the free energy of the amorphous phase is higher than that of solid solutions' at all compositions. Thus the mechanism of amorphous phase formation due to compositional superheating does not account for the present observations. Nevertheless the equilibrium free energy diagrams of Fig.[3a] and Fig.[3b] do predict qualitatively metastable crystalline phases produced by IM. For example,the diagrams predict a large compositional range of b.c.c Ru-Ti solid-solution whereas an h.c.p. solid-solution

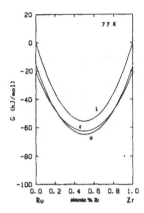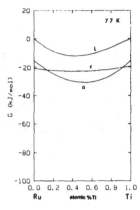

Fig. 3 Free energy
diagrams of Ru-Zr
and Ru-Ti at 77 K.
(L — Liquid (amor-
phous); α — B.C.C.
solid solution; ϵ —
H.C.P. solid solu-
tion).

is predicted to be more stable at high Ru concentration. A
b.c.c. solid solution should form around the 50% composition of
Ru-Zr, etc. These predictions are consistent with the
experiments (Table II). Therefore we may accept that the
equilibrium free energy diagram is relavent to IM processes.
Similar conclusion follows directly from the general discussion
of the applicability of thermodynamics to the late stage of IM
[6]. But why does the free energy diagram not predict amorphous
phase formation in Ru-Zr and Ru-Ti? The existence of a critical
substrate temperature $T^c{}_{sub}$ indicates that a delayed thermally
activated process is occurring after cascade mixing. This kind
of delayed process is usually associated with migration of
defects produced during the cascade. Below $T^c{}_{sub}$ such defects
are traped in the solid and one must take into account their
contribution to the free energy of crystalline phases.

The effect of point defects has been considered as a prime
reason for irradiation induced crystalline phase to amorphous
phase transformaton in Ge,Si [14]. Since the free energy
difference between the crystalline and amorphous phases of Ru-Zr
and Ru-Ti are about 10 and 20 kJ/mol (Fig.[3]), a vacancy
concentration of several percent would reverse this energy
difference (assuming that the vacancy formation energy is 1-2
eV). Although several percent of the vacancy concentrations are
much higher than the equilibrium value at low temperature, it
has been observed by field-ion microscope study of radiation
damage [15]. Furthermore, it has been argued that interstitial
defects may be crucial in destabalizing crystalline phases [16].
In any case, the effect of quenched-in defects seems to be a
necessary condition for amorphous phase formation by IM in
certain systems. Defects should affect the stability of narrow
"line" compound phases more drastically since the narrower the
compound homogeneity range , the greater will be the influence
of atomic defects and site disorder. Thus besides the kinetic
constraints on forming narrow compositional range crystalline
phases, there may also be an absence of thermodynamic driving
forces to form the narrow compounds in a radiation damaged zone.

CONCLUSION

Systematic studies of IM of Ru-Zr and Ru-Ti provide a counterexample to the structural difference rule and the compound structural complexity criteria. This experiment demonstrates that an understanding of the free energy of competing phases including the influence of defects on the free energy of each phase is necessary to explain the results of IM. Furthermore,a more detailed understanding of the kinetic constraints on the nucleation and growth of various phases is required.

ACKNOWLEDGEMENTS

The authors would like to thank A.Ghaffari (Caltech) for sample preparation and Dr.M.Van Rossum for helpful discussions. The authors would like to thank the Department of Energy for partial support under Proj. Agree. No. DE-AT03-81ER10870 under Contract No. DE-AM03-76F00767 and the Office of Naval Research for partial support under Contract No. N00014-84-K-0275.

REFERENCES

1. S.Matteson and M-A.Nicolet, in Am.Rev.Mater.Sci. Vol(18), 339(1983).
2. J.W.Mayer,B.Y.Tsaur,S.S.Lau,andL.S.Hung,Nucl.Instru.Meth. 182/183,1(1981).
3. B.X.Liu,W.L.Johnson,M-A.Nicolet,and S.S.Lau,Appl.Phys.Lett. 42,45(1983).
4. L.S.Hung,M.Nastasi,J.Gyulai and J.W.Mayer,Appl.Phys.Lett. 42,672(1983).
5. Y-T.Cheng,M.Von Rossum,M-A.Nicolet and W.L.Johnson Appl.Phys.Lett.45,185(1984).
6. W.L.Johnson,Y-T.Cheng,M.Van Rossum,and M-A.Nicolet, in IBMM 84 Conference Proceedings,to be published in Nucl.Instr.Meth.B(1985).
7. M.Hanson and K.Anderko,Constitution of Binary Alloy,(McGraw-Hill,New York,1958).
8. R.J.Shunk,Constitution of Binary Alloys. Second Supplement, (McGraw-Hill,New York,1969).
9. J.L.Murray,in Bulletin of Alloy Phase Diagrams,3,216(1982).
10. A.R.Miedema,Philips Tech.Rev.36,217(1976).
11. R.B.Schwarz and W.L.Johnson,Phys.Rev.Lett.51,415(1983).
12. L.Kaufman andH.Bernstein,Computer Calculation of Phase Diagrams,(Acadamic Press,New York,1970).
13. J.Hafner,Phys.Rev.B.21,406(1980).
14. M.L.Swanson,J.R.Parsons and C.W.Hoelke,Radiat.Eff. 9,249(1971).
15. D.Prananik and Seidman,J.Appl.Phys.54,6352(1983).
16. Y.Limoge and A.Barbu,Phys.Rev.B.30,2212(1984).

Schottky Barriers

A COMPARISON OF TRANSIENT ANNEALING METHODS FOR SILICIDE FORMATION

R.E. HARPER, C.J. SOFIELD*, I.H. WILSON AND K.G. STEPHENS
Dept. of Electronic and Electrical Engineering, University of Surrey,
Guildford, Surrey, UK.
*UKAEA Harwell, Didcot, Oxon, UK.

ABSTRACT

Nickel and cobalt silicides have been formed by raster-scanned
electron beam and flash-lamp irradiation of thin metal films on single
crystal (100) and (111) silicon wafers. RBS and channelling measurements
indicate that the $NiSi_2$ is epitaxial and of good crystalline quality
(χ_{min} 4% on (111)); epitaxial $CoSi_2$ was more difficult to form and of
somewhat poorer quality. The elastic recoil technique has been used to
determine bulk and interfacial light element contamination. These
measurements have been correlated with resistivity and SEM studies of the
surface textures.

INTRODUCTION

It is well known that the disilicides of some metals (e.g. Ni, Co,
Fe, Cr) can grow epitaxially on single crystal silicon substrates. Most
research has relied upon conventional furnace annealing [1,2] of thin
metal films on silicon substrates although the best results have been
obtained using sophisticated UHV [3] or liquid phase epitaxy techniques
[4]. However the relatively simple rapid thermal processing methods
which are increasingly being employed for annealing of implanted layers
are also suitable for the formation of both polycrystalline and single-
crystal silicides [5,6].

We have chosen to study the epitaxial growth of two of these
compounds, $NiSi_2$ and $CoSi_2$. Polycrystalline $CoSi_2$ is of course a technologically interesting material because of its low formation temperature
and low resistivity. Furthermore the fact that Co is thought to be the
diffusing species [7] means it would lend itself well to a self-aligned
(salicide) type process [8]. In this work we have used two methods of
rapid annealing, raster-scanned electron beam irradiation and incoherent
light exposure. We have compared the properties of the disilicide layers
thus formed with those already published for furnace-grown material. We
have also investigated the behaviour of light element contaminants
(H,C,O) during silicide growth and have assessed the effect of gross
contamination on both reaction kinetics and the crystallinity and
resistivity of the silicide layer.

EXPERIMENTAL

Commercially available silicon wafers of orientation (100) and
resistivity 4-6Ω cm (p type) or orientation (111) and resistivity 45-50Ω
cm (n type) received a 10 second etch in 10:1 HF followed by rinsing in
deionised water immediately prior to loading into an evaporation system.
Ni films of nominal thickness 300, 660 or 1000Å and Co films of nominal
thickness 500Å were deposited in a vacuum of about 10^{-6} torr. The wafers
were then cleaved into approximately 1 x 1 cm squares for annealing and
analysis.
Annealing was carried out in a vacuum of about 10^{-6} Torr using
either a commercial Lintech electron beam facility or a quartz halogen

flash-lamp system. In the former, the samples were irradiated by a 30 kV electron beam in a vacuum of around 10^{-6} torr. The power density and hence equilibrium temperature was controlled by varying the current. The sample mounting (which has been described elsewhere [9]) was designed to approximate good thermal isolation. Temperatures were measured by IR pyrometry and were in good agreement with calculated values. In the flash-lamp system the mounting was again chosen to minimise conductive heat losses, but here the temperature was controlled using a Pt-Rh thermocouple sandwiched between (but not in contact with) two samples. Some care was taken to ensure that the samples underwent similar thermal cycles in each system. Firstly, since the flash-lamp is limited to about 800°C, only fairly low electron beam power densities were generally used. Secondly, in the electron beam case the rise time depends inversely on the incident power density whereas in the flash-lamp the ramp rate was fixed so the rise time is proportional to the equilibrium temperature; thus annealing times that were long compared to the rise time were used (quoted values refer to the total time, not the dwell time). Finally since two different methods of temperature measurement must lead to some uncertainty a range of thermal cycles were employed in each system in order to ensure some overlap. Typical cycles were 5-10 Wcm^{-2} or 600-750°C with times between 40 and 150 seconds; for such cycles, the rise times in the two systems are less than 10% of the total. In addition some very different electron beam cycles (e.g. 30 Wcm^{-2}, 5 seconds) were tried when studying $CoSi_2$ formation.

The thickness of the deposited metal films and the thicknesses, stoichiometry and crystallinity of the disilicide films were assessed using standard Rutherford Backscattering (RBS) and channelling techniques with 1.5 MeV He ions. Light element analysis was carried out by the elastic reocil (ERDA) technique which has been fully described elsewhere [10]. Briefly, ERDA may be regarded as the inverse of RBS, using a heavy ion incident beam (in this case 30 MeV Cl) and detecting the forward scattered light mass recoil atoms. Selected samples were studied using SEM to examine surface textures. Resistivities were deduced from four-point probe measurements of sheet resistance and the film thicknesses obtained from RBS analysis.

RESULTS

(a) $NiSi_2$ Formation

In the range of thermal cycles studied, two phases of nickel silicide were observed. The formation temperature for disilicide formation depended on substrate orientation: typically, a power density of 5.5 Wcm^{-2} or a flash-lamp temperature of 650°C was sufficient to produce $NiSi_2$ on (111), but a mixture of $NiSi$ and $NiSi_2$ on (100). As soon as the $NiSi_2$ phase was formed there was always some epitaxy, but this improved upon prolonged exposure. For example, using 8 Wcm^{-2} though $NiSi_2$ formed after only 20 s, the χ_{min} only reached its lowest value after 40 s or more. Figure 1 shows our best random and channelled spectra for 1000Å $NiSi_2$ on both (111) and (100) substrates. Similar χ_{min} values (~4% on (111), ~14% on (100)) were obtained using both annealing methods and at a range of temperatures, with the proviso that the lower the temperature the longer the exposure time required to reach this value.

The tail on the low energy side of the Ni signal indicates that the interface is not perfectly sharp. There was some evidence that though prolonged exposure at fixed temperature improved the χ_{min}, the interface

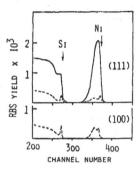

Fig. 1 RBS random and channelled spectra for ~1000Å NiSi$_2$ on <111> and <100> substrates

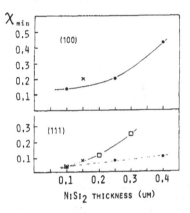

Fig. 2 Dependence of χ_{min} upon film thickness. Solid circles denote this work, crosses are from [6] and open squares from [1]

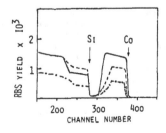

Fig. 3 RBS spectrum for polycrystalline CoSi$_2$ (solid line) and random and channelled spectra (broken lines) for partially epitaxial CoSi$_2$

degraded. Therefore a few electron beam irradiations were carried out using very short times (5-10 seconds) at higher temperatures than were possible in the flash-lamp system. These indeed produced films with sharper interfaces but with poor χ_{min} values. At such high temperatures, increased exposure time again reduced the χ_{min} but caused extremely ragged interfaces and poor stoichiometry: the films appeared silicon-rich with surface textures indicative of island formation.

The dependence of χ_{min} upon film thickness is shown for both substrate orientation in Figure 2. Some points have been added from published work using incoherent light [6] and furnace annealing [1] for comparison. Clearly while the results are similar for thin films the transient annealing methods give much lower χ_{min} values for thicker films on (111).

(b) CoSi$_2$ Formation

The results obtained for CoSi$_2$ formation were considerably different from those presented above. Using a similar range of thermal cycles as for NiSi$_2$ formation, both annealing techniques always produced polycrystalline CoSi$_2$ films as shown by the solid line in Figure 3. The interface is however quite sharp, and the resistivities typically only ~15 μΩcm. Some further electron beam irradiations at very high power densities were more successful in that some epitaxy was obtained with 25 Wcm^{-2} and above. However as can be seen from the spectra represented by broken lines in Figure 3, the films were no longer stoichiometric CoSi$_2$, the interface is very poor and the crystallinity much worse than was observed for NiSi$_2$. The lowest χ_{min} values obtained were 40% on (111) and 70% on (100). In general the higher the power density and/or the longer the irradiation time the more silicon-rich the film appeared and the longer the low-energy tail on the Co signal. Whether this tail truly represents interfacial raggedness is not clear, since these films all had very rough surfaces as shown in Figure 5(c).

(c) Light Element Analysis

ERDA measurements were made on samples from all the wafers before and after compound formation. Figure 4 shows examples for two Ni films

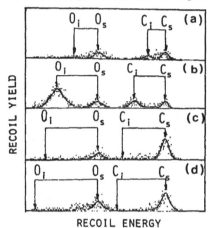

RECOIL ENERGY

Fig. 4 ERDA spectra for (a) and (b) ~300Å Ni films on <111> substrates, and (c) and (d) their corresponding disilicides. Subscripts denote surface and interface O and C positions

of nominal thickness 300Å deposited on (111) substrates and their corresponding disilicides. The calculated positions of surface and interfacial O and C are indicated. The differences are striking; Fig. 4(a) shows a 240Å film with a fairly clean interface and some bulk contamination whereas Fig. 4(b) indicates a heavily contaminated interface with an oxygen level of about 2.7×10^{16} at.cm^{-2}. This is probably a consequence of poor substrate cleaning prior to metal deposition. Unfortunately all the Co films showed similar high contamination levels. Figs. 4(c) and (d) show the ERDA spectra after silicide formation: the light elements have been swept out, even when the original levels were very high, leaving predominantly surface peaks but with some O and C trapped in the

bulk. The total oxygen content has decreased by about 65%; in contrast the total carbon content has increased slightly, presumably owing to surface contamination during annealing. No significant carbon build-up was observed after prolonged exposure to the Cl beam. The greater thickness of the Co films produced some difficulties as the O and C spectra overlapped after silicide formation so that quantitative information was hard to obtain. It was possible to see however that again the total O content decreased while the C surface peak increased.

The effect of oxygen contamination on the formation temperature and properties of $NiSi_2$ may be seen from Table 1. Clearly low levels have little effect on formation temperature, while gross contamination inhibits the reaction and degrades the physical properties. The surface textures also correlated with contamination: Figures 5(a) and (b) show SEM micrographs for $NiSi_2$ grown from the films represented by the spectra shown in Figures 4(a) and (b) respectively. The polycrystalline $CoSi_2$ showed a rather similar surface to that shown in Figure 5(b), again probably because of contamination: the best epitaxial ($\chi_{min} \sim 40\%$) film had similar type of texture but with larger feature sizes as shown in Figure 5(c).

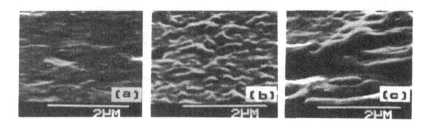

Fig. 5 SEM micrographs for (a) and (b) NiSi$_2$ grown from the films
represented in Fig. 4(a) and (b) respectively, and (c) epitaxial CoSi$_2$

Table I

Properties of NiSi$_2$ Grown from 300Å Ni Films on Substrates Showing

Varying Levels of Light Element Contamination

Substrate Orientation	Interfacial Oxygen 10^{15} at.cm^{-2}	Lowest Formation Temp (°C)	Lowest χ_{min}(%)	Resistivity ($\mu\Omega$cm)
(100)	2.9	700	15	42
(100)	$\lesssim$1	700	14	37
(111)	$\lesssim$1	650	4	34
(111)	27	720	25	70

DISCUSSION AND CONCLUSIONS

The role of light element contamination in silicide formation is
well known [11]. We have demonstrated that (at least for rapid annealing
methods) low levels have little effect on either reaction kinetics or
film properties, gross interfacial contamination not only inhibits NiSi$_2$
growth but degrades the crystallinity, resistivity and surface texture.
We have also observed that both oxygen and carbon are swept out during
both NiSi$_2$ and CoSi$_2$ formation, with the total oxygen content being
reduced by up to 65%. This appears to cast some doubt upon the usefulness of implanted O^{18} as a marker for identification of the dominant
diffusing species during compound formation [7]. The ERDA analysis
further showed that the dependence of NiSi$_2$ upon substrate orientation
was not a consequence of differences in light element contamination.

The fact that while polycrystalline CoSi$_2$ forms at temperatures as
low as 600°C much higher temperatures are required to obtain epitaxial
growth is consistent with other annealing methods [1,4]. The ERDA
analysis suggests that the reason for our high χ_{min} values is the high
level of contamination and we expect therefore to obtain better results
using cleaner films. Comparison with NiSi$_2$ formed from wafers showing
similar high contaminant levels indicate that epitaxial CoSi$_2$ growth is
more susceptible to such contamination than NiSi$_2$.

A comparison of the two annealing techniques for NiSi$_2$ formation
gave essentially identical results in terms of χ_{min}, resistivity and
surface textures, which were comparable to the best obtained using
furnace annealing [1]. It has previously been reported that on (100)

substrates conventional techniques yield discontinuous films with very
ragged interfaces while films on (111) substrates are continuous [2].
Though our films did not have perfectly sharp interfaces the RBS spectra
indicate that there was no difference in the interfaces obtained for the
two orientations, and that particularly for (100) the interfaces are
sharper than obtained using furnace annealing. We conclude therefore
that transient annealing may allow growth of continuous films
even on (100) without resort to UHV conditions [3] or liquid phase
epitaxy [4]. Cross sectional TEM studies are under way to further
investigate this possibility.

ACKNOWLEDGEMENTS

The authors wish to acknowledge Dr. J.M. Shannon, Philips Research
Laboratories, Redhill, Surrey for providing nickel films and for useful
discussions, and Dr. E.A. Maydell-Ondrusz, University of Surrey, for
discussions and help in preparing diagrams. One of us (REH) would like
to thank Philips Research Labs for funding of a research fellowship.

REFERENCES

[1] S. Saitoh, H. Ishiwara, T. Asano and S. Furukawa, Jpn. J. Appl.
     Phys. 20, 1649, 1981.
[2] K.C.R. Chiu, J.N. Poate, J.E. Rowe, T.T. Sheng and A. Cullis, Appl.
     Phys. Lett. 38, 988, 1981.
[3] R.T. Tung, J.M. Gibson and J.M. Poate, Phys. Rev. Lett. 50, 429,
     1983.
[4] R.T. Tung, J.M. Gibson, D.C. Jacobson and J.M. Poate, Appl. Phys.
     Lett. 43, 476, 1983.
[5] R.E. Harper, E.A. Maydell-Ondrusz, I.H. Wilson and K.G. Stephens,
     Mat. Res. Soc. Symp. Proc. 25 51, Elsevier Science Publishing Co.
     Inc., 1984.
[6] A. Nylandsted Larsen, J. Chevallier and G. Sorensen, ibid, 23, 727
[7] C.D. Lien, M. Bartur and M.A. Nicolet, ibid, 25, 51.

[8] F.M. D'Heurle, Proc. 1st Int. Symp. on VLSI Science and Technology,
     Detroit, 1982, 194.
[9] E.A. Maydell-Ondrusz, R.E. Harper, A. Abid, P.L.F. Hemment and
     K.G. Stephens, Proc. Mat. Res. Soc. Symp. Proc. 25, 99.
[10] P.M. Read, C.J. Sofield, M.C. Franks, G.B. Scott and M.J. Thwaites,
     Thin Solid Films 110, 251, 1983.
[11] C. Canali, F. Catelline, G. Ottiavani and M. Prudenziati, Appl.
     Phys. Lett. 33, 187 1978.

REDISTRIBUTION AND INFLUENCE OF ARSENIC IN CHROMIUM SILICIDE FORMATION

L. R. ZHENG, L. S. HUNG AND J. W. MAYER
Department of Materials Science and Engineering
Cornell University, Ithaca, NY 14853

ABSTRACT

The redistribution of arsenic during $CrSi_2$ formation and its influence on the growth rate of the silicide have been investigated with Rutherford backscattering and ion channeling spectroscopy and electron microscopy. Arsenic was introduced by implantation in the metal films or in the silicon substrates. When arsenic was initially in chromium, it was incorporated in $CrSi_2$ during silicide formation and significantly reduced the reaction rate; when arsenic was initially in silicon, it accumulated at the silicon/silicide interface with a less pronounced retarding effect than that if arsenic was present in chromium. The redistribution of dopant atoms is attributed to the fact that silicon is the dominant moving species in $CrSi_2$ formation. The influence of dopant atoms is related to their chemical and physical state.

INTRODUCTION

The redistribution of implanted dopant atoms in silicide formation has attracted much interest during the past few years due to its important implication for shallow junction device technology [1-3]. Silicides are used as ohmic contacts, Schottky barriers and interconnects. In numerous technological solutions, metal films are deposited on previously implanted silicon surfaces. The dopant atoms may segregate or may be snowplowed during silicide formation. This redistribution alters the doping profile and affects shallow junction device characteristics.

Previous experimental results showed that during Pd_2Si formation a fraction of the dopant atoms have been pushed into silicon by the advancing silicon/silicide interface and part of the redistributed dopant atoms occupies substitutional lattice sites [4]. The snowplow process has not been found in the formation of refractory silicides (such as $TaSi_2$ and $TiSi_2$), where arsenic atoms diffuse out and accumulate near the surface.

In the present work we investigated the redistribution of arsenic during $CrSi_2$ formation with arsenic initially located in chromium or in silicon for a better understanding of the dopant behavior and its effect on silicide formation. Unimplanted and krypton implanted samples were used for comparison.

EXPERIMENTAL PROCEDURE

Commercially available n-type Si<100> wafers of 1 Ωcm resistivity were used throughout this work. Following standard cleaning, half of the wafers were immediately loaded into an oil-free e-beam evaporation system for depositing chromium films of 1100 Å. The rest of the samples were implanted with As+ at 200 keV to a dose of $1x10^{16}$ ions cm^{-2}. The wafers were then plasma-oxidized to remove a carbon layer formed during implantation. Lattice damage was removed by thermal annealing in dry oxygen ambient at 950°C for 30 min. Prior to the chromium deposition, the wafers were dipped in diluted HF to etch off oxide. Chromium films of 850 Å in thickness were evaporated on the wafers at pressures in the low 10^{-7} Torr range. Some chromium films on undoped silicon substrates were implanted with As+ and Kr+ at 285 keV to doses of $5x10^{15}$ and $1x10^{16}$ ions cm^{-2}. This energy corresponds

580

to a projected range of Rp ~680 Å with ΔRp ~160 Å for the arsenic concentration profile in the chromium film and Rp ~630 Å with ΔRp ~140 Å for krypton.
Silicide formation took place by thermal annealing in a vacuum furnace with
5×10^{-8} Torr pressure. Analysis of the samples were performed with Rutherford backscattering and ion channeling spectroscopy. Transmission electron
microscopy was employed to identify the growing phases and examine their
microstructures. The thickness of the silicide layer as a function of annealing time was determined with computer simulation of the corresponding backscattering spectra.

RESULTS AND DISCUSSION

1) Dopant Redistribution

The backscattering spectra of a sample implanted with 5×10^{15} As+ cm^{-2}
in chromium before and after anneal at 475°C for 30 min are shown in Fig. 1,
where only the chromium and arsenic profiles are present for clarity. The
arsenic was initially imbedded in the chromium film with a peak concentration of 0.7 at.%. After annealing, a $CrSi_2$ layer of 1250 Å has formed. As
the chromium was consumed in the silicide, a large fraction of the arsenic
originally located in that chromium layer was incorporated in the silicide
and some arsenic moved toward the silicon/silicide interface.

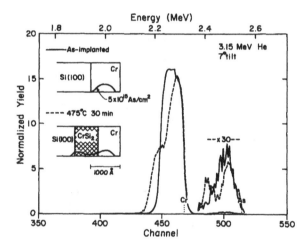

Fig. 1. Ion backscattering spectra of Cr/Si(100) samples implanted with
5×10^{15} As cm^{-2} in the chromium films before and after annealing at 475°C for
30 min. For clarity, the silicon signals are omitted.

The redistribution of arsenic during annealing was completely different
when the arsenic initially was in the silicon substrates. The backscattering
spectra in Fig. 2 showed the formation of 1600 Å $CrSi_2$ after 475°C anneal for
30 min. The arsenic profile of the annealed samples indicated that the front
of the implanted arsenic piled up near the silicon/silicide interface. Since
it is difficult to extract information on the exact depth of the arsenic redistribution from Fig. 2, the annealed sample was re-analyzed after removal
of the unreacted chromium and chromium-disilicide. The sample normal was at
75° relative to incident beams in order to improve depth resolution. Backscattering measurements showed that the arsenic was located at the silicon

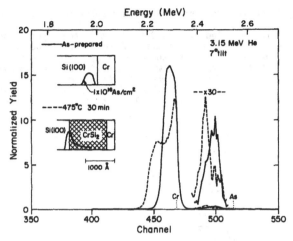

Fig. 2.   Ion backscattering spectra of chromium films on arsenic-doped silicon substrates before and after annealing at 475°C for 30 min.

surface, i.e. the silicon/silicide interface in the unetched sample. To extract information on arsenic lattice location before and after silicide formation, channeling measurements were carried out in an unannealed sample
after stripping off the chromium film. The depth of the arsenic profile extended over 1500 Å in silicon with a peak concentration of 0.7 at.% (Fig.
3a). It was found that the initial amount of arsenic had dropped to about
60% of the implant dose of 1×10^{16} ions cm^{-2} due to the arsenic loss during
950°C anneal and the etching in diluted HF. The amount of substitutional
arsenic was deduced to be around 75%. Channeling measurements of the annealed sample revealed that a total amount of 2×10^{15} arsenic cm^{-2} accumu-

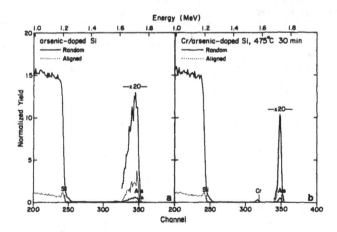

Fig. 3.   Channeling effect measurement of chromium films on arsenic-doped
silicon substrates under various processing conditions: a) before annealing, b) after annealing at 475°C for 30 min. The surface layers were moved
before channeling analysis.

582

lated at the silicon/silicide interface with a substitutionality less than 5% (Fig. 3b).

For the sample implanted with 1×10^{16} kr+ cm^{-2}, the krypton concentration reached 1.4 at.% of chromium at the peak of its distribution. After anneal at 475°C for 30 min, the krypton originally in the reacted chromium film was incorporated into the $CrSi_2$ with a lowering and some broadening of its initial distribution (Fig. 4). When all the chromium film was converted into $CrSi_2$ upon prolonged annealing, the krypton was uniformly distributed in the entire silicide layer and the total amount of krypton was conserved.

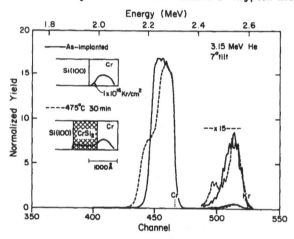

Fig. 4. Ion backscattering spectra of Cr/Si(100) samples implanted with 1×10^{16} Kr cm^{-2} in the chromium films before and after annealing at 475°C for 30 min.

The understanding of the impurity redistribution can be approached based on the moving species during silicide formation and the diffusivity of the impurity in silicides. It is known that in $CrSi_2$ formation, silicon is the dominant moving species [5]. Our recent observation of implanting arsenic in a completely reacted $CrSi_2$ layer revealed low diffusivities of arsenic in $CrSi_2$ at temperatures up to 500°C [6]. Therefore, during $CrSi_2$ formation, the arsenic originally in the silicon substrates is left behind by the moving silicon. It consequently segregated and remained at the silicon/silicide interfaces. When arsenic or krypton was initially present in chromium, the silicon moved into the chromium film and diluted the chromium as well as the impurity concentration. The redistribution, however, is slightly different between arsenic and krypton. Krypton was uniformly distributed in $CrSi_2$, whereas arsenic accumulated near the silicon/silicide interface along with its incorporation in $CrSi_2$. This difference is apparently due to the fact that inert gas has a negligible diffusivity in refractory silicides. This argument is consistent with our observation that argon in the cosputtered refractory silicide remained stationary upon annealing at temperatures up to 900°C [6].

The accumulation of doped arsenic differs from the previous observations in Ti/Si and Ta/Si systems where the arsenic diffuses out the silicide and accumulates at the surface [4]. We believe that the arsenic in Ti- and Ta-silicides is mobile at silicide formation temperatures of above 650°C. Therefore, as soon as the arsenic is released from the substrate, it diffuses throughout the silicide layer without interfacial accumulation. The interfacial segregation of dopant arsenic in the present case also differs from the results reported by Wittmer who found that part of the redistributed arsenic atoms has been pushed into silicon during Pd_2Si formation and the occupancy of substitutional sites is around 50% [4]. It was suggested that the enhanced diffusion is attributed to the generation of point defects in silicon due to the interstitial metal-atoms motion.

2) **Effect on Growth Kinetics**

The growth of $CrSi_2$ at 475°C has been studied for all implanted and unimplanted samples. The results are presented in Fig. 5, where the thickness of the silicide is plotted as a function of annealing time. The initial $CrSi_2$ growth rate in the unimplanted sample was consistent with that reported in ref. 7, whereas the presence of impurities significantly decreased reaction rates. The effect of arsenic on the growth rate of $CrSi_2$ is more pronounced when the arsenic is in the chromium films than when it is in the silicon substrates. An experiment was carried out to determine the chemical state of the arsenic in chromium or in silicon. Single crystal silicon wafers and thin chromium films on SiO_2 substrates were subjected to an arsenic ion implantation with a doping level of ∿4%, and then annealed at 500°C for 30 min. For chromium, we observed the superimposition of two patterns by electron diffraction: one in agreement with the pure chromium pattern and the other corresponding to CrAs reflections. In contrast, no Si-arsenide was formed in the implanted silicon. It appears that the extent of the retarding effect is possibly related to the chemical state of the impurity.

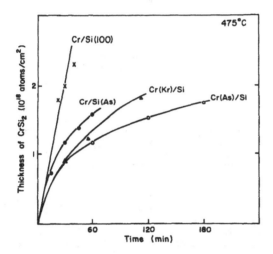

Fig. 5. The thickness of $CrSi_2$ versus anneal time at 475°C for unimplanted and implanted samples.

It has been suggested that two major parameters determine the effect of impurities on metal silicide formation [8]. The first parameter is the initial location of the impurity relative to the moving species, and the other is the chemical affinity of the impurity to the two species. Based on that model, one would not expect a slow down effect on $CrSi_2$ formation in the presence of krypton in chromium, since krypton does not build up at an interface and it is inert with respect to its surrounding. However, the reaction in Kr-implanted samples exhibited a lower rate than that in unimplanted samples. A similar observation has been reported by Lien et al. [9], who

584

found that the growth rate of Co-silicides was reduced for cobalt films on Xe-implanted silicon substrates and speculated that the effect is possibly due to the microstructure changes of Co-silicides resulting from Xe-implanted silicon. The microstructures of $CrSi_2$ formed with pure or Kr-implanted chromium films on undoped silicon substrates were examined with transmission electron microscopy after removal of the unreacted chromium surface layer. We found that the difference in microstructures was not obvious. In the previous study of Ar-implanted amorphous silicon, argon bubbles were formed and grew in size upon annealing [10]. The formation of argon bubbles significantly reduced the epitaxial growth rate. Therefore we speculate that the krypton in $CrSi_2$ may form bubbles and acts as a sink for defects resulting retardation of silicon transport. Further experiments are in progress to determine the growth kinetics of $CrSi_2$ in the presence of krypton and the mechanisms responsible for the retardation.

SUMMARY

When the arsenic was implanted in the silicon substrate, the impurity accumulated at the Si/silicide interface and decreased the growth rate of $CrSi_2$. When the impurity (arsenic or krypton) was imbedded in the chromium film, it was incorporated in the silicide and more significantly reduced the growth rate of $CrSi_2$. The redistribution of impurities was interpreted in terms of the moving species in $CrSi_2$ formation and the diffusivity of impurities in $CrSi_2$. The retarding effect of impurities is related to their chemical state and their physical state at microstructural level.

ACKNOWLEDGMENTS

The work was supported in part by Semiconductor Research Corporation (SRC). Ion implantation was carried out at the National Submicron Facility (NSF).

REFERENCES

1. M. Wittmer and T. E. Seidel, J. Appl. Phys. 49, 5827 (1978).
2. I. Ohdomari, K. N. Tu, K. Suguro, M. Akiyama, I. Kimura, and K. Yoneda, Appl. Phys. Lett. 38, 1015 (1981).
3. L. R. Zheng, L. S. Hung, J. W. Mayer and K. W. Choi, Nucl. Instr. and Meth. (in press).
4. M. Wittmer and K. N. Tu, Phys. Rev. B 29, 2010 (1984).
5. M.-A. Nicolet and S. S. Lau, in VLSI Electronics; Microstructure Science, series edited by E. Einspruch (G. Larrabee, Guest Editor, Academic, NY, 1983), Vol. 6, Chap. 6.
6. L. R. Zheng, L. S. Hung, and J. W. Mayer (unpublished).
7. J. O. Olowolafe, M.-A. Nicolet, and J. W. Mayer, J. Appl. Phys., 47, 5182 (1976).
8. D. M. Scott and M.-A. Nicolet, Nucl. Instr. and Meth. 182/183, 655 (1981).
9. C.-D. Lien and M.-A. Nicolet, in Thin Films and Interfaces II, edited by J.E.E. Baglin, D. R. Campbell and W. K. Chu (Elsevier, NY 1984), p. 131.
10. P. Revesz, M. Wittmer, J. Roth, and J. W. Mayer, J. Appl. Phys. 49, 5199 (1978).

METAL RICH SILICIDE FORMATION BETWEEN THIN FILMS
OF VANADIUM AND AMORPHOUS SILICON

P.A. Psaras, M. Eizenberg[†] and K.N. Tu
IBM Thomas J. Watson Research Center
Yorktown Heights, N.Y. 10598

†Dept. of Materials Engineering and Solid State Institute, Technion, Israel
Institute of Technology, Haifa 32000, Israel.

EXTENDED ABSTRACT:

The formation of silicides via a solid state reaction of silicon and transition
metal thin film is characterized by the sequential appearance of the various
compounds rather than the simultaneous formation of all possible compounds
[1]. Thin films of refractory metals will form disilicides on a single crystal
silicon substrate (unlimited supply of silicon) and since they are stable with
silicon, no other silicides will form. This situation is illustrated in Fig. 1a) by the
formation of VSi_2 in the case of vanadium on silicon. On the other hand, which
silicide will form in the case where there is an excess amount of refractory metal
is unknown. The objective of this work was to study this case utilizing the
vanadium silicon binary system [2].

Thin films of vanadium (1900 and 2500Å in thickness) followed by a 900Å
layer of amorphous silicon were evaporated consecutively via an electron beam
on thermally grown SiO_2 and sapphire (Al_2O_3) substrates, respectively. The
corresponding atomic ratios of vanadium to silicon were 3:1 and 4:1. Heat
treatment was carried out in a purified helium tube furnace. Composition
profiles were measured by Rutherford backscattering spectroscopy (RBS)
utilizing 2.3MeV $^4He^+$ ions and the various compound crystal structures were
determined by Seemann Bohlin glancing angle x ray diffraction [3,4].

The first compound, VSi_2, started to form at 475°C and grew in a planar
form at the vanadium/amorphous silicon interface until all the silicon was
consumed. This process is schematically illustrated in the upper part of Fig. 1b).

The thickness of the VSi_2 layer as obtained from RBS analysis increased linearly
as a function of heat treatment time in the temperature range of 475 to 525°C.
An Arrhenius plot of the growth rate dependence on the reciprocal temperature
(Fig. 2) yields an activation energy of 2.3 ± 0.4eV.

At temperatures higher than 750°C a second silicide (V_5Si_3) grew in a
planar form at the interface between VSi_2 and vanadium until all the VSi_2 was
consumed. (see Fig. 1b). Here again the growth was linear with time, and the
Arrhenius plot (Fig. 2) yields an activation energy of 2.5 ± 0.1eV. For heat
treatment temperatures higher than 725°C, the V_5Si_3 layer interacted with the
excess unconsumed vanadium layer to form a second metal rich compound, V_3Si.
This interaction was nonplanar and thus the growth kinetics of V_3Si could not be
determined by RBS.

It has been observed that the vanadium silicides are formed sequentially in the presence of excess metal: VSi_2, V_5Si_3 and V_3Si. The first compound VSi_2 is the same as that obtained for reactions of vanadium on single crystal silicon (unlimited supply of silicon). All other vanadium rich compounds in the phase diagram of vanadium crystalline silicon [5] except V_6Si_5 have been observed in the later stages of reaction. In previous cases of vanadium on crystalline silicon both linear [6] and parabolic [7] growth rates for VSi_2 were reported. Our results for the vanadium and amorphous silicon system indicate a linear rate which is typically attributed to interface reaction limited growth processes. In this case it is probably due to the limited release of silicon atoms from the amorphous silicon layer. The activation energy for VSi_2 growth (2.3eV) has recently been confirmed by a different technique, the dependence of electrical resistivity on heat treatment temperature [8].

FIGURES

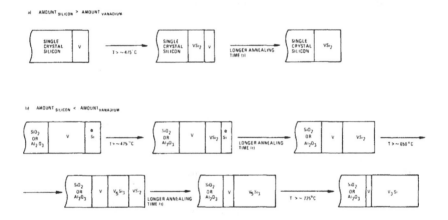

Fig. 1. Schematic diagram depicting reaction progression of a) Thin film vanadium on single crystal silicon (unlimited silicon supply case) and b) Thin film bilayer of amorphous silicon on a thicker vanadium layer (limited silicon supply case).

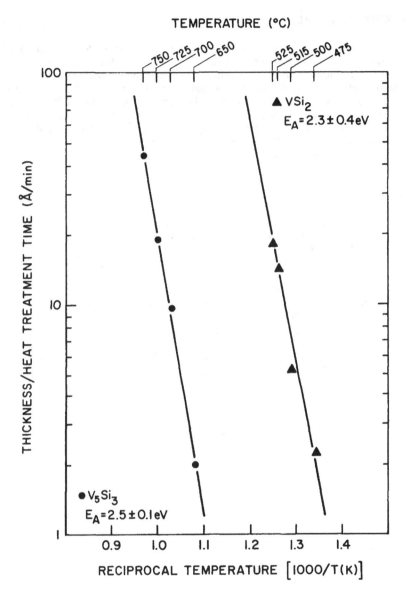

Fig. 2.   Arrhenius plots of (thickness/heat treatment time) vs. reciprocal temperature for VSi$_2$ and V$_5$Si$_3$ growth.

ACKNOWLEDGEMENTS

The authors gratefully acknowledge the Central Scientific Service Materials Laboratory staff for specimen preparation, P.A. Saunders (Ion Beam Technology Group) for Rutherford backscattering spectroscopy exposures, A.A. Levi (Microstructural Studies Group) for various computer programs, C. Palmerstrom (Materials Science Department, Cornell University) for additional Rutherford backscattering spectroscopy exposures, R.D. Thompson, (Submicron Structure Group), G. Ottaviani (Department of Physics, University of Modena, Italy) for helpful discussions and G.B. Stephenson (Microanalysis and Microscopy Group) for reviewing the manuscript.

REFERENCES

1. K.N. Tu and J.W. Mayer, Chap. 10, "Thin Films Interdiffusion and Reactions", ed. by J.M. Poate, K.N. Tu and J.W. Mayer, John Wiley and Sons, New York, (1978).
2. P.A. Psaras, M. Eizenberg and K.N. Tu, J. Appl. Phys., in press.
3. W.K. Chu, J.W. Mayer and M.A. Nicolet, "Backscattering Spectrometry", (Academic Press, New York, 1978).
4. R. Feder and B.S. Berry, J. Appl. Crystallogr. *3*, 372 (1970).
5. W.G. Moffatt, "The Handbook of Binary Phase Diagrams", (General Electric, new York, 1978), Diagram revised in 1983.
6. H. Kraütle, M.A. Nicolet and J.W. Mayer, J. Appl. Phy., *45*, 3304 (1974).
7. K.N. Tu, J.F. Ziegler and C.J. Kircher, Appl. Phys. Lett. *23*, 493 (1973).
8. F. Nava, P.A. Psaras, H. Takai and K.N. Tu, unpublished.

SOLID-PHASE OHMIC CONTACTS TO GaAs WITH TiN DIFFUSION BARRIERS

H. P. KATTELUS*+, J. L. TANDON**, A. H. HAMDI*, and M-A. NICOLET
*California Institute of Technology, Pasadena, California 91125
**Applied Solar Energy Corporation, City of Industry, California 91749
+Permanent Address: Semiconductor Laboratory, Technical Research Centre
of Finland, Otakaari 5A, SF-02150 Espoo 15, Finland

ABSTRACT

We report on contacts to p-type GaAs formed by a GaAs/Pt/TiN/Ag system.
Ohmic behavior in this system is believed to be accomplished by the solid-
state reaction of Pt with GaAs. This reaction is confined by the TiN film
which is thermally stable. In addition, the TiN film acts as an excellent
diffusion barrier in preventing the intermixing of the top Ag layer with
GaAs or Pt. Contacts formed with such controlled reaction have important
implications for the stability of shallow p-n junction devices.

INTRODUCTION

Ohmic contacts to GaAs are generally made by using eutectic-type
metallization compositions. A typical scheme involves a metal (e.g. Au, Ag)
combined with a dopant element (e.g. Ge, Si, Te, etc. for n-type and Zn, Be,
etc. for p-type) [1,2], and possibly other metals as well. The ohmic
behavior in these systems is claimed to be realized by the incorporation of
the dopant in GaAs upon heating above the eutectic temperature. The alloyed
contacts so formed possess inherent problems because of the possible melting
involved, and the subsequent quenching of the reaction. Lateral inhomogeneities are typical in such contacts. The transient nature of the formation
process promotes the irreproducibility and instability of these contacts. In
addition, in the liquid phase, the penetration of the metals in GaAs cannot
be controlled precisely. Thus in this perspective, alloyed contacts often
cause the shorting of p-n junctions.

To obtain stable and reproducible ohmic contacts, a process based on
solid-phase reaction offers obvious advantages [3]. Ohmic contacts to p-
type GaAs formed by GaAs/Pt/TiN/Ag system are discussed in this paper. In
general, the presence of an interposed barrier layer such as TiN is important
in two respects: (a) the layer confines the reaction between Pt and GaAs,
and (b) the layer acts as an excellent diffusion barrier in preventing the
top Ag layer from interfering with the Pt/GaAs reaction.

EXPERIMENTAL

Contacts were made to Zn-doped p-type layers. The layers were grown by metalorganic chemical vapor deposition (MOCVD) on <100> n-type GaAs substrates. The hole concentrations in the layers are $\sim 2 \times 10^{18}$ cm^{-3} ("lightly" doped layers), and their thicknesses are 1-2 μm. In certain cases, the top ~ 1000 Å of the layers were additionally doped to a level of $\sim 5 \times 10^{19}$ cm^{-3} ("heavily" doped layers). Platinum was chosen as a contact metal because of its known reaction with GaAs in the solid-phase [4]. To attempt an enhancement of the surface doping of GaAs, in some cases thin (~ 100 Å) layers of Mg were also interposed between GaAs and Pt. The films of Pt were deposited by e-beam evaporation, whereas Mg and Ag were deposited by RF magnetron sputtering in Ar ambient. Films of TiN were prepared by reactively sputtering Ti in a premixed gas of 80% Ar and 20% N$_2$. During TiN deposition, the substrates were negatively biased with respect to the sputtering chamber to improve the quality of the films [5]. The thicknesses of the films were calibrated by backscattering spectrometry (BS), assuming bulk density values [6].

For contact resistivity measurements, patterns were made photolithographically by lift-off, conforming to the circular transmission line model [7]. Parallel samples were also prepared to monitor the reactions in the contact system by BS. The annealings of the contacts were carried out in vacuum (7×10^{-7} torr) in the temperature range 350-550°C, and for times up to 30 min.

RESULTS

Backscattering Spectrometry Measurements

The reaction of Pt with GaAs in GaAs/Pt/TiN/Ag contacts was monitored as a function of annealing cycles by BS. In Fig. 1, spectra obtained from these contacts are shown before and after annealings for 30 min at 350°C and 550°C. An important feature of Fig. 1 is that no significant change is observed after annealing in the high energy step of the Ag signal and in the low energy step of the Ti signal. This fact implies thermal stability of the TiN film. The role of the TiN film in preventing gross reaction of the top Ag layer with Pt is also exemplified by the absence of a shift in the leading edge of the Pt signal. However, a small amount of Ag penetration into the TiN layer cannot be ruled out from Fig. 1, because the Pt signal interferes with the trailing edge of the Ag signal. To further substantiate the absence of significant Ag movement, measurements were carried out on samples similarly prepared without Pt. The results obtained from these samples confirmed the diffusion barrier properties of the TiN films.

Referring back to Fig. 1, upon heating, the Pt signal is found to shrink progressively in height and broaden toward low energies. Knowing

that the Ag film remains unchanged, this result can be attributed to the
reaction of Pt with GaAs. To further understand this reaction, the top
unreacted Ag layer was selectively etched with a solution of 1 H_2O_2: 1 NH_4OH:
15 H_2O. Upon removal of Ag the characteristic golden color of the stoichiometric TiN below was revealed. BS measurements were then made on these
samples without Ag. The spectra obtained are shown in Fig. 2. Because
removing the Ag film improves the resolution, the steps in the signals of
the reacted Pt and GaAs are much clearer in this figure, when compared to
Fig. 1. These steps are due to the formation of compounds $PtAs_2$ and PtGa
[4]. The similarity between the spectra shown here and those given in
Ref. [4] implies that the reaction of Pt with GaAs is not affected by the
presence of TiN and Ag overlayers. The TiN film thus simply isolates the
reaction of Pt with GaAs from interference by the top Ag layer.

Contact Resistivity Measurements

Contact resistivity measurements were made on patterns prepared conforming to circular transmission line model [7]. To eliminate measurement
errors due to excessive probe resistance, separate pairs of probes were
used for the voltage and the current measurements. End resistance measurements were also made to calculate the modified sheet resistances below the
contacts, which were considered in the calculations of contact resistivity.

The as-deposited GaAs/Pt/TiN/Ag contacts on lightly doped layers are
non-ohmic. Upon heating at 350°C for 30 min, the contacts become ohmic.
The transition so observed can be attributed to the onset of the reaction
of Pt with GaAs. Measurements made on the resistivity of these contacts as
a function of annealing temperature are shown in Fig. 3 (solid circles).
The contact resistivity is $\sim 10^{-3}$ Ωcm^2 at 350°C and increases somewhat with
increasing annealing temperature. This rise may be correlated with the
effects of reacting Pt, to be discussed later.

The as-deposited GaAs/Pt/TiN/Ag contacts made on heavily doped layers
are ohmic. This is in contrast to the contacts formed on lightly doped
layers, as just described. A substantial decrease in the resistivity of
these contacts is observed after annealing at 350°C for 30 min, as shown by
the band in Fig. 3. Further annealing at higher temperatures does not
reduce the resistivity of the contacts significantly, and the value stabilizes somewhat at $\sim 10^{-4}$ Ωcm^2. These data show that the resistivity of GaAs/
Pt/TiN/Ag contacts can be decreased by raising the surface doping of GaAs.

A similar reduction in the contact resistivity is also observed in the
lightly doped samples when a thin layer of Mg (~ 100 Å) is interposed between
GaAs and Pt. The resistivity of these contacts, also shown in Fig. 3 (stars),

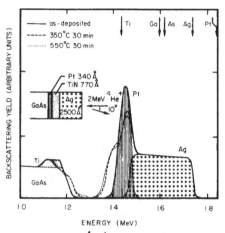

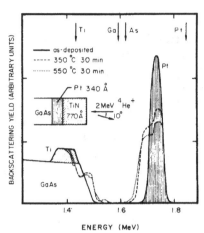

Fig. 1 2 MeV ^{4}He$^+$ backscattering spectra obtained from GaAs/Pt/TiN/Ag contacts before and after annealing.

Fig. 2 2 MeV ^{4}He$^+$ backscattering spectra obtained from the same samples as in Fig. 1 after removing the Ag layer.

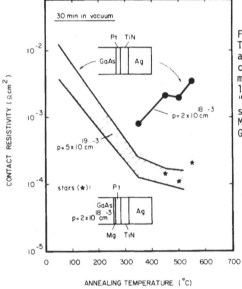

Fig. 3 Resistivity of GaAs/TiN/Ag contacts as a function of annealing temperature. The solid circles represent measurements made on "lightly" doped GaAs layers, the hatched band on "heavily" doped layers, and the stars on contacts with interposed Mg layers between lightly doped GaAs and Pt films.

is comparable to those measured on samples with heavily doped grown layers.

DISCUSSION

The electrical transport properties across a metal-semiconductor contact depend on details of the atomic arrangements at the interface. This well-known fact is convincingly demonstrated by recent studies of the Au-Ge-Ni contact [8] for GaAs, and of the epitaxial $NiSi_2$ contact [9] for Si. Our experimental techniques do not provide the detailed compositional and structural information that is needed to explain the observations reported. However, the reaction of Pt films with GaAs has been characterized by TEM studies before [4]. With this information, we hypothesize that the evolution of the reaction relates to our observations as follows.

The transition from non-ohmic to ohmic behavior after annealing at 350°C for lightly doped GaAs layers is a consequence of the onset of the reaction of Pt with GaAs, resulting in a mixture of $PtAs_2$ and PtGa compounds at the GaAs interface [4] (see Fig. 2). This transition reduces the barrier height between GaAs and the newly formed compounds [3]. Annealing at high temperatures results in the segregation of $PtAs_2$ at the GaAs interface [4], which increases the contact resistivity. These observations suggest that with metal reactions, the barrier height can be changed. The resistivity thus depends on the reaction products directly in contact with the GaAs surface. A very similar correlation between the average composition of compound phases at the interface and the contact resistivity has been reported for the Au-Ge-Ni contact to GaAs [8]. Further evidence about reaction products determining resistivity is provided by comparing GaAs/Ti [10] and GaAs/Pt contacts made on p-type layers with comparable doping concentrations. An order of magnitude higher contact resistivity values are measured for the GaAs/Ti system than for the GaAs/Pt system.

The ohmic character of contacts made on heavily doped layers can be explained by tunneling dominating the current transport across the narrow barrier. The reduction in the contact resistivity by approximately two orders of magnitude upon annealing (see the band shown in Fig. 3) can again be attributed to the decrease in the barrier height. The combined effect of narrowing the barrier and decreasing its height thus results in lowered contact resistivity values. An increase in the contact resistivity upon annealing above 350°C is not observed because the current flow is dominated by the low resistance paths of tunneling.

Upon annealing, a reduction in the resistivity is also observed in contacts made on lightly doped p-type layers with interposed Mg films between Pt and GaAs (see stars in Fig. 3). This reduction can be explained by a doping action of Mg, once again resulting in narrowing the barrier, as well as in reducing its height.

Note that for all ohmic contacts which involve reaction of the metal(s) with a thin conducting surface layer of the substrate, a modification in the sheet resistance below the contacts will take place. This change necessitates a correction which raises the absolute lowest values of the contact resistivities reported here ($\sim 10^{-4}$ Ωcm^2) and may explain the difference with reported values for typical alloyed contact systems, e.g. Au-Zn ($\sim 10^{-5}$ Ωcm^2) [1].

CONCLUSIONS

Stable ohmic contacts to p-type GaAs have been formed using the GaAs/Pt/ TiN/Ag system. The stability of the contacts is accomplished by using the TiN films which act as excellent diffusion barrier layers in confining the solid-phase reaction of Pt with GaAs and in preventing the penetration of Ag. The contact resistivity value is believed to depend on the barrier height of the resulting reacted compound(s) in contact with the GaAs surface. In addition, it is also affected by the surface doping concentration in GaAs. Low values of contact resistivities are achieved on layers grown with high doping concentrations near the surface. Similar reduction in the resistivity values is also accomplished by interposing thin layers of Mg in the metallization system.

ACKNOWLEDGMENTS

The authors would like to thank A. Ghaffari (Caltech) for his assistance with evaporations. The encouragement provided by P. A. Iles (Applied Solar Energy Corporation) is appreciated. The work was supported in part by Sandia National Laboratories under Contract 47-3967 at the California Institute of Technology.

REFERENCES

1. A. Piotrowska, A. Guivarc'h, and G. Pelous, Solid-State Electron. 26, 179 (1983).
2. V. L. Rideout, Solid-State Electron. 18, 541 (1975).
3. A. K. Sinha, Thomas E. Smith, and J. J. Levinstein, IEEE Trans. Electron Devices, 22, 218 (1975).
4. A. K. Sinha and J. M. Poate in, Thin Films - Interdiffusion and Reactions, J. M. Poate, K. N. Tu, and J. W. Mayer, eds. (Wiley, New York, 1978), p. 407.
5. J.-E. Sundgren, B.-O. Johansson, H. T. G. Hentzell, and S.-E. Karlsson, Thin Solid Films, 105, 385 (1983).
6. CRC Handbook of Chemistry and Physics, 62nd edition, (CRC Press, Boca Raton, Florida, 1981-1982).
7. G. K. Reeves, Solid-State Electron. 23, 487 (1980).
8. T. S. Kuan, P. E. Batson, T. N. Jackson, H. Rupprecht, and E. L. Wilkie, J. Appl. Phys. 54, 6952 (1983).
9. R. T. Tung, Phys. Rev. Lett. 52, 461 (1984).
10. M. F. Zhu, A. H. Hamdi, M-A. Nicolet, and J. L. Tandon, Thin Solid Films, (in press).

PHASE TRANSITIONS IN GOLD CONTACTS TO GALLIUM ARSENIDE

EDWARD BEAM III AND D.D.L. CHUNG
Department of Metallurgical Engineering and Materials Science,
Carnegie-Mellon University, Pittsburgh, PA 15213

ABSTRACT

X-ray diffraction was used in situ to study the phase transitions which occurred in 1500 Å Au/GaAs(100) upon heating and cooling. The reaction between Au and GaAs took the form Au + Ga → α Au-Ga. Upon heating, α Au-Ga completely dissolved in liquid Au-Ga. Upon subsequent cooling, β Au-Ga (or Au_7Ga_2) formed. In 1 atm of nitrogen, phase transitions were observed reversibly at $525 \pm 25°C$ (due to the complete dissolution of α Au-Ga upon heating) and $415 \pm 5°C$ (due to the peritectic transformation of β Au-Ga to $α_2$Au-Ga and liquid Au-Ga upon heating). In a vacuum of 425 μ (0.031 Kg/^{2}m) similar phase transitions were observed at $425 \pm 25°C$ and $387 \pm 13°C$, respectively.

INTRODUCTION

Nakanisi's internal friction results on gold thin films on GaAs suggested the occurrence of three melting transitions upon heating in a vacuum of 1×10^{-4} torr [1]. These transitions were labeled T_1, T_2 and T_3. Transitions T_1 and T_2 were observed upon first heating at 440°C and 500°C, respectively; transition T_3 was observed upon second and subsequent heating at 340°C [1]. Moreover, the internal friction results showed that transition T_2 was absent for gold film thicknesses less than 3000 Å so that only transition T_1 was observed upon first heating [1]. As the gold film thickness increased, transition T_2 became more and more significant until T_1 was buried in the shoulder of T_2 in the plot of internal friction vs. temperature [1]. In second and subsequent heating, transition T_3 appeared strongly while transitions T_1 and T_2 were not observed [1].

By using in situ x-ray diffraction during heating and cooling in a pressure of 1 atm, Zeng and Chung [2] observed in 1000 Å Au/GaAs a transition at $478 \pm 22°C$ during first heating and a transition at $413 \pm 5°C$ during second and subsequent heating. The x-ray diffraction results showed that the transition at $478 \pm 22°C$ was due to the melting of a Au-Ga reaction product (tentatively AuGa, 50 at. % Ga) which was formed in the solid state, whereas the transition at $413 \pm 5°C$ was due to the melting of another Au-Ga phase, namely β (or Au_7Ga_2) [2].

In this work, we have used the same x-ray diffraction technique and the same sample preparation technique as Zeng and Chung to study Au/GaAs with a gold film thickness of 1500 Å. Together with the findings reported in Ref. 3, we found that, for 1500 Å Au/GaAs, the reaction took the form Au + Ga → α Au-Ga, where α Au-Ga is the terminal Au-rich solid solution. By in situ x-ray diffraction, we observed that α Au-Ga completely dissolves in liquid Au-Ga (L) upon heating, i.e., α + L → L. This phase transition, though reversible, degrades the contact uniformity, induces phase formation and changes the crystallographic orientation of α [3]. It is therefore of practical importance to investigate the phase transitions in gold contacts to GaAs.

EXPERIMENTAL

Au/GaAs(100) wafers were kindly provided by P. Lindquist of Hewlett-Packard Corporation. The GaAs substrates were Te doped ($2x10^{17}$ -- $3x10^{17}$

596

cm^{-3}), and gold was boat evaporated on to the substrate at 3 Å/sec at a temperature of less than 100°C. The gold thickness was 1500 Å. X-ray diffraction was performed by using a Rigaku D/MAX II powder x-ray diffractometer system and its 1500°C high temperature attachment. The sample chamber was either purged with nitrogen gas or under a dynamic vacuum of 425 ± 25 μ (purged with argon prior to evacuation) during heating. The accuracy of the temperature measurement was ± 5°C. A fine-focus Cu x-ray tube was used. Detection was provided by a scintillation counter. Because a much thicker Al window was used in vacuum than at 1 atm, the diffraction intensities were much lower in vacuum than at 1 atm. Scanning electron microscopy (SEM) was performed at room temperature by using a Cambridge SEM.

Because of experimental limitations, we did not distinguish between Au and α diffraction peaks in this work. However, this distinction was made in our ex situ x-ray diffraction work, which showed that α formed at 350°C or below and had a lattice constant about 0.997 of that of Au [3]. Since β and Au$_7$Ga$_2$ have almost the same lattice constants, distinction between these two phases was not made in this work.

At 1 atm (56 Kg/m^2)

Figure 1 shows x-ray diffraction patterns obtained at room temperature before heating and obtained in situ at 450°C after being held at 450°C for 2 hr. The two patterns are essentially identical. Upon further heating to 550°C, the Au (or α) peaks were observed to vanish completely, as shown in Fig. 2. Although Fig. 1 shows the Au (or α) 111 peak and not the Au (or α) 200 peak, the latter was observed outside the 2θ range of Fig. 1. Upon heating to 550°C, both Au (or α) peaks disappeared, while the GaAs peaks remained. This is attributed to the complete dissolution of α at 525 ± 25°C.

Figure 2 shows a series of x-ray diffraction patterns obtained in situ at different temperatures upon cooling after the first heating. The specimen was cooled stepwise from 550°C to 500, 450, 400 and 350°C, with the time at each temperature being 30 min. The Au (or α) peaks appeared at 500°C and became strong at 450°C. This reversible phase transition which occurs between 500 and 550°C is attributed to the precipitation of α, i.e., L → α + L. Upon further cooling to 400°C, the Au (or α) peaks diminished in intensity, while the β (or Au$_7$Ga$_2$) peaks appeared. At 350°C, the β (or Au$_7$Ga$_2$) peaks became strong. This transition is attributed to the peritectic transformation, i.e., α + L → β. Upon second or subsequent heating, the peritectic transformation of β to α + L and the dissolution of α were again observed at the same respective temperatures.

The reversibility of the peritectic transformation is further shown in Figures 3 and 4. Upon heating (Fig. 3) the β (or Au$_7$Ga$_2$) peaks vanished while the Au (or α) peaks grew; upon cooling (Fig. 4) the β (or Au$_7$Ga$_2$) peaks appeared while the Au (or α) peaks diminished in intensity. That the

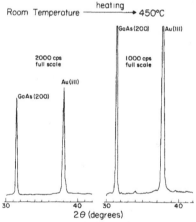

Room Temperature ——heating——→ 450°C

Fig. 1 X-ray diffraction patterns obtained at room temperature before heating and obtained in situ at 450° C and 1 atm after being held at 450° C for 2 hr.

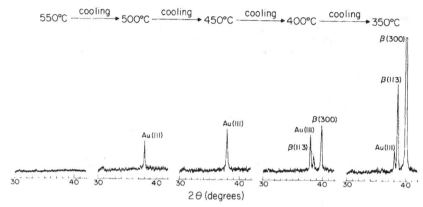

Fig. 2 X-ray diffraction patterns obtained in situ at different temperatures at 1 atm upon cooling after the first heating. The time at each temperature step was 30 min. The GaAs 200 peak is absent due to slight misalignment of this sample.

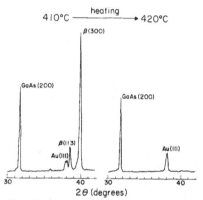

Fig. 3 X-ray diffraction patterns obtained in situ upon heating at 1 atm to show the peritectic transformation of β (or Au₇Ga₂) to α (or Au) and liquid Au-Ga.

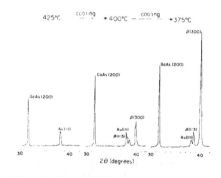

Fig. 4 X-ray diffraction patterns obtained in situ at 1 atm upon cooling to show the reverse of the transformation shown in Fig. 3.

peritectic transformation α + L → β did not cause the Au (or α) peaks to vanish totally was due to the remaining Au (or α), which did not participate in the peritectic transformation. Figure 3 shows that the peritectic temperature is between 410 and 420°C upon heating. Upon cooling, the transformation was observed between 425 and 400°C, as shown in Fig. 4.

In Vacuum (425 μ or 0.031 Kg/m²)

Figure 5 shows x-ray diffraction patterns obtained in situ at 400 and 450°C upon stepwise heating (20-25 min at each temperature step) and at 400 and 350°C upon subsequent stepwise cooling. The Au (or α) peaks observed at room temperature persisted at 400°C, but vanished at 450°C. This is attributed to the complete dissolution of α at 425 ± 25°C. Upon subse-

quent cooling to 400°C, the Au (or α) peaks were still absent. Upon
further cooling to 350°C, the Au (or α) peaks remained absent, but the β
113 and 300 peaks appeared. The irreversibility of the dissolution of α
is in contrast to the reversibility of this transition at 1 atm. Furthermore, Fig. 5 shows that β forms from L without the participation of α,
again in contrast to the peritectic transformation (α + L → β) observed
at 1 atm.

After the appearance of β at 350°C (Fig. 5), the sample was heated
again. At 375°C, the β peaks were still present, but at 400°C they vanished. Subsequent cooling to 375°C caused the reappearance of the β
peaks. Thus, the melting of β (β → L) occurs reversibly at 387 ± 13°C.

After the reappearance of β at 375°C, the sample was further cooled
to 350°C, which caused the growth of the β peaks. Further cooling to
325°C caused the appearance Au$_2$Ga 511 and/or 203 peak, which coexisted
with the β peaks. The Au$_2$Ga peak grew with respect to the β peaks upon
further cooling.

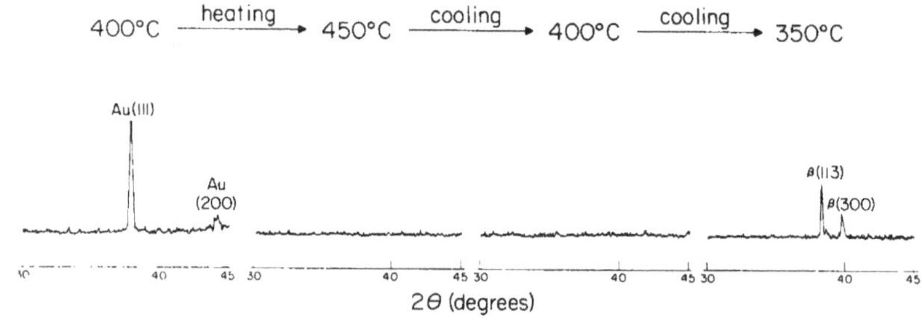

$$400°C \xrightarrow{\text{heating}} 450°C \xrightarrow{\text{cooling}} 400°C \xrightarrow{\text{cooling}} 350°C$$

Fig. 5   X-ray diffraction patterns obtained in situ in vacuum to show the
dissolution of α upon heating and the formation of β upon cooling.

DISCUSSION

As indicated in the Au-Ga phase diagram, the dissolution of α occurs
over a range of temperatures such that the dissolution is complete at
510°C for an overall composition of 21 at. % Ga. The fact that β (21 at.
% Ga) was formed upon subsequent cooling suggests that α and L together
have a composition of 21 at. % Ga. Thus, if one assumes an overall composition of 21 at. % Ga, at 500°C α and L coexist, whereas at 550°C only
L is present. This is just what we observed in 1500 Å Au/GaAs at 1 atm,
during heating as well as cooling.

For 1500 Å of gold in 1 atm, we observed the peritectic transformation
of β to α + L. The transition temperature is between 410 and 420°C. The
Au-Ga phase diagram shows two peritectic transformations: β → α' + L (at
409.8°C) and α' → α + L (at 415.4°C). We did not observe α'. On the other
hand, in Au/Ni/Au-Ge/GaP(111), α' was observed as the reaction product and
the peritectic transformation of α' to α + L was observed [5].

Upon reducing the hydrostatic pressure from 1 atm to 425 μ, the temperature for the complete dissolution of α decreased by ∿ 100°C, while
that for the melting of β decreased about 28°C. A low pressure enhances
the arsenic evolution that accompanies the Au-Ga phase formation, thereby
affecting the amount of As dissolved in the "Au-Ga" phases. This might in
turn affect the equilibrium transition temperatures. The sharp variation
of the Au-Ga liquidus temperature with the Ga concentration (0-22 at. % Ga)
suggests a sharp variation with the As concentration as well. This argument is consistent with the large pressure dependence of the temperature
for complete dissolution of α. Because the transition temperatures
reported here were determined by stepwise heating, with 20 min or more at

each temperature step, the transition temperatures obtained are equilibrium transition temperatures. This means that the dependence of the
transition temperature on the pressure is due to thermodynamics rather
than kinetics.

Other than the effect on the transition temperature, pressure had
three additional effects. Firstly, a low pressure caused the dissolution
of α to be irreversible, probably due to the evacuation of the arsenic
vapor. Secondly, a low pressure caused the melting of β to occur in the
form $\beta \rightarrow L$, whereas it occurred in the form $\beta \rightarrow \alpha + L$ at 1 atm. Thirdly,
Au_2Ga coexisted with β below $325^\circ C$ in vacuum, but was not observed in 1 atm.
The origins of these effects are still to be understood.

Further decrease of the pressure below 425 μ is expected to decrease
the phase transition temperatures further. This is suggested by the
observation that the gold-to-silver transition temperature and the arsenic
evolution temperature decreased with decreasing pressure below 425 μ (0.031
Kg/m^2) [6]. For example, in situ mass spectrometry showed that 2100 Å Au/
GaAs(100) underwent arsenic evolution at $358^\circ C$ at a pressure of $\sim$ 5 X 10^{-6}
torr (3.7X$10^{-7}Kg/m^2$) [6].

Our phase transition temperatures obtained for 1500 Å Au at 425 μ are
in agreement with those (T_1 and T_3) obtained by internal friction on < 3000
Å Au at 1 X 10^{-4} torr (7.4 X $10^{-6}Kg/m^2$) [1]. T_1 was determined by internal
friction to be $440^\circ C$, while our x-ray diffraction results showed it to be
$425 \pm 25^\circ C$ (corresponding to the complete dissolution of α). T_3 was determined by internal friction to be $340^\circ C$, while our x-ray diffraction results
showed it to be either $387 \pm 13^\circ C$ (corresponding to the melting of β) or
$337 \pm 13^\circ C$ (corresponding to the appearance of Au_2Ga upon cooling).
Because we did not heat up Au_2Ga after its formation, we have not yet
observed the melting of Au_2Ga. However, it is likely that the melting of
Au_2Ga occurs reversibly at a temperature close to $337 \pm 13^\circ C$; the Au-Ga
phase diagram suggests its occurrence at $348.9^\circ C$; the dissolution of Au_2Ga
had been observed reversibly at $320^\circ C$ in 600 Å Au/GaAs(100) by in situ
TEM [7].

Based on internal friction results, Nakanisi [1] interpreted T_1 as due
to the melting of a portion of Au, and T_3 as the Au-Ga-As ternary eutectic.
These interpretations are in contrast to those we arrived at by using
x-ray diffraction.

In 1000 Å Au/GaAs, Zeng and Chung [2] observed the vanishing of the β
(or Au_7Ga_2) peaks upon heating at $413 \pm 5^\circ C$ and 1 atm, in agreement with the
temperature found in this work for 1500 Å Au/GaAs at 1 atm. However, in
1000 Å Au/GaAs, the vanishing of the β (or Au_7Ga_2) peaks was not accompanied
by any significant growth of the Au (or α) peaks. The absence of the
transition associated with the dissolution of α in this material suggests
the absence of $\alpha + L$, and this in turn suggests the absence of the peritectic transformation of β to $\alpha + L$. Indeed, for 1000 Å Au, the melting of
β takes the form $\beta \rightarrow L$, rather than that of a peritectic transformation.
It occurs at $412 \pm 5^\circ C$ at 1 atm [2].

The behavior for 1500 Å of gold at 1 atm (this work) agrees with the
Au-Ga phase diagram better than that for 1000 Å of gold at 1 atm [2].
This is reasonable since the phase diagram is for the bulk material system.

Ex situ x-ray diffraction of 1500 Å Au/GaAs (same sample as used in
this work) showed that the reaction between Au and GaAs took the form Au +
Ga $\rightarrow \alpha$ and that this reaction occurred at $350^\circ C$ or below [3]. In this work,
in situ observation of the dissolution of α gives additional evidence for
the formation of α.

The difference between the results for 1500 Å Au at 1 atm (this work)
and 1000 Å Au at 1 atm [2] originates from the fact that the interfacial
reaction takes the form Au + Ga $\rightarrow \alpha$ for 1500 Å Au/GaAs [3], whereas the
interfacial reaction takes the form Au + Ga $\rightarrow$ Au-Ga (tentatively AuGa) for
1000 Å Au/GaAs [2]. It should be mentioned that the behavior observed here
for 1500 Å Au was not due to insufficient time for the formation of Au-Ga

(tentatively AuGa) to take place. After 2 hr at 450°C, no additional peak in the x-ray diffraction pattern was observed for 1500 Å Au, whereas the reaction to form AuGa had gone to completion for 1000 Å Au. Furthermore, the absence of AuGa phase formation in 1500 Å Au/GaAs after 2 hr at 450°C was not due to the hindrance of the thick Au film to As evolution, because As evolution had been observed from 2100 Å Au/GaAs at $\sim 5 \times 10^{-6}$ torr (3.7 $\times 10^{-7}$Kg/m^2) [6]. Therefore, we believe that the thickness dependence observed by Nakanisi and us is not due to kinetic reasons, but is due to thermodynamic reasons.

Both x-ray diffraction and internal friction [1] indicated that the gold film thickness affects the phase transitions. However, internal friction [1] showed a change from "$T_1 - T_3$ behavior" to "$T_2 - T_3$ behavior" at a gold film thickness of 3000 Å, whereas x-ray diffraction suggested a similar change at a gold film thickness of 1250 ± 250 Å. The details of this effect of the gold film thickness remains to be investigated. Nevertheless, our x-ray diffraction results show that this effect stems from the effect of thickness on the form of the reaction between Au and GaAs, as previously discussed.

It is quite probable that yet different reactions occur for Au films considerably thinner than 1000 Å. For Au film of thickness 600 Å, ex situ transmission electron microscopy (TEM) revealed the formation of Au$_2$Ga precipitates at 250°C [7]. For Au film of thickness $\sim$100 Å in situ TEM revealed the formation of rectangular features upon heating to $\sim380^{\circ}$C, the vanishing of the Au diffraction rings upon further heating to 470°C, and the formation of β (or Au$_7$Ga$_2$) upon subsequent cooling to room temperature [8].

CONCLUSION

In situ x-ray diffraction was used to observe the phase transitions in 1500 Å Au/GaAs. At 1 atm (56 Kg/m^2) of nitrogen, we observed transitions at 525 ± 25 and 415 ± 5°C; at 425 μ (0.031 Kg/m^2), we observed similar transitions at 425 ± 25 and 387 ± 13°C, respectively. Thus the decrease in pressure decreased the transition temperatures, as also observed by Zeng and Chung for the transition which they observed at 478 ± 22°C at 1 atm in 1000 Å Au/GaAs [2]. For 1500 Å Au/GaAs, we found that the higher temperature transition (525 ± 25°C at 1 atm, 425 ± 25°C at 425 μ) is due to the completion of the dissolution of α, i.e. α + L → L, and that the lower temperature transition (415 ± 5°C at 1 atm, 387 ± 13°C at 425 μ or 0.031 Kg/m^2) is due to the melting of β (or Au$_7$Ga$_2$). At 1 atm, the melting takes the form of a reversible peritectic transformation, i.e., β → α + L; at 425 μ or 0.031 Kg/m^2, the melting is also reversible, but is simply of the form β → L.

REFERENCES

1. T. Nakanisi, Japan. J. Appl. Phys. 12, 1818 (1973).
2. X.-F. Zeng and D. D. L. Chung, Solid-State Electron. 27, 339 (1984).
3. D. D. L. Chung and Edward Beam III, Thin Solid Films, in press.
4. X.-F. Zeng and D. D. L. Chung, Thin Solid Films 93, 207 (1982).
5. R. A. Ginley, D. D. L. Chung and D. S. Ginley, Solid-State Electron. 27, 137 (1984).
6. S. Leung, L. K. Wong, D. D. L. Chung and A. G. Milnes, J. Electrochem. Soc. 130, 462 (1983).
7. T. Yoshiie, C. L. Bauer and A. G. Milnes, Thin Solid Films 111, 149 (1984).
8. K. Kumar, Japan. J. Appl. Phys. 18, 713 (1979).

Pd_2Si on Si(111) GROWTH KINETICS
STUDIED BY X-RAY DIFFRACTION

Betty Coulman[*] and Haydn Chen[**]
*, **Dept. of Metallurgy and Mining Engineering and the Materials
Research Laboratory, University of Illinois at Urbana-Champgaign, IL
61801.
*Present address: Philips Research Laboratory Sunnyvale, Signetics
Corp., Sunnyvale, CA 94088.

ABSTRACT

Results are presented for the kinetics of growth of Pd_2Si interfacial layers obtained by an X-ray diffraction technique. Epitaxial
Pd_2Si films were grown on Si(111) substrates over a temperature range
of 160-222°C. The parabolic rate law observed is in qualitative agreement with those reported by investigators using other techniques (RBS,
AES, Electron Microprobe). There appear to be two kinetics regimes dis-
tinquished by diffusion paths with different activation energies (1.35±
0.10 eV vs. 1.05±0.10 eV). The presence of impurities and the detailed
Pd_2Si microstructure will influence how the reacting species are transported through the lattice.

INTRODUCTION

Pd_2Si is among the group of silicides categorized as epitaxial.
It can be grown readily from evaporated Pd films on chemically cleaned
Si(111) substrates of moderate doping. Thin interfacial layers of epitaxial Pd_2Si have been observed to form during Pd deposition at room
temperature. Subsequent anneals at temperatures up to 700°C result in
continued growth of the single product phase [1]. Relative misorientations between different regions of the Pd_2Si films amount to less than
2° [2]. TEM cross-sections reveal abrupt interfaces with misfit
dislocations at the interfaces [3].

Pd_2Si was considered an ideal prototypical system for an in situ
X-ray diffraction thin film kinetics study because of its favorable characteristics for delivering adequate diffracted intensity from a small
amount of material. In particular, Pd_2Si films are single phase, of
mosaic structure with highly preferred orientation and contain a strong
X-ray scatterer (Pd). In addition, the formation of Pd_2Si at relatively low temperatures (200°C) enabled use of a simple furnace design for
in situ measurements.

Results from an X-ray diffraction study are compared to previous results. A discussion of the influences of impurities and structure on the
rate of Pd_2Si formation on Si(111) is presented.

GROWTH KINETICS BY X-RAY DIFFRACTION

The desired outcome of a kinetics experiment is a mathematical description of the progress of a reaction with time. Thin film growth kinetics equations usually relate the thickness of the product phase to
time. Standard techniques for probing thin film thickness are RBS, AES
with sputter profiling and the electron microprobe. Although not usually
regarded as a thin film thickness guage, X-ray diffraction offers some

unique advantages for this purpose. The non-destructive nature and sim-
plicity of X-ray diffraction experiments, as well as their phase specificity, deserve consideration when these attributes can be beneficial.

In situ growth of kinematically diffracting material with constant
structure and negligible X-ray attenuation (e.g. thin film) is characterized by the proportionality of integrated intensity to the volume of
material irradiated and, hence, to thin film thickness [4]. Thicker
films require an additional factor to correct integrated intensity for
absorption.

EXPERIMENTAL METHODS

Palladium films were deposited onto one-inch diameter chemically
cleaned Si(111) substrates by electron beam evaporation of Marz grade Pd
in a vacuum of 10^{-8} Torr. Film thicknesses were 100 nm as determined
by a quartz monitor. Substrate doping was n-type to a resistivity of
10 Ωcm.

Samples were enclosed in a mini-furnace which was mounted on a diffractometer in place of a goniometer head. The furnace was evacuated to
10^{-6} Torr. Cu Kα X-rays were supplied by a sealed tube generator operated at 45kV and 22.5 mA. A singly bent LiF monochromator was used.

The integrated intensity of the Pd_2Si (00.2) reflection was measured as a function of time. The constant of proportionality to Pd_2Si
film thickness was determined at reaction completion. A separate RBS
study of film thicknesses of progressively transformed samples confirmed
porportionality throughout growth at 200°C [4].

RESULTS

The growth of Pd_2Si on Si(111) was observed to exhibit a parabolic
time dependence, in agreement with prevous studies. The rate constant,
k, was determined over a temperature range of 160-222°C. An activation
energy for growth, E_a, of 1.06± 0.10 eV was found with the prefactor,
k_0, equal to 7 x 10^{-4} cm²/sec. [5]. For comparison, the various
growth rate parameters which have been measured for the Pd_2Si on
Si(111) system are given in Table I [1,5,6,7,8,9]. Activation energies
range from 1.05 to 1.5 eV. The spread of the prefactors is quite large,
from 5.7 x 10^{-4} to 31 cm²/sec (nearly 5 orders of magnitude).
Evaluation of the various rate constants at 200°C ($k_{200°C}$) reveals
that they fall within about an order of magnitude of each other. Thus,
the effect of larger activation energies is partially offset by the associated larger prefactors. Consideration of the authors' stated uncertainties in E_a (Table I), the results appear to be divided into two
groups: the first with E_a=1.35± 0.10eV and the second with E_a= 1.05±
0.10eV.

The data from Table I have been plotted in Figure 1. Individual
data points are not indicated; the authors' best linear fits over the
temperature ranges studied are given.

TABLE I : TABULATION OF Pd_2Si KINETICS STUDIES ON Si(III)

$(\text{FILM THICKNESS})^2 = k \times \text{TIME}$

$k = k_0 \ \text{EXP} \ (-E_a/RT) \ cm^2/sec$

INVESTIGATORS	TECHNIQUE	TEMPERATURES	E_a	k_0	$k_{200°C}$
		(°C)	(eV)	(cm^2/sec)	(cm^2/sec)
Hutchins and Shepela, 1973	e⁻ Microprobe	125, 200, 250	1.3	7×10^{-2}	1×10^{-15}
	Visual Timing	240-400	1.2-1.3	—	—
Bower, Sigurd and Scott, 1973	RBS	225, 250, 275[a]	1.5±0.1	30.9	3.3×10^{-15}
Fertig and Robinson, 1976	AES	176, 200, 225, 248	1.4±0.2	2.4	3.0×10^{-15}
Wittmer and Tu, 1983	RBS	200, 225 250, 275	1.35±0.10[b]	0.72[b]	3.0×10^{-15}[b]
			1.05±0.10[c]	5.7×10^{-4}[c]	3.8×10^{-15}[c]
Cheung, Lau, Nicolet, Mayer and Sheng, 1980	RBS	190, 200, 227, 250, 270	1.05±0.10	2.4×10^{-3}	1.5×10^{-14}
Coulman and Chen, 1984	XRD	160, 180, 188, 200, 210, 222	1.06±0.10	7×10^{-4}	4×10^{-15}

a. Data at 200°C measured on Si(100)
b. $4.6 \times 10^{14} \ cm^{-3}$ P-doped Si
c. $5.0 \times 10^{20} \ cm^{-3}$ As-doped Si

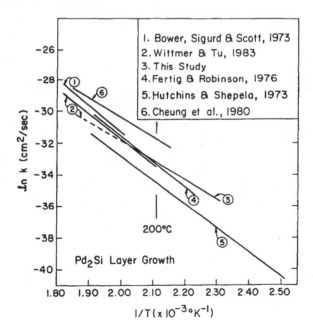

1. Bower, Sigurd & Scott, 1973
2. Wittmer & Tu, 1983
3. This Study
4. Fertig & Robinson, 1976
5. Hutchins & Shepela, 1973
6. Cheung et al., 1980

200°C

Pd_2Si Layer Growth

FIGURE 1: Arrhenius plots of kinetics results for Pd_2Si growth on Si(lll).

DISCUSSION

The spread of the kinetics results of different investigators for Pd_2Si growth on Si(111) suggests a sensitivity to the detailed sample conditions. Three influences will be considered here: impurities, structure and implantation damage.

The intentional introduction of impurities into either the Pd film or the Si substrate has been carried out for a limited number of possible contaminants. Most of this data was collected from samples of Pd on Si(100), however, where the diffusion behavior is known to be different from that in the Pd on Si(111) system. Two effects have been noted for Si(111) substrate samples.

The presence of an interfacial oxide layer caused a delay in the start of the transformation to Pd_2Si, but had no other effect on the kinetics [6]. The activation energy for growth in this study was found to be 1.5 ± 0.1 eV. It is not known if oxide has the same effect on growth proceeding by a pathway with $E_a = 1.05$ eV. Implantation of As into the unreacted substrate, followed by an anneal at 900°C to intro-duce a high doping level, influences the growth kinetics [8]. Relative to substrates of low to moderate P doping, the high As samples reacted with a lower activiation energy and lower k_0. The implantation and anneal processes associated with the As doping undoubtedly introduce structural changes as well as the high doping. The implanted and anneal-ed high As doped samples behaved in the same way as the moderately doped ones used in this study did. It is not clear why the two apparently have growth mechanism in common. Certainly there is a need for controlled impurity studies which do not introduce the complication of implantation related changes.

The effect of structure on the reaction kinetics is not yet clear. The silicide microstructure, through which diffusants move, is expected to influence the kinetics. Surface accumulation studies using Ni indi-cate that the boundaries of polycrystalline silicide serve as diffusion pathways [10]. It may be that the two E_a regimes correspond to bound-ary dominated vs. bulk diffusion dominated growth.

Two studies comparing Si(100) substrate samples and Si(111) samples indicate opposite effects. In one case, the activation energy for growth on Si(111) was found to be 10% higher [9] while in the other, it was 5% lower [8] than for Si(100). Since the nature of the Pd_2Si boundaries will be a function of the degree of relative misorientation of small regions (i.e. high vs. low angle), it is expected that changes in the growth kinetics may reveal the misorientation of boundaries necessary for them to serve as effective diffusion pathways. Thus, an experiment in which misorientation alone is varied would be enlightening. Data on mis-orientation could be collected at the same time in situ kinetics are mea-sured using an X-ray diffraction technique.

The effect of ion implantation without a post-anneal at the inter-face with Si on Pd_2Si growth at 250°C is to accelerate growth [11]. This would be consistent with growth with a higher activation energy. Post-implant anneals appear to alter the kinetics in favor of the lower activation energy.

SUMMARY

X-ray diffraction is an attractive alternative for thin film growth kinetics studies. Its capabilities for controlled structural studies are yet to be exploited.

Two activation energy regimes appear to exist for Pd_2Si growth on Si(111). Impurities and structure are likely to play important roles in defining the diffusion pathways for growth. The search for clear trends has just begun.

ACKNOWLEDGEMENTS

This work was carried out under DOE grant #DE-AC02-76ER01198.

The cooperation of Paul Ho of IBM in providing samples is appreciated.

REFERENCES

1. G. A. Hutchins and A. Shepela, Thin Solid Films 18, 343 (1973).

2. H. Chen, G. White and S. Stock, Thin Solid Films 93, 161 (1982).

3. D. Cherns, D. Smith, W. Krakow and P. Batson, Philos. Mag. A 45, 107 (1982).

4. B. Coulman, H. Chen and L. Rehn, to appear in December 15, 1984, issue J. Appl. Phys.

5. Betty Ann Coulman, Ph.D. thesis, University of Illinois at Urbana-Champaign, 1984.

6. R. Bower, D. Sigurd and R. Scott, Solid-State Electron. 16, 1461 (1973).

7. D. Fertig and G. Robinson, Solid-State Electron. 19, 407 (1976).

8. M. Wittmer and K. Tu, Phys. Rev. B 27, 1173 (1983).

9. H. Cheung, S. Lau, M-A. Nicolet, J. Mayer and T. Sheng in Proceedings of Thin Film Interfaces and Interactions, edited by J. Baglin and J. Poate. (The Electrochemical Society, Princeton, 1980) Vol. 80-2, p. 494.

10. E. Zingu and J. Mayer in Thin Films and Interfaces II, edited by J. Baglin, D. Campbell and W. Chu (North-Holland, New York, 1984), p. 45.

11. I. Ohdomari, K. Tu and W. Hammer, Radiat. Eff. 49, 1 (1980).

SELF ALIGNED NITRIDATION OF TiSi$_2$: A TiN/TiSi$_2$ CONTACT STRUCTURE

PAUL J. ROSSER* AND GARY J. TOMKINS**
*Standard Telecommunication Laboratories Limited, London Road,
Harlow, Essex, UK
**Standard Telephones and Cables, Maidstone Road, Foots Cray, Kent, UK.

ABSTRACT

   Titanium nitride is an ideal barrier to silicon migration into
aluminium metallisations in MOS devices. Various means by which a
TiN,TiSi$_2$ contact structure can be achieved to give low resistivity
interconnects and stable contacts are outlined. It is shown that
TiN/ TiSi$_2$ contact structures self aligned to contact holes cut down to
the TiSi$_2$ can be simply achieved. Results showing the stability of the
nitride film against silicon diffusion are presented.

Introduction

   Titanium disilicide is increasingly being adopted as a replacement
for polysilicon in advanced MOS devices due to its very low resistivity.
In addition, the original process implementations in which the silicide
simply replaced the polysilicon gate interconnect are now being superseded
by more advanced self aligned processes such as the SALICIDE process in
which the source and drain regions are silicided in addition to the gate
interconnect [1,2].
   Aluminium is the most commonly used interconnect material in silicon
technology. However as circuit dimensions are reduced so the reliability
of aluminium contacts to silicon becomes more of a problem. Junction
spiking, in which aluminium can penetrate over a micron into the silicon,
is well reported [3]. The addition of silicon to the aluminium in order
to alleviate this often results in silicon precipitates at the contact
openings which have a deleterious effect on the contact resistance. The
formation of titanium disilicide at the contact regions, as would occur in
the SALICIDE process, is not expected to improve matters significantly as
it is know that silicon diffuses very readily through this film.
   Titanium nitride is an ideal material for use in multilayer contact
structures as it is not only a low resistivity material, but also blocks
the diffusion of silicon and is stable against reation with aluminium up
to approximately 550°C, well in excess of temperatures routinely
encountered in post metallisation anneals [4,5,6]. A metallisation scheme
in which the aluminium is separated from the silicide by a nitride barrier
layer clearly offers the potential of low resistivity interconnects with
low contact resistance and protection against junction spiking.
   Several techniques for incorporating a titanium nitride diffusion
barrier in a silicide process have been considered. One approach is to
deposit the film after depositing the titanium disilicide. By
cosputtering titanium and silicon and then reactively sputtering titanium
in a nitrogen ambient it is possible to deposit both layers in the same
vacuum system, thereby ensuring a clean interface [4] (fig.1). This
approach entails more complicated deposition processes which could
compromise high throughput and machine reliability. More important is the
fact that it is incompatible with any of the self aligned processes in
which a selective wash is used, as this wash removes titanium nitride very
efficiently.
   To ensure compatibility with the self aligned silicide processes the
nitride layer could alternatively be deposited immediately prior to
aluminium deposition (fig.2). The nitride could then be etched at the

608

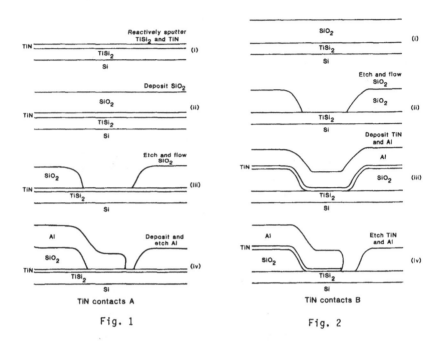

Fig. 1   Fig. 2

same time as the aluminium to result in a nitride layer only where there
is an aluminium interconnect. This approach again involves more
complicated deposition and etch processes. In addition, aluminium suffers
not only from junction spiking but also from electromigration.
Electromigration could result in aluminium bridging across the nitride
barrier and causing precipitation where it contacts with the underlying
silicide.

A third approach would be to deposit and define the nitride layer in
order to completely cover the contact region prior to aluminium deposition
(fig.3). This approach entails the addition of extra deposition,
photolithographic and etch steps, with the yield hazard entailed by each.

The approach favoured by this laboratory, which is presented in this
paper, entails the addition of no extra steps to an advanced MOS process.
The nitride layer is formed only in those regions where silicide is
exposed prior to aluminium deposition (fig.4).

Experimental detail

The substrates used were either (100) single crystal silicon wafers
or 0.5μm thick degenerately phosphorous doped polycrystalline silicon
films deposited at 600°C in an LPCVD system. The polysilicon films were
deposited over oxidised silicon wafers.

Two systems have been used to deposit thin films of titanium; an
Electrotech electron-beam evaporator with a base pressure of 4.10^{-5} Pa
and a Leybold Heraeus load locked magnetron sputterer with a base pressure
of 10^{-5} Pa. Results from the films deposited in these two systems were
in all cases very similar. Those from sputtered titanium will be
presented where possible.

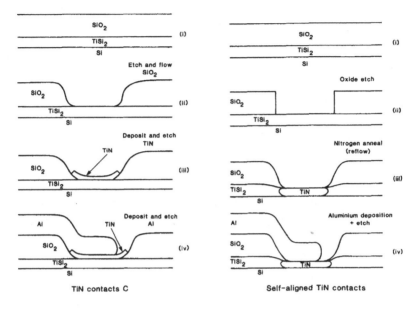

Fig. 3   Fig. 4

Immediately prior to the titanium deposition the substrates were dipped in 10:1 HF for 10 seconds in order to remove the native oxide layer. After loading the wafers into the load lock it was evacuated to better than 4.10^{-3} Pa with a turbomolecular pump. The substrates were then transferred into the main chamber which is maintained at its base pressure of 10^{-5} Pa with a cryopump. The chamber was backfilled to 7.10^{-1} Pa with argon. Titanium was then deposited at 10 nm per minute to a total thickness of 100 nm.

The films were annealed in an A.G. Associate's Heatpulse 210-T rapid thermal processing furnace. Typically the films were annealed in a nitrogen ambient at 850°C for 30 seconds. During this anneal the underlying silicon diffuses into and reacts with the titanium to form a low resistivity titanium disilicide layer. Impurities in the titanium, such as oxygen and nitrogen, are not incorporated into the silicide but are pushed toward the surface of the titanium resulting in an 'impurity layer' of titanium oxides and nitrides approximately 40 nm thick. An AES spectrum of an annealed film is shown in fig.5.

After the silicide formation anneal the impurity layer was removed using the selective wash consisting of hydrogen peroxide and ammonium hydroxide. Fig.6 shows an AES spectrum of a silicide film after the selective etch. No significant impurities are detected within the resolution of the AES equipment except for a thin (~5 nm) silicon oxide layer at the surface. RBS analysis confirms the presence of stoichiometric titanium disilicide.

The titanium disilicide films were then annealed in a nitrogen ambient at 1000°C for 30 minutes. After this anneal the sheet resistivity of the film rose from 1.4 ohms/□ to approximately 6 ohms/□, and its appearance changed from that of metallic looking titanium disilicide to a yellow gold colour characteristic of titanium nitride.

610

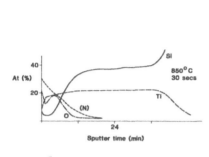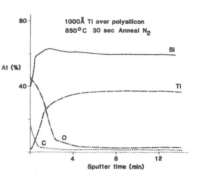

Fig.5 A.E.S.* analysis of 1000Å Ti over silicon annealed
in the heatpulse

Fig.6 A.E.S.* of selectively etched film

Analysis of the film using AES and RBS showed it to be a mixture or compound of titanium and nitrogen with a very low level of impurities other than silicon. Subsequent RHEED analysis confirmed that titanium nitride was indeed formed, (fig.7).

In order to demonstrate the effectiveness of the titanium nitride film formed by this technique as a barrier to silicon diffusion the film was annealed in a wet oxidising ambient at 850°C for 20 minutes. If silicon were able to diffuse from the substrate to the surface it would be oxidised to form silicon dioxide, as occurs during the oxidation of titanium disilicide. If the titanium nitride layer acted as an effective barrier then only titanium oxides would be expected. AES analysis, fig. 8, showed this latter to be the case despite the presence of some silicon in the nitride film.

As device geometries continue to be reduced so does the importance of minimising the diffusion of dopants during subsequent processing. This is clearly demonstrated by the trend toward lower temperatures and more rapid processing. Although the anneal used to form the nitride would be acceptable in most existing processes, being comparable to the P.S.G. reflow anneals at present used prior to aluminium metallisation, it could not be considered ideally suited to many proposed small geometry processes. Work in this laboratory and others has demonstrated the feasibility of using rapid thermal processing to flow the phosphorous

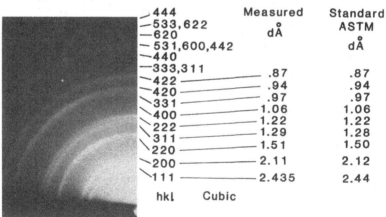

hkl	Cubic	Measured dÅ	Standard ASTM dÅ
444			
533,622			
620			
531,600,442			
440			
333,311		.87	.87
422		.94	.94
420		.97	.97
331		1.06	1.06
400		1.22	1.22
222		1.29	1.28
311		1.51	1.50
220		2.11	2.12
200		2.435	2.44
111			

Fig.7   RHEED analysis of TiN film

*In Fig. 5 and 6 the N signal due to overlap of the N and Ti Auger electron energy peaks has been subtracted.

611

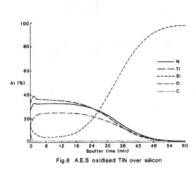

Fig.8 A.E.S oxidised TiN over silicon

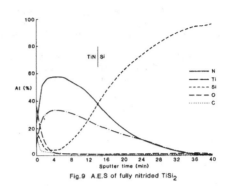

Fig.9 A.E.S of fully nitrided TiSi$_2$

doped glass [7] and the use
of a similar anneal to form
the nitride diffusion barrier
would be very advantageous.
Fig.9 shows an AES spectrum
of a film annealed for 10
seconds at 1100°C in the
Heatpulse in a nitrogen
ambient. Even in this short
time it has proved possible
to convert the silicide completely into titanium nitride.
By careful control of the anneal,
profiles as shown in fig.10
are possible in which only part
of the silicide is nitrided.

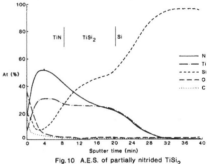

Fig.10 A.E.S. of partially nitrided TiSi$_2$

A proposed process implementation of this self aligned nitriding
of TiSi$_2$ is given in fig.4.

Conclusion

A self-aligned contact metallisation process has been proposed
and preliminary results presented. The process combines the low
resistivity of aluminium for long interconnects with the diffusion
barrier properties of titanium nitride and the relatively low resistivity
of titanium disilicide for gate level interconnects. The process outlined
does not involve any additional photolithographic or etch steps.

Acknowledgement

The authors are grateful for the assistance of Carol A. Mayston
in carrying out this work, Barry Lamb and Tony Hawkridge for AES analysis
and A.R. Abid of the University of Surrey for RHEED analysis. The authors
thank STC plc for permission to publish this paper.

612

References

(1) G.H. Osborn, M.Y Tsai, S. Roberts, C.J. Lucchese, C.Y. Ting, in Proceedings of the ULSI Science and Technology Symposium, 82-7, 213 (Electrochem. Soc., Detroit, 1982).

(2) C.Y. Ting, S.S. Iyer, C.M. Osborn, G.J. Hu, A.M. Schweighart, in Proceedings of the VLSI Science and Technology Symposium, 82-7, 224 (Electrochem. Soc., Detroit, 1982)

(3) R.B. Marcus, T.T. Sheng, Transmission Electron. Microscopy of Silicon VLSI Circuits and Structures, 100, (New York, J. Wiley a Sons, 1983).

(4) H. Norstrom, T. Doucher, M. Ostling, C.S. Peterson, Physica Scripta, 28, 633 (1983.

(5) M. Wittmer, J. Appl. Phys. 53, 1007 (1982)

(6) M. Wittmer, H. Melchior, Thin Solid Films 93, 397 (1982)

(7) J.S. Mercier, in Proceedings of the VLSI Science and Technology Symposium 82-7, 377, (Electrochem. Soc., New Orleans, 1984).

SURFACE CHARACTERIZATION OF ARSENIC IMPLANTED SILICON (100): A NEW INSIGHT INTO THE INHIBITION OF ALUMINUM/SILICON INTERDIFFUSION

NICOLE HERBOTS,* D. GLOESENER,+ E. J. VAN LOENEN**, A. E. M. J. FISCHER++
*Guest scientist at the Solid State Division, Oak Ridge National
Laboratory,*** Oak Ridge, TN 37830, from FAI-UCL, 3, pl. du Levant,
B-1348, Louvain-La-Neuve, Belgium.
+FAI-UCL, 3, pl. du Levant, B-1348, Louvain-La-Neuve, Belgium.
**FOM-Instituut, Kruislaan 407, 1098-SJ Amsterdam, The Netherlands.

ABSTRACT

Arsenic segregation at Si(100) surfaces during annealing (890-970°C) has been studied by medium energy ion scattering (MEIS), Rutherford backscattering spectrometry (RBS), ion scattering spectrometry (ISS), and Auger electron spectroscopy (AES). The unique depth resolution of MEIS revealed that arsenic segregated in a two-dimensional layer at the clean Si surface during annealing. For a surface with a native oxide the arsenic piled up at the Si/oxide interface. This segregation peak was no longer present on the Si surface after conventional contact opening. By metallizing arsenic junctions with Al:Si, 1% it was found that a segregation annealing step inhibited Al/Si interdiffusion. Diodes as shallow as 180 nm could be metallized without spiking.

INTRODUCTION

In previous studies,[1,2] it was shown that the presence of segregated impurities at the interface between two thin films could inhibit the inter-diffusion between the layers. In this work the same method is applied to the interface between Si(100) and a metallic film, Al:Si, 1%. The inter-diffusion at this interface is a well-known problem in contact technology [3]. The experiment was performed in two phases. First, the segregation of arsenic at the clean Si (100) surface was demonstrated and measured with nearly monolayer resolution and with good elemental sensitivity. Special attention is given to the effect of native oxide present on the Si surface during annealing, because this native oxide is always present in standard IC processing. In the second phase, the effect of arsenic segregation on I-V characteristics of shallow diodes was measured. A detailed study of the electrical properties will be published elsewhere.

EXPERIMENTAL PROCEDURE

Three groups of samples were prepared: (1) samples for in situ segregation studies in the MEIS UHV chamber, (2) test-circuits using dif-ferent processes to modify the arsenic surface concentration on top of N^+P junctions, and (3) samples processed as in the second group, but with omission of the patterning by photolithography, in order to obtain planar structures.

The samples analyzed by MEIS consisted of Si(100) wafers implanted with 4.69×10^{15} arsenic/cm^2 at 40 keV. One of the samples was cleaned inside the UHV chamber, by sputtering with 2×10^{15} argon/cm^2. Annealing at 970°C for as little as three minutes reordered the surface to the well-known (2x1) LEED pattern. Surface cleanliness was checked by in-situ AES analysis.
***Operated by Martin Marietta Energy Systems, Inc. under Contract No. DE-AC05-84OR21400 for the U.S. Department of Energy.

614

The two other groups of samples underwent an 8-level N-MOS process at 950°C. After the contact window opening, two of those three samples underwent an additional annealing under a nitrogen flow of 2 l/min at 950°C, called here "segregation annealing". One of them was then etched with HF,2% The samples of the second group were then metallized with Al:Si, 1% and sintered at 465°C for 30 min. MEIS was used on the first and third group in order to analyze the composition of the first atomic layers with high depth resolution. The energy of the analyzing helium beam was 174.5 keV. A toroidal electrostatic analyser with an energy resolution of $\Delta E/E =$ 4×10^{-3} was used. For the presently used scattering conditions, the detector resolution corresponds to a depth resolution of 0.5 nm. The pressure was less than 1.10-10 Torr during analysis.

High energy ion scattering was used to measure arsenic depth profiles deeper in the substrate than MEIS allowed, at the expense of poorer depth resolution.

ISS was performed with a 2 keV Neon beam having an intensity of 60 nA. The purpose of this analysis was to compare ISS sputter profiles with MEIS.

I-V characteristics were taken from a diode having a 340 x 340 μ^2 area, with 100 contacts opening of $3.5 \times 3.5 \ \mu^2$. The junction depth was 0.18 μ. The voltage range was -5 to 1 V.

RESULTS AND DISCUSSION

Atomically clean Si(100):

Arsenic was detected at the surface immediately after the last step of the cleaning procedure, that is after 3 min annealing at 970°C (Fig. 1).

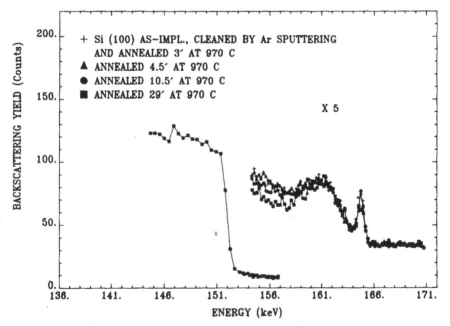

Fig.1. MEIS random spectrum of Si (100) arsenic implanted with 4.69 x 10[15]at/cm[2] at 40 keV, cleaned in UHV and sequentially 970⁰C.

Thus the segregation of arsenic at the Si surface takes place during a short transient process. The width of the arsenic peak was not larger than the resolution of the analyzer, providing evidence that the segregation is confined to less than two atomic layers. It can therefore be concluded that this small amount of arsenic does not form three-dimensional clusters, but segregates in a two-dimensional layer.

The amount of segregated arsenic is 0.05 monolayers (Fig. 1), where one monolayer corresponds to 6.8×10^{14} at/cm^2. Sequential annealing up to 29 min. did not modify this concentration. The fact that the amount of segregated arsenic remains constant is due to the equilibrium between segregation and desorption. This is consistent with thermal desorption spectroscopy results [4], where the amount of tightly bounded arsenic monomers on Si (100) was found to be about 10% in the 800-1000°C range. There seems to be disagreement with the AES measurements of ref. [5], showing the amount of segregated arsenic to vary strongly with time. However, no UHV cleaning was done in the latter study. As we will show below the presence of a native oxide dramatically influences the surface segregation.

The profile underneath the surface presented interesting features. A depletion of arsenic is observed directly below the segregated layer. This is consistent with the theoretical segregation profiles calculated by J. Kirschner [6]. Thus, the MEIS measurement provides evidence of strong concentration fluctuations within a few monolayers. The concentration of arsenic increases under the depleted layer up to a level of about 8×10^{20} at/cm^3.

Si (100)/native oxide:

If the native oxide was left intact on the Si (100) surface, strong arsenic segregation was observed, as shown in Fig. 2. The arsenic piles up

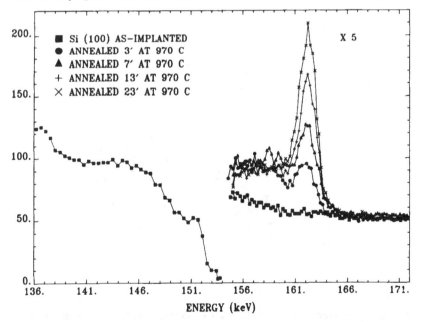

Fig. 2. MEIS random spectrum of Si (100) arsenic implanted and sequentially annealed at 970°C as in Fig. 1, with native oxide left on the surface.

below the oxide at the Si/oxide interface, as can be judged from the shift of the arsenic peak.

The native oxide acted as a barrier for the outdiffusion, so that the time evolution of the profile differed strongly from the clean surface case. The amount of segregated arsenic increased as the square root of time showing that the segregation is diffusion limited by the supply of arsenic from the bulk. After 23 min. of annealing, the amount of arsenic at the Si/oxide interface was 5.4×10^{14} at/cm^2. The MEIS measurement in Fig. 2 also provides information on the lateral non-uniformity of the native oxide. The shape of the step on the Si edge directly reflects the oxide thickness distribution. From the average width of the step and from the shift of the arsenic peak an average oxide thickness of 2.6 nm is found. This rather large value is likely due to the fact that the Si surface had been amorphized by the ion implantation.

In the case of the Si wafers having undergone the steps of the modified NMOS process, the following results were obtained. When the last step was an oxidation or an annealing under a thick oxide, MEIS revealed after the contact opening etch a flat arsenic profile from the surface into the bulk. If a "segregation annealing" was made in a non-oxidizing ambient after the contact window opening, so that no moving Si/oxide interface was present during the annealing, arsenic segregation took place at the Si/oxide interface. MEIS energy spectra were taken on a non-metallized 0.18 μ diode manufactured during a 3.5 μ N-MOS process. Fig. 3 shows the spectra measured in a 20 range of exit angles scattering centered around the Si [111] direction. Whereas ions backscattered from Si are strongly blocked along Si crystal directions, no blocking effects are observed on the arsenic signal. This indicates that the arsenic is either present at the very interface or is not coherent with the Si lattice.

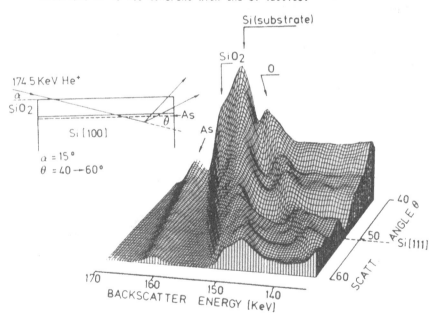

Fig. 3. MEIS spectra of an 0.18 μ diode having undergone a "segregation annealing".

If the surface was submitted to a cleaning procedure prior to metalli-zation, as is usually the case in standard IC processing, an etch with diluted hydrofluoric acid (2%) for 60 sec removed the native oxide layer and the segregated arsenic. AES was not sensitive enough to arsenic to pro-vide a measure of the dopant concentration. ISS revealed an accumulation of As on the surface, but was unable to resolve the second (depleted) mono-layer because of the intermixing associated with sputter-profiling. A com-parison of the measured profiles with computer simulations was made. SUPREM [7] underestimates by about one order of magnitude the arsenic con-centration through the first 7 nm from the surface.

I-V characteristics were taken with three different surface conditions: a) surfaces which underwent a segregation annealing before metallization, b) surfaces which underwent the segregation annealing, and were submitted thereafter to a cleaning procedure that did remove the native oxide and the segregated arsenic and c) surfaces without any segregation annealing after the contact opening. The three junctions were all 0.18 µ. Diodes from the first and second group show good diode behavior with ideality factors below 1.07, whereas diodes from the third group showed the typical behavior of a destroyed junctions with breakdowns all along the curve. Completely flat characteristics were obtained under reverse bias for the diodes of the first group because of the presence of the residual oxide. The second group exhibited true ohmic contact. Saturation currents averaged 2.5×10^{-13} A for both diodes. Those values are excellent as compared with other attempts to metallize shallow diodes with Al [8].

CONCLUSIONS

Several conclusions can be drawn: 1) Arsenic segregates in a two-dimensional layer at the clean Si (100) surface upon annealing. 2) The presence of native oxide influences the arsenic segregation profile, inducing a pile-up at the Si/oxide interface. 3) Spiking at the Si/Al interface can be prevented by segregation annealing immediately before metallization.

REFERENCES

[1] N. Herbots, et al., J. of the Electrochem. Soc., Vol. 131, N. 3, March 1983.
[2] N. Herbots, et al., Proc. of the Intern. Conf. on Ion Beam Modif. of Mater., July 1984, Ithaca, NY, to be publ. in Nucl. Instrum. Meth. 8 (1985).
[3] R. Rosenberg, et al., Thin films interdiffusion and reactions, ed. by J. M. Poate, K. N. Tu, and J. W. Mayer, pp.14-34 (1978).
[4] S. C. Perino, et al., Grain Boundaries in semicond., Proc. of the the Mater. Res. Soc. ann. Meet., Nov. 1981, Boton, MA, ed. by Leamy, Pike et al, pp. 147-151 (1982).
[5] M. Tabe, et al. J. Appl. Phys. 50(8), August 1979, pp. 5292-5295.
[6] J. Kirschner, ibid. as [2].
[7] Stanford University Process Engineering Models Program, Version 0-05.
[8] L. S. Hung, J. W. Mayer, M. Zhang, and E. D. Wolf, Appl. Phys. Lett, Vol 43(12), pp.1123-1125(1983).

TITANIUM SILICIDE FILMS DEPOSITED BY
LOW PRESSURE CHEMICAL VAPOR DEPOSITION

PRABHA K. TEDROW, VIDA ILDEREM, AND R. REIF
Massachusetts Institute of Technology, Dept. of Electrical Engineering and
Computer Science, Cambridge, MA 02139

ABSTRACT

Smooth titanium silicide films have been deposited using a Low
Pressure Chemical Vapor Deposition (LPCVD) process. A system has been
designed and built for the LPCVD of titanium silicide. It is a cold wall
reactor with the wafer being heated externally by infrared lamps. Sequential deposition of polycrystalline silicon (polysilicon) and titanium
silicide films, and in-situ annealing of these films, if required, can
be performed in this system. A turbomolecular pump is used to provide a
contaminant free environment with a base pressure of $< 10^{-7}$ torr. SiH_4
and $TiCl_4$ are used as silicon and titanium sources, respectively.
Titanium silicide films with resistivities ranging from 22 to 39 $\mu\Omega$-
cm have been obtained. At low deposition rates, these films have surface
roughnesses ranging from 50 to 250 Å. From X-ray diffractometry, it was
determined that the as-deposited titanium silicide films were polycrystalline, and $TiSi_2$ was the predominant phase. Si/Ti ratios of 1.8 to 2.3
were obtained from Rutherford Backscattering Spectroscopy (RBS). Auger
analyses did not show any impurities such as oxygen, carbon or chlorine
in these films.

INTRODUCTION

Advances in integrated circuit technology have resulted in smaller
and faster devices. In order to be able to continue the trend of scaling
down of device dimensions, it has become necessary to develop new materials
which can replace or can be used in conjunction with polysilicon as
interconnection and gate materials. The high resistivity of polysilicon
(heavily doped polysilicon has a resistivity > 300 $\mu\Omega$-cm [1]) imposes a
limit on its applicability, thus the new materials must have lower resistivity than polysilicon and must be compatible with present integrated
circuit processes. Refractory metal silicides meet these requirements
with their metallic conductivity and high temperature stability.

Refractory metal silicides are most commonly prepared by some form of
physical vapor deposition process such as sputtering [2] and co-evaporation [3]. However, as the newer generation of circuits evolve, side walls
are getting steeper and the contact areas are getting smaller, which makes
it imperative to explore a Chemical Vapor Deposition (CVD) process that
will offer good side wall coverage. In addition to conformal coverage,
CVD also offers the advantage of high throughput.

We have designed and built a reactor for low pressure chemical vapor
deposition of titanium silicide. The reactor and the deposition procedure
are described elsewhere [4]. A schematic of the reaction chamber is shown
in figure 1.

RESULTS AND DISCUSSION

Sequential films of polysilicon and titanium silicide were deposited
at temperatures of 650 to 700°C, at pressures of 50 to 460 mtorr, and at
various $SiH_4/TiCl_4$ flow rates.

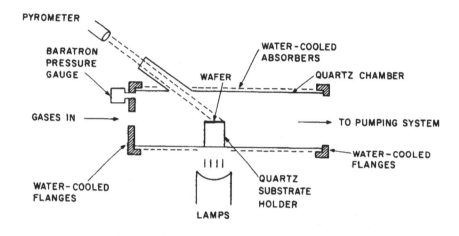

Figure 1. Schematic of the reaction chamber

Several analytical techniques have been used to characterize these films. RBS was used to determine the Si/Ti ratio, and the film thicknesses of the titanium silicide and the underlying polysilicon. A surface profilometer was also used to confirm the film thicknesses and to measure the surface roughness of the films. Nomarsky optical microscopy was also used to evaluate the surface roughness. Auger analysis was used to determine the presence of any impurities and X-ray diffractometry was used to identify the silicide phases and to check the crystallinity of the films. Sheet resistances of titanium silicide films were measured by a four point probe technique.

As-deposited titanium silicide films (figure 2) were polycrystalline with TiSi$_2$ being the predominant phase.

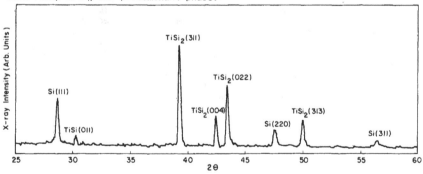

Figure 2. X-ray spectrum of an as-deposited LPCVD TiSi$_2$ film

In this X-ray spectrum, along with the $TiSi_2$ peaks, the underlying polysilicon peaks and a weak TiSi peak can also be identified.
   Smooth titanium silicide films were obtained at low deposition rates (Table I).

TABLE I PROPERTIES OF THE AS-DEPOSITED LPCVD TITANIUM SILICIDE FILMS.

DEPOSITION RATE (Å/min)	SURFACE ROUGHNESS (Å)	THICKNESS (Å)	RESISTIVITY ($\mu\Omega$-cm)
——	50	2600	26
117	250	2000	22
253*	900	3800	152
>2000	2300	15500	39
840	550	17000	81

*This film contained a high oxygen concentration.

The surface roughness of these silicide films ranged from 50 to 250 Å. These films had resistivities ranging from 22 to 26 $\mu\Omega$-cm for film thicknesses of 2000 to 2600 Å. No impurities such as oxygen, carbon, or chlorine were detected in these films.
   High deposition rates (in excess of 2000 Å/min) had resulted in rough surface films; a surface roughness of 2300 Å for a 15500 Å thick film was measured. RBS studies showed these films to have a large transition region between the polysilicon films and the silicide films which made it difficult to estimate the silicide film thickness and hence its resistivity. These thick films tended to have higher resistivities (39-80 $\mu\Omega$-cm) due to the difficulty in thickness measurements. None of these films except for one (See Table I) showed any impurities. This film had a high oxygen contamination which is believed to be responsible for its high resistivity (152 $\mu\Omega$-cm). This contamination was caused by a leak in the vacuum system which points to the importance of having a clean environment during silicide deposition. Preliminary results also indicate that it is important to deposit polysilicon just before the silicide deposition. One may be able to utilize this property for selective deposition of $TiSi_2$.
   All the as-deposited films were annealed at 850°C for 15 minutes in argon ambient. No changes were observed in the already present titanium silicide phases which can indicate that the as-deposited films were already in their final stable form. Table I summarizes some of the results for the deposition rate, surface morphology, thickness and the resistivity of the as-deposited LPCVD titanium silicide films.

CONCLUSION

   A reactor for LPCVD of titanium silicide has been designed and built. Sequential films of polysilicon and titanium silicide have been deposited successfully in this reactor. Titanium silicide films with resistivities ranging from 22 to 39 $\mu\Omega$-cm were obtained. At low deposition rates, these films had surface roughnesses ranging from 50 to 250 Å. High deposition rates resulted in films with surface roughnesses of the order of 2000 Å. From X-ray diffractometry, it was determined that the as-deposited titanium

622

silicide films were "polycrystalline", and TiSi$_2$ was the predominant phase, no impurities such as oxygen, carbon and chlorine were detected in the films.
This work is supported by Analog Devices, Inc. (Res. Agmt. 8-3-82), and by the Semiconductor Research Corp. (Contract No. 83-01-033).

REFERENCES

1. M. Y. Tsai et al., Semiconductor Silicon 1981, edited by H. R. Huff, R. J. Kriegler and Y. Takeishi, (Electrochem. Soc., Pennington, New Jersey, 1981), p.573.

2. B. L. Crowder and S. Zirinsky, IEEE Trans. Elec. Dev., ED-26, 369(1979).

3. S. P. Murarka, J. Vac. Sci. Tech., 17, 775(1980).

4. P. Tedrow, V. Ilderem and R. Reif, to be published in APL, Jan 15(1985).

PRE-ANNEALING OF TiN BARRIERS IN Al METALLIZATION OF SILICON

W. SINKE, P.K. STOUT AND F.W. SARIS
FOM-Institute for Atomic and Molecular Physics, Kruislaan 407, 1098 SJ Amsterdam, The Netherlands

ABSTRACT

We have studied the influence of a high-temperature pre-anneal on the barrier performance of TiN in Al metallization of Si. The results show that barrier failure is shifted towards a higher temperature by 20°C - 40°C when the barrier is pre-annealed at 600°C - 800°C. In addition, we studied the failure mechanism and found that the barrier breaks down by compound formation.

INTRODUCTION

In VLSI and solar cell processing there is a need for reliable contact metallization on silicon. Aluminium, which is a widely used contact material, causes pits and spikes upon thermal treatment which can lead to short-circuiting of the device. A common concept to solve this problem is to use a thin, conductive barrier film between Si and Al [1-5]. Lately, transition-metal nitrides such as TiN have received great interest because of their attractive properties, which make them promising as barrier material. It was found that TiN is impermeable to Si atoms at temperatures up to 700°C for several hours [6]. However, with Al on top, the Si/TiN/Al structure degrades between 500°C and 600°C [1,6-9].

Although it is recognized that the deposition method influences the barrier properties, this has not been thoroughly investigated. In most cases the TiN film was prepared by sputtering, only in a very few cases by evaporation [7]. Krusin-Elbaum et al. [10] found improvement of their ZrN barrier when it was annealed at a high temperature prior to Al deposition, but their starting material was practically amorphous. The role of oxygen contamination with respect to the barrier properties is not well understood either.

In this paper we compare the barrier properties of TiN made by sputtering and evaporation and study the effect of a high-temperature pre-annealing of the TiN layer in the Si/TiN/Al system, using different analysing techniques. In addition, we discuss the degradation mechanism of the structure.

SAMPLE PREPARATION

Two different methods were used to deposit a 50 nm film of TiN on an Si substrate. The first method was e-beam evaporation of Ti in an N_2 ambient. Before deposition, the UHV evaporation chamber was pumped down to a total pressure of 2×10^{-8} Torr, using a turbo-molecular pump in combination with a Ti sublimation pump and a $1 - N_2$ trap. Subsequently, 99.999% pure N_2 was let into the system and the pressure was kept constant at $3 - 5 \times 10^{-5}$ Torr. During evaporation, the substrate temperature was 200°C. Before opening the shutter, Ti was evaporated during several minutes to remove possible contaminants from the surface of the Ti source and to stabilize the evaporation rate at $0.08 - 0.11$ nm/s.

The second method to deposit TiN was reactive sputtering of Ti in a mixed N_2/Ar atmosphere using a planar diode r.f. sputtering system. The base pressure before the sputtering gases were introduced was $3 - 4 \times 10^{-7}$ Torr. Prior to deposition, the substrates were back-sputter etched. Then sputter deposition took place at a total sputtering pressure of 1×10^{-2} Torr. The input power was 1 kW. No negative substrate bias was applied. During sputtering, the substrate temperature remained below 100°C. Two deposition experiments were carried out in two slightly different systems. In the first experiment (A) the

N_2-partial pressure was 2×10^{-5} Torr and the deposition rate was 0.24 nm/s. In the second experiment (B) a rotating substrate holder was used in order to enhance the thickness uniformity of the film; the N_2-partial pressure was 5×10^{-5} Torr and the rate was 0.07 nm/s.

TiN FILM CHARACTERIZATION

The thickness and composition of the TiN films was determined using Rutherford backscattering spectrometry (RBS) and channeling. In all samples we calculated a Ti:N atomic ratio of $1.0 : 1.0 \pm 0.1$ (Table 1).

TABLE 1. Properties of as-deposited TiN films

METHOD OF DEPOSITION	COMPOSITION Ti : N : O	FILM THICKNESS (nm)	GRAIN SIZE (nm)	RESISTIVITY (μΩ cm)
	(±0.03)(±0.03)			
evaporation	1.00 : 0.94 : 0.26	50	> 25	370
sputtering (A)	1.00 : 1.10 : 0.19	37	> 44	170
sputtering (B)	1.00 : 0.89 : 0.29	41	> 44	830

The colour of the TiN films, which is a sensitive indicator of the atomic composition [11], was golden-yellow in all cases, confirming the proper stoichiometry as found by RBS.

Oxygen was found in all samples (Table 1) and appeared to be uniformly distributed in the film. In the case of sputter-deposition, we found a higher oxygen concentration in the samples which were deposited at the lower rate (B). During evaporation, O_2 is only present at a very low partial pressure ($< 10^{-9}$ Torr). Furthermore, we could not detect any form of titaniumoxide in the evaporation beam using a quadrupole mass spectrometer. Therefore we assume that oxygen comes in the evaporated samples during exposure to air [12].

The specific electrical resistivity of TiN films deposited on high-resistivity Si was measured using a four-point probe. The measured values are much higher than the bulk resistivity of TiN ($22 \mu\Omega$ cm, [3]), which is probably due to the oxygen content. The resistivity is a strong function of the amount of oxygen which is incorporated [13]; such a dependence is also found in our results (see Table 1).

The phases were identified using Cu Kα X-ray diffraction (XRD) in a Siemens/Philips diffractometer. In all as-deposited films the only phase which is observed is that of fcc TiN. Whereas the evaporated TiN films show no preferred crystalline orientation, the sputtered films show a strong preference for the <111> orientation. The width of the diffraction peaks is to some extent indicative of grain size. However, inhomogeneous strain in the film can also cause broadening of the diffraction peaks, therefore, only a minimal grain size can be calculated from the peak width (Table 1).

Before Al deposition all TiN samples were divided into four sets: three sets were annealed for 30 minutes at the following temperatures: 600°C, 700°C and 800°C. The fourth set was used as a reference. These annealings as well as the annealings after Al deposition were performed in a diffusion pumped vacuum furnace with a h.f. heated cylindrical carbon susceptor and a base pressure of 1×10^{-5} Torr or less. A chromel-alumel thermocouple attached to a control sample provided

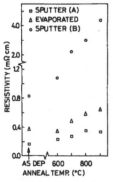

Fig.1. Film resistivity of TiN vs. annealing temperature.

the temperature readings. After the annealings, the TiN samples were examined by RBS, XRD and resistivity measurements.

A comparison of the RBS spectra of annealed films and as-deposited films shows that no silicide has been formed during annealing.

XRD measurements of the annealed films reveal no significant changes with respect to the reference samples.

In all cases the film resistivity increases upon annealing (Fig.1). This can be caused by a small amount of extra oxygen which comes in during annealing or by a redistribution of oxygen. However, using RBS, this cannot be detected. The drastic increase of the resistivity of the sputtered (B) film is remarkable.

RESULTS ON BARRIER PERFORMANCE

In order to examine the barrier properties of the different TiN layers we investigated the thermal stability of the Si/TiN/Al structure. The structures were prepared as follows: (1) TiN deposition on an Si wafer (see SAMPLE PREPARATION), (2) exposure to air, (3) pre-annealing in high vacuum (see TiN-FILM CHARACTERIZATION), (4) exposure to air, (5) e-beam evaporation of a 110 nm Al film onto TiN, (6) exposure to air, (7) heating in high vacuum (30 min), 500°C to 580°C, steps of 20°C.

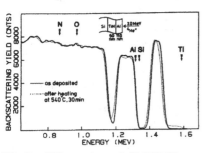

According to RBS measurements all structures remain intact upon a heat treatment at temperatures up to 520°C

Heating at 540°C results in a significant change in the RBS spectra for the samples which were not pre-annealed (= reference samples); an example of this is shown in Fig.2. Ti appears at the surface, while the width of the main Ti peak decreases. However, there is no Si found at the surface and the Si/TiN interface seems unaffected. The pre-annealed samples show very little change

Fig.2. RBS spectra of Si/TiN/Al structures. TiN prepared by reactive evaporation, not pre-annealed.

upon heating at 540°C (see Fig.3a). In this case the foot in the Ti RBS signal is explained by a local contraction of the aluminium, leaving spots (dimension > 10 μm) of uncovered TiN, as was seen by optical microscopy. This contraction of Al is found on almost all samples which were annealed at 500°C or higher, the estimated total area of uncovered TiN is 1 - 2%.

After a heat treatment at 560°C, also the samples which were pre-annealed show degradation (see Fig.3b), although less than the degradation of reference

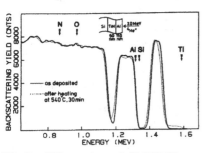

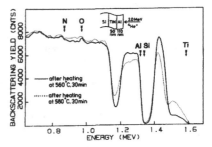

Fig.3a. RBS spectra of Si/TiN/Al. TiN prepared by reactive evaporation, pre-annealed at 800°C.

Fig.3b. RBS spectra of Si/TiN/Al. TiN prepared by reactive evaporation, pre-annealed at 800°C.

samples after heating at 540°C.
Heating at 580°C for 30 minutes
causes a further and severe degradation of all structures (see Figs.
2 and 3b). XRD measurements on the
degraded structures (540°C for the
reference (not pre-annealed) samples and 560°C for the pre-annealed
samples) reveal the formation of a
compound which can be identified
as Al$_3$Ti. This is illustrated by
Fig.4, which shows in addition to
the Al and TiN diffraction lines
also five diffraction lines of tetragonal Al$_3$Ti. Reaction of the top
part of TiN with Al resulting in
Al$_3$Ti formation agrees with the RBS
picture of Fig. 2. Splitting of the RBS Ti signal into two separate peaks,
characteristic for almost all structures which underwent a change due to the
heat treatment, will be explained below.

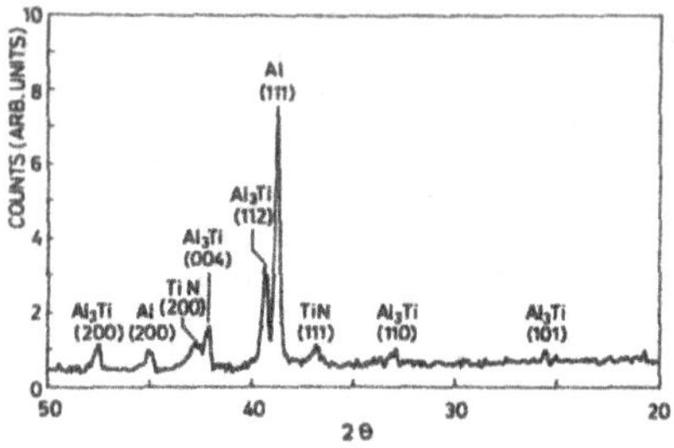

Fig.4. XRD spectrum of a Si/TiN/Al
structure after heating at 580°C, TiN
pre-annealed at 800°C.

With scanning electron microscopy (SEM) it was observed that in most samples after heating at 540°C or higher, the Al surface roughens and that the
Al layer shows holes (the size of these holes is of a different order of magnitude than the size of the craters described above, namely < 1 μm).

An example of this is shown in Fig. 5.
Owing to those holes the underlying material is locally almost uncovered, which
may result in the splitting of the original
Ti RBS signal into two peaks and a lowering
of the height of the main Ti peak. From
SEM micrographs the total hole area is estimated to be less than 10% of the surface area in most cases. Therefore this
surface roughening alone cannot be fully
responsible for the observed lowering and
splitting of the Ti RBS signal. We assume
there is an intermediate region in which
Ti is almost absent that causes splitting
of the Ti RBS peak. As the N RBS signal is
almost unchanged after heating, it is likely that this intermediate region consists
of AlN. This would fit with Wittmer's
suggestion [6] that it is the formation of
AlN with a heat of formation which is comparable to TiN, that initiates the decomposition of the TiN film, thus enabling
the released Ti atoms to diffuse to the
surface and react with Al to form Al$_3$Ti.

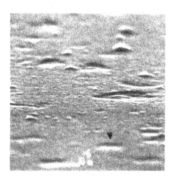

Fig.5. SEM picture of a Si/TiN/Al
structure after heating 540°C,
TiN prepared by sputtering, not
pre-annealed.

In order to evaluate this reaction path we have performed computer simulations of the RBS spectra assuming the final structure to consist of Si/TiN/
AlN/Al$_3$Ti-Al. To obtain quantitative agreement we first simulated an as-deposited Si/TiN/Al structure (Fig.6), which fits the measured spectrum. Simulations
of Si/TiN/AlN/Al$_3$Ti/Al and Si/TiN/AlN/Al$_x$Ti structures (Fig.6) were performed
to account for local reactions of the form: TiN + 4Al --→ AlN + Al$_3$Ti, while
assuming conservation of atomic species. From Fig.6 it is clear that a mixed
structure, with top layers consisting of Al$_3$Ti/Al and Al$_x$Ti, can fit the measured spectra in Fig.2 and 3 for the higher temperatures.

Further support for the degradation model comes from analysis of samples
which were etched after the heat treatment in dilute (10%), warm (50°C) HCl to
remove Al. After etching, the characteristic golden-yellow colour of TiN re-
appeared. No surface irregularities are observed anymore by SEM, indicating

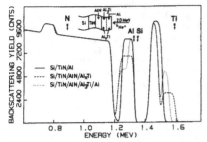

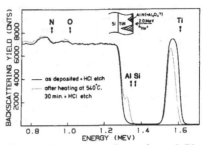

Fig.6. Simulations of RBS measurements.

Fig.7. RBS spectra of samples of Fig.2
after etching in HCl.

that holes as seen in Fig.5 do not penetrate into the remaining TiN layer. XRD
shows the etch-removal of Al₃Ti, as only the two broad peaks of TiN are present
in the spectrum. Fig.7 shows a comparison of RBS spectra taken after etching
of a reference (unheated) sample and the 540°C sample of Fig.2. The reduced
width of the Ti peak corresponds with the width of the main Ti peak of Fig.2.
The presence of Al at the surface and the shifting of the Ti RBS signal to lower energies indicate that the TiN layer of the heated sample contains some Al
compound. As N is still detected at the surface the major part of this compound
is likely to be AlN. From the change in the slope of the front edge of the Ti
RBS peak and from the fact that the Al peak is not separated from the Si signal we conclude that there is not a uniform layer of AlN on top of the TiN,
but a 'graded' structure (see inset Fig.7). The reason we did not detect the
AlN by XRD can be that it is only present in very small quantities.
    We notice that the 540°C RBS spectrum of Fig.7 shows some extra oxygen at
the surface. This suggests, that besides AlN, also Al₂O₃ is formed.
    Both RBS and XRD measurements suggest that pre-annealing of the TiN layers
improves the barrier properties with respect to Al. However, these methods are
not sensitive to small changes in the structures. On the other hand, such small
changes can have tremendous effects on the electrical characteristics of the
device. Therefore we tested this conclusion by measuring the I(V) characteristics of Si shallow-junction diodes which were simultaneously prepared, TiN-
deposited (sputtered (A) and evaporated) and pre-annealed. A 2 µm evaporated
Al layer was used as a top contact. With the evaporated TiN as a barrier the
junction was short-circuited after heating at 560°C for 30 minutes when the

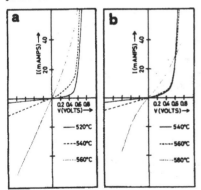

layer was not pre-annealed (see Fig.8a),
whereas for the diodes with the pre-
annealed TiN layer this was the case only
after heating at 580°C (see Fig.8b). The
same holds for the sputtered (A) film,
but already at 20 degrees lower temperatures. Short-circuiting of the junction
coincides with severe pitting of the
surface. The pits are rectangular in
shape and oriented along the main Si
crystal axes, indicating interaction of
the Si substrate and Al.
    Comparing the barrier properties
of the TiN layers made by the described
techniques it is found by RBS that the
sputtered (B) film has the best barrier
performance, followed by the evaporated
film. In comparison with the sputtered
(A) film, the sputtered (B) film and
the evaporated film contain more oxygen.
Further, there is an excess of Ti instead of an excess of N.

Fig.8. Diode characteristics of a
Si(p-n⁺)/TiN/Al structure after
heating; (a) TiN not pre-annealed,
(b) TiN pre-annealed at 800°C.

628

An interesting question is why the barrier quality improves upon annealing. Different from Ting [7] or Krusin-Elbaum et al. [10] our X-ray data show that the starting material is already poly-crystalline and annealing does not significantly alter grain size. We suggest that annealing improves the stoichiometry of the TiN film. Upon annealing, excess Ti (or N) and possibly O segregate at the grain boundaries. In the case of the sputtered (B) film and the evaporated film the excess Ti might segregate at the grain boundaries and react with the O (which at least partly came in during exposure to air) to form TiO_x. In this way, fast diffusion paths are blocked and reaction of Al and Si during heating is retarded. Unfortunately, best barrier performance is found for the films which contain most oxygen, and have the highest resistivity.

In this model, the TiN film not only acts as a diffusion barrier but also as a sacrificial barrier, which is broken down by compound formation. The forming of AlN is highly undesired because of its high resistivity. Therefore, it has been suggested [6,9] to apply a layer of Ti between TiN and Al. In such a structure only Al_3Ti would be formed, no AlN.

CONCLUDING REMARKS

In studying the thermal stability of the Si/TiN/Al structure we found that the structure degrades after heat treatments for 30 minutes at 540°C and above, due to chemical decomposition of the TiN layer by Al, thereby forming Al_3Ti and probably AlN. The characteristic RBS spectrum of the partially degraded structure can be explained by assuming a layered structure consisting of Si/[TiN/AlN] and a mixture of Al_3Ti/Al and Al_xTi on top. The onset of degradation shifts towards higher temperatures by 20°C - 40°C when the TiN is annealed prior to Al deposition, both for TiN layers made by sputtering and by evaporation.

ACKNOWLEDGEMENTS

The authors wish to thank C. Bakker of the University of Amsterdam for taking SEM pictures, N.M. van der Pers of the Technical University of Delft for performing XRD, J. Haisma of the Philips Research Laboratories in Eindhoven for preparing the sputtered TiN films and Gerrit Frijlink for preparing the evaporated TiN films.
This work is part of the research program of FOM and was financially supported by ZWO.

REFERENCES

1. M.Wittmer, J.Vac.Sci.Technol. A, 2 (1984) 273.
2. C.Y.Ting and M.Wittmer, Thin Solid Films 96 (1982) 327.
3. M.A.Nicolet, Thin Solid Films 52 (1978) 415.
4. M.A.Nicolet and M.Bartur, J.Vac.Sci.Technol. 19 (1981) 786.
5. R.S.Nowicki and M.A.Nicolet, Thin Solid Films 96 (1982) 317.
6. M.Wittmer, J.Appl.Phys. 53 (1982) 1007.
7. C.Y.Ting, J.Vac.Sci.Technol. 21 (1982) 14.
8. R.J. Schutz, Thin Solid Films 104 (1983) 89.
9. M.Wittmer, Appl.Phys.Lett. 37 (1980) 540.
10. L.Krusin-Elbaum, M.Wittmer, C.Y.Ting and J.J.Cuomo, Thin Solid Films 104 (1983) 81.
11. K.Y.Ahn, M.Wittmer and C.Y.Ting, Thin Solid Films 107 (1983) 45.
12. P.J.Martin, R.P.Netterfield and W.G.Sainty, Vacuum 32 (1982) 359.
13. H.von Seefeld, N.W.Cheung, M. Mäenpää and M.A.Nicolet, IEEE Trans.Electron Dev. 27 (1980) 873.

CROSS-SECTIONAL TEM STUDY OF RHODIUM ON SINGLE CRYSTAL AND AMORPHOUS SILICON

S.R. Herd, P.A. Psaras, I.J. Fisher[a] and K.N. Tu
IBM Thomas J. Watson Research Center, Yorktown Heights, N.Y. 10598

ABSTRACT:

A crystalline interfacial bilayer, consisting of about 5nm each of RhSi and Rh_2Si was found after E-beam deposition of 95nm of Rh onto either Si[100] or α-Si substrates. Cross sectional TEM of the as-deposited and annealed specimens showed no change occurred after 24hrs at 200°C. With infinite supply of Si, as on the Si[100] substrate, RhSi was found as a major growing phase, although Rh_2Si also grew at a much slower rate. With a limited supply of Si, as in the α-Si case, RhSi first formed until all α-Si was consumed (2 hrs at 400°C) and then transformed partially to Rh_2Si after 4 hrs at 400°C. This transformation could be confirmed by RBS since Rh_2Si layer thickness exceeded 30nm.

INTRODUCTION

It is generally accepted that silicide phases form sequentially during metal silicon reactions at elevated temperatures. First phase formation in a large number of metal silicon systems has been the subject of much investigation [1,2]. The first phase observed to grow may not necessarily be the first or only phase which nucleates. The theory, that during thin film preparation atomic intermixing at the interface takes place, which is very deposition parameter sensitive, is quite reasonable. The existence of such interfacial layers, including contamination and oxides, have been shown by spectroscopic surface methods for extremely thin depositions [3,4]. Conventional flat on transmission electron microscopy (TEM) although capable of high lateral resolution has not been able to show such layers in superposition and the conventional analytical methods of x-ray diffraction and Rutherford Backscattering (RBS) are depth resolution limited to 20-30nm.

RESULTS AND DISCUSSION

We have used cross sectional TEM to investigate the as deposited and reacted layers (at 400°C) of 90nm of Rhodium (Rh) on single crystal (Si[100]) and amorphous silicon (α-Si). The α-Si was deposited before the Rh in the same pump down (at 3×10^{-7} torr) and with only a 65nm thickness (on 500nm of SiO_2 on Si[111]). This represents a specimen with a limited supply of Si as compared to the unlimited supply of Si on the [100] substrate.

In the as deposited state we found a crystalline interfacial layer of 11-12nm in thickness. This layer is actually composed of two distinctly separate silicide layers each 5-6nm thick. The crystal size is many times larger (hundreds of nm) than the thickness and there is no epitaxial relationship between them or with the Si[100] surface. In fact the layers on Si[100] and α-Si are quite similar and remain unchanged by annealing at

[a] I.J. Fisher, Fairleigh Dickenson University, Rutherford, N.J. 07080.

200°C for 24 hrs. Fig. 1 shows a brightfield (a) and darkfield (b) photomicrograph of the entire cross section of the Rh film on [100]Si. The bilayer interface is clearly visible at I. Bright (a) and darkfield (b) micrographs are taken of different sample areas. Fig. 1c shows a typical electron diffraction (E.D.) pattern from an interfacial region. The pattern contains additional reflections (other than Rh and Si) that were used to identify the silicide phase, which is close to Rh, as Rh_2Si. Darkfield microscopy with small objective apertures and small area limiting apertures was similarly used to identify the phase close to Si as RhSi.

The growth of both phases could be followed by TEM observation after various annealing times at 400°C and the data are given in Table I. We note that the values are average layer thickness values, gathered from numerous micrographs of multiple cross sections. This is due to the fact that it is difficult to obtain evenly thinned (i.e. across the film thickness) cross sections. Also, there exists a genuine variation in grain sizes, together with grain overlap due to large section thickness, and inaccuracies occur due to specimen tilt positioning. RhSi is the major growing phase, although Rh_2Si also grows very slowly. Fig. 2 shows a bright (a) and darkfield (b) micrograph of Rh on α-Si after 0.5 hrs at 400°C. Figure 2c shows the E.D. pattern with the selected reflection, which was used to obtain the Rh_2Si crystal contrast in the darkfield image (b), in the center of the high contrast objective aperture. After 2 hrs at 400°C all the α-Si has been consumed and a trilayer structure (RhSi, Rh_2Si and Rh) is observed. The conversion of RhSi to Rh_2Si occurs after 4 hrs at 400°C. Figure 3 shows a 4-layer like structure after the 4 hrs at 400°C anneal. Figure 3b and 3c are darkfield micrographs with Rh and Rh_2Si reflections respectively. The central Rh_2Si layer consists of a two-grained structure and there is a thin layer of RhSi (less than after 2 hrs) near SiO_2 and a very small amount of Rh (possibly Rh_5Si_3) at the surface. This behavior is of course not observed for Rh on Si[100], where RhSi remains the major growing phase.

The conversion of RhSi to Rh_2Si with limited supply of Si is confirmed by the RBS data. Note the shift upwards in Fig. 4 at the low energy side of the Rh spectrum after the 4 hrs anneal (curve d) compared to the 2 hrs anneal (curve c). This shift is not present in Fig. 5 which is for Rh on Si [100]. In conclusion, we show that the formation of a crystalline multiphase interface forms during deposition of Rh onto Si. Both Rh_2Si and RhSi coexist in the as-deposited state and upon annealing to 400°C. This is different from the so-called "single phase growth" in thin films. The width of interfacial layers (in our case about 10nm) most likely depends on the film deposition parameters as well as the reactivity of the atomic species involved.

REFERENCES

1. K.N. Tu and J.W. Mayer, "Thin Films Interdiffusion and Reaction", J.M. Poate, K.N. Tu and J.W. Mayer Eds., (Wiley, New York, 1978) p. 359.

2. R.M. Walser and R.W. Bene, Appl. Phys. Lett. *28*, 624 (1976).

3. G.W. Rubloff, Surface Science *132*, 268 (1983).

4. J.C. Tsang, R. Matz, Y. Yokota and G.W. Rubloff, J. Vac. Sci. Technol. A2 (1984) 556.

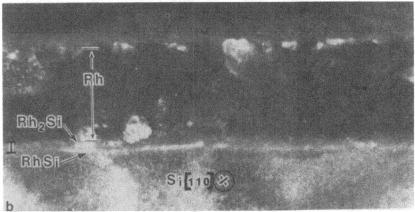

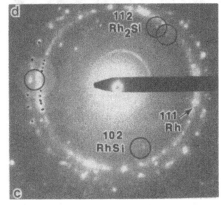

FIG. 1.

a) brightfield

b) darkfield micrograph of entire cross-section of 95nm of Rh on Si [100] as deposited.

c) Electron diffraction of interface with extra reflection due to Rh_2Si and RhSi.

FIG. 2.

a) brightfield
b) darkfield micrograph of interface of Rh
on α-Si after 0.5 hrs at 400°C.
c) Electron Diffraction of Rh$_2$Si crystal.
Si[110] pattern of substrate by super positioning.

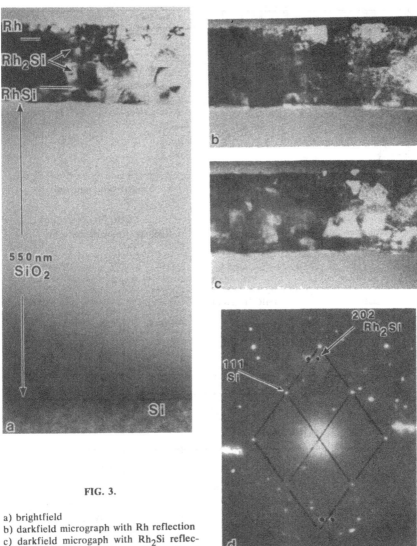

FIG. 3.

a) brightfield
b) darkfield micrograph with Rh reflection
c) darkfield microgaph with Rh$_2$Si reflection of entire cross section on SiO$_2$ after 4 hrs at 400°C

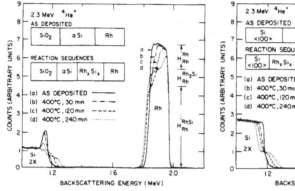

Fig. 4.

RBS spectra of Rh on Si [100]

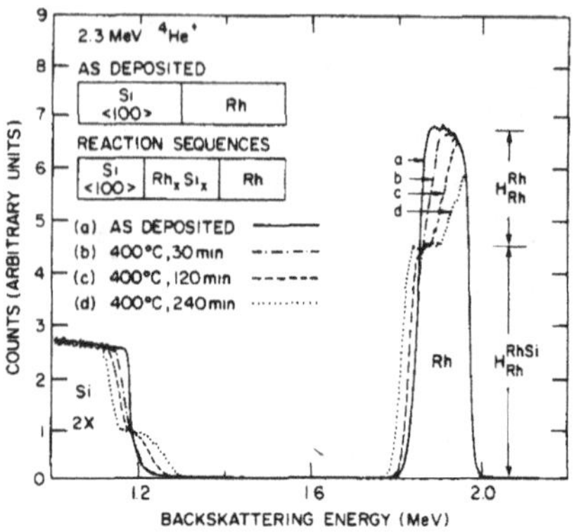

FIG. 5.

RBS spectra of Rh on α-Si

TABLE I

Anneal Time (hrs)	THICKNESS (nm)			
	Si	RhSi	Rh$_2$Si	Rh
As dep. and 24hrs 200°C	α-Si 66	5.5	5.5	90
	Si[100] infinite	5.0	6.0	90
0.5 400°C	α-Si 48	30	15	75
	Si[100] infinite	--	--	--
1 400°C	α-Si 30	40	20	60
	Si[100] infinite	60	10	60
2 400°C	α-Si 19 in spots	68	33	45(Rh$_5$Si$_3$?)
	Si[100] infinite	90	20	30
4 400°C	α-Si 0	38	86(2x43)	23
	Si[100] infinite	120	25	10

STRUCTURE ANALYSIS OF Ni-SILICIDES FORMED IN LATERAL DIFFUSION COUPLES

S.H. CHEN, J.C. BARBOUR, L.R. ZHENG, C.B. CARTER, AND J.W. MAYER
Department of Materials Science and Engineering
Cornell University, Ithaca, NY 14853

ABSTRACT

The microstructures of the silicide Ni_5Si_2, which formed in self-supporting Ni-Si lateral-diffusion couples has been studied using high-resolution electron microscopy. Two different polymorphs (or polytypes) for Ni_5Si_2 have been observed. The actual composition of one polytype is confirmed to be $Ni_{31}Si_{12}$, while the other one has not yet been identified. Variations in the distribution of the two polytypes, as observed in the present study, may account for the composition range of Ni_5Si_2 in the Ni-Si phase diagram.

INTRODUCTION

Multiphase formation in lateral Ni-Si diffusion couples has been observed using a new specimen preparation technique for transmission electron microscopy (TEM) studies [1,2]. Each of the compounds Ni_3Si, Ni_5Si_2, Ni_2Si, Ni_3Si_2, and NiSi has been found to form across the Ni-Si interdiffusion region. The different phases were identified using a combination of energy-dispersive x-ray spectroscopy (EDS), selected-area electron diffraction and microdiffraction. The specimen preparation has also been applied to other systems including Pd-Si [1] and Ni-GaAs [3].

It has been predicted [4] that concentration gradients may exist across some of the phases in Ni/Si diffusion couples because their growth is controlled by diffusion rather than by the movement of the phase boundary. However, no experimental evidence has been reported for the existence of such concentration gradients. If such concentration gradients do exist, they are too small to be detected by EDS [1,2]. It has been observed that Ni_5Si_2 is the only phase that has a high density of planar defects in the lateral Ni-Si diffusion couples [2]. The Ni-Si phase diagram indicates a small, but finite, composition range for Ni_5Si_2, in contrast to the contiguous phases, i.e. Ni_2Si and Ni_3Si_2. Such a small variation in composition is not easily detected using EDS. Therefore, high-resolution electron microscopy (HREM) has been used in the present study to examine the possible relation between these planar defects and composition variation within the Ni_5Si_2. A local variation of the crystal structure of the "Ni_5Si_2" phase may itself accommodate the concentration gradient. (See, for example the discussion of 'chemical twinning' in ref. 5.)

EXPERIMENTAL DETAILS

A layer of amorphous Si, $\sim$ 30 nm thick (see the schematic diagram in Fig. 1), was electron-beam deposited on a NaCl substrate at a base pressure of 2×10^{-7} Torr. A layer of Ni 100 nm thick, was then deposited through square openings (400 μm x 400 μm) in a mask produced by a photolithographic technique, so that a well-defined area of Ni was formed with relatively sharp edges on the amorphous Si film. The NaCl substrate was then removed by dissolving it in water and the films, which were floating on the surface of the water, were picked up on molybdenum TEM grids. The specimens were then annealed in a vacuum at a pressure of 2×10^{-8} Torr. Transmission electron microscopy was carried out using a Siemens 102 TEM, operating at 125 kV.

636

As-deposited

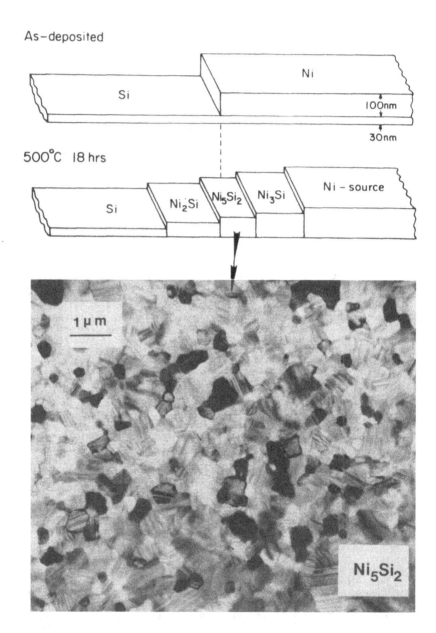

500°C 18 hrs

Fig. 1. Schematic diagram showing the side view of the lateral diffusion couple before and after annealing at 500°C for 18 h. The incident electron beam is perpendicular to the Si and Ni layers during TEM observations. The micrograph shows a typical bright-field TEM image for Ni_5Si_2 where the "twin-like" microstructure can be seen within many of the grains.

RESULTS AND DISCUSSION

Fig. 1 shows a pair of schematic diagrams illustrating the initial geometry of the diffusion couple and the main features of the sample after annealing for 18 hours at 500°C. A very thin region of NiSi also actually forms [2] at this temperature. The ratio of Ni:Si layer thickness was chosen to ensure a supply of excess Ni. The TEM image shows the equiaxed grain structure of a typical region of the polycrystalline Ni_5Si_2 formed in the lateral Ni-Si diffusion couple. Within the grains, many planar defects are clearly visible. In order to carry out a high-resolution study, grains in particular orientations, where direct structure imaging would be possible with the limited point resolution (3.4 Å) of the Siemens microscope, were selected using the double-tilting stage in the TEM.

Fig. 2 shows electron diffraction patterns and the corresponding high-resolution lattice-fringe images of two different "Ni_5Si_2" polymorph structures which have been found in the present study. The diffraction pattern in (a) indicates that the structure of this polymorph is consistent with that proposed for $Ni_{31}Si_{12}$ [6,7], which has a hexagonal crystal lattice with a = 6.68 Å and c = 12.28 Å. The Ni:Si ratio is clearly very close to 5:2. The crystal structure of $Ni_{31}Si_{12}$ was proposed by Frank and Schubert [7] based on their x-ray data. Using this structure, a computer simulation of the high-resolution images has been performed [8], and the results compared to the experimental images (see Fig. 3). Excellent agreement between the experimental and simulated images has been found.

The other polymorph [Fig. 2(b)] has a hexagonal lattice a = 6.68 Å and c = 18.42 Å (i.e. a appears to be the same but c is 50% larger than that found for $Ni_{31}Si_{12}$): its actual crystal structure and composition have not yet been determined. In Fig. 2, the electron-beam direction is <11$\bar{2}$0> and the fringes in both images are parallel to the "basal" planes of both poly-morphs. The ratio of the "basal"-plane spacing for the two polytypes is 1.5 and the existence of one polymorph within the other (arrowed) can be seen in the high-resolution image in Fig. 2b. The polytypism, also referred to as one-dimensional polymorphism, is commonly observed in minerals such as ZnS, or in semiconductors such as SiC. It occurs when polymorphs differ only in the stacking sequence of identical two-dimensional layers. A possible relation between the two polytypes illustrated in Fig. 2 is that the first (Fig. 2a) corresponds to an abab stacking sequence and the second (Fig. 2b) corresponds to an abcabc stacking sequence: a, b and c represent the same layers as are present in Fig. 2a but arranged in different sequence relative to one another.

The determination of the crystal structure of "Ni_5Si_2" or $Ni_{31}Si_{12}$ which was reported [7,9,10] to give a hexagonal unit cell with lattice parameters a = 6.68 Å and c = 12.28 Å, was based on single-crystal or powder x-ray diffraction studies. This silicide structure was believed to be a thermodynamically stable one because of the very long annealing time involved in its formation. In the present study, the occurrence of the other polytype and of many planar defects within the Ni_5Si_2 region of the diffusion couples therefore suggests that the microstructure of the Ni_5Si_2 is related to its growth during the phase transformation.

CONCLUSION

The sequence of multiphase formation in thin film lateral diffusion couples can be examined in situ by TEM. The present paper has demonstrated that the Ni_5Si_2 'phase' formed during the thin film reaction actually consists of two polytypes. The structure of one of these polytypes has been shown, by high-resolution TEM, to be consistent with that based on x-ray diffraction studies and having a composition $Ni_{31}Si_{12}$. The precise composition and structure of the other polytypes have not yet been determined. The existence of the two polytypes within the 'Ni_5Si_2' region of the diffusion

658

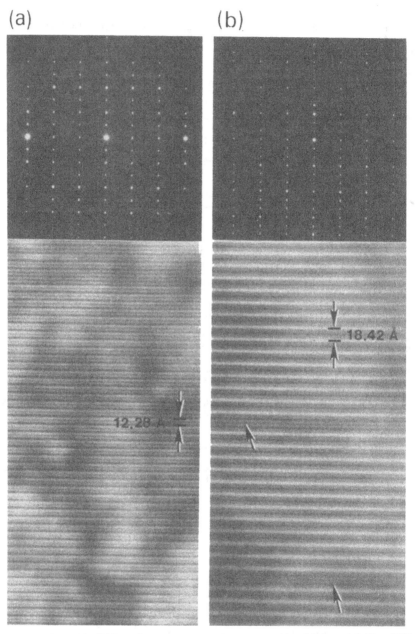

Fig. 2. Electron diffraction patterns and corresponding high-resolution TEM images showing two polytypes for "Ni$_5$Si$_2$". 'a' corresponds to the known Ni$_{31}$Si$_{12}$ structure.

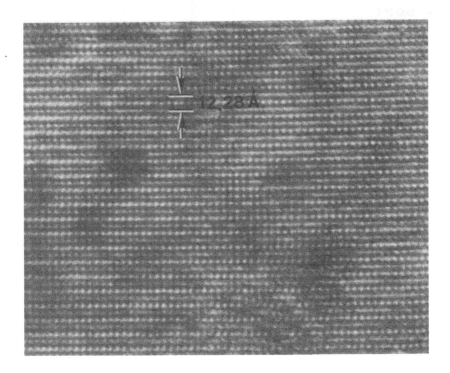

Fig. 3. High-resolution TEM image of $Ni_{31}Si_{12}$.

couples may either occur in order to accommodate a local variation in composition or it may be associated with the mechanism of the phase transformation.

ACKNOWLEDGMENTS

The authors are grateful to Mr. R. Coles for maintaining the microscopes which are part of Materials Science Center Facility at Cornell which is supported by NSF. This research is supported by SRC Microscience and Technology Program at Cornell.

640

REFERENCES

1. S.H. Chen, L.R. Zheng, J.C. Barbour, E.C. Zingu, L.S. Hung, C.B.
    Carter, and J.W. Mayer, Materials Letters 2, 469 (1984).
2. S.H. Chen, L.R. Zheng, C.B. Carter, and J.W. Mayer, J. Appl. Phys., 57,
    258 (1985).
3. S.H. Chen, C.J. Palmstrom, T. Ohashi, and C.B. Carter, to be submitted
    for publication (1984).
4. U. Gosele and K.N. Tu, J. Appl. Phys. 53, 3252 (1982).
5. S. Andersson and B. G. Hyde, J. Solid State Chem. 9, 92 (1974).
6. M.-A. Nicolet and S.S. Lau, in VLSI Electronics: Microstructure
    Science, Vol. 6, edited by N.G. Einspruch and G.B. Larrabee (Academic
    Press, New York) (1983), Chap. 6.
7. K. Frank and K. Schubert, Acta Cryst. B27, 916 (1971).
8. S.H. Chen, Z. Elgat, L.R. Zheng, C.B. Carter, and J.W. Mayer, to be
    presented at the ASU Conference on High Resolution Electron Microscopy,
    proceedings to be published in Ultramicroscopy.
9. G. Pilstrom, Acta Chem. Scand. 15, 893 (1961).
10. G.S. Saini, L.D. Calveert, and J.B. Taylor, Canadian J. Chem., 42,
    1511 (1964).

EPITAXIAL PHASES FORMATION DUE TO INTERACTION BETWEEN
Ni THIN FILMS AND GaAs

A. LAHAV, M. EIZENBERG AND Y. KOMEM
Dept. of Materials Engineering, Technion, Haifa 32000, Israel

ABSTRACT

Solid state reactions between Ni thin films and (100) GaAs were studied by transmission electron microscopy, X-ray diffraction, Auger electron spectroscopy and Rutherford backscattering. The hexagonal ternary phase Ni_2GaAs is formed at the temperature range of 150 to 300°C with the following epitaxial relations to the substrate: $(10\bar{1}1)//(001)$ and $[\bar{1}101]//[\bar{1}00]$. The gradual decrease with temperature of the $(10\bar{1}1)$ interplanar spacing of this phase may be related to strain relaxation associated with microtwins formation. At the temperature range of 350 to 550°C precipitation of NiAs in Ni_2GaAs matrix and an increase in the twin size are observed. In this range the $(10\bar{1}1)$ interplanar spacing of the ternary phase increases with temperature, possibly due to stoichiometric changes resulting from NiAs precipitation. At 600°C the ternary phase decomposes into NiGa and NiAs with average grain size of about 1 micron in lateral dimension. The cubic phase, NiGa, has simple epitaxial relations with GaAs: $(001)//(001)$ and $[010]//[010]$. Ni atoms are the dominant diffusing species during the ternary phase growth process as was observed in a marker experiment utilizing a very thin Ta Layer interposed between the Ni film and the substrate.

1. INTRODUCTION

Similarly to silicides [1-3], intermetallic compounds consisting of metal and gallium and/or arsenic, formed by solid state reaction between metal thin films and GaAs substrate may be used as stable ohmic or Schottky contacts to GaAs devices. In particular, the intermetallic compounds that grow epitaxially to GaAs are attractive since they may be utilized for preparation of heterostructures or for channeling implanted ions into the substrate through the contact. In this paper we present a brief survey of metallurgical investigation of the Ni-GaAs interaction in the temperature range of 100-600°C. Nickel was chosen as a contact metal as a representative of the near noble metals which are characterized by relatively low reaction temperatures. In combination with gold and germanium nickel is used for ohmic contacts preparation to n-GaAs [4]. Nickel is also very attractive because it was reported to form an epitaxial compound Ni_2GaAs with (111) GaAs [5].

2. EXPERIMENTAL DETAILS

Nickel films ranging in thickness from 35 to 250 nm were e-gun deposited in a vacuum of 1×10^{-7} torr on cleaned and chemically etched n-type GaAs (001) oriented wafers. Annealing was performed in a flow of forming gas (88% N_2 and 12% H_2) in the temperature range of 100-600°C. The reaction products were characterized by Auger Electron Spectroscopy (AES), X-Ray Diffraction (XRD), Nomarski microscopy, Scanning Transmission Electron Microscopy (STEM), Energy Dispersive X-Ray Spectroscopy (EDS) combined with STEM, and by Rutherford Backscattering Spectroscopy (RBS). Specimens for TEM examination were prepared by chemical thinning from the back side of the samples.

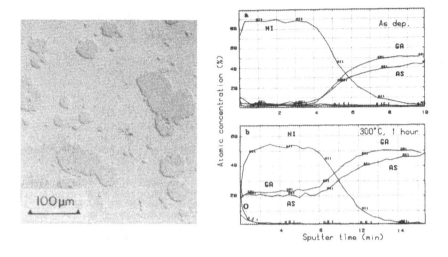

Fig.1. Nomarski micrograph of the sample annealed at 250°C for 1 hour.

Fig.2. AES profiles of Ni/GaAs: a) as-deposited, b) annealed at 300°C for 1 hour

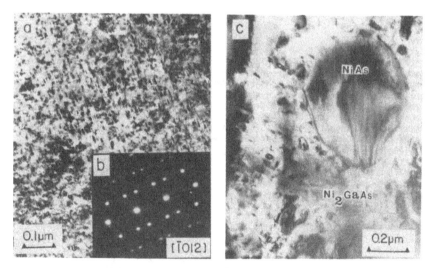

Fig.3. a) Bright field TEM image and b) SAD pattern from Ni_2GaAs micro-twins after annealing at 250°C for 1 hour, c) Bright field image of NiAs precipitate in Ni_2GaAs matrix after annealing at 500°C for 1 hour.

3. RESULTS AND DISCUSSION

Nomarski micrograph of the sample annealed at 250°C for 1 hour (Fig.1) shows reacted regions in the form of dark islands on the grey background of the Ni-film. After annealing at 350°C for 1 hour almost all the surface was reacted. The AES depth profiles of the as-deposited and reacted samples are shown in Fig.2. The calculated atomic ratio of Ni:Ga:As = 2:1:1 for the reacted sample confirms the ternary phase stoichiometry of Ni2GaAs, the compound which was identified earlier [5]. The bright field TEM image of the reacted area in the sample annealed at 250°C for 1 hour is shown in Fig.3a. In contrast to the as-deposited Ni film which is polycrystalline with average grain size of 20 nm, selected area diffraction (SAD) of the reacted area is characterized by single crystal twinned diffraction patterns (Fig.3b). The microtwins of rhomboid shape, 5 x 10 nm in size can be seen in Fig.3a. The SAD patterns from different zone axes of Ni2GaAs obtained by sample tilting were indexed and fitted to a hexagonal system with a ratio of c/a = 1.2248. In all cases the $[10\bar{1}1]$ zone axis of Ni2GaAs was parallel to [001] zone axis of GaAs. Therefore, one can deduce the following epitaxial relation (Fig.4):

$$(10\bar{1}1)_{Ni_2GaAs}//(001)_{GaAs} \text{ and } [\bar{1}101]_{Ni_2GaAs}//[100]_{GaAs}$$

The four twin planes of Ni2GaAs - $(\bar{1}101)$, $(0\bar{1}11)$, $(\bar{1}012)$ and $(1\bar{2}10)$ are parallel to the (100), (01$\bar{0}$), (110) and ($\bar{1}$10) planes of GaAs, respectively. The described above epitaxial relation between hexagonal (10$\bar{1}$1) and cubic (001) planes is rather unusual. The epitaxial relation which is often found between hexagonal and cubic phases is $(0001)_h//(111)_c$. However in our case the c/a ratio of the ternary phase (equal to $\sqrt{3}$ / $\sqrt{2}$) results in a hexagonal structure which is pseudocubic [6] and allows the epitaxy along $(10\bar{1}1)_h//(001)_c$ planes in addition to the epitaxy of the usual type reported by M. Ogawa [5]. In this case the symmetry difference between the four fold cubic direction and the non-symmetric hexagonal direction is compensated by the quadruple twinning in Ni2GaAs [7].

The reaction mechanism at the initial stage was studied using a marker of very thin (about 0.5 nm) tantalum film e-gun deposited on the chemically etched GaAs surface prior to nickel evaporation. Ta does not react with GaAs up to annealing temperatures of 600°C [8]. On the other hand, such a thin layer can not be continuous and would consist of Ta-islands which do not prevent the reaction between Ni and GaAs. RBS spectra of the marker samples in the as-deposited state and after annealing at 320°C for 2 hours are shown in Fig.5. One can see that in the reacted sample the Ta peak has moved towards the surface, thus indicating that the reaction is governed by indiffusion of Ni atoms rather than by As and Ga outdiffusion. The appearance of a large number of reacted islands on the sample surface leads to the conclusion that the reaction is controlled by nucleation of the ternary phase at the interface with GaAs.

XRD pattern of the as-deposited sample with 250 nm thich Ni film is given in Fig.6a. Samples annealed in the temperature range of 100-300°C (Fig.6b) show besides peaks of Ni and GaAs, two additional peaks whose intensity increases with temperature. These peaks were indexed as $(10\bar{1}1)$ and (2022) of Ni2GaAs. In higher temperature range 350-550°C the diffractogram shows a third additional peak which is attributed to $(10\bar{1}1)$ of NiAs (Fig.6c). The intensity of Ni peaks decreases with the annealing temperature and they finally disappear at 450°C. TEM and EDS examination of the sample annealed at 500°C for 1 hour shows presence of NiAs precipitates in the Ni2GaAs matrix (Fig.3c). The twin size of Ni2GaAs in this sample has increased up to 20 x 40 nm. The NiAs

644

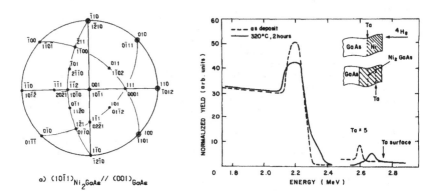

a) $(10\bar{1}1)_{Ni_2GaAs} // (001)_{GaAs}$

Fig.4. Superimposed stereographic projection of Ni2GaAs (four indices) and GaAs (three indices) showing epitaxial relations for (001) GaAs substrate.

Fig.5. RBS spectra of Ni/GaAs with Ta marker: as deposited and annealed at 320°C for 2 hours.

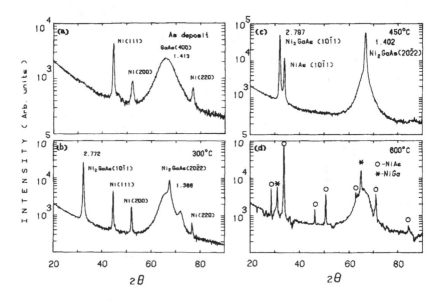

Fig.6. 6 XRD patterns of Ni/GaAs: a) as-deposited and after 4 hour annealing at b) 300°C, c) 450°C, d) 600°C.

precipitates were found to be coherent with the Ni_2GaAs matrix at the early stage of their formation. With the increase of their size they lose the coherence and become random.

The interplanar spacings of the $(10\bar{1}1)$ and $(20\bar{2}2)$ reflections of the ternary phase Ni_2GaAs change with annealing temperature. At the first reaction stage in the range of 100-300°C they decrease and at the second stage of 350-550°C they increase with increasing of annealing temperature. The early decrease of the interplanar spacings can be explained by adjustment of the lattice parameters of the ternary phase to those of GaAs (d(200) = 2.832 A°) due to epitaxy. As the thickness of the ternary phase increases, the twin formation releases strain at the interface and the ternary phase acquires its equilibrium lattice parameters which are those measured after 300°C annealing. Using the value of $c/a = 1.2248$ obtained by electron diffraction for the samples annealed at this temperature, the calculated lattice parameters of Ni_2GaAs are $a_0 = 3.925$ and $c_0 = 4.807$ Å . These values slightly differ from those reported by Ogawa [5] $a_0 = 3.84$, $c_0 = 4.96$ Å and $c/a = 1.292$. The increase of $(10\bar{1}1)$ interplanar spacing in the temperature range of 350-550°C can be associated with the ternary phase decomposition from Ni_2GaAs to NiGa due to NiAs precipitation.

After annealing at 600°C XRD pattern (Fig.6d) shows all peaks of NiAs and two peaks, (100) and (200), of NiGa. The bright field TEM image of this sample (Fig.7a) shows rounded isometrical grains of NiAs and NiGa about 1 μm in lateral dimension. The TEM and EDX analysis indicate that the dark grains belong to NiGa and the bright ones to NiAs. The colour distinction in bright field image originates from thickness difference which is apparently a result of higher etching rate of NiAs relative to that of NiGa during the sample preparation. The dark field image taken from the (110) reflection of NiGa (Fig.7b) shows that all the NiGa grains have the same orientation and are epitaxially related to GaAs:

$$(001)_{NiGa}//(001)_{GaAs} \quad \text{and} \quad [100]_{NiGa}//[100]_{GaAs}$$

The NiAs grains, on the other hand, have random orientation.

The decomposition of the ternary phase due to NiAs precipitation at temperatures higher than 300°C lead to the assumption that Ni_2GaAs is metastable. Since the phase diagram of Ni-Ga-As is not known and there is no data on Ni_2GaAs phase in bulk form, this assumption needs still to be proved. However, if it is so, then the formation of Ni_2GaAs at the interface with GaAs at low temperatures could be stabilized by the free energy gain due to epitaxy [9], and due to a less extent of atomic diffusion required to form this phase rather than separate grains of NiAs and NiGa.

4. CONCLUSION

The solid state reaction between Ni thin films and GaAs results in epitaxial formation of Ni_2GaAs in the temperature range of 100-550°C and NiGa at 600°C. The fact that both these phases make epitaxial contacts to (001) and (111) GaAs substrates can be used for buried contacts formation.

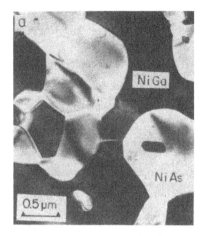
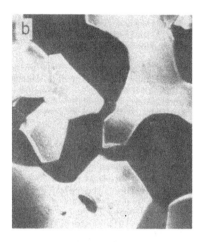

Fig.7. a) Bright field TEM image of sample annealed at 600°C for 1 hour.
b) Dark field image taken from (110) reflection of NiGa.

ACKNOWLEDGEMENTS

The authors gratefully acknowledge the assistance of C. Cytermann and
R. Brener in AES analysis and the help of Z. Brat in the early TEM measurements. R. Fastow's (Cornell University) help in carrying out the RBS
analysis is very much appreciated.

REFERENCES

1.   K.N. Tu in "Preparation and Properties of Thin Films" - Treatise
     on Materials Science and Technology, V.24, Ed. by K.N. Tu and
     R. Rosenberg, Academic Press, Chapter 7, 237, (1982).

2.   S.P. Murarka, "Silicides for VLSI Applications", Academic Press,
     (1983).

3.   M. Eizenberg in "VLSI Science and Technology 1984", Ed. by K.E. Bean
     and G.A. Rozgonyi, The Electrochemical Society, Pennington, 348,
     (1984).

4.   N. Braslau, J. Vac. Sci. Technol. 19, 803 (1981).

5.   M. Ogawa, Thin Solid Films, 70, 181, (1980).

6.   C. Barrett and T.B. Massalski, "Structure of Metals", 3rd ed.
     Pergamon Press, (1980).

7.   I.H. Scobey, C.A. Wallace and R.C.C. Ward, J. Appl. Cryst., 6, 425,
     (1973).

8.   A. Lahav and M. Eizenberg, Appl. Phys. Lett., 45, 256 (1984).

9.   R.F.C. Farrow, J. Vac. Sci. Technol. B 1 (2), 222 (1983).

Author Index

Subject Index

Printed in the United States
By Bookmasters